Advanced Researches in Optics

Volume I

Advanced Researches in Optics
Volume I

Edited by **Annie Kent**

NY RESEARCH PRESS
New York

Published by NY Research Press,
23 West, 55th Street, Suite 816,
New York, NY 10019, USA
www.nyresearchpress.com

Advanced Researches in Optics: Volume I
Edited by Annie Kent

International Standard Book Number: 978-1-63238-020-3 (Hardback)

Printed in China.

Contents

	Preface	VII
Chapter 1	**Anomalous Propagation in Low Index Contrast Metamaterials: Assessment of the Beam Collimation Condition** Jun Tan, Weiwei Song and Wei Jiang	1
Chapter 2	**Light Combining for Interferometric Switching** Marco Masi, Mattia Mancinelli, Paolo Bettotti and Lorenzo Pavesi	7
Chapter 3	**Ultrafast Plasmonic Nanoantenna-ITO Hybrid Switches** Martina Abb and Otto L. Muskens	24
Chapter 4	**Plasmonic Bandgaps in 1D Arrays of Slits on Metal Layers Excited by Out-of-Plane Sources** Roberto Marani, Marco Grande, Vincenzo Petruzzelli and Antonella D'Orazio	29
Chapter 5	**Coupling Schemes in Terahertz Planar Metamaterials** Dibakar Roy Chowdhury, Ranjan Singh, Antoinette J. Taylor, Hou-Tong Chen, Weili Zhang and Abul K. Azad	41
Chapter 6	**Surface Plasmon Singularities** Gabriel Martínez-Niconoff, P.Martínez-Vara, G. Díaz-González, J. Silva-Barranco and A. Carbajal-Domínguez	53
Chapter 7	**Manipulating the Propagation of Solitons with Solid-Core Photonic Bandgap Fibers** O. Vanvincq, A. Kudlinski, A. Bétourné, A.Mussot, Y. Quiquempois and G. Bouwmans	60
Chapter 8	**Nonlinear Pulse Shaping in Fibres for Pulse Generation and Optical Processing** Sonia Boscolo and Christophe Finot	72
Chapter 9	**Nonlinear Coherent Directional Coupler: Coupled Mode Theory and BPM Simulation** Dharmadas Kumbhakar	86
Chapter 10	**Mie Plasmons: Modes Volumes, Quality Factors, and Coupling Strengths (Purcell Factor) to a Dipolar Emitter** G. Colas des Francs, S. Derom, R. Vincent, A. Bouhelier and A.Dereux	94
Chapter 11	**Diffuse Optical Imaging Using Decomposition Methods** Binlin Wu, M. Alrubaiee, W. Cai, M. Xu and S. K. Gayen	102

Chapter 12	**Zeeman Effect of Sm Atoms by High-Resolution Diode-Laser Spectroscopy** Wei-Guo Jin and Tatsuya Minowa	**114**
Chapter 13	**Polishing Sapphire Substrates by 355nm Ultraviolet Laser** X.Wei, X. Z. Xie, W. Hu and J. F. Huang	**119**
Chapter 14	**Fresnel Lens with Embedded Vortices** Sunil Vyas, Rakesh Kumar Singh, Devinder Pal Ghai and P. Senthilkumaran	**126**
Chapter 15	**Generation of Bessel Surface Plasmon Polaritons in a Finite-Thickness Metal Film** S. N. Kurilkina, V. N. Belyi and N. S. Kazak	**133**
Chapter 16	**Subwavelength Plasmonic Waveguides and Plasmonic Materials** Ruoxi Yang and Zhaolin Lu	**146**
Chapter 17	**Development of Automatic 3D Blood Vessel Search and Automatic Blood Sampling System by Using Hybrid Stereo-Autofocus Method** Eiji Nakamachi, Yusuke Morita, and Yoshifumi Mizuno	**158**
Chapter 18	**Effect of Fluorescent Particle Size on the Modulation Efficiency of Ultrasound-Modulated Fluorescence** Yuan Liu, Baohong Yuan and Joseph Vignola	**169**
Chapter 19	**Bragg-Scattering Four-Wave Mixing in Nonlinear Fibers with Intracavity Frequency-Shifted Laser Pumps** Katarzyna Krupa, Michela Bettenzana, Alessandro Tonello, Vincent Couderc, Philippe Di Bin, Stefan Wabnitz and Alain Barthélémy	**176**
Chapter 20	**Experiments on Linear and Nonlinear Localization of Optical Vortices in Optically Induced Photonic Lattices** Daohong Song, Cibo Lou, Liqin Tang, Zhuoyi Ye, Jingjun Xu and Zhigang Chen	**183**
Chapter 21	**Volumetric Diffuse Optical Tomography for Small Animals Using a CCD-Camera-Based Imaging System** Zi-Jing Lin, Haijing Niu, Lin Li and Hanli Liu	**193**
	Permissions	
	List of Contributors	

Preface

Optics is a sub-disciplinary field of physics, which includes the study of properties of light, interaction of light with matter and the development of methods as well as instruments which are used for the detection of this phenomenon. The origins of the subject of optics can be traced to the ancient Egyptians and Mesopotamians, when they began developing lenses. In fact, glass spheres were filled with water to make lenses by the ancient Romans and Greeks.

Field optics is used to describe concepts of all forms of light waves; visible light (VIBGYOR), ultraviolet light and infrared. Practical optics is illustrated using simplified models because it is practically impossible to describe the electromagnetic behaviour of light. Examples include geometric optics and physical optics. Light exhibits characteristics of both a wave and a particle and this characteristic influences a lot of phenomena related to Optics. Quantum mechanics takes into consideration the fact that light depicts properties of a particle. Classical optics can be categorised into two main branches: geometrical optics and physical optics.

I especially wish to acknowledge the contributing authors, without whom a work of this magnitude would clearly not have been realizable. I thank all the authors for allocating much of their scarce time to this project. Not only do I appreciate their participation, but also their adherence as a group to the time parameters set for this publication. I hope that this book proves to be a resourceful guide for both basic and advanced concepts in Optics. As an Editor and also an Author in this field, I am privileged to be editing a book with such interesting topics, written by a selected group of gifted researchers. I would also like to thank my family for their endless support.

I hope that this book encourages readers to actively explore these broad and exciting research areas.

Editor

Anomalous Propagation in Low Index Contrast Metamaterials: Assessment of the Beam Collimation Condition

Jun Tan,[1] Weiwei Song,[1] and Wei Jiang[1,2]

[1] *Department of Electrical and Computer Engineering, Rutgers University, Piscataway, NJ 08854, USA*
[2] *Institute for Advanced Materials, Devices, and Nanotechnology, Rutgers University, Piscataway, NJ 08854, USA*

Correspondence should be addressed to Wei Jiang, wjiangnj@rci.rutgers.edu

Academic Editor: Georgios Veronis

Anomalous beam propagation in low index-contrast metamaterials has been analyzed. The condition for a well-collimated beam is found to be depending on the beam width and the pertinent Fourier component of the dielectric function. Guided by this condition, an ultra-compact metamaterial structure is designed to deflect a light beam at a wide angle. The structure is tolerant to structural parameter deviation and has a wide bandwidth.

1. Introduction

Metamaterials have aroused wide spread interest in recent years [1–12]. Many metamaterial structures involve metallic and dielectric components with a large refractive index-contrast between two material components. Such a large index-contrast results in the capability of confining and maneuvering light path in a very small scale. On the other hand, it is possible to construct metamaterials with relatively low index-contrast. It would be helpful to understand the capability and limit of metamaterials in the low index-contrast case. In this work, we will study a low index-contrast metamaterial system composed of polymer and air components in such a context. Negative refraction is a topic of wide interest in metamaterials research [1]. Negative refraction can be considered as one aspect of the anomalous characteristics of an optical beam in metamaterials. We will analyze low index-contrast metamaterials that can produce certain anomalous propagation characteristics. This can be utilized to maneuver light beams of sufficiently large widths with potential applications in polymer photonic integrated circuits [13]. Note that the low index-contrast metamaterials discussed here differ from the low-index (or epsilon-near-zero) metamaterials, whose constitutive components have relatively large indices.

Anomalous beam propagation in metamaterials can be traced back to early studies of photonic crystals in 1998 [14]. Note that photonic crystals can be considered as one category of metamaterials whose refractive indices are usually real. The abnormalities include negative refraction, unusually large refraction/deflection angle, and unusual wavelength sensitivity [14]. The proposed applications of these effects include beam deflectors [15], wavelength-division multiplexers [14, 15]. To utilize such anomalous propagation in compact devices, it is generally preferred to use a narrow beam at the entrance so that the device lateral width can be minimized. However, the anomalous propagation is usually accompanied by greater angular dispersion, which significantly exacerbates the angular divergence of a narrow beam in a metamaterial. For high index-contrast systems such as silicon-air photonic crystals, this problem can be solved through engineering the dispersion surface [16]. However, this angular divergence problem has not been analyzed in a general manner, especially for metamaterials of small index-contrasts. Note that many low-index materials such as polymers are attractive in a wide range of photonics applications because of their low fabrication/processing cost and flexible form [13]. In this work, we will start from analyzing the beam collimation condition in low index-contrast metamaterials in Section 2. Then Section 3 will present a design of a beam deflection device and address some other practical issues of low index-contrast metamaterials. Lastly, conclusions will be presented in Section 4.

2. Beam Collimation Condition in Low Index-Contrast Metamaterials

The anomalous propagation characteristics observed in these metamaterials originate from the underlying subwavelength structures. Related problems have been studied in various scenarios in other fields. Particularly, in X-ray diffraction study, a "dynamic theory" has been developed [17] to understand the beam divergence related issue for X-ray propagation in a crystal. Of particular value is the spherical wave part of the dynamic theory of X-ray diffraction. In the limit of zero beam width, the entering beam becomes a completely diverging spherical wave (a cylindrical wave in two dimensional problems), which represents the worst beam divergence scenario. This spherical wave formulation can be easily adapted to help understand the beam divergence issue in a low index-contrast metamaterial. Considering an entering beam with center wavevector $\mathbf{k}_0$ diffracted by a Fourier component ε_G of the dielectric function, the first order diffraction has a wave vector $\mathbf{k}_g = \mathbf{k}_0 + \mathbf{G}$, where $\mathbf{G}$ is a reciprocal lattice vector. The dynamic X-ray diffraction theory gives the wave in the crystal as [17]:

$$E_g(x,y) = A(x,y) \sum_{\sigma=\pm 1} \int_{-\infty}^{+\infty} B(s) \left[\frac{\beta}{\sigma\sqrt{s^2+\beta^2}} \right] \times \exp\left[i\sigma\xi_1\sqrt{s^2+\beta^2} - i\xi_2 s \right] ds, \quad (1)$$

where $A(x,y)$ is a fast oscillating field with the ensuing integral being its envelope, ξ_1 and ξ_2 represent the local coordinates of an oblique two-dimensional (2D) coordinate system (hence, ξ_1 and ξ_2 are two linear functions of x and y) induced by the dispersion surface; $\sigma = 1$ and -1 for two branches of dispersion surface, respectively; $s = \Delta k + C_1 k_0$, with Δk being the wavevector deviation from k_0; $\beta = C_2 k_0 \varepsilon_G$. The geometric constants C_1, C_2 are of order unity. Here the dispersion surface refers to the constant frequency surface in the reciprocal space. The function $B(s)$ is introduced by us in this work to describe the wavevector distribution (note that s is a function of Δk). The usage of $B(s)$ will be discussed shortly. Equation (1) was obtained through lengthy and complicated derivations, where the physical meaning of the quantities s, β, and ξ_i became obscure. However, it is sufficient to note that the phase factor $\exp[i\sigma\xi_1\sqrt{s^2+\beta^2} - i\xi_2 s]$ in (1) is equivalent to $\exp[i(\mathbf{k} - \mathbf{k}_0)\cdot\mathbf{r}]$, where $\mathbf{k}_0$ is constant. Thus, s and $\sqrt{s^2+\beta^2}$ are the projections of wavevector deviation $\Delta\mathbf{k} = \mathbf{k} - \mathbf{k}_c$ on the ξ_1 and ξ_2 axes, respectively. Note that as s varies, β varies accordingly, so that $\mathbf{k}$ remains on the dispersion surface. Therefore, β is a function of s. In the spherical-wave dynamic X-ray theory [17], due to the limitation of analytic study, only the case $B(s) \equiv 1$ is investigated. It was analytically shown that the wave pattern $E_g(x,y)$ spreads within a triangle spanned by $\mathbf{k}_0$ and $\mathbf{k}_g$ and have hook-shaped interference patterns. For polymer photonic integrated circuits applications, such a triangular spread leads to significant beam divergence, which may increase device lateral dimensions and optical loss. In addition, an additional beam focuser or waveguide taper becomes necessary to couple light into an output waveguide. This increases the device complexity and size.

To overcome these issues originated from beam divergence, it is necessary to quantitatively study the beam divergence with varying structure parameters and seek approaches to minimize the beam divergence under the constraints of small index-contrast. To start, we note that in the X-ray theory, the beam is assumed to originate from a point source with a uniform angular intensity spectrum. Such an ideal simplification is not applicable to integrated optics applications where the input beam has a finite width. We have introduced a wavevector distribution function $B(s)$ in (1) to deal with arbitrary beam profiles. Note that for any given lateral beam profile $b(x_\perp)$ in space ($x_\perp$ being the coordinate transverse to the beam propagation direction), the corresponding wavevector distribution function $\widetilde{B}(\Delta k_\perp)$ can be obtained from the Fourier transform of $b(x_\perp)$. Note that for a monochromatic beam in 2D space, the wavevector deviation has only one degree of freedom. Therefore, one can easily link $\widetilde{B}(\Delta k_\perp)$ to $B(s)$, noting that s is a linear function of Δk. Here we address the case where the wavevector is limited to a small angular range such that $\Delta k_{\max} \ll |\beta|$. Assuming a Gaussian beam profile $B(s)$ and expanding (1) in $\Delta s/\beta$ (or equivalently $\Delta k/\beta$), one can readily show that for a sufficiently small $\Delta k/\beta \ll 1$, the beam is well collimated without obvious divergence. This case is of our interest here. Introduce the wavevector angle θ_k, with respect to the beam axis, we find $\Delta k = k_0 \sin\theta_k$. Also, note $\beta = C_2 k_0 \varepsilon_G$, where C_2 is a constant on the order of unity. Hence, the condition ($\Delta k_{\max} \ll |\beta|$) for a well-collimated beam can also be formulated as

$$\theta_{\mathrm{km}} \ll \varepsilon_G, \quad (2)$$

where θ_{km} is the characteristic angular spread of the wavevector (roughly speaking, the beam divergence angle) for a finite beam. Consider a Gaussian beam with a width w propagating in a 2D square lattice of air holes in a medium of relative dielectric constant ε_1. It is well known that the beam angular divergence is related to the beam width as $\theta_{\mathrm{km}} \propto \lambda/w$. One can also readily show that the Fourier coefficient ε_G generally has the form of Bessel functions for a photonic crystal composed of circular holes [18]. Thus, based on (2), we can readily show that the condition for a well-collimated beam is given by

$$w \gg w_c = \frac{2\lambda}{\pi\sqrt{\varepsilon_1}(\varepsilon_1 - 1)} \frac{a}{r J_1(2\pi r/a)}, \quad (3)$$

where r is the hole radius, a is the lattices constant, λ is the wavelength, $J_1(x)$ is a Bessel function of the first kind, and w_c can be regarded a characteristic beam width pertinent to the divergence issue. Assume a metamaterial is composed of air holes in a polymer matrix with typical polymer index $n_{\mathrm{polymer}} = 1.5$, we plot w_c as a function

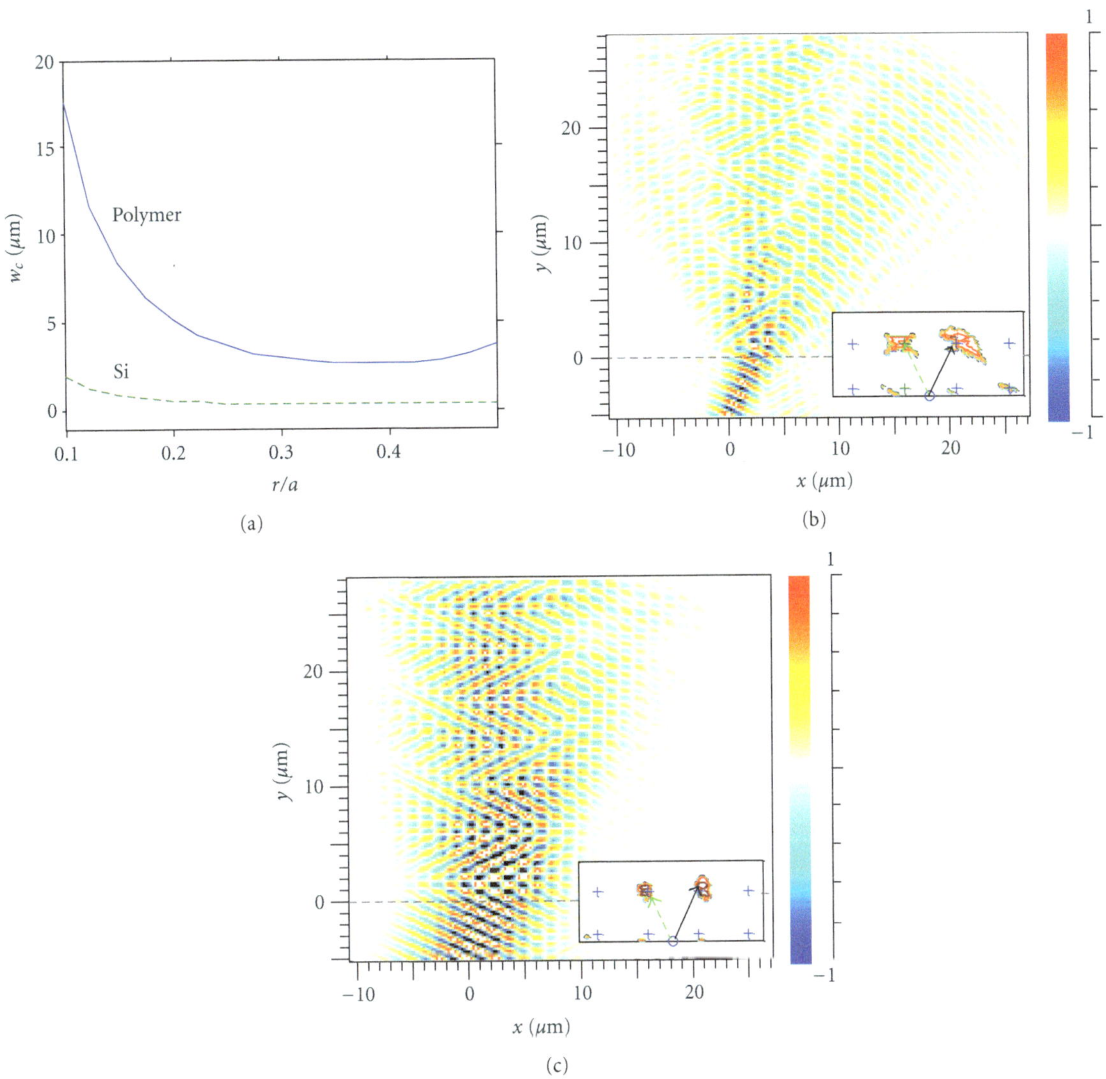

FIGURE 1: (a) w_c versus r/a curve given by (3) for polymer-based and silicon-based metamaterials. (b) Beam propagation in a polymer metamaterial for a Gaussian beam with a width $w = 3\,\mu m$. The metamaterial region starts at $5\,\mu m$ from the bottom of the figure (the interface is indicated by a dashed line) and extends upwards. Inset: 2D Fourier transform of the spatial field in the photonic crystal. The wavevector $\mathbf{k}_0$ and $\mathbf{k}_g$ are indicated by a black arrow and a green dash arrow. The wavevector of various diffraction orders (on a square lattice in the reciprocal space) are marked by crosses. (c) Same as in (b) except $w = 10\,\mu m$.

of r/a in Figure 1(a) for $\lambda = 1.55\,\mu m$. For comparison, the w_c for a high index-contrast metamaterial composed of air holes in a silicon matrix is also shown. Evidently, the polymer-based metamaterial system requires substantially wider beams than the silicon-based metamaterial system. In this work, the low index-contrast regime pertaining to metamaterial beam collimation refers to $|\Delta n|$ substantially less than unity (e.g., $|\Delta n| \sim 0.5$), which is in contrast to high index-contrast metamaterial systems such as silicon-air or silver-air. Note that w should usually be at least 3 to 10 times larger than w_c in order to satisfy (3). For the polymer-based metamaterial, the minimal $w_c \sim 2.7\,\mu m$ appears at $r/a \sim 0.37$. For $r/a \geq 0.3$, the curve is relatively flat, indicating w_c is insensitive to hole radius change in this range. Overall, Figure 1(a) shows that the beam width must be much larger than $3\,\mu m$ for a polymer-based metamaterial at $\lambda = 1.55\,\mu m$. Figures 1(b) and 1(c) compare two beams of $3\,\mu m$ and $10\,\mu m$ wide, propagating in a $r/a = 0.3$ lattice as simulated by the Finite Difference Time-Domain (FDTD) method for polarization normal to the plane. The metamaterial region starts at $5\,\mu m$ from the bottom (each division on the axis is a micron) and extends upward. For the $10\,\mu m$ wide beam case in Figure 1(c), the optical wave inside the metamaterial has the majority of its power propagating in a collimated beam (vertically upwards and slightly tilting to the left), whereas the 3-μm wide beam in Figure 1(b) results in a diverging field spreading over a triangle whose left and right edges are roughly parallel to $\mathbf{k}_0$ and $\mathbf{k}_g$. Note that in Figure 1(c), the incident beam points towards the right hand side whereas the beam propagation direction inside the metamaterial slightly tilts towards the left side. This indicates that the beam experiences negative refraction

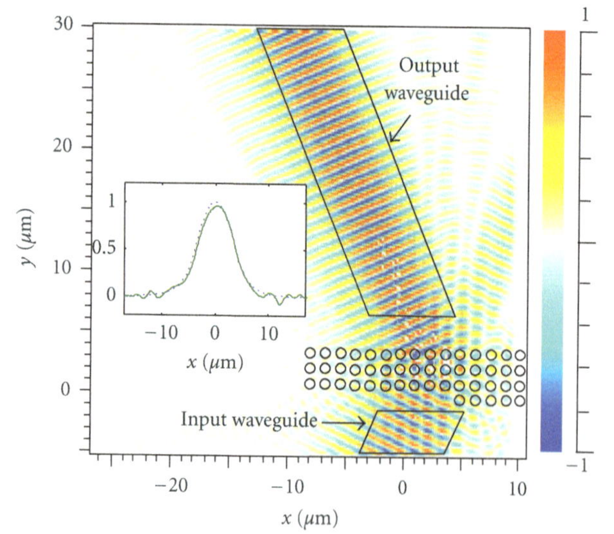

FIGURE 2: An ultra-compact 47.3° beam deflector (effectively a wide-angle waveguide bend). Inset indicates the output beam profile (solid curve) is well matched to the waveguide mode profile (dotted line).

at the metamaterial surface, which is anomalous compared to ordinary refraction. For Figure 1(b), the diverging beam wipes out the signature of negative refraction. These FDTD simulation results confirm that the beam width w must be sufficiently larger than the characteristic beam width w_c in order to form a well-collimated beam and observe negative refraction in a low index-contrast metamaterial. The insets in Figures 1(b) and 1(c) show the power spectrum of the 2D spatial Fourier transforms of the respective spatial fields. The un-diffracted wavevector $\mathbf{k}_0$ is shown by an arrow with the origin $\mathbf{k} = 0$ noted by a circle. Because the lower half of the Fourier spectrum is symmetric to the upper half, only the upper half the 2D Fourier spectrum is shown. The wavevectors of various diffraction orders (on a square lattice in the reciprocal space) are marked by crosses. Their spatial Fourier spectra are dominated by two Fourier components (centered around $\mathbf{k}_0$ and $\mathbf{k}_g$).

3. Potential Device Application: A Beam Deflector

A closer look of Figure 1(c) reveals an alternation of two different wave components, whose (center) wavevectors are different from the beam direction. When the beam is collimated as Figure 1(c), this can be utilized to realize low-loss wide angle beam deflection. Figure 2 presents the 2D FDTD simulation result of a wide-angle (47.3°) deflector, which essentially has the functionality of a wide-angle waveguide bend. The incident beam is a waveguide mode matched to a typical fiber mode, which has a mode width around 10 μm. The 2D square lattice metamaterial has a lattice constant $a = 1.3\,\mu$m, and a hole radius $r = 0.3a$. The lattice constant and hole radius are relatively large to facilitate fabrication. More discussion on practical implementation and fabrication will be presented at the end of this section. High transmission and excellent mode match (see the inset of Figure 2) are achieved at the output end, giving an insertion loss less than 0.5 dB. Note that for the noncollimated case shown in Figure 1(b), because of widely spread beam (whose profile quickly deviates from Gaussian after a short propagation length) inside the metamaterial, it is very difficult to achieve a collimated deflected beam at the output and the loss is usually significantly higher. Note that for the case shown in Figure 2, the beam profile quickly stabilizes after exiting the metamaterial region. The bottom layer of the structure is intentionally designed to be incomplete. Our simulation indicates that such an incomplete layer enhances the transmission. One possible explanation is that when the beam is making a left turn, the right portion of the beam is on the "outer track" and travels a longer distance. Thus, the right portion of the beam also needs a longer interaction length in the metamaterial to complete the turn on the "outer track." Therefore, the right portion of the metamaterial shall be relatively thicker. It is worth noting that at the output surface of the metamaterial region in Figure 2, the beam experiences a second negative refraction. The negative refractions at the front and back surfaces both contribute substantially to producing a large total beam deflection angle.

Figure 3 shows the transmission loss variation with respect to two essential parameters (λ and n). For simulation results in Figure 3, the baseline structure has all parameters same as the optimized structure in Figure 2 except that we have used a generic polymer refractive index $n_{\text{polymer}} = 1.5$ (the structure in Figure 2 is optimized to have $n_{\text{polymer}} = 1.42$). This baseline structure has an insertion of 0.7 dB. Additionally, the loss variation with 50 nm hole radius deviation is less than 0.1 dB. These results show that this device is wide band (whole C-Band) and relatively insensitive to slight deviation of polymer index and fabrication tolerance of the hole radius. Note that a hole radius tolerance of 50 nm is readily achievable using electron beam lithography. Also, the relative insensitivity to polymer index allows a wide range of polymers to choose from. Note that we have intentionally designed our structure with a relatively large a, and an intermediate r/a to facilitate the fabrication. Note the traditional waveguide bends for polymer integrated photonic devices are usually several millimeters long due to the small index-contrast of polymers [13]. The metamaterial structures obviously have advantages in device miniaturization.

For practical implementation of this structure, each waveguide will be a channel waveguide composed of higher index polymer core and lower index polymer cladding, as shown in Figure 4. In the metamaterial region, the core layer will be etched away; thus, the conformal nature of the top polymer cladding will likely lower the top cladding surface to the same height as the top surface of the core layer. To ensure the functionality of the metamaterial, the metamaterial region will be etched to a depth that is greater than the depth the polymer core. Note that polymer waveguides generally have fairly small refractive index-contrasts ($\Delta n \sim 0.01$) between core and cladding,

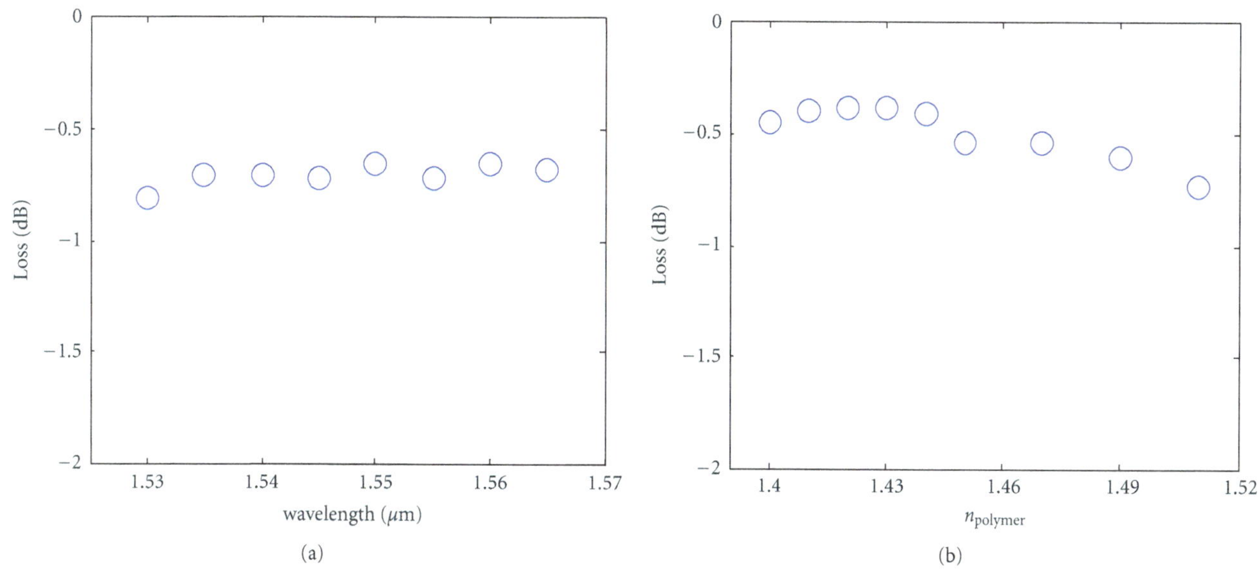

FIGURE 3: Loss variation with (a) wavelength and (b) polymer index.

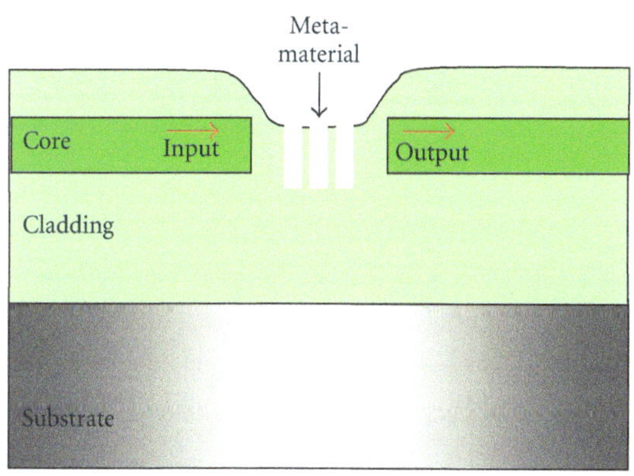

FIGURE 4: Schematic of an example device structure (side view, not drawn to scale).

and relatively large cross-sections. Here we assume a core thickness of 3 μm and a top cladding thickness of 2 μm. As such, it is estimated that we need to etch ~3 μm deep into the polymer in the metamaterial region. Although there is no vertical confinement mechanism in the metamaterial region, optical loss due to beam divergence in the vertical direction is expected to be insignificant due to the short length of the metamaterial region (~5 μm long) through which the beam travels.

4. Conclusions

In summary, the condition for achieving well-collimated beams in a metamaterial has been analyzed. Guided by this analysis, we have designed ultra-compact, wide-angle beam deflection devices that are significantly smaller than conventional polymer integrated photonic devices. The simulations also show that these devices have a wide bandwidth and are insensitive to fabrication tolerances. A wide range of polymers can be chosen to fabricate these structures. Such polymer photonic devices are attractive for their compact size, low cost and flexible form.

Acknowledgment

This work is supported in part by AFOSR Grant No. FA9550-08-1-0394.

References

[1] D. R. Smith, J. B. Pendry, and M. C. K. Wiltshire, "Metamaterials and negative refractive index," *Science*, vol. 305, no. 5685, pp. 788–792, 2004.
[2] V. M. Shalaev, W. Cai, U. K. Chettiar et al., "Negative index of refraction in optical metamaterials," *Optics Letters*, vol. 30, no. 24, pp. 3356–3358, 2005.
[3] S. Zhang, W. Fan, N. C. Panoiu, K. J. Malloy, R. M. Osgood, and S. R. J. Brueck, "Experimental demonstration of near-infrared negative-index metamaterials," *Physical Review Letters*, vol. 95, no. 13, Article ID 137404, 4 pages, 2005.
[4] S. Linden, C. Enkrich, M. Wegener, J. Zhou, T. Koschny, and C. M. Soukoulis, "Magnetic response of metamaterials at 100 terahertz," *Science*, vol. 306, no. 5700, pp. 1351–1353, 2004.
[5] D. Schurig, J. J. Mock, B. J. Justice et al., "Metamaterial electromagnetic cloak at microwave frequencies," *Science*, vol. 314, no. 5801, pp. 977–980, 2006.
[6] H. J. Lezec, J. A. Dionne, and H. A. Atwater, "Negative refraction at visible frequencies," *Science*, vol. 316, no. 5823, pp. 430–432, 2007.
[7] N. Engheta, "Circuits with light at nanoscales: optical nanocircuits inspired by metamaterials," *Science*, vol. 317, no. 5845, pp. 1698–1702, 2007.
[8] J. A. Schuller, R. Zia, T. Taubner, and M. L. Brongersma, "Dielectric metamaterials based on electric and magnetic resonances of silicon carbide particles," *Physical Review Letters*, vol. 99, no. 10, Article ID 107401, 4 pages, 2007.

[9] J. Valentine, S. Zhang, T. Zentgraf et al., "Three-dimensional optical metamaterial with a negative refractive index," *Nature*, vol. 455, no. 7211, pp. 376–379, 2008.

[10] S. E. Kocabaş, G. Veronis, D. A. B. Miller, and S. H. Fan, "Transmission line and equivalent circuit models for plasmonic waveguide components," *IEEE Journal on Selected Topics in Quantum Electronics*, vol. 14, no. 6, pp. 1462–1472, 2008.

[11] W. Zhao, X. Huang, and Z. Lu, "Super Talbot effect in indefinite metamaterial," *Optics Express*, vol. 19, no. 16, pp. 15297–15303, 2011.

[12] J. X. Chen, P. Wang, C. C. Chen, Y. H. Lu, H. Ming, and Q. W. Zhan, "Plasmonic EIT-like switching in bright-dark-bright plasmon resonators," *Optics Express*, vol. 19, no. 7, pp. 5970–5978, 2011.

[13] R. T. Chen, "Polymer-based photonic integrated circuits," *Optics and Laser Technology*, vol. 25, no. 6, pp. 347–365, 1993.

[14] H. Kosaka, T. Kawashima, A. Tomita et al., "Superprism phenomena in photonic crystals," *Physical Review B*, vol. 58, no. 16, pp. 10096–10099, 1998.

[15] T. Baba and M. Nakamura, "Photonic crystal light deflection devices using the superprism effect," *IEEE Journal of Quantum Electronics*, vol. 38, no. 7, pp. 909–914, 2002.

[16] B. Momeni, J. Huang, M. Soltani et al., "Compact wavelength demultiplexing using focusing negative index photonic crystal superprisms," *Optics Express*, vol. 14, no. 6, pp. 2413–2422, 2006.

[17] L. V. Azaroff, R. Kaplow, N. Kato, R. J. Weiss, A. J. C. Wilson, and R. A. Young, *X-Ray Diffraction*, McGraw-Hill, New York, NY, USA, 1974.

[18] W. Jiang and R. T. Chen, "Symmetry-induced singularities of the dispersion surface curvature and high sensitivities of a photonic crystal," *Physical Review B*, vol. 77, no. 7, Article ID 075104, 8 pages, 2008.

2

Light Combining for Interferometric Switching

Marco Masi,[1] Mattia Mancinelli,[2] Paolo Bettotti,[2] and Lorenzo Pavesi[2]

[1] *Institut des Nanotechnologies de Lyon, Université de Lyon, INL-UMR5270, CNRS, INSA de Lyon, 69621 Villeurbanne, France*
[2] *Nanoscience Laboratory, Department of Physics, University of Trento, Via Sommarive 14, 38121 Trento, Italy*

Correspondence should be addressed to Marco Masi, marco.masi@gmail.com

Academic Editor: Francesco Prudenzano

Interferometric switching as a routing method in sequence of coupled optical microresonators is explored. Mach-Zhender interferometry is extended to systems of side-coupled integrated sequences of resonators (SCISSORs) and coupled resonators optical waveguides (CROWs). We generalized Coupled Mode Theory (CMT) to a system of three coupled waveguides. The two bus interferometric switching functions of SCISSOR and CROW resonant structures are investigated. A novel switching device based on three input phase modulation ports is presented. This device displays a wide range of switching behaviors which might lead to new interesting applications.

1. Introduction

The tremendous growth of communication services and information technologies demands new and enhanced networking capabilities. Switching is one of the main functions of communication systems and networks. In particular, novel optical switching technologies have the potential to play a decisive role in future telecommunication and information processing systems. The extensive deployment of wavelength division multiplexing (WDM) technologies demands for high-speed, scalable, and rapidly reconfigurable network switching. Several optical switching technologies exist, like electrooptic (EO), acoustooptic (AO), thermooptic (TO), optomechanical (OM), and optical amplifier (OA) based switching. Refractive index modulation, inducing phase differences for switching functionalities, resorts especially to EO or TO effects. These are widely used in directional couplers, Mach-Zehnder interferometers, and multi-mode interference (MMI) switches (for a good review see, e.g., [1]). The present paper focuses on some possible extensions of these interferometric techniques by means of sequences of microoptical resonator systems, like SCISSORs (side-coupled integrated spaced-sequences of resonators) and CROWs (coupled resonator optical waveguides) [2–6]. A particular combination of these allows to use three input waveguides for amplitude and phase modulation. The advantage of a system of resonators over conventional single-resonator scheme coupling is its larger spectral band, the possibility to operate through interferometry on several channels and its higher design flexibility. In particular, this latter aspect allows for a diversity of behaviors compared to usual directional coupling switching.

The practical realization of such kind of phase switching devices is, however, still limited by present photonic fabrication tolerances. Phase shifting of light requires the same interferometric precision as the device operation itself. For example, the phase difference between two interfering light beams needs to be tightly controlled at the subwavelength level to provide efficient switching. Even more sensitive to small deviations from the nominal parameters are devices where coherent addition of signals is achieved by arranging complex systems of resonator chains. These require nanometer level accuracy in the fabrication of the single resonators and their mutual spacings. For instance, coupled resonator induced transparency (CRIT) effects emerge easily only for few nanometer deviations [7–12]. However, while present optical lithography still suffers of few nanometers imprecision needing further improvements, the next generation optical lithography is likely to achieve an order of magnitude leap in accuracy paving the way to novel interferometric devices [13]. Also, the fabrication of electrooptic silicon-photonic modulators which are capable of providing smooth and uniform phase shifts over a broad spectrum is a topic of intense research and promises to find a wide area of applications [14]. Elsewhere we have demonstrated the practical realization of a novel silicon photonic interferometric switching device showing that resonator phase switching is indeed feasible [12]. Therefore, it is interesting to explore from the theoretical point of view some extensions of conventional directional coupling methods which could have potential applications in upcoming interferometric devices.

The next section will briefly summarize the basics of coupled mode theory (CMT) which will be particularly useful in our context. Section 2 will set the stage for interferometric switching between three waveguides which in turn will be used in Section 3 where interferometric switching with a dual bus waveguide resonator system is described. Section 4 will extend this to a particular arrangement of three-bus resonator systems.

2. CMT and Phase Switching for the Double-Sided Symmetric Codirectional Coupler

CMT has grown to a vast subject in the last three decades (for a good introduction see e.g., [15]). Let us quickly recall some standard equations which describe in particular the coupling between two waveguides.

We assume that the two waveguides have the same geometry and propagation constants β, and they are homogenous and isotropic. Material or other types of losses are negligible. Within these assumptions, the fields in the first and second waveguide are given by the solution of the following coupled differential equations:

$$\frac{\partial A_1(z)}{\partial z} = icA_2(z),$$
$$\frac{\partial A_2(z)}{\partial z} = icA_1(z), \tag{1}$$

with $A_1(z)$ and $A_2(z)$ are the z dependent parts of the fields, and c a mutual mode coupling coefficient. c is obtained from perturbation theory by the amplitudes cross-sectional integral over the section of the two waveguides of the codirectional coupler and has the dimension of an inverse length. If we impose as initial conditions

$$A_1(0) = A_1 e^{i\phi_1}, \qquad A_2(0) = A_2 e^{i\phi_2}, \tag{2}$$

then the solutions at z are

$$A_1(z) = tA_1 e^{i\phi_1} + i\kappa A_2 e^{i\phi_2},$$
$$A_2(z) = i\kappa A_1 e^{i\phi_1} + tA_2 e^{i\phi_2}, \tag{3}$$

with

$$\kappa = \sin(cz)e^{i\beta z}, \qquad t = \cos(cz)e^{i\beta z}, \tag{4}$$

being κ and t the cross- and through-coupling coefficients, respectively, and $e^{i\beta z}$ accounts for the phase shift. κ and t are given by

$$\kappa = \sin\left(\frac{\pi L}{2L_c}\right)e^{i\beta L}, \qquad t = \cos\left(\frac{\pi L}{2L_c}\right)e^{i\beta L}, \tag{5}$$

where

$$L_c = \frac{\lambda}{2(n_{\text{effe}}(\lambda) - n_{\text{effo}}(\lambda))} \tag{6}$$

is the coupling length, λ the wavelength, L the length of the coupling section, while n_{effe} and n_{effo} are the even and odd mode wavelength-dependent effective indexes, respectively. Then, from (4) and (5), one obtains

$$c = \frac{\pi}{2L_c}. \tag{7}$$

From (3), it follows that

$$|A_1(z)|^2 = |t|^2|A_1|^2 + |\kappa|^2|A_2|^2 + 2|\kappa||t||A_1||A_2|\sin\Delta\phi,$$
$$|A_2(z)|^2 = |\kappa|^2|A_1|^2 + |t|^2|A_2|^2 - 2|\kappa||t||A_1||A_2|\sin\Delta\phi, \tag{8}$$

with $\Delta\phi = \phi_1 - \phi_2$. For simplicity in (8), the dependence of the coupling coefficients on z is omitted. Power conservation condition for the lossless system follows straightforwardly from (5), or (8), as $|t|^2 + |\kappa|^2 = 1$.

If we set $A_2 = 0$, we obtain the well-known power exchange expressions for the single-sided codirectional coupler made of two equal waveguides

$$|A_1(z)|^2 = \cos^2(cz)|A_1|^2 = |t|^2|A_1|^2,$$
$$|A_2(z)|^2 = \sin^2(cz)|A_1|^2 = |\kappa|^2|A_1|^2. \tag{9}$$

The Mach-Zehnder modulator is obtained when two signals with same amplitude ($A_1 = A_2 = A$) are coupled into the waveguides. In fact, (8) becomes

$$|A_1(z)|^2 = |A|^2(1 + 2|\kappa||t|\sin\Delta\phi),$$
$$|A_2(z)|^2 = |A|^2(1 - 2|\kappa||t|\sin\Delta\phi), \tag{10}$$

which shows that, when $\Delta\phi = 0$, no power exchange between the two waveguides occurs, while, for $\Delta\phi = m(\pi/2)$ (m an integer), the power oscillates between the two waveguides depending on the coupling section length, L. If $cL = \pi/4$ and $\Delta\phi = \pm\pi/2$, then all the power is transferred into one or the other waveguide.

Let us now extend (1) to the situation of Figure 1, where three waveguides couple. We call the central waveguide the Drop waveguide for a reason that will be clear in the following. We consider the case of a Drop signal A_D excited by the fields A_1 and A_2 in the upper and lower waveguides. Then, neglecting the direct coupling between the two outer waveguides, we write a set of three coupled differential equations:

$$\frac{\partial A_1(z)}{\partial z} = icA_D(z),$$
$$\frac{\partial A_D(z)}{\partial z} = icA_1(z) + icA_2(z), \tag{11}$$
$$\frac{\partial A_2(z)}{\partial z} = icA_D(z).$$

We impose the initial conditions

$$A_1(0) = A_1 e^{i\phi_1},$$
$$A_D(0) = A_D e^{i\phi_D}, \tag{12}$$
$$A_2(0) = A_2 e^{i\phi_2},$$

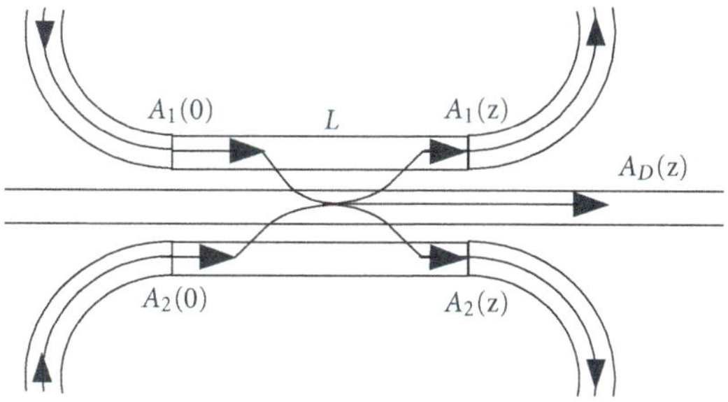

FIGURE 1: Double sided symmetric codirectional coupler.

where ϕ_1, ϕ_1, and ϕ_D express the phase of A_1, A_2, and A_D, respectively. The solutions of (11) are

$$\begin{aligned} A_1(z) &= t' A_1 e^{i\phi_1} + i\kappa_D A_D e^{i\phi_D} - \kappa' A_2 e^{i\phi_2}, \\ A_D(z) &= i\kappa_D A_1 e^{i\phi_1} + t_D A_D e^{i\phi_D} + i\kappa_D A_2 e^{i\phi_2}, \\ A_2(z) &= -\kappa' A_1 e^{i\phi_1} + i\kappa_D A_D e^{i\phi_D} + t' A_2 e^{i\phi_2}, \end{aligned} \tag{13}$$

with

$$t' = \cos^2\left(\frac{cz}{\sqrt{2}}\right)e^{i\beta z}, \qquad t_D = \cos\left(\sqrt{2}cz\right)e^{i\beta z}, \tag{14}$$

$$\kappa' = \sin^2\left(\frac{cz}{\sqrt{2}}\right)e^{i\beta z}, \qquad \kappa_D = \frac{1}{\sqrt{2}}\sin\left(\sqrt{2}cz\right)e^{i\beta z}. \tag{15}$$

The term $1/\sqrt{2}$ in (15) is a consequence of the symmetry of the system (symmetric coupling and geometry). For simplicity, let us consider $A_D = 0$, then

$$\begin{aligned} |A_1(z)|^2 &= |t'|^2|A_1|^2 + |\kappa'|^2|A_2|^2 \\ &\quad - 2|\kappa'||t'||A_1||A_2|\cos\Delta\phi, \end{aligned} \tag{16}$$

$$|A_D(z)|^2 = |\kappa_D|^2\left(|A_1|^2 + |A_2|^2 + 2|A_1||A_2|\cos\Delta\phi\right), \tag{17}$$

$$\begin{aligned} |A_2(z)|^2 &= |\kappa'|^2|A_1|^2 + |t'|^2|A_2|^2 \\ &\quad - 2|\kappa'||t'||A_1||A_2|\cos\Delta\phi. \end{aligned} \tag{18}$$

If the system is lossless, power conservation implies $|k'|^2 + |k_D|^2 + |t'|^2 = 1$. Despite not being directly coupled to each others, (16) and (18) show a possible crosstalk between the outer waveguides through the intermediate Drop port (crossing of arrows through the Drop port shown in Figure 1). This can be quantified by using as initial condition, for example, $A_2 = 0$ which implies from (16)–(18)

$$\begin{aligned} |A_1(z)|^2 &= |A_1|^2\cos^4\left(\frac{cz}{\sqrt{2}}\right), \\ |A_D(z)|^2 &= \frac{|A_1|^2}{2}\sin^2\left(\sqrt{2}cz\right), \\ |A_2(z)|^2 &= |A_1|^2\sin^4\left(\frac{cz}{\sqrt{2}}\right). \end{aligned} \tag{19}$$

Equation (19) is plotted in Figure 2 in normalized intensity and distance. These show that the signal from the upper (lower) waveguide is never completely recovered in the Drop port because part of it recouples towards the lower (upper)

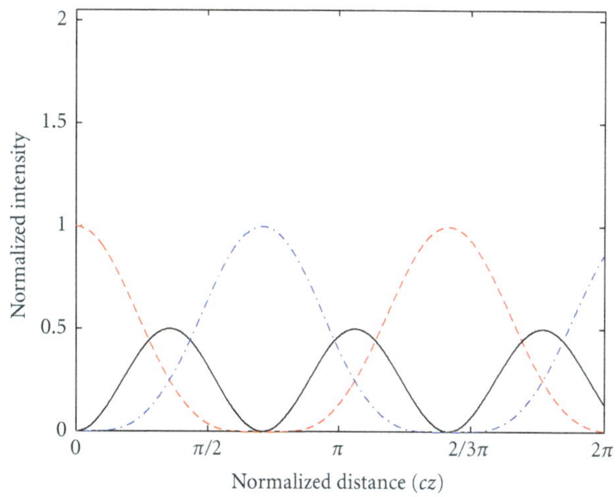

FIGURE 2: Power transfer in the waveguides of the ds-coupler for one input signal only ((16), (17), and (18), with initial conditions $A_1 = 1$, $A_2 = 0$). Red dashed line: amplitude $A_1(cz)$; blue dotted-dashed line: amplitude $A_2(cz)$; black solid line: amplitude $A_D(cz)$.

waveguide. The Drop behaves like a "power transiting" port, and the amount of light that crosses the structure can be considered as a measure of crosstalk. However, we will see that this interpretation must be taken with caution.

In fact, if we consider a same input signal ($A_1 = A_2 = A$) on the top and bottom waveguides, (16)–(18) simplify to

$$\begin{aligned} |A_1(z)|^2 = |A_2(z)|^2 &= |A|^2\left(1 - \frac{1}{2}\sin^2\left(\sqrt{2}cz\right)(1 + \cos\Delta\phi)\right), \\ |A_D(z)|^2 &= |A|^2(1 + \cos\Delta\phi)\sin^2\left(\sqrt{2}cz\right). \end{aligned} \tag{20}$$

The case for in phase signals ($\Delta\phi = 0$) is shown in Figure 3(a). In this case, the energy transfer between the Drop and external ports occurs harmonically. A 100% power transfer to the Drop port occurs when

$$L = \frac{\pi}{2\sqrt{2}c} = \frac{L_c}{\sqrt{2}}. \tag{21}$$

Figures 3(b) and 3(c) show this case for a phase difference of $\Delta\phi = \pi/4$ and $\Delta\phi = (3/4)\pi$: the power in the outer waveguides 1 and 2 (magenta dotted-dashed line) oscillates around the initial input value but is never zero, while the Drop signal (black solid line) oscillates according to the propagation length like in the previous case, but with less maximum intensity. If instead a phase difference of $\Delta\phi = \pi$ is applied (Figure 3(d)), there is no power in the Drop port.

Therefore, a particularly interesting situation occurs when a ds-coupler has a coupling section long as in (21) ($cz = (1/\sqrt{2})(\pi/2)$ in Figure 3). In fact, for two equal input signals the ds-coupler acts as an interferometric switch: for $\Delta\phi = 0$, all the power goes into the drop port (Figure 4), while, for $\Delta\phi = \pi$, all the power stays in waveguides 1 and 2 (Figure 5). Under these circumstances, the crosstalk between waveguide 1 and 2 has been suppressed. At first, this seems to be at odds with what shown in Figure 2 where a

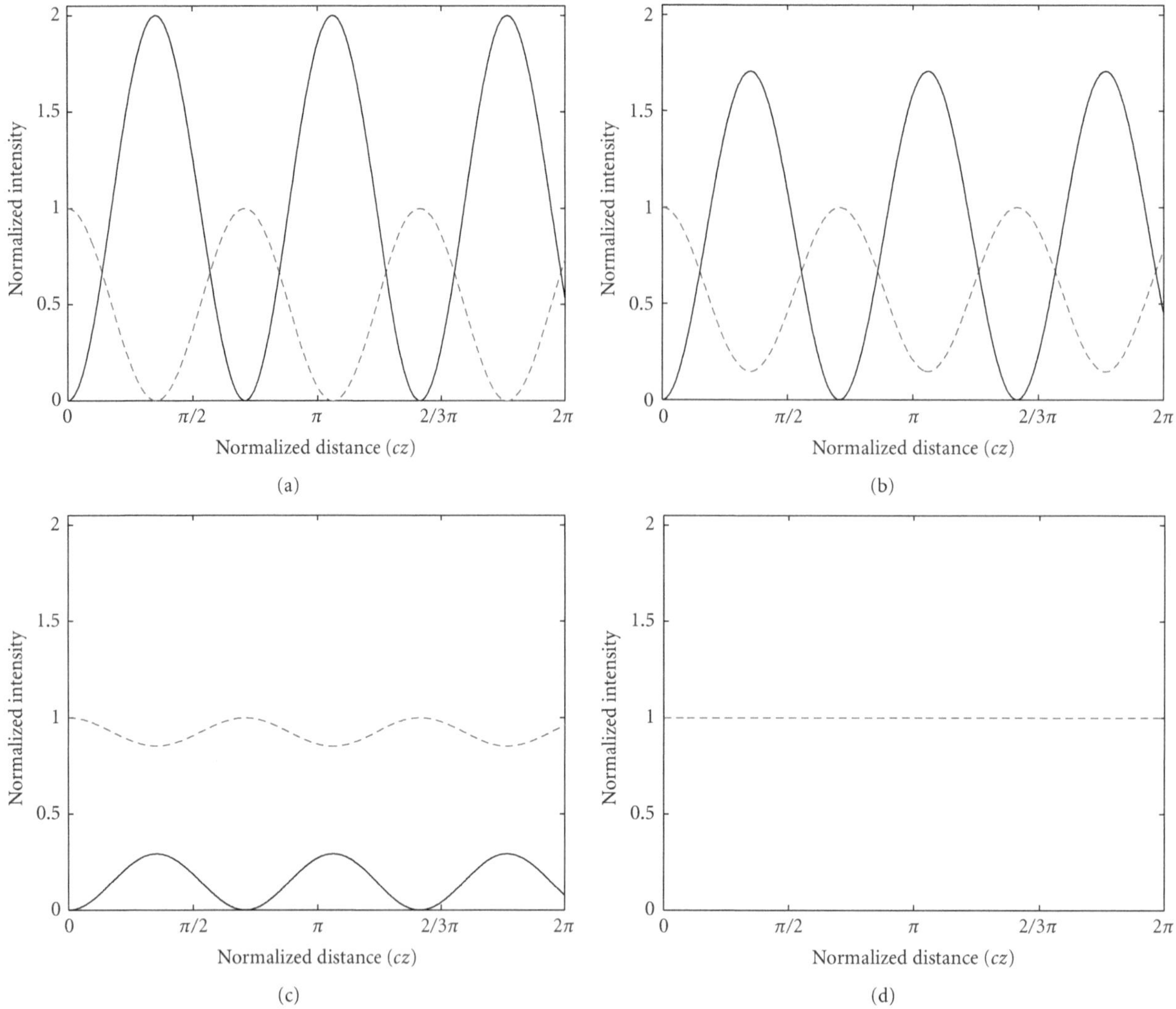

FIGURE 3: Power transfer in the waveguides of the ds-coupler for two equal input signals ((16), (17), and (18), with initial conditions $A_1 = 1$, $A_2 = 1$, $\Delta\phi = (0, \pi/4, (3/4)\pi, \pi)$ in (a), (b), (c), (d), resp.). Magenta dotted-dashed line: amplitude $A_1(cz)$ and $A_2(cz)$; black solid line: amplitude $A_D(cz)$.

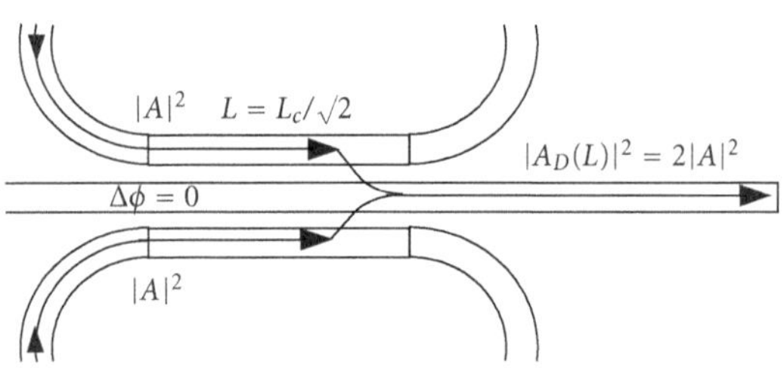

FIGURE 4: The ds-coupler switch in phase: all the power goes into the Drop port.

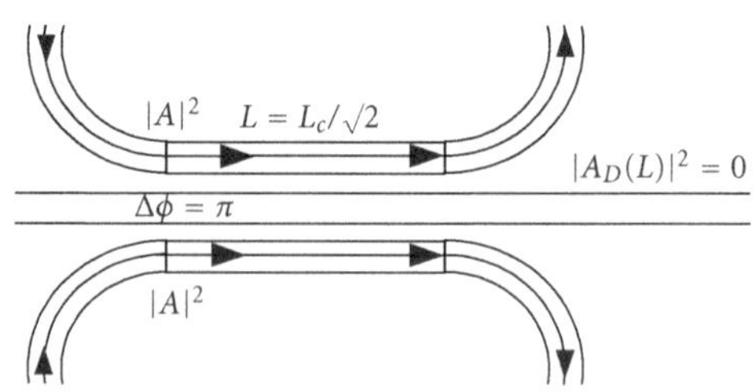

FIGURE 5: The ds-coupler switch out of phase: no power couples into the Drop port.

significant power transfer from waveguide 1 to waveguide 2 is observed. This apparent contradiction is explained by the fact that the signal in waveguide 1 (waveguide 2) undergoes twice a phase change of $\pi/2$ (evanescent waves have always a $\pi/2$ phase difference with respect to the waveguide core field phase) and the net π dephased cross coupled wave interferes destructively in waveguide 2 (waveguide 1). This interference effect leads to a mutual subtraction of the cross-coupled powers.

For completeness, we mention also the case when the light is injected only in the Drop waveguide, that is,

$$A_1(0) = 0, \qquad A_D(0) = A_D, \qquad A_2(0) = 0, \tag{22}$$

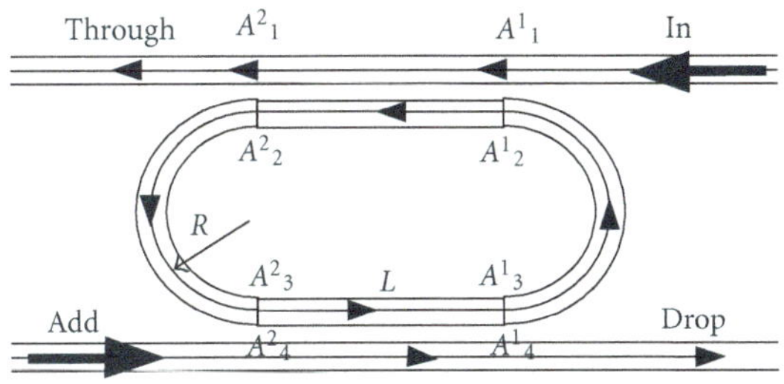

FIGURE 6: The single racetrack switch.

then

$$A_1(z) = A_2(z) = \frac{iA_D \sin(\sqrt{2}cz)}{\sqrt{2}} e^{i\beta z},$$
$$A_D(z) = A_D \cos\left(\sqrt{2}cz\right) e^{i\beta z}, \tag{23}$$

whereby of course

$$|A_1(z)|^2 = |A_2(z)|^2 = \frac{|A_D|^2}{2} \sin^2\left(\sqrt{2}cz\right) = |A_D|^2 \frac{|\kappa_D|^2}{2},$$
$$|A_D(z)|^2 = |A_D|^2 \cos^2\left(\sqrt{2}cz\right) = |A_D|^2 |t_D|^2. \tag{24}$$

These show that the power is equally distributed among the outer waveguides. Note that by imposing a coupling section length equal to the value given by (21), it leads to a complete power transfer into waveguide 1 and 2, that is, the ds-coupler can be used as a splitter. Its advantage over other splitting devices, as Y-branches or MMI splitters, is that it is less sensitive to imbalances, provided that the spacing difference between the waveguides arising due to fabrication imperfections is negligible. Its disadvantage is obviously that it is wavelength dependent.

3. Dual Bus Resonator Interferometric Switch

A further extension of the formalism is to consider a single resonator as in Figure 6. We consider a racetrack resonator so that we can apply the CMT extension developed in Section 2. For simplicity, the bus-waveguide gaps and the coupling section lengths are the same everywhere. To apply the transfer matrix approach [16–21], a first step is to relate the amplitudes (A_1^1, A_2^1) with (A_1^2, A_2^2) (see Figure 6 for the definitions). From (3), we have:

$$\begin{pmatrix} A_1^2 \\ A_2^2 \end{pmatrix} = \mathbf{M} \begin{pmatrix} A_1^1 \\ A_2^1 \end{pmatrix} = \begin{pmatrix} t & i\kappa \\ i\kappa & t \end{pmatrix} \begin{pmatrix} A_1^1 \\ A_2^1 \end{pmatrix}. \tag{25}$$

In order to "transfer" the signal, it is more appropriate to write the relationships $(A_1^1, A_1^2) \rightarrow (A_2^1, A_2^2)$. For which one obtains the coupling matrix $\mathbf{K}$:

$$\begin{pmatrix} A_2^1 \\ A_2^2 \end{pmatrix} = \mathbf{K} \begin{pmatrix} A_1^1 \\ A_1^2 \end{pmatrix} = \frac{1}{i|k|} \begin{pmatrix} -|t| & e^{-i\beta L} \\ -e^{i\beta L} & |t| \end{pmatrix} \begin{pmatrix} A_1^1 \\ A_1^2 \end{pmatrix}, \tag{26}$$

where, for the straight sections, the lossless case ($|\kappa_1|^2 + |t_1|^2 = 1$) was considered. We instead assume bending losses in the racetrack. The (forward and backward) internal propagation matrix $\mathbf{P}$ is

$$\begin{pmatrix} A_3^2 \\ A_3^1 \end{pmatrix} = \mathbf{P} \begin{pmatrix} A_2^1 \\ A_2^2 \end{pmatrix} = \begin{pmatrix} 0 & a e^{i\beta\pi R} \\ \frac{1}{a} e^{-i\beta\pi R} & 0 \end{pmatrix} \begin{pmatrix} A_2^1 \\ A_2^2 \end{pmatrix}, \tag{27}$$

where $a = e^{-\alpha\pi R}$ is the half round trip loss factor, α the total loss per unit length, and R the curvature radius.

Equations (26) and (27) define a transfer matrix $\mathbf{T}^{\mathrm{PK}}$ as

$$\begin{pmatrix} A_3^2 \\ A_3^1 \end{pmatrix} = \mathbf{T}^{\mathrm{PK}} \begin{pmatrix} A_1^1 \\ A_1^2 \end{pmatrix} = \mathbf{PK} \begin{pmatrix} A_1^1 \\ A_1^2 \end{pmatrix}. \tag{28}$$

With the definition of the Input, Through, Drop, and Add ports as in Figure 6, we define A_{In}, A_{Th}, A_D, and A_{Ad} the respective field amplitudes. Therefore,

$$\begin{pmatrix} A_{\mathrm{Ad}} \\ A_D \end{pmatrix} = \mathbf{T}^1 \begin{pmatrix} A_{\mathrm{In}} \\ A_{\mathrm{Th}} \end{pmatrix} = \mathbf{KT}^{\mathrm{PK}} \begin{pmatrix} A_{\mathrm{In}} \\ A_{\mathrm{Th}} \end{pmatrix}. \tag{29}$$

By rearranging the elements of the transfer matrix $\mathbf{T}^1$ into the scattering matrix $\mathbf{S}^1$, one obtains

$$\begin{pmatrix} A_{\mathrm{Th}} \\ A_D \end{pmatrix} = \mathbf{S}^1 \begin{pmatrix} A_{\mathrm{In}} \\ A_{\mathrm{Ad}} \end{pmatrix}. \tag{30}$$

To simulate the device of Figure 6, we use the following parameters: bent curvature radius of 3.25 μm, gap spacing of 0.2 μm, silicon waveguides width of 0.45 μm, silicon waveguide core height of 0.22 μm, SiO_2 cladding of 0.75 μm, and −0.03 dB/90° bent loss in TE polarization [22].

The first interesting situation is when we inject two input signals, one in the Input port and the other in the Add port such that $A_{\mathrm{In}} = e^{i\phi_1}$ and $A_{\mathrm{Ad}} = e^{i\phi_2}$, with ϕ_1 and ϕ_2 their phases. For $\Delta\phi = \phi_2 - \phi_1 = \pi/2$, we obtain Figure 7. The parameter γ represents the round-trip dephasing of the resonator given as $\gamma = 2\pi p_{\mathrm{opt}}(\Delta\lambda/\lambda^2)$, where $p_{\mathrm{opt}} = (2\pi R + 2L)\, n_{\mathrm{eff}}$ is the optical path of the resonator, $\Delta\lambda$ the wavelength shift from a racetrack resonance, and n_{eff} the effective index (at resonance wavelength) of the waveguides (see Appendix A).

The γ dependencies of the Drop and Through signals are asymmetric. This is the manifestation of the Fano resonance

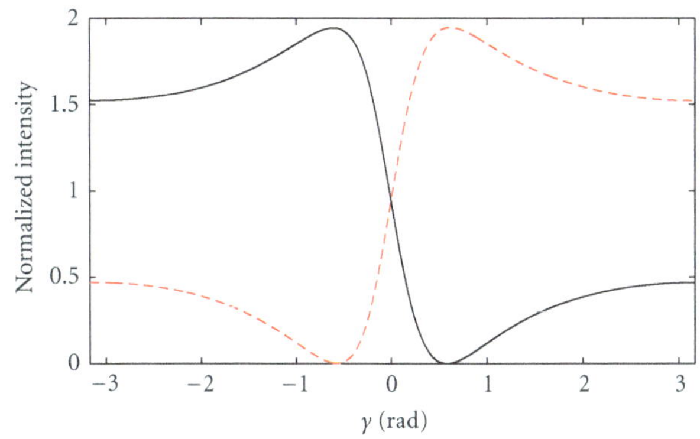

FIGURE 7: The dual-bus single resonator Fano resonance spectrum for $\Delta\phi = \pi/2$ (Drop: black solid line, Through: red dashed line).

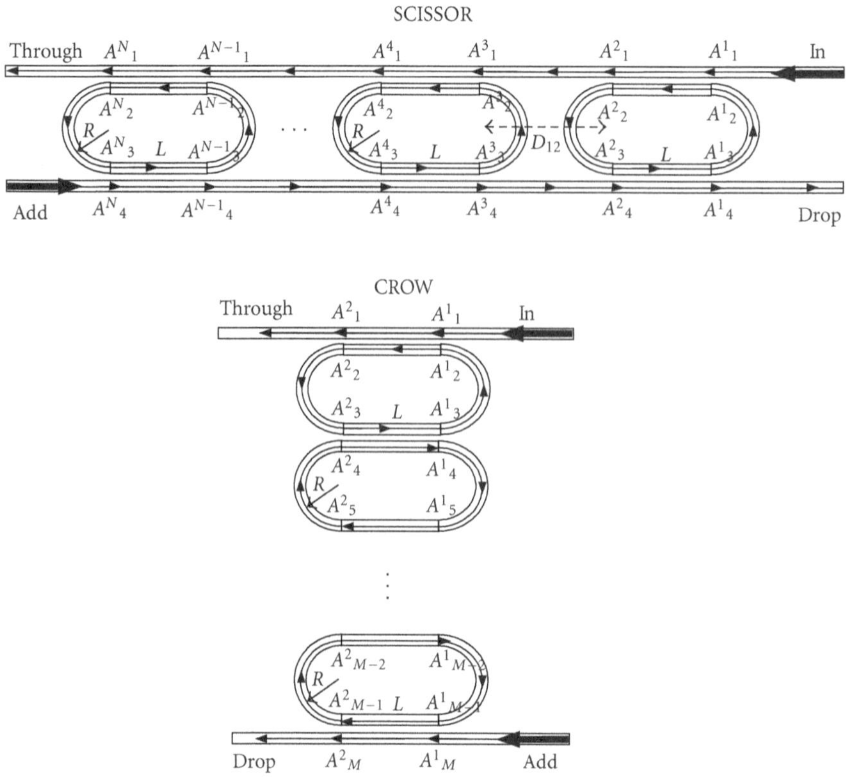

Figure 8: The SCISSOR and CROW resonant structures.

that arises from the interference between the resonant modes in the racetrack, where a narrow band superimposes on the waveguide flat spectral background [23, 24].

Devices where two phase modulated incoming signals are injected can be used for light switching functions. In this case, a high extinction ratio is obtained. For instance, at $\gamma = \pm 0.6$, in Figure 7, the drop and through signals have a maximum or vanish (the maximum intensity does not reach 2 because bent losses are considered). Now, if $\Delta\phi$ is changed from $\pi/2$, as in Figure 7, to $3/2\pi$, a similar spectrum is found with the two signals changed (the black solid line would go into the red dashed line). This means that, by phase modulation, the signal output can be switched in the Through port or in the Drop port.

It is possible to extend this interferometric switching functions to a parallel or serial chain of coupled resonators, that is, in CROW or SCISSOR, Figure 8.

To model a SCISSOR, we introduce an external propagation matrix $\mathbf{Q}^j$ connecting the jth to the $(j+1)$th resonator

$$\begin{pmatrix} A_1^{j+1} \\ A_4^{j+1} \end{pmatrix} = \mathbf{Q}^j \begin{pmatrix} A_1^j \\ A_4^j \end{pmatrix} = \begin{pmatrix} e^{i\beta D^{jj+1}} & 0 \\ 0 & e^{-i\beta D^{jj+1}} \end{pmatrix} \begin{pmatrix} A_1^j \\ A_4^j \end{pmatrix}, \quad (31)$$

where D^{jj+1} is the distance which separates the jth to $(j+1)$th resonator. Then, (30) can be generalized to the SCISSOR structure with a scattering matrix $\mathbf{S_h}$:

$$\begin{pmatrix} A_{\text{Th}} \\ A_D \end{pmatrix} = \mathbf{S_h} \begin{pmatrix} A_{\text{In}} \\ A_{\text{Ad}} \end{pmatrix} = \mathbf{S}^N \mathbf{Q}^{N-1} \mathbf{S}^{N-1} \cdots \mathbf{Q}^1 \mathbf{S}^1 \begin{pmatrix} A_{\text{In}} \\ A_{\text{Ad}} \end{pmatrix}. \quad (32)$$

For the CROW, we simply generalize (29) to

$$\begin{pmatrix} A_{\text{Ad}} \\ A_D \end{pmatrix} = \mathbf{T}^N \cdots \mathbf{T}^2 \mathbf{T}^1 \begin{pmatrix} A_{\text{In}} \\ A_{\text{Th}} \end{pmatrix}, \quad (33)$$

where $\mathbf{T}^k$ $(k = 1, \ldots, N)$ is the transfer matrix for the kth resonator and is given by (28). From this we obtain, the scattering matrix $\mathbf{S_v}$

$$\begin{pmatrix} A_{\text{Th}} \\ A_D \end{pmatrix} = \mathbf{S_v} \begin{pmatrix} A_{\text{In}} \\ A_{\text{Ad}} \end{pmatrix}. \quad (34)$$

Figure 9 shows a spectrum of a SCISSOR made of a chain of eight racetrack resonators. Figure 9(a) represents the spectrum when only one signal is injected: high-order filter flat-box spectrum appears. When a second signal is injected at the Add port with the same amplitude but with a phase difference $\Delta\phi = \pi/2$ with respect to the signal injected in the In port, Figure 9(b) is obtained. We call this situation, where we have two input signals, the double bus situation. The point $\gamma = \pm 1.5$ can be used to switch the signal from the Through to the Drop port by controlling the phase at the Add signal. Moving the Add signal phase from $\pi/2$ to $3/2\pi$ switches the output from the Through to the Drop port. Possibly, a wavelength band can be switched by using the $\gamma = \pm 1.75$ point, though with low extinction rates. Therefore, with this respect, the SCISSOR geometry presents limited advantages over the single resonator case of Figure 7.

An eight racetrack CROW structure has a behavior shown in Figure 10. As in the previous case, the left and right

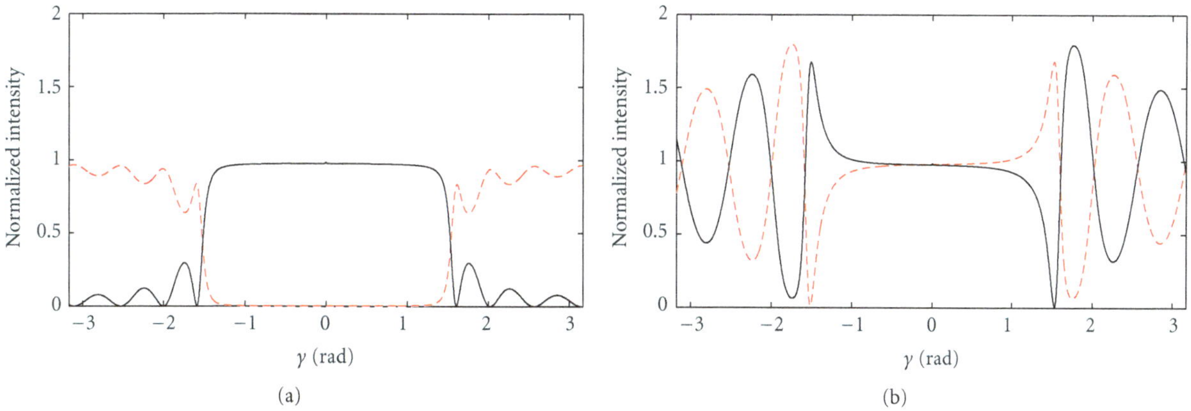

FIGURE 9: The single (a) and dual-bus (b) eight resonator SCISSOR spectrum for $\Delta\phi = \pi/2$. (Drop: black solid line, Through: red dashed line).

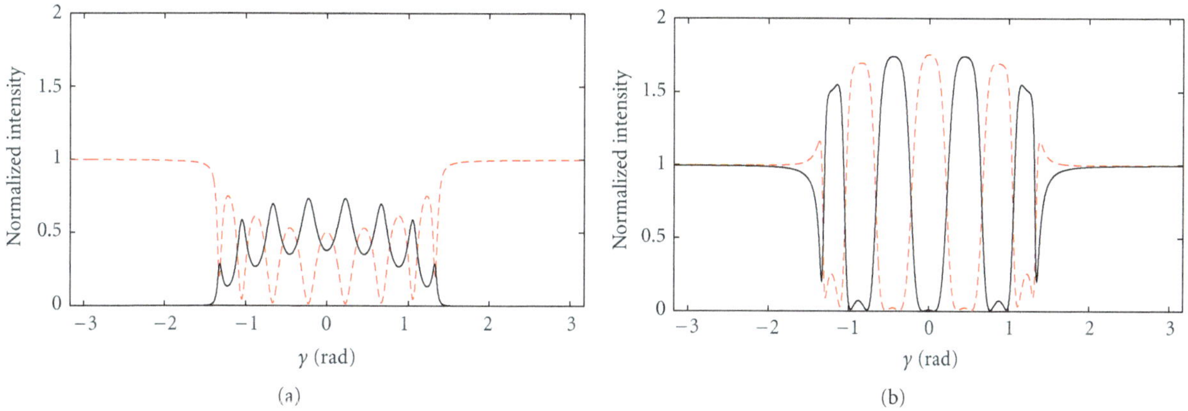

FIGURE 10: The single (a) and dual-bus (b) eight resonator CROW spectrum for $\Delta\phi = \pi/2$. (Lines as in Figure 9).

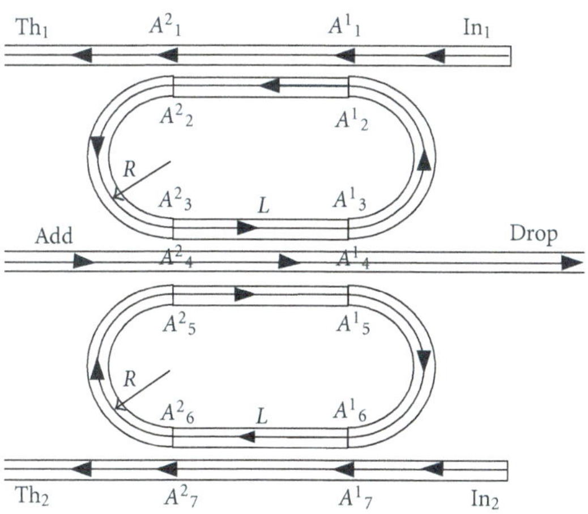

FIGURE 11: The resonator pair ds-coupler switch.

graphs represent the single- and double-bus spectra for the same dephasing. CROW are very different from SCISSOR: given a N resonator CROW, the single bus device has N Drop resonances (Figure 10(a)) while the double bus device has $N - 1$ bands when a $\Delta\phi = \pi/2$ is considered (Figure 10(b)). Also, in this case, for a dephasing of $\Delta\phi = (3/2)\pi$, one can flip the Through ↔ Drop port outputs. Not all bands have the same extinction efficiency, only the central ones exhibit an almost ideal on-off. The CROW has the advantage with respect to the SCISSOR and single resonator that it allows band routing or it can be used as an optical interleaver, that is, as a device that separates a set of channels into two sets, routing them separately towards the Drop and Through ports. The width of each band depends on the structural parameters.

4. Three-Bus Resonator Interferometric Switch

A further interesting situation is to model a structure like that of Figure 11. It shows a single pair of racetrack resonators coupled to three-bus waveguides. This interferometric switch works as follows: the input signals In_1 and In_2 are tuned to a resonance wavelength and have the same amplitude. Then, depending on their relative phase, they interfere constructively or destructively in the Drop waveguide. In the former case, the signal is expected to couple into the Drop port, while, in the latter case, the signal is transmitted to the Through ports. Experiments demonstrate this principle [11, 12].

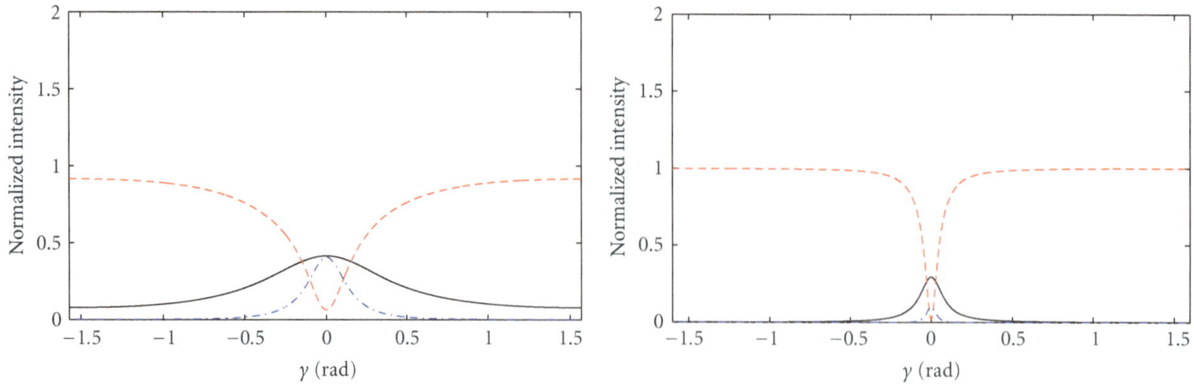

(a) (b)

FIGURE 12: Spectral response of the device in Figure 11, with $A_{\mathrm{In}_1} = 1, A_{\mathrm{Ad}} = 0, A_{\mathrm{In}_2} = 0, L = 10\,\mu\mathrm{m}$ (a), and $L = 3\,\mu\mathrm{m}$ (b) (Drop port: black solid line, Through1 port: red dashed line, Through2 port: blue dotted-dashed line).

In this section, we model the device as a three-bus device, that is, we drive it with three input signals whose phase is independently controlled. To relate the known signals $A_1^1 = A_{\mathrm{In}_1}$, $A_4^2 = A_{\mathrm{Ad}}$, $A_7^1 = A_{\mathrm{In}_2}$ to the unknown $A_1^2 = A_{\mathrm{Th}_1}$, $A_4^1 = A_D$, and $A_7^2 = A_{\mathrm{Th}_2}$ (Figure 11), we use a general transfer matrix $\mathbf{T}$:

$$\begin{pmatrix} A_{\mathrm{Th}_1} \\ A_D \\ A_{\mathrm{Th}_2} \end{pmatrix} = \begin{pmatrix} T_{11} & T_{21} & T_{31} \\ T_{21} & T_{22} & T_{21} \\ T_{31} & T_{21} & T_{11} \end{pmatrix} \begin{pmatrix} A_{\mathrm{In}_1} \\ A_{\mathrm{Ad}} \\ A_{\mathrm{In}_2} \end{pmatrix} = \mathbf{T} \begin{pmatrix} A_{\mathrm{In}_1} \\ A_{\mathrm{Ad}} \\ A_{\mathrm{In}_2} \end{pmatrix}, \quad (35)$$

whose elements are reported in Appendix B. The device (and therefore $\mathbf{T}$) is symmetric: the paths $\mathrm{In}_1 \to \mathrm{Through}_1$, $\mathrm{In}_1 \to$ Drop, $\mathrm{In}_1 \to \mathrm{Through}_2$ are equivalent to the paths $\mathrm{In}_2 \to \mathrm{Through}_2$, $\mathrm{In}_2 \to$ Drop, $\mathrm{In}_2 \to \mathrm{Through}_1$.

Figure 12 shows the spectral response of the device when used in single bus configuration, that is, only one input signal is injected ($A_{\mathrm{In}_1} = 1, A_{\mathrm{Ad}} = 0, A_{\mathrm{In}_2} = 0$), for two coupling section lengths of $L = 10\,\mu$m and $L = 3\,\mu$m. The drop signal (black line) shows a typical resonant behavior with a sharper feature for the $L = 3\,\mu$m device than for the $L = 10\,\mu$m device because of its lower coupling efficiency. It is also interesting to note that the crosstalk is much higher for $L = 10\,\mu$m than for $3\,\mu$m. At resonance, the power is almost equally distributed between the Drop and $\mathrm{Through}_2$ ports.

Figure 13 shows the case of a device operated in a duals bus configuration. We used $L = 10\,\mu$m for the coupling section. The input signals have the same amplitude and a phase difference of $\Delta\phi = 0, \pi/4, 3/4\pi, \pi$ for Figures 13(a), 13(b), 13(c), 13(d), respectively. When the two input signals are in phase, and the wavelength is resonant with the two racetracks, then most of the power is transferred to the Drop port. This situation was experimentally confirmed in [11, 12].

The Through ports will not exhibit complete extinction due to crosstalk effects (Figure 13(a)). Similarly, when the two input signals are in antiphase (Figure 13(d)), a complete destructive interference is achieved in the Drop waveguide and no Drop signal is observed. Most of the power goes straight in the Through ports. Note that the minima at resonance in the Through signals are due to the resonator round trip losses. With no dephasing, the Through signal minima coincide, with the Drop signal maximum at resonance λ_0 (Figure 13(a)). In the other case, the Through signals show the behaviors of a Fano resonance (Figures 13(b) and 13(c)). When there is a phase shift between the input signals, the Through signals show characteristics of Fano resonance with a different minimum at $\lambda_F = \lambda_0 \pm 0.8$ nm. This occurs already for a small $\Delta\phi = \pi/10$. Moreover, the Through signal resonance depth depends on the phase shift and vanishes almost completely at $\Delta\phi = \pi/4$ (Figure 13(b)). This is potentially useful to build interferometric switching devices with good on-off extinctions.

To highlight the phase shift dependence, we repeated the calculations at the resonant or at the Fano wavelengths (λ_0 or λ_F, resp.) by varying $\Delta\phi$. At the same time, we varied the intensity of the input signals. The results are shown in Figure 14 for λ_0 and in Figure 15 for λ_F, for $|A_{\mathrm{In}_2}|^2 = \delta \times |A_{\mathrm{In}_1}|^2$ with $\delta = (1, 0.75, 0.25, 0)$. Several features can be observed. First, for perfect input balance (Figure 14(a) and Figure 15(a)), the Drop signal goes to zero at $\Delta\phi = \pi$, while the Through signals vanish only for λ_F and not for λ_0. For this reason, interferometric switching is at λ_F than at λ_0. Second, for λ_F (Figure 15), the three output signals (Drop, $\mathrm{Through}_1$ and $\mathrm{Through}_2$) vanish for three different values of $\Delta\phi$: $\Delta\phi = \pi/4$ the $\mathrm{Through}_2$, $\Delta\phi = \pi$ the Drop, and $\Delta\phi = 7/4\pi$ the $\mathrm{Through}_1$. Third, the transmission of the input signals to the through ports is more effective for λ_F than for λ_0, which might be useful for practical purposes. Fourth, from Figures 14 and 15, it is clear that an interferometric switch device must be robust against input signal imbalance.

Similar features are obtained for a device with $L = 3\,\mu$m. The main difference is that λ_F is more near to λ_0 which makes difficult the use of the Fano concepts for interferometric routing.

A further extension of the ds-router concept is a double SCISSOR structure as shown in Figure 16. To model it, we transform $\mathbf{T}$ (35) into the "scattering matrix" $\mathbf{S}$ such that

$$\begin{pmatrix} A_{\mathrm{Th}_1} \\ A_{\mathrm{Ad}} \\ A_{\mathrm{Th}_2} \end{pmatrix} = \mathbf{S} \begin{pmatrix} A_{\mathrm{In}_1} \\ A_D \\ A_{\mathrm{In}_2} \end{pmatrix}. \quad (36)$$

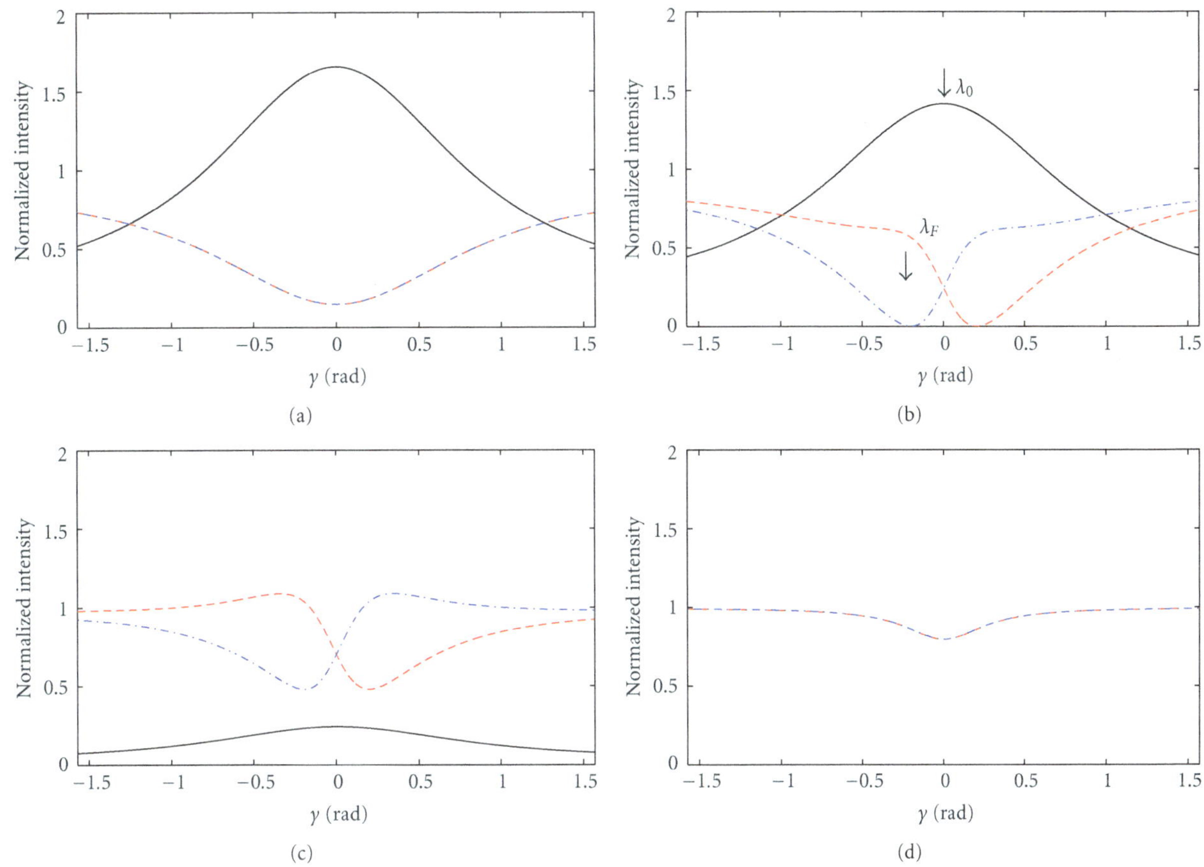

FIGURE 13: Spectral response of the device in Figure 11, with $A_{\mathrm{In}_1} = (1, e^{i(\pi/4)}, e^{i(3/4)\pi}, e^{i\pi})$ ((a), (b), (c), (d), resp.), $A_{\mathrm{Ad}} = 0$, $A_{\mathrm{In}_2} = 1$. (Lines as in Figure 12.)

After simple algebraic manipulation, one finds

$$\mathbf{S} = \frac{1}{T_{22}} \begin{pmatrix} T_{11}T_{22} - T_{12}T_{21} & T_{12} & T_{31}T_{22} - T_{12}T_{21} \\ -T_{21} & 1 & -T_{21} \\ T_{31}T_{22} - T_{12}T_{21} & T_{12} & T_{11}T_{22} - T_{12}T_{21} \end{pmatrix}, \quad (37)$$

where the lower indexes indicate the row × column element of $\mathbf{T}$.

Then, by introducing an external propagation matrix $\mathbf{Q}^k$ which connects the Through and Add ports of the kth pair of resonators with the In and Drop ports of the $(k+1)$th pair as

$$\begin{pmatrix} A_{\mathrm{In}_1^{k+1}} \\ A_{D^{k+1}} \\ A_{\mathrm{In}_2^{k+1}} \end{pmatrix} = \mathbf{Q}^{\mathbf{k}} \begin{pmatrix} A_{\mathrm{Th}_1^k} \\ A_{\mathrm{Ad}^k} \\ A_{\mathrm{Th}_2^k} \end{pmatrix} = \begin{pmatrix} e^{i\beta D^k} & 0 & 0 \\ 0 & e^{-i\beta D^k} & 0 \\ 0 & 0 & e^{i\beta D^k} \end{pmatrix} \begin{pmatrix} A_{\mathrm{Th}_1^k} \\ A_{\mathrm{Ad}^k} \\ A_{\mathrm{Th}_2^k} \end{pmatrix} \quad (k = 1, \ldots, N), \quad (38)$$

we can express the general equations for the device of Figure 16 with a total transfer matrix $\mathbf{S}_{\mathrm{Tot}}$:

$$\begin{pmatrix} A_{\mathrm{Th}_1^N} \\ A_{\mathrm{Ad}^N} \\ A_{\mathrm{Th}_2^N} \end{pmatrix} = \mathbf{S}_{\mathrm{Tot}} \begin{pmatrix} A_{\mathrm{In}_1^1} \\ A_{D^1} \\ A_{\mathrm{In}_2^1} \end{pmatrix} = \mathbf{S}^N \mathbf{Q}^{N-1} \mathbf{S}^{N-1} \cdots \mathbf{Q}^1 \mathbf{S}^1 \begin{pmatrix} A_{\mathrm{In}_1^1} \\ A_{D^1} \\ A_{\mathrm{In}_2^1} \end{pmatrix}. \quad (39)$$

And, finally, after inverting $\mathbf{S}_{\mathrm{Tot}}$ (37), we get

$$\begin{pmatrix} A_{\mathrm{Th}_1^N} \\ A_{D^1} \\ A_{\mathrm{Th}_2^N} \end{pmatrix} = \mathbf{T}_{\mathrm{Tot}} \begin{pmatrix} A_{\mathrm{In}_1^1} \\ A_{\mathrm{Ad}^N} \\ A_{\mathrm{In}_2^1} \end{pmatrix}, \quad (40)$$

which is the generalized version of (35).

Figure 17 shows the spectral response for the 1,2,4, and 8 pair SCISSOR interferometric switch (Figures 17(a), 17(b), 17(c), 17(d), resp.) with only one input signal. The $(N-1)$ central dips are *not* due to fabrication imperfections [11, 12] but are intrinsic features due to the existence of the central Drop waveguide between the two resonator chains. The system behaves like a single SCISSOR row with asymmetric gaps which introduce a correspondent asymmetric phase shift at each resonator gap leading to CRIT-like features. It is in this spectral region, near the center of the resonance band, where light couples more efficiently into the resonators and travels repeatedly through them producing slow light effects. For this reason, it is also in this central CRIT-like region where most of the attenuation occurs due to radiation and roughness losses on the resonator bent. Note how, aside from the flat-box spectrum, despite strong input imbalance, the two Through ports response (red dashed and blue dotted-dashed lines) tend to merge together inside the mode's spectral range with increasing the number of

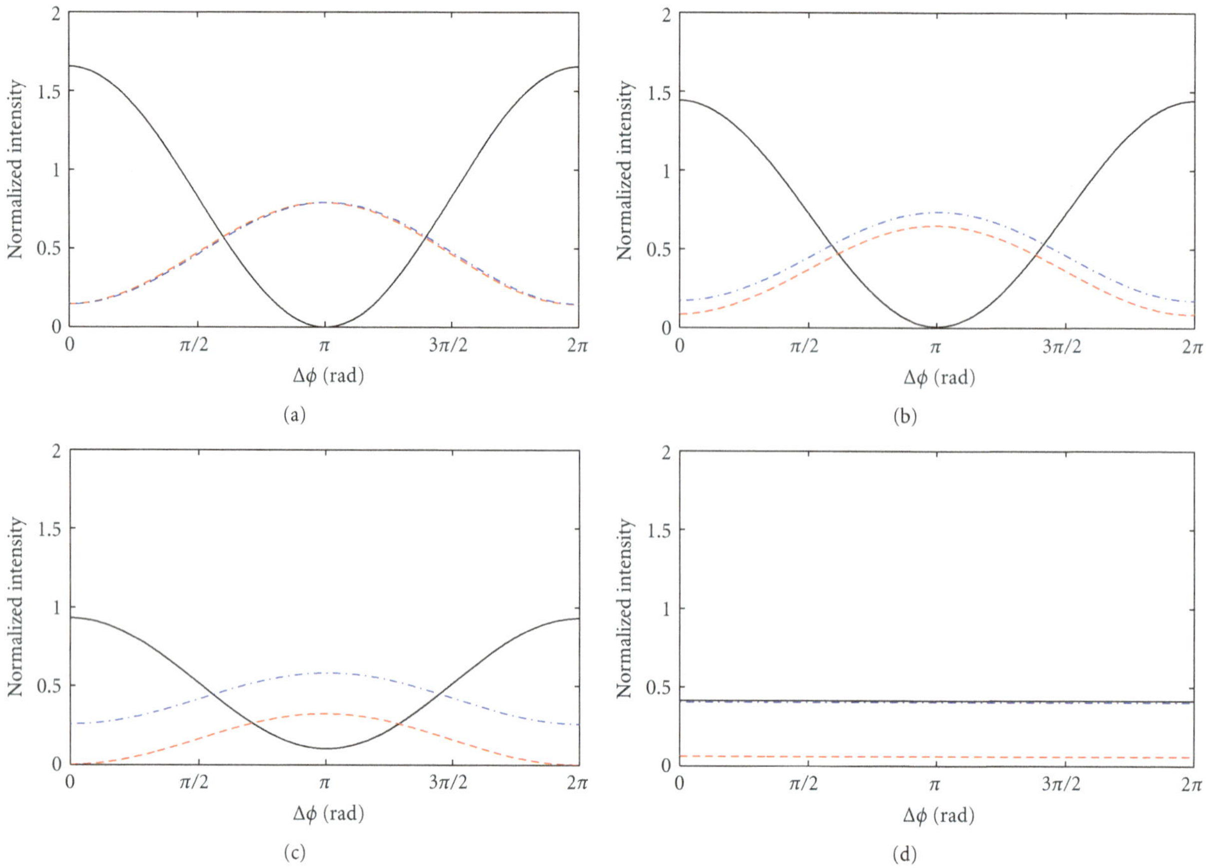

FIGURE 14: Phase response of the device in Figure 11 at λ_0, with $A_{\mathrm{In}_1} = e^{i\Delta\Phi}$, $A_{\mathrm{Ad}} = 0$, $|A_{\mathrm{In}_2}|^2 = (1,\ 0.75,\ 0.25,\ 0)$ ((a), (b), (c), (d), resp.) (Lines as in Figure 12.)

resonator pairs. This is because in an N-pair SCISSOR the input signals couple to the various resonator pairs, and, while they are propagating along the sequence, their power is progressively balanced due to the large crosstalk. This robustness of SCISSOR devices against input imbalances could have practical advantages where a power injection is uncertain or difficult to tailor.

Figure 18 illustrates the use of a dual bus SCISSOR interferometric switch with two equal input signals. Increasing the number of resonator, the response gets a flat-box shape and a good rejection of the input signal at the two Through ports is obtained.

Figure 19 shows the effect of a phase shift between the two input signals. As $\Delta\phi$ increases, the Drop signal decreases and vanishes for $\Delta\phi = \pi$. The Through signals are coincident within the stop-band and differ outside this wavelength region when $\Delta\phi \neq 0$.

We studied also the effect of an imbalance between the two input signals. A phase sweep at $\gamma = 1$ for four different intensities is shown in Figure 20. Contrary to the single pair interferometric switch affected by Fano resonance asymmetries and unequal Through port signal intensities along the phase sweep, the 2×8 SCISSOR interferometric switch flattens out the differences on the Through signals and shows its robustness against imbalances.

5. Three-Bus Interferometric Switch with Three Input Signals

Another potentially interesting application is to operate the ds-coupler interferometric switch of Figure 11 with three input signals, that is, to use the Add port as a third input port. Figure 21 shows its spectral response when $A_{\mathrm{In}_1} = e^{i\phi_1}$, $A_{\mathrm{Ad}} = e^{i\phi_{\mathrm{Add}}}$, $A_{\mathrm{In}_2} = e^{i\phi_2}$. Figure 21(a) shows the case of all three equal input phases $((\phi_1, \phi_{\mathrm{Add}}, \phi_2) = (0,0,0))$: the Drop and the two Through signals are equal with slight differences caused by the bending losses. Figure 21(b) reproduces the case $(\phi_1,\ \phi_{\mathrm{Add}},\ \phi_2) = (0, \pi/4, \pi/2)$: at the resonance wavelength, the two Through signals are high while the Drop signals are low. The Through signal resonances are Fano resonances. Figure 21(c) represents the alternative case where only a π-shift on the Add signal is applied $((\phi_1, \phi_{\mathrm{Add}}, \phi_2) = (0, \pi, 0))$: the Drop signal maximizes at resonance, while the two Through signals are low. Figure 21(d) shows the case when a π-shift is applied to both the Add and one Input signal $(\phi_1, \phi_{\mathrm{Add}}, \phi_2) = (0, \pi, \pi)$: all the input signals are addressed to the $\mathrm{Through}_1$ port. The symmetric case where the signals are directed to the $\mathrm{Through}_2$ port is obtained by applying the same phase shift to the other Input signal.

This three input signal configuration allows using the Add signal as a control signal to drive the interferometric

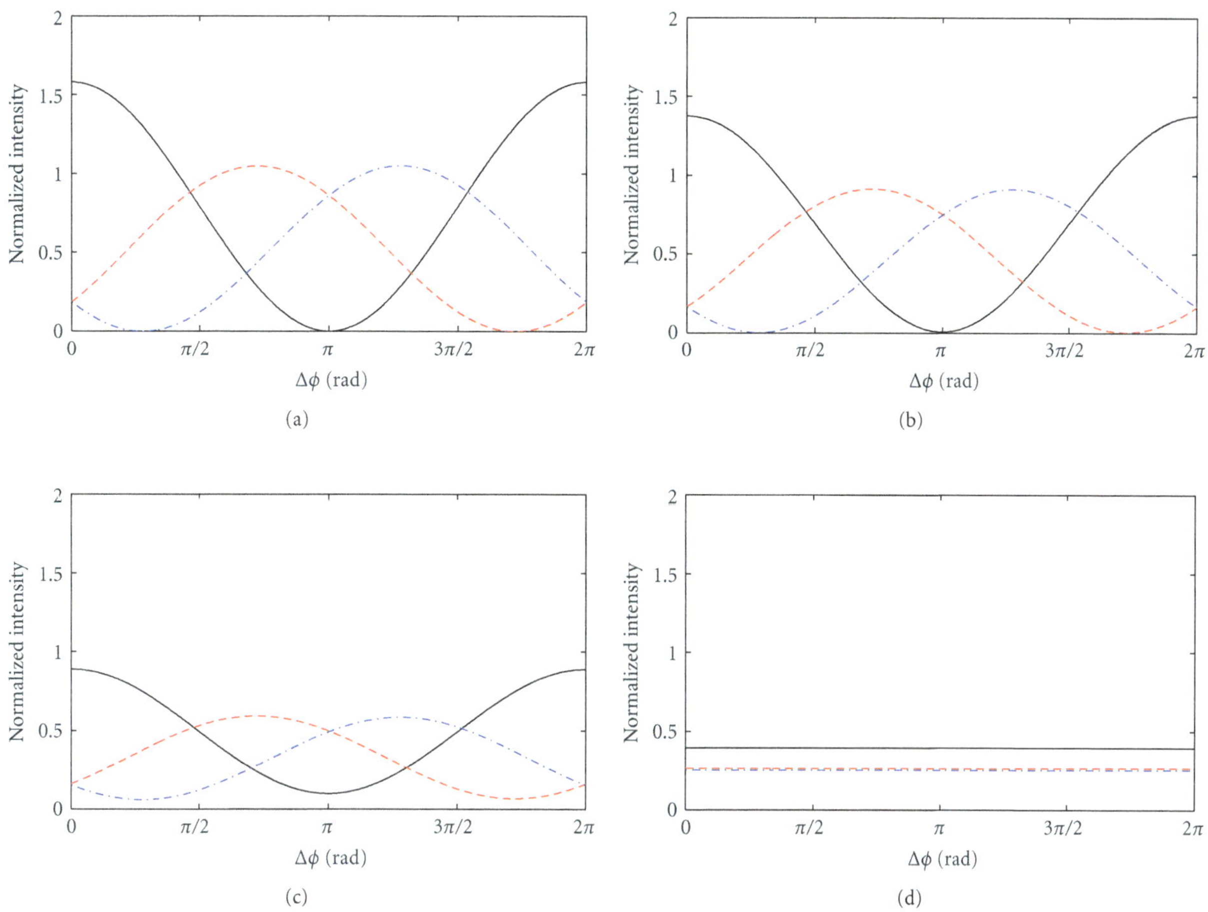

FIGURE 15: Phase response of the device in Figure 11 at λ_F, with $A_{\mathrm{In}_1} = e^{i\Delta\Phi}$, $A_{\mathrm{Ad}} = 0$, $|A_{\mathrm{In}_2}|^2 = (1,\ 0.75,\ 0.25,\ 0)$ ((a), (b), (c), (d), resp.). (Lines as in Figure 12.)

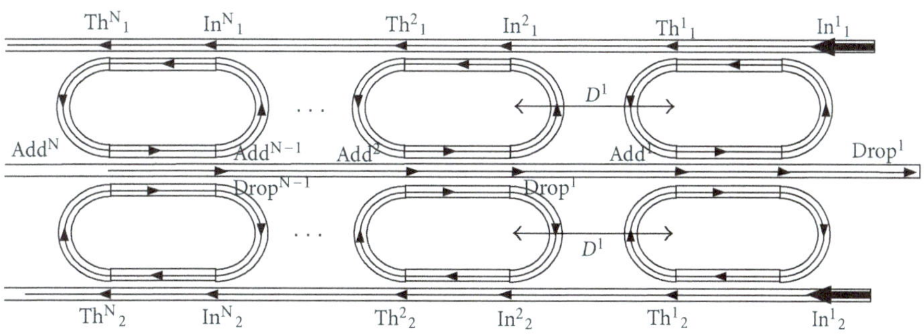

FIGURE 16: The $2 \times N$ SCISSOR phase switch.

switch to different functionalities (compare Figure 13 with Figure 21). In fact, if the phase of the Add signal is changed from 0 to π, the Drop signal is switched from 1 to 3 (in absence of bending losses); while, if the phase of one of the input is changed from 0 to π and $\phi_{\mathrm{Add}} = \pi$, one can switch off and on the Through signals. Many other sorts of combinations in phase and intensity at the Input ports, resonator numbers, and overall resonant device geometry for phase switching applications could be imagined in the most diverse configurations.

6. Conclusions

A three-bus waveguide multiple resonator-based interferometric switching device was discussed which is able to route light over a broad range of wavelengths by phase

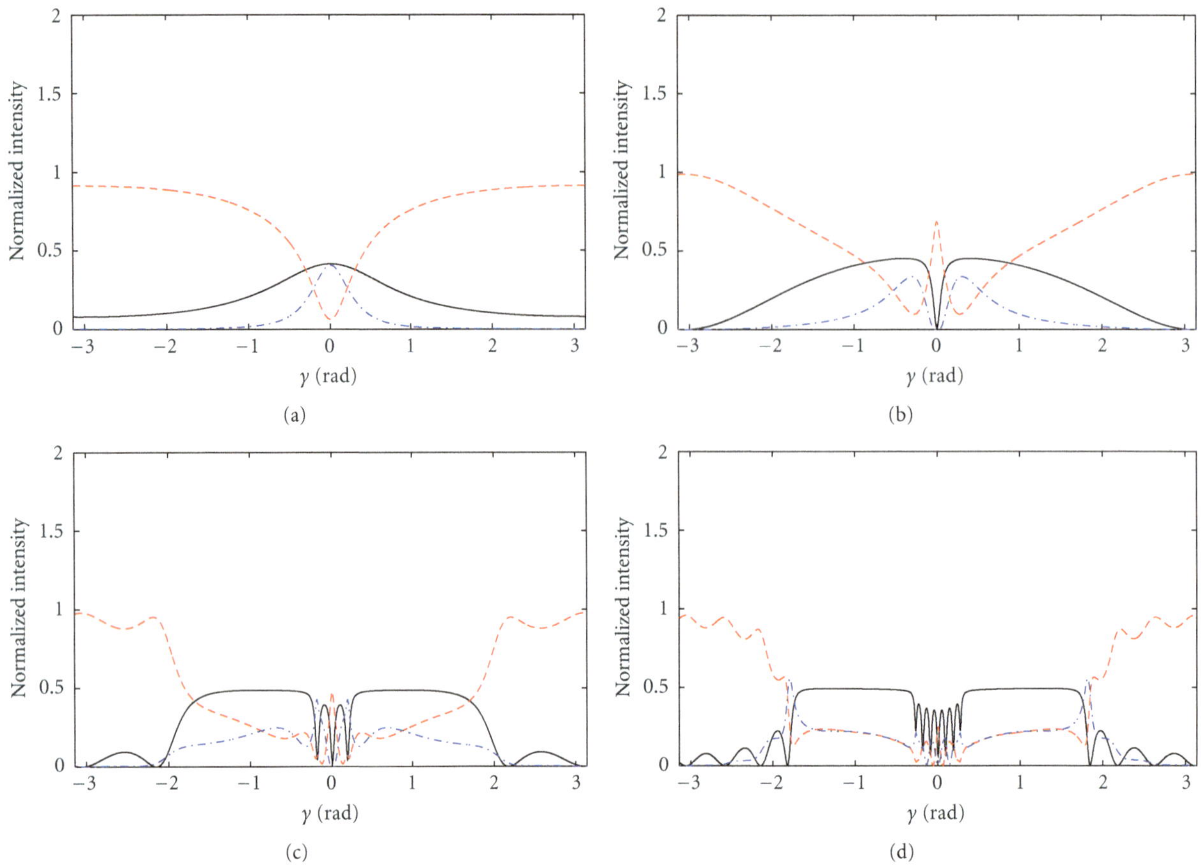

FIGURE 17: Spectral response of the SCISSOR switch with 1, 2, 4, 8 pairs of resonators ((a), (b), (c), (d), resp.) and inputs $A_{\text{In}_1} = 1$, $A_{\text{Ad}} = 0$, $A_{\text{In}_2} = 0$. (Lines as in Figure 12.)

modulation. To obtain its behavior, it was necessary to analyze first the performance of the two bus waveguide single resonator and SCISSOR and CROW systems. We showed that extending side coupling and Mach-Zehnder phase switching techniques to multiple waveguides and to chains of resonators could be useful to enhance their switching capabilities. The proposed interferometer device shows the possibility to control a lightwave signal with the phase of another signal. This allows to add a further degree of freedom in designing reconfigurable optical routers and complex networks. Simple examples can be foreseen in single channel switches where the switch redirects the signal in two different channels without absorbing the light (Figure 7), or in a dual channel interleaver where the signal bands can be alternatively directed on one or the other channels by phase control (Figure 10(b)), or in a three channel router where the signal is routed along three different directions by controlling the wavelength and the phase (Figure 13(b)). It is clear, though, that the potentiality of the scheme proposed in this work is not exhausted by these examples and more can be envisaged by optical network specialists. However, we believe that herewith we outlined some proposals which display how this is a field with vast possibilities, hopefully inspiring further investigations of possible applications of novel interferometric switching devices.

Appendices

A.

Instead of displaying the spectral response of a resonant structure in terms of the wavelength, it is possible to express it as a resonator round trip dephasing, clearly with the assumption of a nondispersive system. The resonant wavelength of the mth mode, λ_m, is thus in general

$$\lambda_m = \frac{p_{\text{opt}}(\lambda_m)}{m}, \tag{A.1}$$

with the wavelength-dependent optical path of the racetrack resonator $p_{\text{opt}} = (2\pi R + 2L)n_{\text{eff}}(\lambda_m)$, $n_{\text{eff}}(\lambda_m)$ the effective index of the resonator waveguides at the mth resonant wavelength, and m the resonance azimutal number. In the case of dispersionless systems, (A.1) is no longer an implicit function, and any two modes at λ_1 and λ_2 are determined solely by the optical path of the resonator ($p_{\text{opt}}(\lambda_m) = p_{\text{opt}}$ for all m) and by the mode azimutal number m, as

$$\lambda_1 = \frac{p_{\text{opt}}}{m}, \qquad \lambda_2 = \frac{p_{\text{opt}}}{m+1}. \tag{A.2}$$

This means that the difference (i.e., the FSR) is

$$\lambda_2 - \lambda_1 = -\frac{p_{\text{opt}}}{m(m+1)} = -\frac{\lambda_1}{m+1} = -\frac{\lambda_1\lambda_2}{p_{\text{opt}}}. \tag{A.3}$$

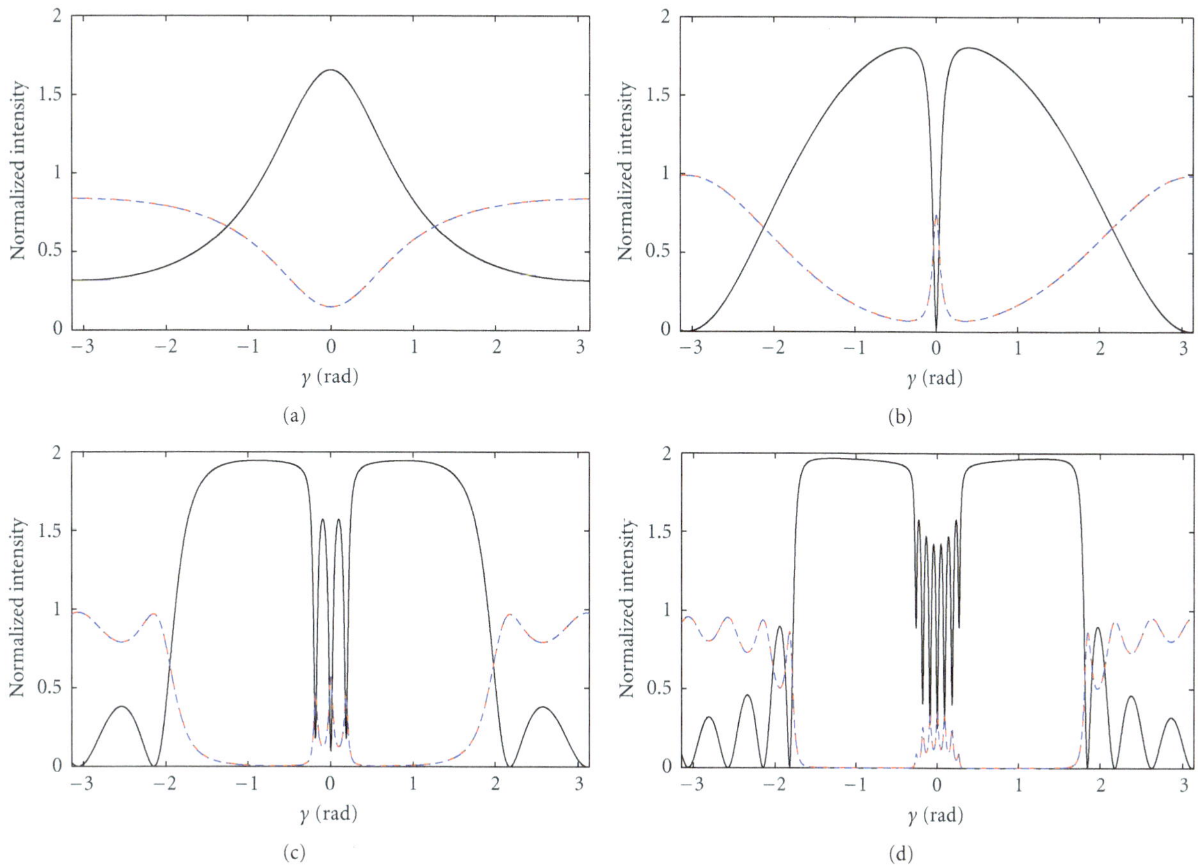

FIGURE 18: Spectral response of the SCISSOR switch with 1, 2, 4, 8 pairs of resonators ((a), (b), (c), (d), resp.), and inputs $A_{\text{In}_1} = 1, A_{\text{Ad}} = 0, A_{\text{In}_2} = 1$, with same phase. (Lines as in Figure 12.)

Then, the parameter γ representing the round-trip dephasing of the resonator can be defined as an "angular deviation" from resonance at some wavelength λ as

$$\lambda = \lambda_1 + \frac{\gamma}{2\pi}(\lambda_2 - \lambda_1) = \lambda_1 - \frac{\gamma}{2\pi}\frac{\lambda_1\lambda_2}{p_{\text{opt}}}, \tag{A.4}$$

and from which

$$\gamma = 2\pi p_{\text{opt}}\frac{\lambda_1 - \lambda}{\lambda_1\lambda_2} \approx 2\pi p_{\text{opt}}\frac{\Delta\lambda}{\lambda^2}. \tag{A.5}$$

B.

In Figure 11, we can relate $(A_3^2, A_4^2, A_5^2) \rightarrow (A_3^1, A_4^1, A_5^1)$ through (13) as

$$\begin{pmatrix} A_3^1 \\ A_4^1 \\ A_5^1 \end{pmatrix} = \mathcal{M}\begin{pmatrix} A_3^2 \\ A_4^2 \\ A_5^2 \end{pmatrix} = \begin{pmatrix} t' & i\kappa_D & -\kappa' \\ i\kappa_D & t_D & i\kappa_D \\ -\kappa' & i\kappa_D & t' \end{pmatrix}\begin{pmatrix} A_3^2 \\ A_4^2 \\ A_5^2 \end{pmatrix}. \tag{B.1}$$

Proceeding in the same way as we have done with matrix **M** of (25), we look for the coupling matrix that relates in the interference section the upper to the lower waveguide through the middle one, that is, $(A_3^2, A_4^2, A_3^1) \rightarrow (A_5^2, A_4^1, A_5^1)$, and obtain the second transfer matrix $\mathbf{T}^{\mathbf{b}}$:

$$\begin{pmatrix} A_5^2 \\ A_4^1 \\ A_5^1 \end{pmatrix} = \mathbf{T}^{\mathbf{b}}\begin{pmatrix} A_3^2 \\ A_4^2 \\ A_3^1 \end{pmatrix} = \frac{1}{\kappa'}\begin{pmatrix} t' & i\kappa_D & -1 \\ i\kappa_D(\kappa' + t') & t_D\kappa' - \kappa_D^2 & -i\kappa_D \\ t'^2 - \kappa'^2 & i\kappa_D(\kappa' + t') & -t' \end{pmatrix}\begin{pmatrix} A_3^2 \\ A_4^2 \\ A_3^1 \end{pmatrix}. \tag{B.2}$$

Due to a symmetric coupling, **K** of (26) can be applied also at $(A_6^1, A_6^2) \rightarrow (A_7^1, A_7^2)$. For the same reason, the same internal propagation matrix of (27) can be applied also at $(A_5^2, A_5^1) \rightarrow (A_6^1, A_6^2)$. Therefore, the third transfer matrix connects $(A_5^2, A_5^1) \rightarrow (A_7^1, A_7^2)$ in the same way as (28) (but with the order of the propagation and coupling matrixes inverted):

$$\begin{pmatrix} A_7^1 \\ A_7^2 \end{pmatrix} = \mathbf{T}^{\mathbf{KP}}\begin{pmatrix} A_5^2 \\ A_5^1 \end{pmatrix} = \mathbf{KP}\begin{pmatrix} A_5^2 \\ A_5^1 \end{pmatrix}. \tag{B.3}$$

To proceed in the inverse direction, that is, from ports (In_2, Through_2) to (In_1, Through_1), note that, again because of coupling symmetry, $\mathbf{T}^{\mathbf{PK}}$ relates also $(A_7^1, A_7^2) \rightarrow (A_5^2, A_5^1)$, $\mathbf{T}^{\mathbf{KP}}$ does the same with $(A_3^2, A_3^1) \rightarrow (A_1^1, A_1^2)$, and $\mathbf{T}^{\mathbf{b}}$ connects also $(A_5^2, A_4^2, A_5^1) \rightarrow (A_3^2, A_4^1, A_3^1)$.

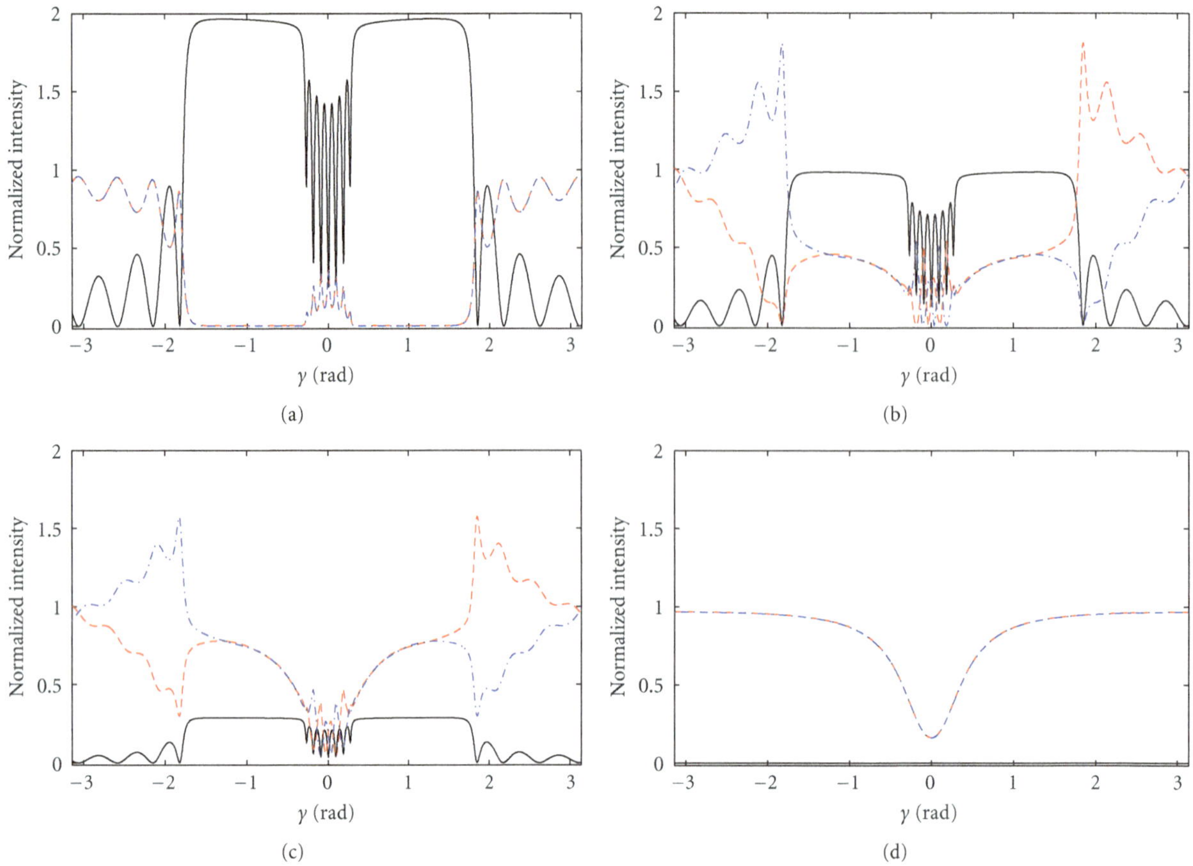

FIGURE 19: Spectral response of the 2 × 8 SCISSOR switch of Figure 16 with inputs, $A_{\mathrm{In}_1} = (1,\ e^{i(\pi/2)}, e^{i(3/4)\pi}, e^{i\pi})$ ((a), (b), (c), (d) resp.), $A_{\mathrm{Ad}} = 0$, $A_{\mathrm{In}_2} = 1$ (Lines as in Figure 12.)

Now, we can obtain the overall device response of Figure 11 by separating the contributions to the two Through and the Drop ports as the sum of three different terms corresponding to the three following device states.

State (I). Input signal in In_1 only is injected and propagated from the top to the bottom of the device

$$\begin{pmatrix} A_{\mathrm{In}_1} = |A_{\mathrm{In}_1}| e^{i\phi_1} \\ A_{\mathrm{Ad}} = 0 \\ A_{\mathrm{In}_2} = 0 \end{pmatrix} \longmapsto \begin{pmatrix} A_{\mathrm{Th}_1^{(\mathrm{I})}} \\ A_{D^{(\mathrm{I})}} \\ A_{\mathrm{Th}_2^{(\mathrm{I})}} \end{pmatrix}. \tag{B.4}$$

Schematically, this goes as follows (from the right to the left):

$$\longleftarrow \begin{pmatrix} A_3^2 \\ A_3^1 \end{pmatrix} = \mathbf{T}^{\mathrm{PK}} \begin{pmatrix} A_{\mathrm{In}_1} \\ A_{\mathrm{Th}_1^{(\mathrm{I})}} \end{pmatrix} \tag{B.5}$$

$$\longleftarrow \begin{pmatrix} A_5^2 \\ A_{D^{(\mathrm{I})}} \\ A_5^1 \end{pmatrix} = \mathbf{T}^{\mathrm{b}} \begin{pmatrix} A_3^2 \\ 0 \\ A_3^1 \end{pmatrix} \longleftarrow \tag{B.6}$$

$$\begin{pmatrix} 0 \\ A_{\mathrm{Th}_2^{(\mathrm{I})}} \end{pmatrix} = \mathbf{T}^{\mathrm{KP}} \begin{pmatrix} A_5^2 \\ A_5^1 \end{pmatrix} \longleftarrow . \tag{B.7}$$

This means that the first set of equations from the first transferral of (B.5)

$$\begin{aligned} A_3^2 &= T_{11}^{\mathrm{PK}} A_{\mathrm{In}_1} + T_{12}^{\mathrm{PK}} A_{\mathrm{Th}_1^{(\mathrm{I})}}, \\ A_3^1 &= T_{21}^{\mathrm{PK}} A_{\mathrm{In}_1} + T_{22}^{\mathrm{PK}} A_{\mathrm{Th}_1^{(\mathrm{I})}}, \end{aligned} \tag{B.8}$$

have to be inserted into (B.6) to obtain

$$\begin{aligned} A_5^2 = {} & \left(T_{11}^b T_{11}^{\mathrm{PK}} + T_{13}^b T_{21}^{\mathrm{PK}}\right) A_{\mathrm{In}_1} \\ & + \left(T_{11}^b T_{12}^{\mathrm{PK}} + T_{13}^b T_{22}^{\mathrm{PK}}\right) A_{\mathrm{Th}_1^{(\mathrm{I})}}, \end{aligned} \tag{B.9}$$

$$\begin{aligned} A_{D^{(\mathrm{I})}} = {} & \left(T_{21}^b T_{11}^{\mathrm{PK}} + T_{23}^b T_{21}^{\mathrm{PK}}\right) A_{\mathrm{In}_1} \\ & + \left(T_{21}^b T_{12}^{\mathrm{PK}} + T_{23}^b T_{22}^{\mathrm{PK}}\right) A_{\mathrm{Th}_1^{(\mathrm{I})}}, \end{aligned} \tag{B.10}$$

$$\begin{aligned} A_5^1 = {} & \left(T_{31}^b T_{11}^{\mathrm{PK}} + T_{33}^b T_{21}^{\mathrm{PK}}\right) A_{\mathrm{In}_1} \\ & + \left(T_{31}^b T_{12}^{\mathrm{PK}} + T_{33}^b T_{22}^{\mathrm{PK}}\right) A_{\mathrm{Th}_1^{(\mathrm{I})}}. \end{aligned} \tag{B.11}$$

Proceeding further with the next transfer, we have from (B.7) that

$$\begin{aligned} 0 &= T_{11}^{\mathrm{KP}} A_5^2 + T_{12}^{\mathrm{KP}} A_5^1, \\ A_{\mathrm{Th}_2^{(\mathrm{I})}} &= T_{21}^{\mathrm{KP}} A_5^2 + T_{22}^{\mathrm{KP}} A_5^1, \end{aligned} \tag{B.12}$$

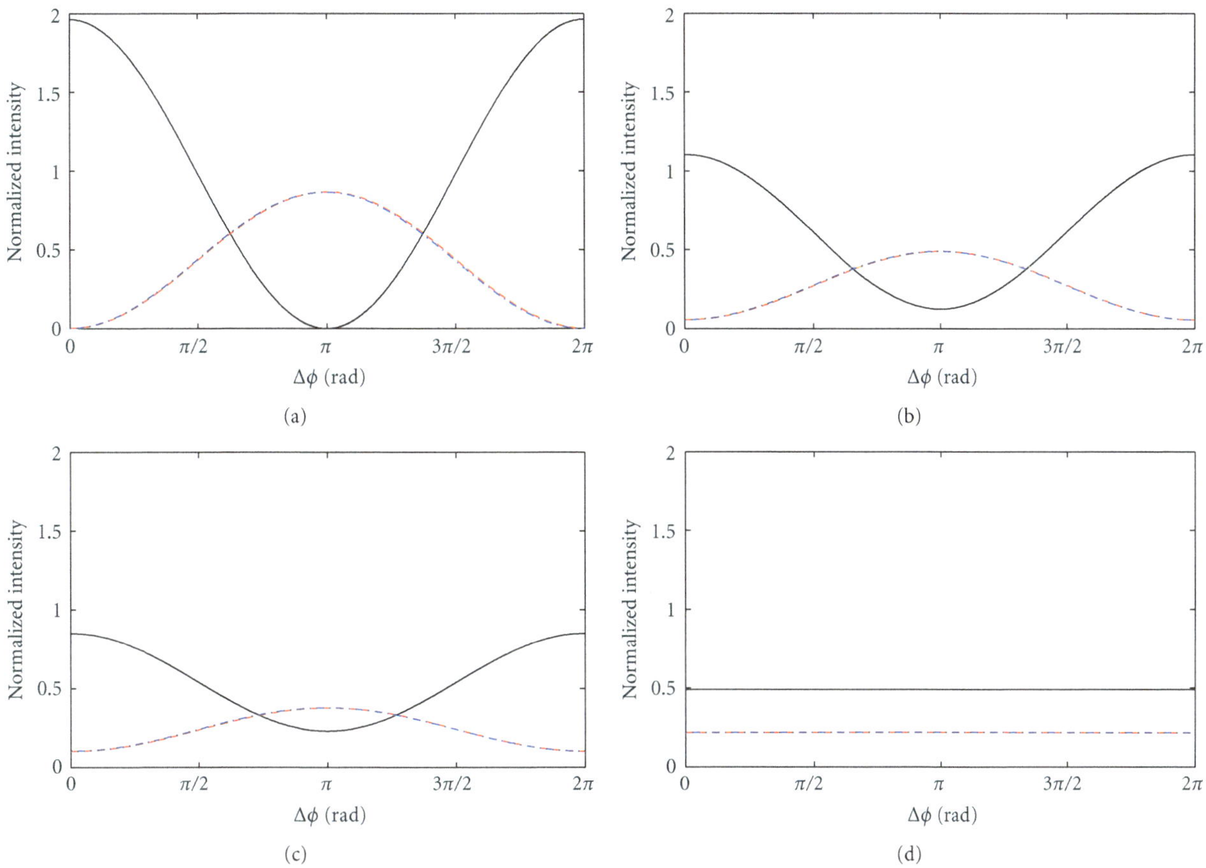

FIGURE 20: Phase response of the 2 × 8 SCISSOR switch of Figure 16 with input imbalances $|A_{\text{In}_1}|^2 = (1, 0.25, 0.1, 0)$ ((a), (b), (c), (d), resp.), $A_{\text{Ad}} = 0$, $A_{\text{In}_2} = 1$, at phase round-trip shift $\gamma = 1$ (Lines as in Figure 12.)

which, through (B.9), (B.10), and (B.11), leads to

$$\begin{aligned} A_{\text{Th}_1^{(\text{I})}} &= T_{11} A_{\text{In}_1}, \\ A_{D^{(\text{I})}} &= T_{21} A_{\text{In}_1}, \\ A_{\text{Th}_2^{(\text{I})}} &= T_{31} A_{\text{In}_1}, \end{aligned} \tag{B.13}$$

with

$$\begin{aligned} T_{11} &= -\frac{T_{11}^{\text{KP}} A + T_{12}^{\text{KP}} B}{T_{11}^{\text{KP}} C + T_{12}^{\text{KP}} D}, \\ T_{21} &= E + F T_{11}, \\ T_{31} &= G + H T_{11}, \\ A &= T_{11}^{b} T_{11}^{\text{PK}} + T_{13}^{b} T_{21}^{\text{PK}}, \\ B &= T_{31}^{b} T_{11}^{\text{PK}} + T_{33}^{b} T_{21}^{\text{PK}}, \\ C &= T_{11}^{b} T_{12}^{\text{PK}} + T_{13}^{b} T_{22}^{\text{PK}}, \\ D &= T_{31}^{b} T_{12}^{\text{PK}} + T_{33}^{b} T_{22}^{\text{PK}}, \\ E &= T_{21}^{b} T_{11}^{\text{PK}} + T_{23}^{b} T_{21}^{\text{PK}}, \\ F &= T_{21}^{b} T_{12}^{\text{PK}} + T_{23}^{b} T_{22}^{\text{PK}}, \\ G &= T_{21}^{\text{KP}} A + T_{22}^{\text{KP}} B, \\ H &= T_{21}^{\text{KP}} C + T_{22}^{\text{KP}} D. \end{aligned} \tag{B.14}$$

State (II). Input signal in In_2 only is injected and propagated from the bottom to the top of the device.

$$\begin{pmatrix} A_{\text{In}_1} = 0 \\ A_{\text{Ad}} = 0 \\ A_{\text{In}_2} = |A_{\text{In}_2}| e^{i\phi_2} \end{pmatrix} \longmapsto \begin{pmatrix} A_{\text{Th}_1^{(\text{III})}} \\ A_{D^{(\text{III})}} \\ A_{\text{Th}_2^{(\text{III})}} \end{pmatrix}. \tag{B.15}$$

One proceeds exactly as for State (I), but in the opposite direction. The only difference in the final result is that in (B.13) the ports (In_1, $\text{Through}_1^{(\text{I})}$, $\text{Through}_2^{(\text{I})}$) have to be exchanged with (In_2, $\text{Through}_2^{(\text{III})}$, $\text{Through}_1^{(\text{III})}$):

$$\begin{aligned} A_{\text{Th}_1^{(\text{III})}} &= T_{13} A_{\text{In}_2}, \\ A_{\text{Drop}^{(\text{III})}} &= T_{23} A_{\text{In}_2}, \\ A_{\text{Th}_2^{(\text{III})}} &= T_{33} A_{\text{In}_2}, \end{aligned} \tag{B.16}$$

with $T_{13} = T_{31}$, $T_{23} = T_{21}$, and $T_{33} = T_{11}$.

State (III). Add port signal only is injected and propagated from the central Add waveguide towards the Drop and upper and lower Through ports.

In the transfer matrix of (35), only one coefficient is missing, namely, T_{22}. This is fixed by the other matrix elements and power conservation criteria. Expressing explicitly

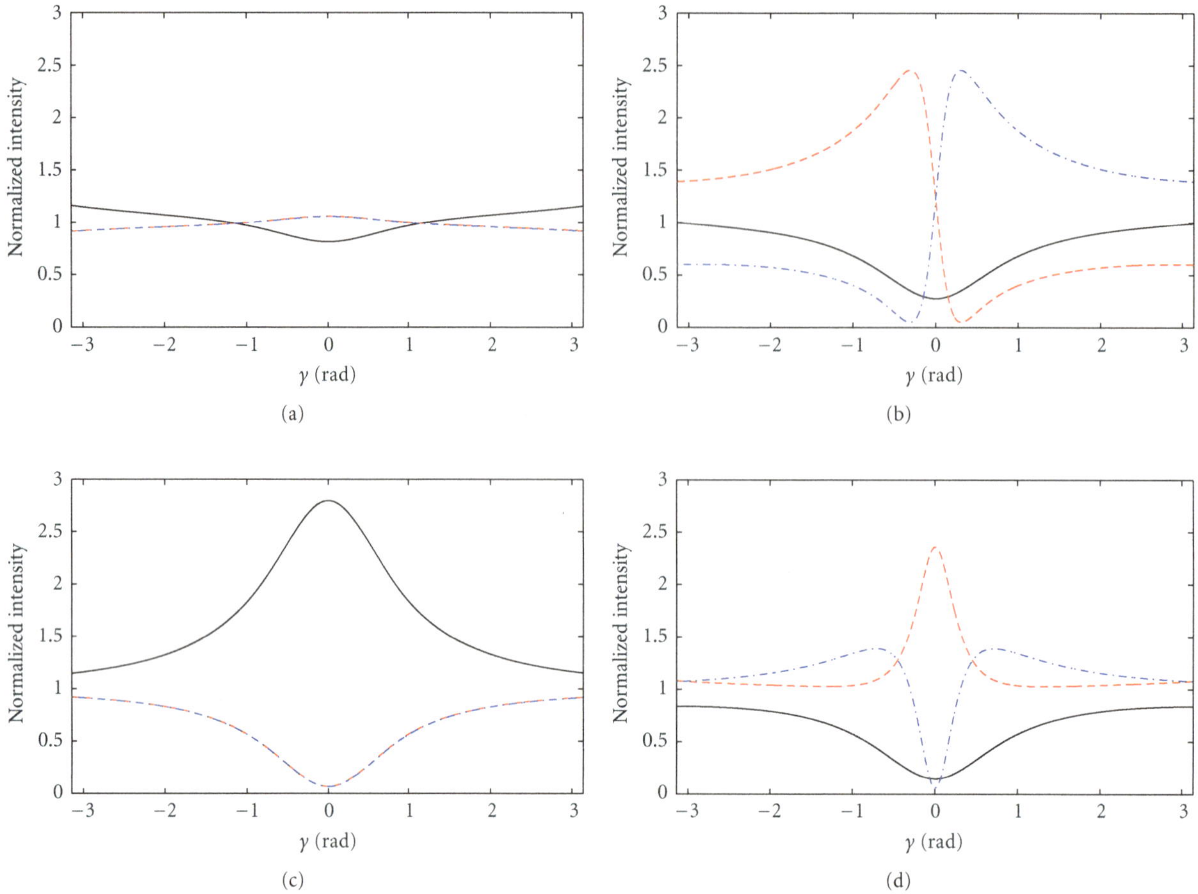

FIGURE 21: Spectral response of the 2 × 1 SCISSOR switch of Figure 16 with inputs $A_{\text{In}_1} = e^{i\phi_1}$, $A_{\text{Ad}} = e^{i\phi_{\text{Add}}}$, $A_{\text{In}_2} = e^{i\phi_2}$, and $(\phi_1, \phi_{\text{Add}}, \phi_2) = (0, 0, 0)$; $(\phi_1, \phi_{\text{Add}}, \phi_2) = (0, \pi/4, \pi/2)$; $(\phi_1, \phi_{\text{Add}}, \phi_2) = (0, \pi, 0)$; $(\phi_1, \phi_{\text{Add}}, \phi_2) = (0, \pi, \pi)$ for figures (a), (b), (c), (d), respectively (Lines as in Figure 12.)

with the matrix elements of **T** the power balance between the input and output intensities of the waves, it must hold:

$$\begin{aligned}
&|A_{\text{In}_1}|^2 + |A_{\text{In}_2}|^2 + |A_{\text{Ad}}|^2 \\
&= |A_{\text{Th}_1}|^2 + |A_D|^2 + |A_{\text{Th}_2}|^2 + \text{Loss} \\
&= \left(|T_{11}|^2 + |T_{21}|^2 + |T_{31}|^2\right)\left(|A_{\text{In}_1}|^2 + |A_{\text{In}_2}|^2\right) \\
&\quad + \left(2|T_{21}|^2 + |T_{22}|^2\right)|A_{\text{Ad}}|^2 \\
&\quad + \left(T_{11}T_{31}^* + |T_{21}|^2 + T_{31}T_{11}^*\right)\left(A_{\text{In}_1}A_{\text{In}_2}^* + A_{\text{In}_1}^*A_{\text{In}_2}\right) \\
&\quad + \left(T_{11}T_{21}^* + T_{21}T_{22}^* + T_{31}T_{21}^*\right)\left(A_{\text{In}_1} + A_{\text{In}_1}\right)A_{\text{Ad}}^* \\
&\quad + \left(T_{11}^*T_{21} + T_{21}^*T_{22} + T_{31}^*T_{21}\right)\left(A_{\text{In}_1}^* + A_{\text{In}_2}\right)A_{\text{Ad}} + \text{Loss},
\end{aligned} \tag{B.17}$$

with the last term indicating the losses of the structure. This implies the following conditions:

$$|T_{11}|^2 + |T_{21}|^2 + |T_{31}|^2 = 1, \tag{B.18}$$

$$2|T_{21}|^2 + |T_{22}|^2 = 1, \tag{B.19}$$

$$T_{11}T_{31}^* + |T_{21}|^2 + T_{31}T_{11}^* = 0, \tag{B.20}$$

$$T_{11}^*T_{21} + T_{21}^*T_{22} + T_{31}^*T_{21} = 0. \tag{B.21}$$

Conditions (B.18) and (B.20) are already satisfied by the coefficients found in the two previous cases (very cumbersome and long calculations). From (B.21), we finally obtain the last coefficient for matrix (35):

$$T_{22} = -\frac{T_{21}}{T_{21}^*}\left(T_{11}^* + T_{31}^*\right), \tag{B.22}$$

which satisfies also (B.19) through (B.20).

Acknowledgment

The authors acknowledge support of the EU through the FP7 ICT-(216405) Project Wavelength Division Multiplexed Photonic Layer on CMOS.

References

[1] T. S. El-Bawab, *Optical Switching*, Springer, 2006.

[2] A. Yariv, Y. Xu, R. K. Lee, and A. Scherer, "Coupled-resonator optical waveguide: a proposal and analysis," *Optics Letters*, vol. 24, no. 11, pp. 711–713, 1999.

[3] J. E. Heebner, R. W. Boyd, and Q. H. Park, "SCISSOR solitons and other novel propagation effects in microresonator-modified waveguides," *Journal of the Optical Society of America B*, vol. 19, no. 4, pp. 722–731, 2002.

[4] J. E. Heebner, P. Chak, S. Pereira, J. E. Sipe, and R. W. Boyd, "Distributed and localized feedback in microresonator sequences for linear and nonlinear optics," *Journal of the Optical Society of America B*, vol. 21, no. 10, pp. 1818–1832, 2004.

[5] J. Capmany, P. Muñoz, J. D. Domenech, and M. A. Muriel, "Apodized coupled resonator waveguides," *Optics Express*, vol. 15, no. 16, pp. 10196–10206, 2007.

[6] S. Y. Cho and R. Soref, "Apodized SCISSORs for filtering and switching," *Optics Express*, vol. 16, no. 23, pp. 19078–19090, 2008.

[7] D. U. Smith, H. Chang, K. A. Fuller, A. T. Rosenberger, and R. W. Boyd, "Coupled-resonator-induced transparency," *Physical Review A*, vol. 69, no. 6, Article ID 063804, 6 pages, 2004.

[8] Q. Xu, S. Sandhu, M. L. Povinelli, J. Shakya, S. Fan, and M. Lipson, "Experimental realization of an on-chip all-optical analogue to electromagnetically induced transparency," *Physical Review Letters*, vol. 96, no. 12, Article ID 123901, 4 pages, 2006.

[9] Y. F. Xiao, X. B. Zou, W. Jiang, Y. L. Chen, and G. C. Guo, "Analog to multiple electromagnetically induced transparency in all-optical drop-filter systems," *Physical Review A*, vol. 75, no. 6, Article ID 063833, 4 pages, 2007.

[10] X. Yang, M. Yu, D. L. Kwong, and C. W. Wong, "All-optical analog to electromagnetically induced transparency in multiple coupled photonic crystal cavities," *Physical Review Letters*, vol. 102, no. 17, Article ID 173902, 4 pages, 2009.

[11] M. Mancinelli, R. Guider, M. Masi et al., "Optical characterization of a SCISSOR device," *Optics Express*, vol. 19, no. 14, pp. 13664–13674, 2011.

[12] M. Masi, M. Mancinelli, A. Battarelli et al., "A silicon photonic interferometric router device based on SCISSOR concept," *Journal of Lightwave Technology*, vol. 29, no. 18, pp. 2747–2753, 2011.

[13] "Technology focus on Lithography," *Nature Photonics*, vol. 4, no. 1, 2010.

[14] G. T. Reed, G. Mashanovich, F. Y. Gardes, and D. J. Thomson, "Silicon optical modulators," *Nature Photonics*, vol. 4, article 660, 2010.

[15] K. Okamoto, *Fundamentals of Optical Waveguides*, Academic Press, 2nd edition, 2006.

[16] Y. M. Landobasa, S. Darmawan, and M. K. Chin, "Matrix analysis of 2-d microresonator lattice optical filters," *IEEE Journal of Quantum Electronics*, vol. 41, no. 11, pp. 1410–1418, 2005.

[17] J. K. S. Poon, J. Scheuer, S. Mookherjea, G. T. Paloczi, Y. Huang, and A. Yariv, "Matrix analysis of microring coupled-resonator optical waveguides," *Optics Express*, vol. 12, no. 1, pp. 90–103, 2004.

[18] C.-S. Ma, X. Yan, and X.-Y. Wang, "Matrix analysis of 2D microring resonator arrays," *Journal of Modern Optics*, vol. 55, no. 1, pp. 143–154, 2008.

[19] J. K. S. Poon, P. Chak, J. M. Choi, and A. Yariv, "Slowing light with Fabry-Perot resonator arrays," *Journal of the Optical Society of America B*, vol. 24, no. 11, pp. 2763–2769, 2007.

[20] S. Y. Cho and R. Soref, "Interferometric microring-resonant 2×2 optical switches," *Optics Express*, vol. 16, no. 17, pp. 13304–13314, 2008.

[21] F. Xia, L. Sekaric, M. O'Boyle, and Y. Vlasov, "Coupled resonator optical waveguides based on silicon-on-insulator photonic wires," *Applied Physics Letters*, vol. 89, no. 4, Article ID 041122, 3 pages, 2006.

[22] M. Masi, R. Orobtchouk, G. Fan, J. M. Fedeli, and L. Pavesi, "Towards a realistic modelling of ultra-compact racetrack resonators," *Journal of Lightwave Technology*, vol. 28, no. 22, pp. 3233–3242, 2010.

[23] U. Fano, "Sullo spettro di assorbimento dei gas nobili presso il limite dello spettro d'arco," *Nuovo Cimento*, vol. 12, no. 3, pp. 154–161, 1935.

[24] U. Fano, "Effects of configuration interaction on intensities and phase shifts," *Physical Review*, vol. 124, no. 6, pp. 1866–1878, 1961.

3

Ultrafast Plasmonic Nanoantenna-ITO Hybrid Switches

Martina Abb and Otto L. Muskens

SEPnet and the Department of Physics and Astronomy, University of Southampton, Highfield, Southampton SO17 1BJ, UK

Correspondence should be addressed to Otto L. Muskens, o.muskens@soton.ac.uk

Academic Editor: Alexandre Bouhelier

We present here a new type of device using nonlinear hybrid antenna-semiconductor (ITO) interaction. We observe a picosecond transient response from the antenna that cannot be explained by either pure ITO or antenna nonlinearities independently. We study the dependence of the hybrid interaction on several experimental parameters, including the polarization of excitation and detection.

Nanoscale plasmonic components such as nanoantennas [1, 2] are of enormous interest for their capabilities of locally enhancing electromagnetic fields and controlling emission. Active control of such components will enable a new generation of tunable devices. We recently introduced a new concept of antenna switches relying on photoconductive loading of the gap between the arms of a dimer antenna [3]. Experimentally, modulation of localized plasmon modes has been achieved using the refractive index of liquid crystals [4]. Active tuning of the antenna gap has also been demonstrated using mechanical means, such as using stretchable elastomeric films [5] to reversibly engineer the distance between particle dimers. The temporal response of these changes is limited by the gap loading mechanisms that are used.

As a potentially ultrafast implementation of our concept [3], we demonstrated experimentally picosecond all-optical control of a plasmonic nanoantenna on ITO, referred to as an nanoantenna-ITO hybrid [6]. For ITO, unity-order changes of the refractive index have been achieved [7] by applying an electric field. We have shown that optical pumping of ITO leads to a local reduction of the free-carrier density in the substrate. The transient nonlinear response of the antenna-ITO hybrid shows a redshift of the plasmon resonance, as opposed to transient bleaching on SiO_2.

Here, we present a detailed study of the response of antennas on 20 nm thick ITO film (low cond.), both with and without ITO cover layer. We find an effect for both perpendicular and parallel modes of the antenna, although the parallel mode shift is more pronounced, as well as a polarization dependence of the pump beam. Additionally, we look at the interaction of a larger cylindrical goldpad with the ITO surrounding it to further investigate the injection mechanism of fast electrons from the gold into the ITO.

Nanoantennas were fabricated with e-beam lithography on low-conductivity (70–100 Ω/sq, 20 nm thick) and high-conductivity (8–12 Ω/sq, 120 nm thick) ITO substrates from Sigma-Aldrich using a bilayer of photoresist. After exposure and development, 25 nm of gold was deposited, followed by liftoff. To stay in focus when moving away from the edge of the sample, pads were burned along the way, which turned out useful as a test area later. The nanostructures were covered by sputtering a matching ITO layer of 25 nm on either top or left without cover. This did not produce a measurable difference, as we shall see.

For ultrafast spectroscopy, pulses of 4 ps length with a repetition rate of 40 MHz from an ytterbium fiber laser amplifier (Fianium, Ltd.) were used. The linear antenna response was spectrally characterized using the spatial modulation technique [8, 9], selecting wavelengths between 500–1800 nm using a subtractive mode double prism monochromator. The tunable probe beam was focused onto the sample from the front with a Mitutoyo 100x NIR objective with 0.5 NA, giving a spot size of 0.7 μm FWHM for 900 nm wavelength. As pump beam, we split off part of the original 1064 nm beam and frequency-double it with a 5 mm KTP

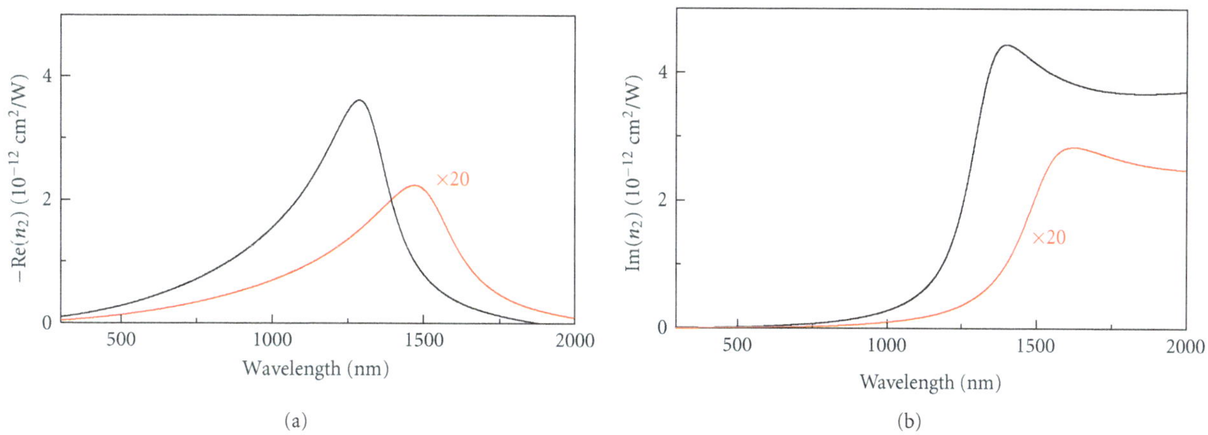

FIGURE 1: Free-carrier nonlinearity described by the nonlinear coefficient for 20 nm low-conductivity (red) and 120 nm high-conductivity (black) ITO.

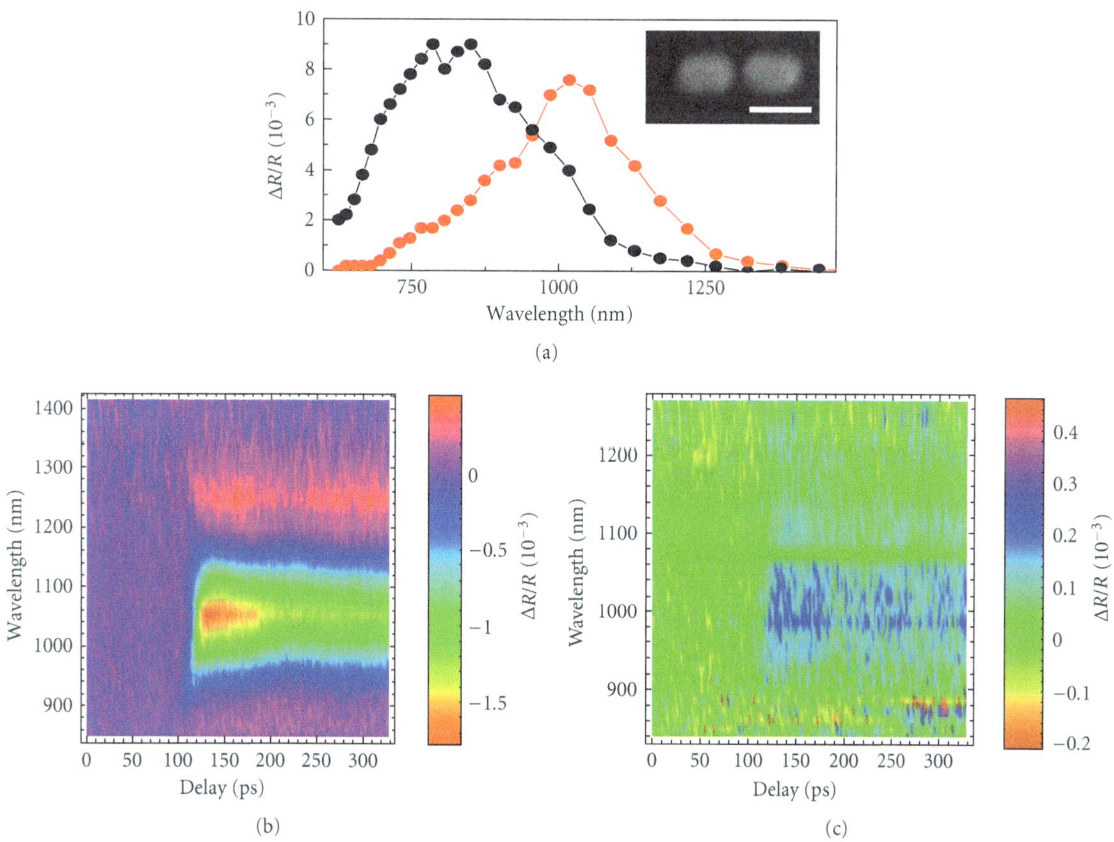

FIGURE 2: (a) Plasmonic resonances along the long (red circles) and the short (black circles) axis of our dimer antenna on low-conductivity ITO, covered with a 20 nm layer of low-conductivity ITO. The inset shows a SEM picture of the antenna, with the scale bar being for 200 nm. (b) and (c) depict the time traces of the pump probe nonlinear response, (b) for the longitudinal and (c) for the transversal mode.

crystal to obtain a 532 nm beam which was focused on the backside of the sample by a 0.6 NA aspheric lens, yielding a spot of 2 μm FWHM. To characterize the nonlinearity of the ITO and the antenna-ITO hybrids, we took time-resolved scans for different wavelengths to give a clear picture of the dynamics.

The nonlinearity of ITO is governed by the free-carrier density, as shown in [6]. This third-order nonlinearity can be expressed by a nonlinear coefficient n_2, representing the resulting refractive index modulation normalized to the optical peak intensity of the excitation beam: $n_2 = \Delta n/I$.

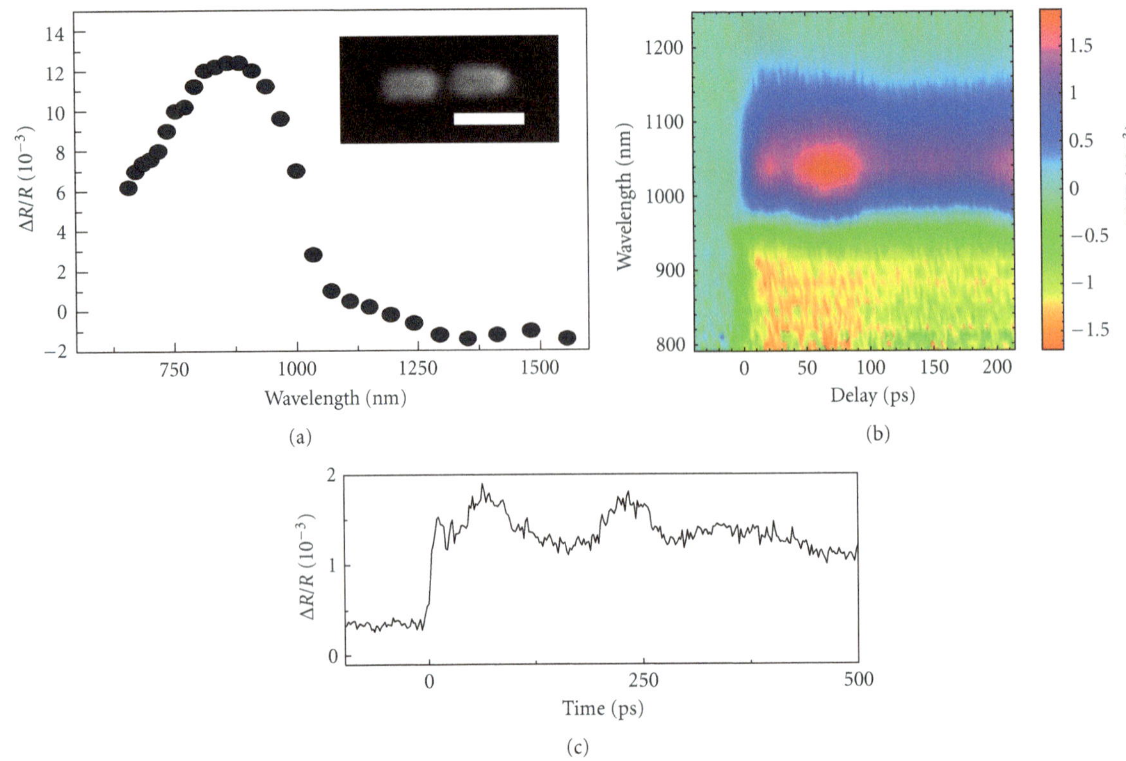

FIGURE 3: Spectral resonance of an antenna with a gap of 40 nm on low-conductivity ITO without cover. Optical pumping leads to a redshift of the original resonance. To the left of the plasmonic resonance and on the maximum of the pump probe signal, we observe an acoustic mode.

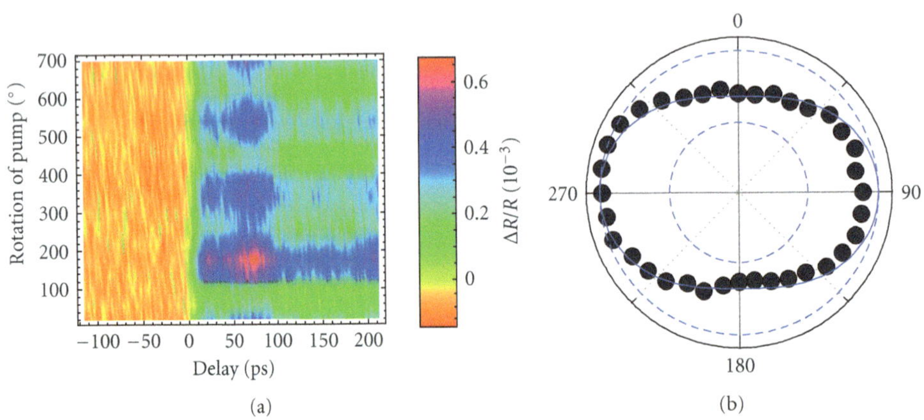

FIGURE 4: Dependence of the pump probe signal from Figure 3 on the polarization of the pump.

Resulting values of n_2 for both the real and the imaginary part of this nonlinear refractive index coefficient over the spectral range are shown in Figure 1 for 20 nm of low-conductivity and 120 nm of high-conductivity ITO. These are obtained from fits to the experimental reflectivity data shown in [6], where we found initial carrier concentrations of $N = 9.2 \times 10^{20}\,\text{cm}^{-3}$ for the high-conductivity ITO and $N = 7.3 \times 10^{20}\,\text{cm}^{-3}$ for the low-conductivity ITO. Optical pumping with 532 nm and 180 pJ pump energy was shown to deplete the ITO, and we observed changes in carrier concentration of $\Delta N = (2.1 \pm 0.1) \times 10^{18}\,\text{cm}^{-3}$ for the high-conductivity ITO and $\Delta N = (5.4 \pm 0.2) \times 10^{16}\,\text{cm}^{-3}$ for the low-conductivity ITO. We find a much larger nonlinearity (more than a factor 20) for the high-conductivity ITO, as can be expected. Our values for the Kerr nonlinearity are in excellent agreement with literature values of similar ITO films [10], taken at a single wavelength around 720 nm.

On the antennas, we measure the extinction spectra of an antenna with a gap of 40 nm on low-conductivity ITO, covered with a matching layer of ITO. We find a resonance at around 800 nm for the perpendicular mode

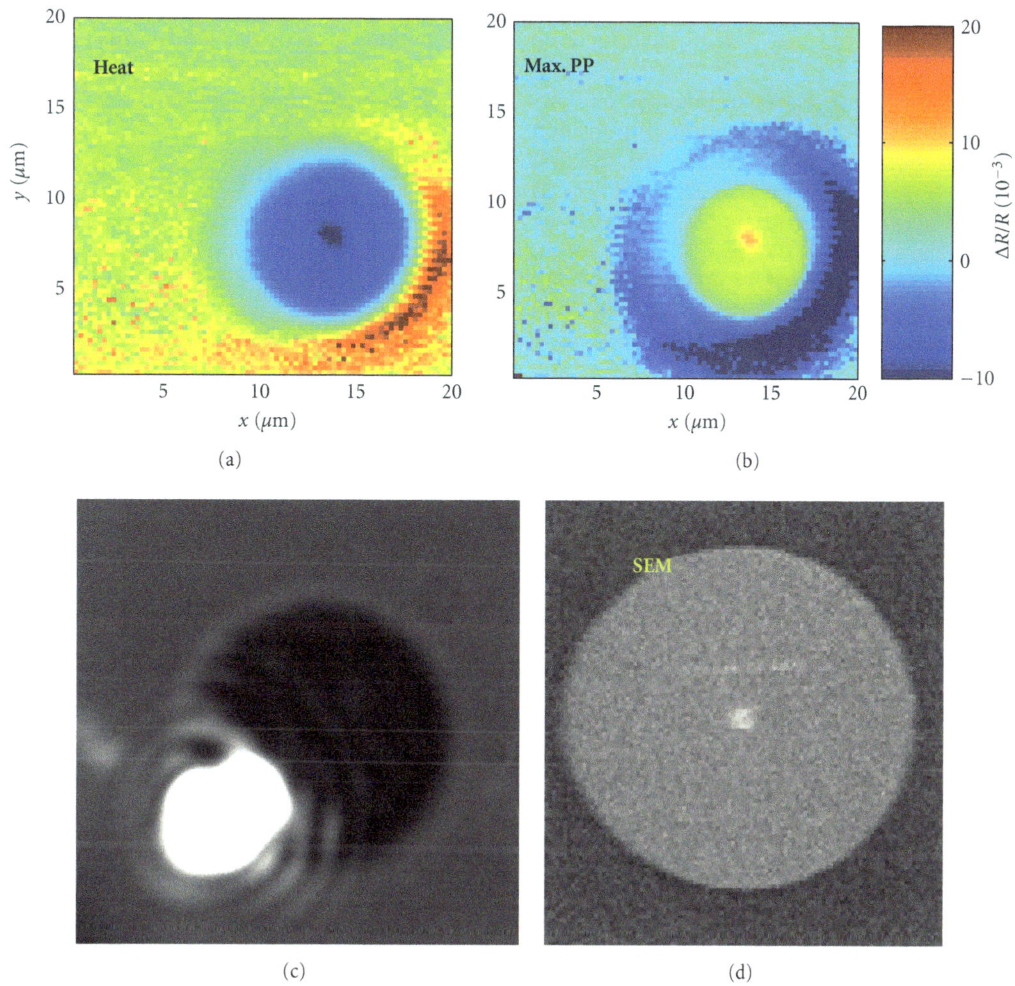

FIGURE 5: Pump probe signal around a gold pad at 985 nm probe wavelength and a pump energy of 180 pJ, for the heat background (a) and the maximum pump probe signal with the heat background subtracted (b). Below are a microscope image with the beam on the edge of the contamination dot (c) and a SEM image of the dot (d).

and another at around 1000 nm for the parallel antenna mode, see Figure 2(a). The corresponding that spectrally resolved nonlinear time trace is shown in Figures 2(b) and 2(c), with the heat backround already subtracted. The spectral bipolar form is different from the well-known transient bleaching found for all kinds of nanoparticles [11, 12], which corresponds to a decrease and broadening of the plasmonic resonance due to fast electron generation in the particle which leads to a change in the refractive index of the gold itself. Instead, we observe a bipolar transient response for both polarizations, although the effect is more pronounced for polarization parallel to the antenna. The negative pump probe signal can be identified with the spatial modulation extinction spectra, with a new redshifted resonance appearing next to it.

When modelling single antenna extinction spectra with FEM simulations (Comsol v3.5), the refractive index change and the concomitant resonance shift predicted for changes in carrier density of the order of our measured value for just the ITO film, $\Delta N = (2.1 \pm 0.1) \times 10^{18}\,\text{cm}^{-3}$ cannot account for the large shift. This leads us to deduce that this must be an effect of the hybrid ITO-antenna system, as described in [6]. We propose that the shift is caused by hot electron injection from the gold into the surrounding ITO. This proposed energy transfer mechanism consists of two parts: first, hot electrons are injected from the antennas into the ITO substrate, leading to a temperature increase in the close vicinity. Apart from the heating effect, the hot electrons are not accountable for changes in the refractive index, since they are swiftly replaced by cold electrons transferring back to the antenna. But the heating effect is big enough to give way to a secondary effect, that of migration of free carriers from the heated region to the surrounding ITO, thus effectively depleting the ITO. This depletion changes the local refractive index of the ITO and therefore shifts the plasmon resonance of the antenna. The nonlinear coefficients needed to describe the Kerr ITO-antenna hybrid nonlinearity is a factor 37 larger than the pure substrate in the case of high-conductivity ITO and as large as a factor of 1475 in the case of low-conductivity ITO (see [6]), which makes the results for low-conductivity ITO-antenna hybrids even more remarkable.

Another notable difference between antennas on low- and high-conductivity ITO that supports our proposition is the fact that we observe acoustic modes in the fast nonlinear response of hybrids on low-conductivity ITO, as shown in Figure 3, but not for hybrids on high-conductivity ITO [6]. This was the case for 11 out of 17 antennas on low-conductivity ITO, for 15 out of 24 antennas on SiO_2, and for none of the 19 antennas on high-conductivity ITO that we have studied. These vibrations are present only on the maximum signal of the new redshifted mode and have oscillation periods of 100–300 ps for our antennas, depending on size. The reason for this position of the acoustic modes is still unclear. Their occurrence, however, indicates that the fast electrons that are excited by the pump beam transfer their energy to the lattice which starts the vibrations in the nanostructure. On high-conductivity ITO, on the other hand, we did not observe these lattice vibrations [6], which leads us to believe that the energy is not transferred to the lattice, but instead is dissipated into heat after the hot electrons are injected into the surrounding semiconductor.

In Figure 4, we show measurements of the excitation anisotropy for an antenna on low-conductivity ITO obtained by rotating a half-wave plate in the pump beam. We find an excitation anisotropy $(\Delta R_{\parallel} - \Delta R_{\perp})/(\Delta R_{\parallel} + \Delta R_{\perp})$ of 0.19 ± 0.02, which indicates that absorption takes place partly through the nanoantenna [13].

To further investigate the hybrid interaction at the gold-ITO interface, we take advantage of one of the gold pads that was written into the photoresist to optimize the focus before exposure. Such a gold pad is shown in Figure 5, as a SEM picture as well as a picture taken in the home-built microscope. We performed pump probe scans on an area of $20\,\mu m \times 20\,\mu m$, both for the slow signal typical for heating of the sample (Figure 5(a)) as well as for the fast pump probe signal. Interestingly, for the heat scan, we observe a clear minimum on the gold pad itself, corresponding to a local heating of the gold surface. For the fast scan (Figure 5(b)), however, the gold pad itself has a clearly positive signal unlike its immediate surroundings. We observe a corona of negative signal up to $2\,\mu m$ around the contamination dot, which is even further than the 126 nm hot electron diffusion length that we suggest in [6]. This is partly due to the size of the pump beam (around $2\,\mu m$ FWHM), which generates heat and hot electrons in areas around the smaller probe spot.

In conclusion, we have shown ultrafast all-optical control of ITO-nanoantenna hybrids. We identify a picosecond nonlinear response that can be described by an enormous corresponding Kerr-nonlinearity that is a factor of 1475 bigger than for the pure ITO. This huge nonlinearity originates in free carriers being injected from the antenna whose energy generates substantial heat in the surrounding substrate. This is very promising indeed for exploitation in creating nanophotonic switches in other designs and warrants further investigations. Hybrid plasmonic components are of great interest for active control of optical fields and integration of photonic and electronic functionalities. Our antenna-ITO hybrid in particular holds great promise for this since electrically controlled refractive index changes for ITO have already been demonstrated [7].

Acknowledgments

The authors thank Z. Webber and M. A. Bampton for their technical support. O. L. Muskens thanks support from EPSRC through Grant EP/H019669/1.

References

[1] W. L. Barnes, A. Dereux, and T. W. Ebbesen, "Surface plasmon subwavelength optics," *Nature*, vol. 424, no. 6950, pp. 824–830, 2003.

[2] O. L. Muskens, V. Giannini, J. A. Sánchez-Gil, and J. G. Rivas, "Optical scattering resonances of single and coupled dimer plasmonic nanoantennas," *Optics Express*, vol. 15, no. 26, pp. 17736–17746, 2007.

[3] N. Large, M. Abb, J. Aizpurua, and O. L. Muskens, "Photoconductively loaded plasmonic nanoantenna as building block for ultracompact optical switches," *Nano Letters*, vol. 10, no. 5, pp. 1741–1746, 2010.

[4] J. Berthelot, A. Bouhelier, C. Huang et al., "Tuning of an optical dimer nanoantenna by electrically controlling its load impedance," *Nano Letters*, vol. 9, no. 11, pp. 3914–3921, 2009.

[5] F. Huang and J. J. Baumberg, "Actively tuned plasmons on elastomerically driven Au nanoparticle dimers," *Nano Letters*, vol. 10, no. 5, pp. 1787–1792, 2010.

[6] M. Abb, P. Albella, J. Aizpurua, and O. L. Muskens, "All-optical control of a single plasmonic nanoantenna-ITO hybrid," *Nano Letters*, vol. 11, no. 6, pp. 2457–2463, 2011.

[7] E. Feigenbaum, K. Diest, and H. A. Atwater, "Unity-order index change in transparent conducting oxides at visible frequencies," *Nano Letters*, vol. 10, no. 6, pp. 2111–2116, 2010.

[8] A. Arbouet, D. Christofilos, N. del Fatti, F. Vallée, J. R. Huntzinger, and M. Broyer, "Direct measurement of the single-metal-cluster optical absorption," *Physical Review Letters*, vol. 93, Article ID 127401, 2004.

[9] O. L. Muskens, N. del Fatti, F. Vallée, J. R. Huntzinger, and M. Broyer, "Single metal nanoparticle absorption spectroscopy and optical characterization," *Applied Physics Letters*, vol. 88, Article ID 063109, 3 pages, 2006.

[10] H. I. Elim, W. Ji, and F. Zhu, "Carrier concentration dependence of optical Kerr nonlinearity in indium tin oxide films," *Applied Physics B: Lasers and Optics*, vol. 82, no. 3, pp. 439–442, 2006.

[11] S. Link and M. A. El-Sayed, "Spectral properties and relaxation dynamics of surface plasmon electronic oscillations in gold and silver nanodots and nanorods," *Journal of Physical Chemistry B*, vol. 103, no. 40, pp. 8410–8426, 1999.

[12] C. Voisin, D. Christofilos, P. A. Loukakos et al., "Ultrafast electron-electron scattering and energy exchanges in noble-metal nanoparticles," *Physical Review B*, vol. 69, no. 19, Article ID 195416, 2004.

[13] O. L. Muskens, G. Bachelier, N. del Fatti et al., "Quantitative absorption spectroscopy of a single gold nanorod," *Journal of Physical Chemistry C*, vol. 112, no. 24, pp. 8917–8921, 2008.

4

Plasmonic Bandgaps in 1D Arrays of Slits on Metal Layers Excited by Out-of-Plane Sources

Roberto Marani, Marco Grande, Vincenzo Petruzzelli, and Antonella D'Orazio

Dipartimento di Elettrotecnica ed Elettronica, Politecnico di Bari, Via E. Orabona 4, 70125 Bari, Italy

Correspondence should be addressed to Antonella D'Orazio, dorazio@poliba.it

Academic Editor: Zhaolin Lu

We analyze the effective opening of finite bands of inhibited transmission in realistic systems excited by actual out-of-plane sources. We first observe how the excitation of surface plasmon polaritons in one-dimensional arrays of metal slits depends on the angle of incidence of the source field. Then, the well-known grating-coupling equation is revised in order to find an asymmetric structure with equivalent parameters which, under perfectly normal excitation, is able to exhibit surface plasmon polariton modes at the same wavelengths of the original structure which undergoes a nonorthogonal incidence of the light. In this way we demonstrate through finite-element simulations that a realistic system, probed by a source beam in a finite light-cone, can be effectively decomposed in several equivalent systems with different physical and geometrical parameters, with results in the enlargement of the theoretically expected punctual minimum of transmission.

1. Introduction

In the last years, the excitation of collective oscillations of free electrons over metal interfaces has been deeply investigated in literature since it allows the successful control of light in terms of localization and enhancement of electromagnetic fields. As already known, the free-electron oscillations can assume several forms, ranging from localized surface dielectric oscillations on metal nanoparticles (localized surface plasmons, LSPs) to freely propagating electron density waves on metal surfaces (surface plasmon polaritons, SPPs). Surface plasmons show unique properties when they propagate over metal gratings and interact with free waves in the dielectric surroundings. The results of the fundamental research in the area of plasmonics have demonstrated that SPPs and LSPs can be exploited for many potential applications residing in sensing [1–5], light manipulation [6, 7] and emission [8–10], energy conversion [11–13], and optical nanoantennas [14, 15], just to mention a few.

Many researchers have proved that under particular conditions many effects of enhanced and inhibited transmission emerge. For instance, subwavelength apertures carved on metal layers provide extraordinary optical transmission (EOT) caused by the resonant excitation of SPP modes over the input and the output interfaces between the metal and the dielectric [16, 17]. The in-phase cooperation among modes allows light beams impinging far from the apertures to assist the spectral response with results in normalized-to-area transmissions higher than one. Moreover, a second mechanism of resonance starts arising when the input wavelength is able to excite waveguide modes within the slits which can be exploited to carry energy within the aperture along a direction perpendicular to the grating wavevector, but in the grating plane (gap plasmon, GP) [18]. One-dimensional gratings allow also the existence of minima of transmission, whose properties are directly linked to the geometrical and physical parameters of the whole systems [19]. Moreover, this resonant behavior, due to the formation of propagating SPP modes over the metal-dielectric interface, can be found even in two-dimensional arrays of metal objects [20]. Without loss of generality, the incident electromagnetic field can be decoupled in two components along the grating directions, each one experiencing its own periodicity, so that a two-dimensional array can be simplified in two different one-dimensional gratings, depending on the polarization of the source.

All the considerations made for arrays of slits face with practical problems due to the splitting of surface modes when the system is illuminated by a source with nonorthogonal incidence. As demonstrated with the experimental analysis in [4], whenever a realistic measurement is performed over arrays of metal objects illuminated by white light lamp, the resulting resonance giving inhibited transmission is constituted by a weighted superposition of resonances in a cone of allowed angles of incidence depending on the numerical aperture of the microscope objective used in the optical setup. In other words the expected single resonance broadens since many SPP modes are excited over the metal-dielectric interface, giving rise to an enlargement of the plasmonic bandgap. This feature is useful for applications in thin-film solar cell technology, where such structures are required to act as broadband reflectors for source field impinging with arbitrary incidences.

In this paper we give evidence to the formation of an effective plasmonic bandgap in an array of sub-wavelength slits on a metal layer excited by an out-of-plane actual source. In order to achieve this goal, we start from the analysis of the theoretical effects of the presence of a tilted source on the excitation of SPP modes. Then we define an equivalence of structures able to convert such modes in spatially separated SPP resonances, with the same spectral properties of the original ones, but excited through the application of an orthogonal source. Since the equivalent parameters depend on the angle of incidence of the source field of the original structure, we observe that a practical system, excited by an input beam in a finite light-cone, has a spectral behaviour comparable to the one obtained by a superposition of results coming out by the investigation of several equivalent structures probed by perfectly orthogonal sources. Thus, we demonstrate the effective opening of a finite bandgap in actual systems, although a single minimum of transmission is expected.

The paper is organized as follows: in Section 2 we will show the basis of the analysis of one-dimensional arrays of sub-wavelength slits on metal layers in both symmetric and asymmetric configurations and under different angles of incidence; then, Section 3 will be dedicated to the definition of the parameters of the equivalent structure, together with a comparison of results with those of the original arrangement; conclusions and remarks will be finally reported in Section 4.

2. One-Dimensional Arrays of Slits

A sketch of the structure considered in this paper is reported in Figure 1. It consists of an infinite one-dimensional array of slits carved on a gold thin film. As shown in Figure 1, the geometrical parameters defining the structure are the metal thickness w, the grating period p, and the width of the strips l, whereas a is the extension of the sub-wavelength gaps, easily equal to the difference between the grating period and the strips width. In this first part of analysis, the structure is assumed to be placed in a homogeneous environment made of air.

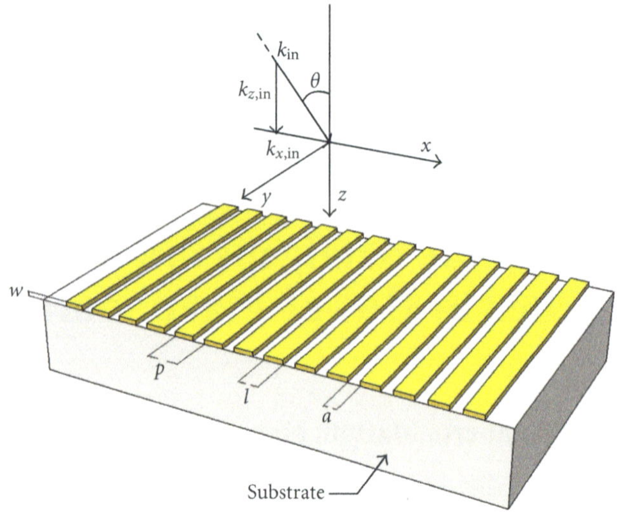

Figure 1: Sketch of an infinite one-dimensional array of metal strips, $(l \times w)$ sized, p is the period of the array, and a is the sub-wavelength extension of the slits.

Maxwell's equations for the propagation of electromagnetic waves have been solved via a finite-element commercial code [21] applied on a computing domain bounded by periodic boundary conditions along the x-direction in Figure 1 and perfectly matched layers in the z-direction. Since Maxwell's equations are rigorously solved depending on the spatial sampling of the window, we impose a maximum size of each subdomain at most equal to the smallest wavelength in each medium divided by a term of 15, whereas the sub-wavelength apertures are sampled with a maximum spatial step of 5 nm to guarantee a sufficient number of elements to model the discontinuities.

Gold regions are then represented through a Drude model assisted by two Lorentz oscillators to fit the experimental data reported in [22]:

$$\varepsilon_m(\omega) = \varepsilon_\infty - \frac{\omega_D^2}{\omega^2 + j\omega\gamma_D} - \sum_{k=1}^{2} \frac{\delta_k \omega_k^2}{\omega^2 - \omega_k^2 + 2j\omega\gamma_k}, \quad (1)$$

where the high-frequency complex dielectric constant of the metal ε_∞ is 5.398, the plasma frequency ω_D and the collision frequency γ_D of the Drude model are $1.398 \cdot 10^{16}$ and $1.033 \cdot 10^{14}$ rad/s, respectively. The parameters describing the contribution of each Lorentz oscillator are the amplitude δ_k, the resonant angular frequency ω_k, and the damping constant γ_k, $k = 1, 2$, which are set according to [23]: $\delta_1 = 0.681$, $\omega_1 = 4.274 \cdot 10^{16}$ rad/s, $\gamma_1 = 4.353 \cdot 10^{14}$ rad/s, $\delta_2 = 1.861$, $\omega_2 = 5.225 \cdot 10^{16}$ rad/s and $\gamma_2 = 6.608 \cdot 10^{14}$ rad/s.

The description of the numerical setup is completed by the definition of the excitation, which is a collection of CW plane waves in the wavelength range between 500 and 1000 nm. This source field emerges from the top boundary of the computing cell with the wavevector $\mathbf{k}_{in}$ depicted in Figure 1, which is tilted of an angle θ with respect to the z-axis. Moreover, the polarization vector is assumed to be parallel to the grating wavevector in order to excite SPPs over the corrugated interfaces. The resulting fields are then collected on transmission and reflection line monitors placed on the bottom and on the top of the domain, 3 μm far from the array, to compute the integral of the Poynting vector flowing through them.

2.1. Orthogonal Incidence. The exhaustive comprehension of the behavior of the free-standing system depicted in Figure 1 can be easily obtained by giving a first look into the transmission and reflection diagrams. Here the geometrical parameters are $w = 200$ nm, $l = 550$ nm, and $p = 600$ nm. Moreover, the simulation has been performed by considering an input pulse illuminating orthogonally the whole system that is, the input wavevector is perpendicular to the grating plane. Transmission and reflection diagrams are reported in Figure 2(a).

It is important to underline that these spectra result by the application of periodic boundary conditions, that is, considering an infinite array of slits. Although this behavior is merely numerical, it is confirmed by experiments since it has been demonstrated that a number of sub-wavelength slits higher than 6 is enough to get the saturation of transmission to its threshold value [24]. The analysis of Figure 2(a) gives evidence to three distinctive regions of the diagram:

(i) a first region at higher wavelengths (800 nm), or lower frequencies where transmission shows a high maximum,

(ii) a deep of transmission localized around 650 nm corresponding to the center of a dip of transmission,

(iii) a sharp peak of transmission at lower wavelengths, or higher frequencies, around 600 nm.

In other words, the presence of a periodicity seems to induce a dip for the light impinging orthogonally on the metal grating as occurs in photonic bandgap structures. Then, this minimum separates two well-defined states. The first state at lower energies ($\lambda = 800$ nm) presents the character of a first-order Fabry-Pérot (FP) resonance acting in a cavity region bounded by two metal walls, defining the aperture, and two dielectric discontinuities between the homogeneous background and the metal-insulator-metal region [19]. The second resonant state of maximum of transmission at higher energies ($\lambda = 611$ nm) is due to the presence of two SPP modes propagating over both the input and the output interfaces, giving rise to extraordinary optical transmission (EOT) [17]. In this case the input plane wave excites first a SPP mode over the input interface which couples with the symmetric SPP mode over the output interface through the array of slits. Finally SPPs start cooperating to the radiation of an outgoing plane wave with the same momentum, that is, wavelength, of the incident one. It is worth noting that the EOT state suffers the presence of absorption within gold, which already occurs around 600 nm. For this reason this choice of parameters, that is, $w = 200$ nm, $p = 600$ nm, and $l = 550$ nm, does not give evidence to a so-well defined maximum of transmission, which would be clear for an equivalent array of slits made on a silver film. Moreover, although this preliminary study is performed on a one-dimensional array, the same behavior can be found also in 2D arrays of gold rectangular patches [20].

More interesting for our purposes is the state of minimum of transmission at $\lambda = 627.5$ nm. The inspection of Figure 2(b), which maps the absolute value of the total magnetic field at this wavelength, reveals that the electromagnetic energy is totally localized on the input interface, where a SPP mode is excited, so that the field is not transmitted to the output. The coupling mechanism between the incident light and the SPP resonating over the grating originates leaky modes, standing over the interface, which reradiate light to the reflection region with effects in the minimum of inhibited transmission through the aperture. It is of relevance to observe that energy is not flowing along the surface, as clearly shown by the inspection of the Poynting vector in Figure 2(b), thus demonstrating the stationary character of this SPP wave under the illumination of an infinite-extended plane wave.

From an analytical point of view, this minimum state occurs whenever the incident wavelength excites the SPP of the unperturbed air-metal interface, that is, for the first resonant order and under normal incidence:

$$\lambda_{\text{SPP}} = p, \qquad (2)$$

where λ_{SPP} is the SPP wavelength given by the dispersion relation of SPPs propagating over a flat interface between the metal and the dielectric. The validity of this formulation is further proven by the initial hypothesis of sub-wavelength apertures, which does not alter significantly the dispersion relation of the SPP propagating on the flat metal layer.

At this stage it is important to point out that each dielectric metal interface acts independently of the other one since the thickness of the metal layer is much higher than the skin depth of gold at the wavelength of interest. This statement can be proven by the examination of Figure 3(a), which details the spectral behavior of the same array of slits placed over a silica substrate with a refractive index n_{sub} equal to 1.45.

In this case it is clear that the matching condition in (2) is individually verified at the input and the output interfaces which are made of the same metal, but of different dielectrics, namely, air and silica. In particular, the minima occurs at $\lambda = 626.5$ nm, almost equal to the one obtained in the previous symmetric case, and $\lambda = 901.5$ nm, which corresponds to the excitation of a SPP mode on the gold-silica interface. This result is also confirmed by Figures 3(b) and 3(c) which plot the xz-section of the profile of the absolute value of the total magnetic field.

The only difference between the two minima of transmission resides on the presence of electromagnetic energy within the sub-wavelength aperture. With reference to Figure 3(c), the operation wavelength of the input field is no longer able to excite a resonant SPP mode over the air-gold interface. The field penetrates through the apertures and coherently excites a SPP mode over the gold-silica interface. Unlike the case in Figure 3(b), where the excitation of a resonant SPP over the first interface cuts off the waveguide mode within the

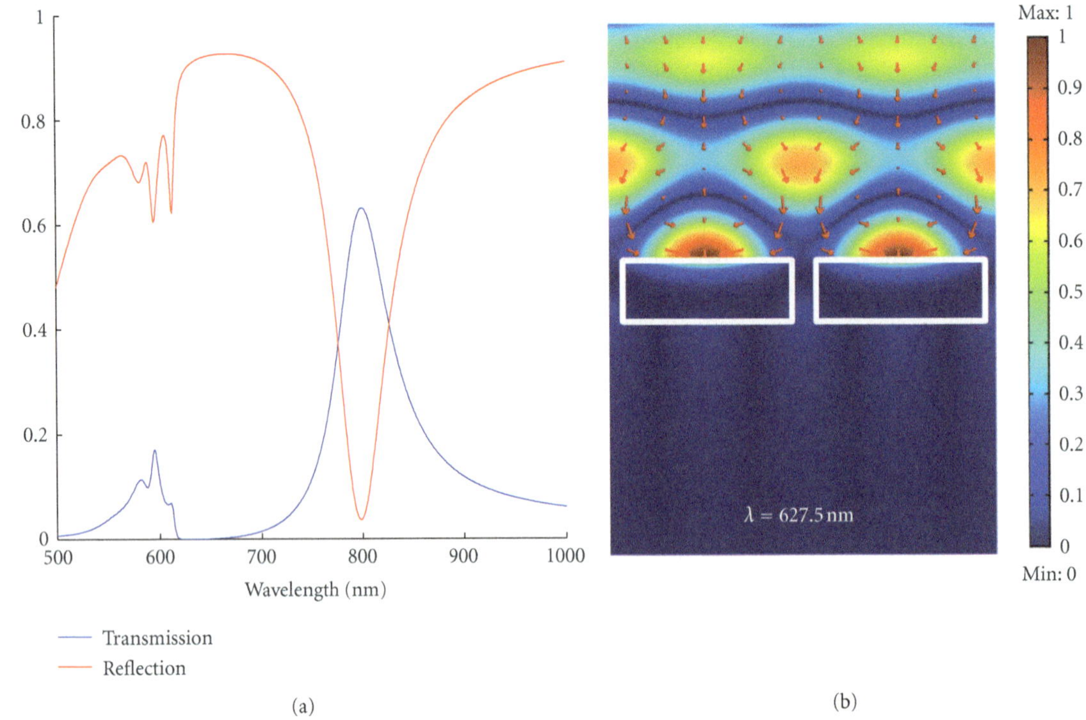

FIGURE 2: (a) Transmission and reflection diagrams of the free-standing array of metal stripes with $w = 200$ nm, $p = 600$ nm, $l = 550$ nm. (b) Map of the transversal section of the absolute value of the magnetic field computed at $\lambda = 627.5$ nm; red arrows display the direction of the Poynting vector.

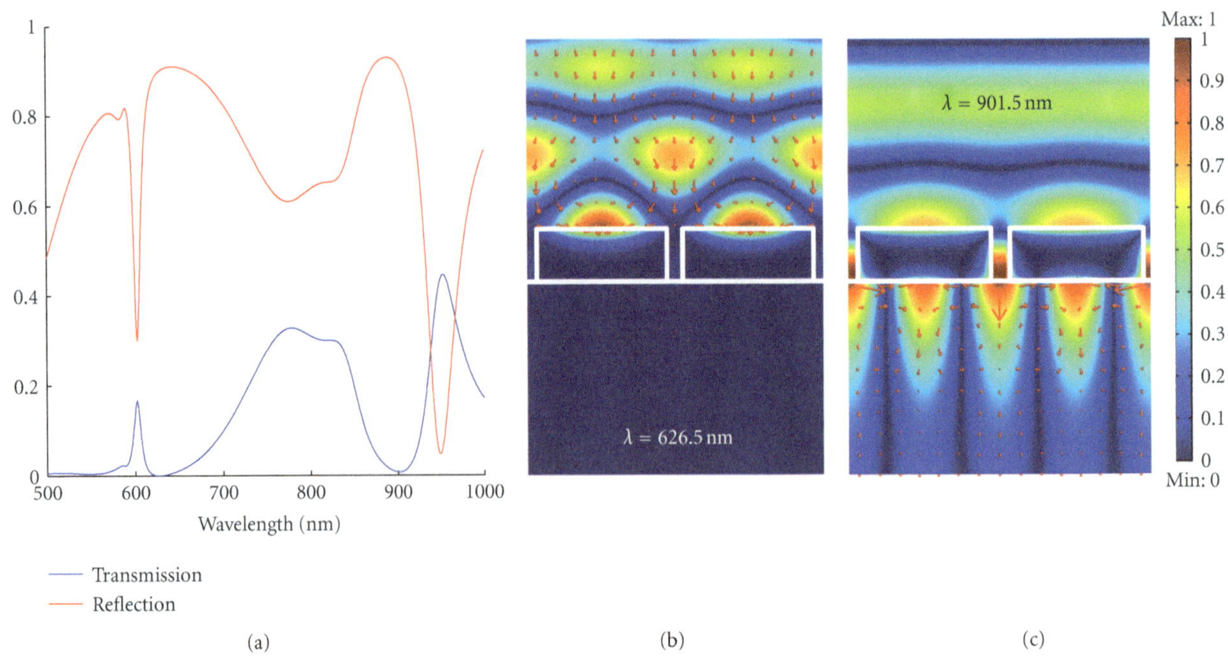

FIGURE 3: (a) Spectral response of an array of slits on the gold layer ($w = 200$ nm, $p = 600$ nm, $l = 550$ nm) placed over a silica substrate ($n_{sub} = 1.45$). Profile of the absolute value of the magnetic field computed at (b) $\lambda = 626.5$ nm and (c) $\lambda = 901.5$ nm; red arrows express the direction of the Poynting vector.

aperture, for $\lambda = 901.5$ nm the presence of electromagnetic energy in this region affects the transmission, which does not reach completely zero. This behavior is even more evident by looking at the direction of the Poynting vector in Figure 3(c), where the lengths of the arrows coming out from the apertures are higher than those related to the resonating SPPs over the output interface.

2.2. Nonorthogonal Incidence. The dependence of the angle of incidence θ on the position of the minimum of transmission is of crucial importance in the spectral behavior of the proposed structure. Equation (2) gives the condition for the excitation of SPP via grating coupling only in the case of orthogonal incidence and for a first order of resonance. If the light impinges with an angle θ (see Figure 1), the matching condition between wavevectors gives [25]

$$k_{\text{SPP}} = k_0 \sin\theta \pm \Delta k_x, \tag{3}$$

where $\triangle\ kx = mG = m \cdot 2\pi/p$ gives the contribution of any perturbation to the smooth metal-dielectric interface mediated by the apertures. Equation (3) can be rearranged to highlight the position of the minimum of transmission. In particular $\lambda_{\text{res},m}$, that is, the free-space wavelength at which the external light excites a SPP mode on the metal grating, can be deduced using the following formulation:

$$\lambda_{\text{res},m} = \pm\frac{p}{m}\left(\sqrt{\frac{\varepsilon_d \varepsilon_m(\lambda_{\text{res},m})}{\varepsilon_d + \varepsilon_m(\lambda_{\text{res},m})}} - \sqrt{\varepsilon_d}\sin\theta\right). \tag{4}$$

The most important effect of the presence of a tilted light source resides in the splitting of the minimum of transmission in two different minima, which are very sensitive to the variation of this angle. An example of application of (4) to the case of an infinite array of slits on gold is reported in Figure 4. Here the minimum position is presented for the first two orders. Moreover the periodicity of the array of slits has been chosen equal to 700 nm in order to give clear evidence to the presence of a minimum of transmission for the wavelengths ranging between 500 and 1000 nm. In addition, the structure is placed in a homogeneous medium, namely air, in order to keep the effect of the variation of the angle of incidence apart from the effect of a nonhomogeneous environment due to the presence of a dielectric substrate.

Since two different SPP modes decouple on the input interface of the metal grating, each mode has an effective refractive index, depending on the corresponding wavelength of excitation of the mode, satisfying (4). In particular

$$n^{\pm}_{\text{SPP},m}(\theta) = m\frac{\lambda^{\pm}_{\text{res},m}(\theta)}{p}. \tag{5}$$

Although this representation of the SPP mode is always valid, n_{SPP} can range between well-defined limits. The lowest limit is due to the presence of ohmic losses within gold, since $\lambda^{-}_{\text{res},m}$ often occurs in the bulk-absorption region of the gold itself when θ increases. On the contrary, $n^{+}_{\text{SPP},m}$ can reach asymptotically the condition $n_{\text{SPP}}(\theta = 90\circ) =$

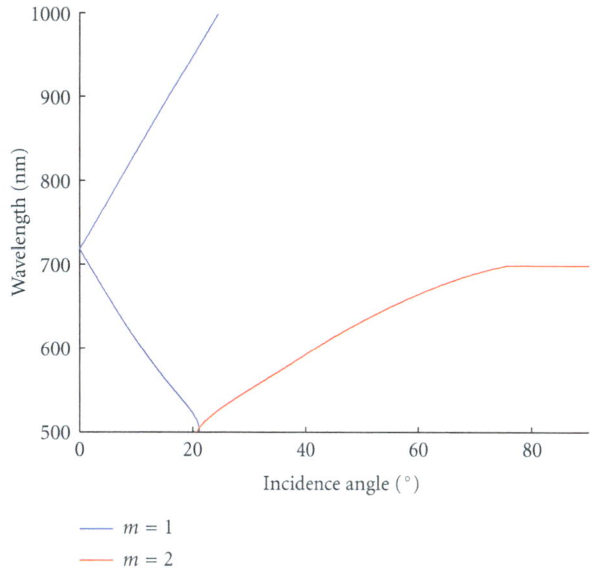

FIGURE 4: Spectral position of the first two orders of the grating-coupling state as a function of the angle of incidence θ of the source field. The period of the array is now equal to 700 nm.

$2 \cdot n_{\text{SPP}}(\theta = 0)$. This has results in the in-plane excitation of SPP modes which undergoes the well-known theory of the Bragg reflectors, forming the basis of the photonic crystal technology [26, 27].

Transmission and reflection diagrams computed on line monitors parallel to the metal-dielectric interfaces are reported in Figure 5. Here the gold film is 150 nm thick, the array of slits has a period of 700 nm, and the width of the sub-wavelength apertures is 50 nm. Moreover, we assume to probe the structure applying a plane wave with an infinite-extended wavefront, with angles of incidence equal to 0, 5°, 10°, and 15°, respectively.

The excitation of two modes whose spectral positions depend on the incidence angle has effects in the formation of two different dips of transmission along the z-axis. The relative distance between these modes increases with the angle θ, following the behavior displayed in Figure 4, down to the absorption region of gold, which is almost reached for $\theta = 15°$.

The xz-profile of the absolute value of the total magnetic field is shown in Figure 6. In this case we consider the same structure discussed above ($p = 700$ nm, $w = 150$ nm, $l = 650$ nm) with an angle of incidence $\theta = 10°$ (for its definition in the reference system see the inset of Figure 6.

A first look to Figure 6 reveals that these plots are qualitatively comparable with that in Figure 2(b), valid in the case of orthogonal incidence over a free-standing array of slits with period equal to 600 nm. In both cases the absolute value of the magnetic fields resembles a first order of resonance, since only one maximum of the electromagnetic field is always localized in the center of the metal elements.

Different considerations have to be carried out to explain the actual meaning of the direction of the Poynting vectors. Here the two SPP modes, due to the presence of a nonzero

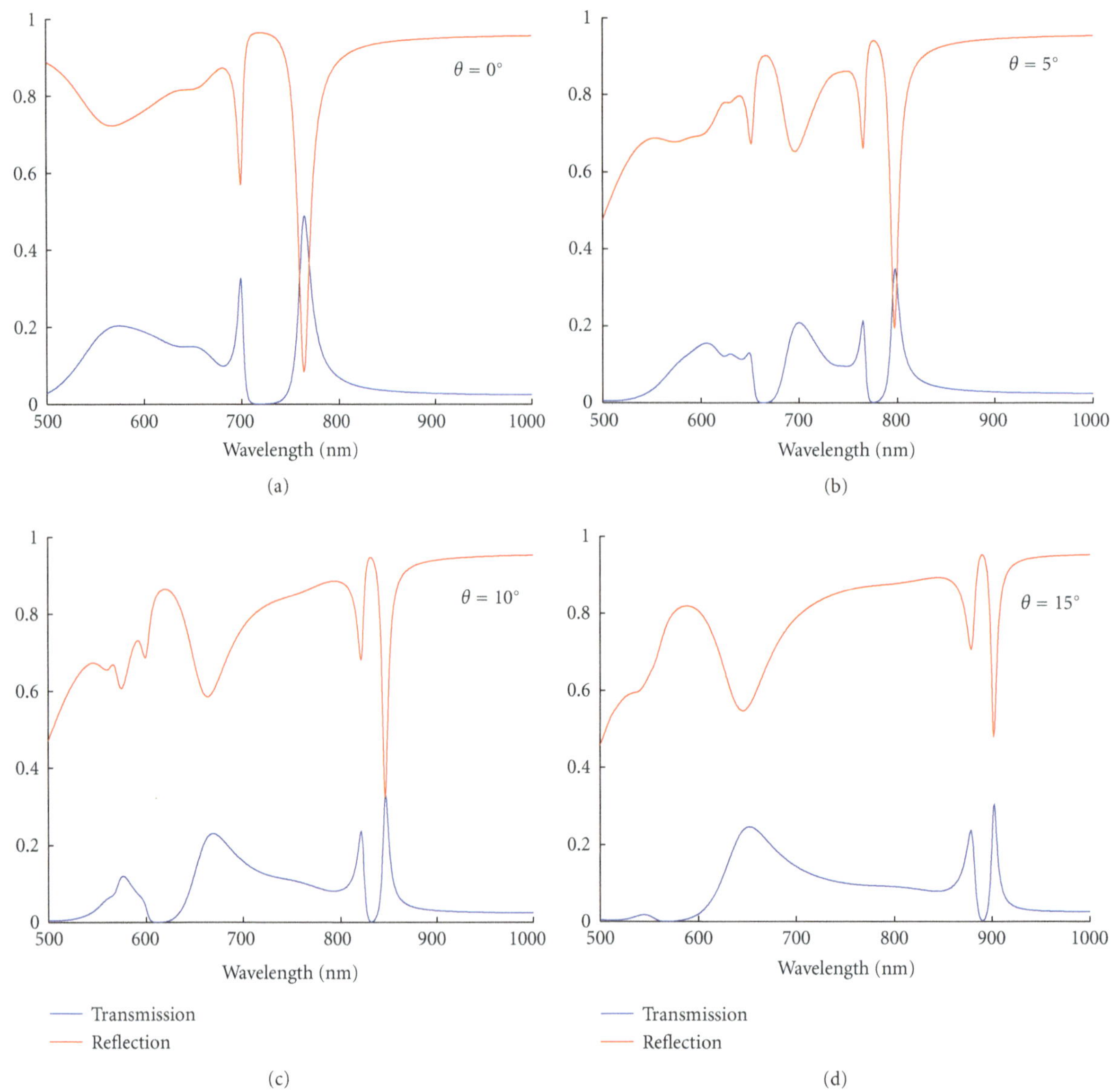

FIGURE 5: Parametric analysis of the transmission and the reflection diagrams of a free-standing array of gold strips computed for (a) $\theta = 0^\circ$, (b) $\theta = 5^\circ$, (c) $\theta = 10^\circ$, and (d) $\theta = 15^\circ$. The geometrical parameters are $p = 700$ nm, $w = 150$ nm, and $l = 650$ nm.

angle of incidence, are counterpropagating over the metal-dielectric interface. The up mode at $\lambda = 833.61$ nm has an opposite direction with reference to the x-component of the exciting wavevector $\mathbf{k}_{\mathrm{in}}$. Taking this mode to the up edge of the range of existence of $n^{+}_{\mathrm{SPP},1}$, that is, for in-plane excitation ($\theta = 90^\circ$), we will find that the SPP is completely reflected accordingly to Bragg's law. As known in SPP bandgap structures [28], the operation wavelength in the medium has to be substituted by the SPP wavelength coming out by the dispersion equation of the SPP propagating over the smooth metal-dielectric interface. Quantitatively, this condition occurs at $\lambda^{+}_{\mathrm{res},1}(\theta = 90^\circ) = 1403.5$ nm. At the same time, the down mode propagates with the same direction of the x-component of $\mathbf{k}_{\mathrm{in}}$, opposite to that of the up mode. As observed previously, the lowest edge for the propagation of this mode, that is, at $\theta = 90^\circ$, falls down to the absorption region of the metal and thus cannot be estimated with accuracy.

3. Equivalent Structure

Once the physics below the excitation of SPPs over a metal gratin has been explained, we go on with the analysis of realistic systems, invoking an equivalence of structures to better display that the punctual minima of transmission broaden to actually form an effective stop band for light.

With reference to Figure 7, the original structure, illuminated by a nonorthogonal light beam, is studied through an equivalent structure with effective period and substrate,

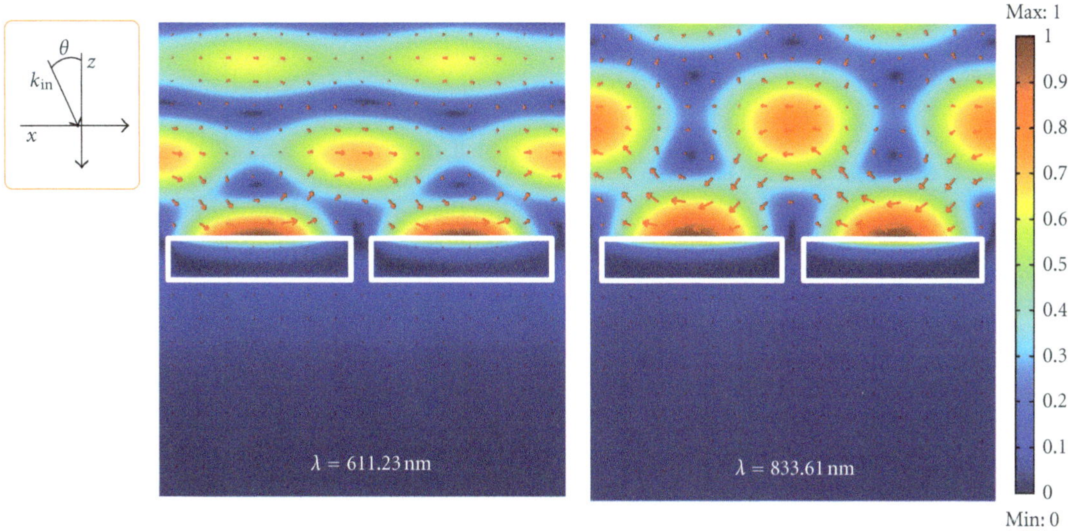

(a) (b)

FIGURE 6: Absolute value of the magnetic field for the structure described in Figure 5 with $\theta = 10°$ computed at (a) $\lambda = 611.23$ nm and (b) $\lambda = 833.61$ nm. Red arrows in both plots express the Poynting vector direction.

but orthogonally excited. The choice of the equivalent parameters depends on the couple of values (m,θ) of the original system: the effective period p_{eq} is chosen to reproduce the SPP excitation of the mth down mode at $\lambda^{-}_{res,m}(\theta)$, whereas the effective substrate is found to model the effect of the mth up mode at $\lambda^{+}_{res,m}(\theta)$. In this way the two modes due to the non-zero angle of incidence behave as two SPP modes resonating over the two interfaces of the asymmetric structure: the down mode at $\lambda^{-}_{res,m}(\theta)$ over the metal-cover interface and the up mode at $\lambda^{+}_{res,m}(\theta)$ over the metal-substrate interface. In this hypothesis, it is clear that the validity of the equivalence showed in Figure 7 is guaranteed at the wavelengths where the up and down modes are excited for a particular value of the angle of incidence θ, that is, at $\lambda^{+}_{res,m}(\theta)$ and $\lambda^{-}_{res,m}(\theta)$. For convenience we suppose to cover the equivalent structure with air.

Since the expression of the equivalent period is merely dependent on $\lambda^{-}_{res,m}(\theta)$, we can make (2) more explicit and merge this result to (4), thus obtaining

$$\frac{p_{eq}}{m} = \frac{\lambda^{-}_{min,m}(\theta)}{\sqrt{\varepsilon_m\left(\lambda^{-}_{min,m}(\theta)\right)/\left(1+\varepsilon_m\left(\lambda^{-}_{min,m}(\theta)\right)\right)}}. \quad (6)$$

In a similar way, but referring to the metal-substrate interface of a p_{eq}-periodic grating, we can find the analytical expression of the dielectric function of the equivalent substrate:

$$\varepsilon_{d,eq} = \frac{\varepsilon_m\left(\lambda^{+}_{min,m}(\theta)\right)\left(m\left(\lambda^{+}_{min,m}(\theta)/p_{eq}\right)\right)^2}{\varepsilon_m\left(\lambda^{+}_{min,m}(\theta)\right) - \left(m\left(\lambda^{+}_{min,m}(\theta)/p_{eq}\right)\right)^2}. \quad (7)$$

It is important to underline that this expression can be applied for each wavelength of the range of interest,

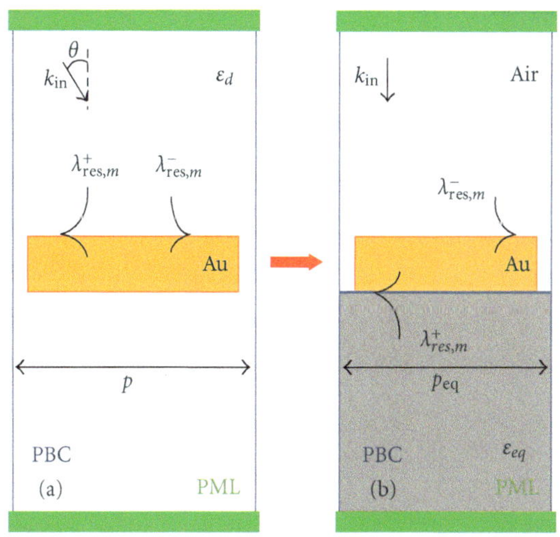

FIGURE 7: Sketch of (a) the original array of slits and (b) the corresponding equivalent structure with orthogonal excitation. SPP modes are also reported on the proper metal-dielectric interfaces.

since it does not interact with the equivalent down mode resonating over the metal-cover interface (see Paragraph 2.1). Nevertheless, it only has a physical meaning for values of λ close to that one assumed by $\lambda^{-}_{res,m}(\theta)$ for a particular θ value. Moreover, the formulations (6) and (7) do not include the metal thickness, since it does not alter considerably the SPP resonance over the grating when the thickness itself is much larger than the skin depth of metal, that is, as in the presented cases.

The analytical formulation of the parameters involved in the model of the equivalent system allows the detailed extraction of the actual values to be implemented. Table 1

Table 1: Resonant wavelengths corresponding to the excitation of the SPP modes over the air-metal interface of the array of gold stripes with $p = 700\,\text{nm}$ and corresponding parameters for the equivalent structures.

θ	$\lambda^{+}_{\text{res},1}$ (nm)	$\lambda^{-}_{\text{res},1}$ (nm)	p_{eq} (nm)	$\varepsilon_{\text{eq}}(\theta)$
0°	718.49	718.49	700	1
5°	775.72	663.08	641.02	1.3814
10°	833.61	611.23	583.88	1.9071
15°	891.29	565.12	530.07	2.6169

reports the resonant wavelengths of the SPP modes of the first order for the cases already presented in Figure 5 and the equivalent parameters obtained by the application of (6) and (7).

The application of these parameters to the equivalent structure allows the comparison of spectra with those already exposed in Figure 5. In particular Figure 8 reports this result in terms of transmission profile for the case in which the original structure is illuminated by a light beam with an angle of incidence equal to 5°, 10°, and 15°.

Before going through the inspection of Figure 8, it is of relevance to notice that the equivalence described by (6) and (7) is suitable only in a wavelength range close to the SPP resonances, so that the comparison of diagrams is expected to show significant differences far from the resonant wavelengths of the two SPP modes. As a result, the diagrams in Figures 8(a) and 8(b) show a good agreement between SPP resonances found by the equivalent structure on those of the original one. Table 2 reports the comparison of the resonant wavelengths resulting from the analysis of the original structure and the equivalent ones with angles of incidence of the source field equal to 5°, 10° and 15°.

In particular, it is evident a good matching between the resonant wavelengths of the systems corresponding to the cases of $\theta = 5°$ and $\theta = 10°$, with a relative error smaller than 0.7%. On the other hand, Figure 8(c) gives evidence to an overestimation of the resonant wavelengths with respect to the one obtained with the original structure. The nature of this phenomenon can be found in the presence of losses induced by the metal on the down mode which raises close to the absorption band of gold. This mode only affects the expression of the equivalent period, according to (6), thus inducing an error in its model. Then, the measurement uncertainty due to high absorption interferes with the definition of the higher resonance, which depends itself on the down mode. These considerations allow us to confirm that the range of validity of θ for the proposed equivalence of structures is determined by the relative position of the down mode with respect to the absorption band of the metal, in a similar way to what noticed in the previous paragraph with reference to (5). This implies that, if perfect metals are considered, that is, any ohmic-loss mechanism is taken into account, the equivalence is always verified. Further simulations prove the equivalence of results under this hypothesis with negligible relative errors (lower than 0.06 %) for each angle of incidence.

Moreover, although this is not of interest for this paper, which is mostly focused on the excitation of SPP modes, it is important to show that the peaks of resonance of high transmission at the lowest edges of the up modes disappear in all the proposed cases. This occurrence has a physical meaning ascribable to the character of this kind of resonance. In other words, since this peak of high transmission is due to the presence of coupled SPPs propagating on both input and output interfaces with the same phase laws, it can be only found in symmetric systems [17]. The presence of an equivalent substrate cuts off this resonant mode with results in the transmission profiles.

Figure 9 maps the absolute value of the resonant magnetic field obtained by the analysis of the equivalent structure related to the case of $\theta = 10°$.

The analysis of Figure 9 highlights the different behaviors of the two resonating modes. As expected, the down mode at $\lambda^{-}_{\text{res},1} = 611.23\,\text{nm}$ resonates as a SPP mode over the air-gold interface, inducing an efficient stop condition for the energy impinging on the grating, in a manner similar to that in Figure 6(a), valid for the original system with $\theta = 10°$. For this reason all the down modes have a depth comparable to the ones obtained by the response of the original structures, within a numerical tolerance. Conversely, the up mode changes its nature becoming a SPP mode resonating over the gold-substrate interface. As already observed in Paragraph 2.1, the excitation of a SPP mode over the output interface does not switch off the resonance within the sub-wavelength aperture, thus allowing light to reach the transmission. This aspect does not allow the achievement of the zero-transmission condition for the up mode, as comes out from Figure 8.

Another important aspect regards the direction of the Poynting vector related to the resonant modes over the two interfaces. As shown previously, the presence of a non-zero angle of incidence induces the excitation of two counterpropagating modes at different wavelengths in the original system. By looking at the red arrows in Figure 9, the energy of the SPP modes stands over the grating without a dominant flow direction. This is the result of the orthogonal excitation of the slit array which produces a stationary mode reradiating to the reflection. Nevertheless, it is important to underline that this is only due to the kind of excitation used in our numerical measurements, that is, an infinite-extended plane wave. When the source is spatially bounded, assuming, for instance, a Gaussian shape, each mode at $\lambda^{-}_{\text{res},m}$ and $\lambda^{-}_{\text{res},m}$ stationary resonates with the profile in Figure 9 only in the bound region effectively covered by the incidence of the input beam. Figure 10 shows the magnetic field response of the structure in Figure 9 under the application of a Gaussian-shaped input pulse able to excite the down mode at $\lambda = 611.23\,\text{nm}$.

The analysis of the Poynting vector directions in Figure 10 proves that, since each of these modes is the superposition of the two counterpropagating modes, these components decouple propagating in opposite direction through the side regions. Moreover, the propagation length of such modes has been found the same of the original structure, since the main loss mechanism suffered by SPPs

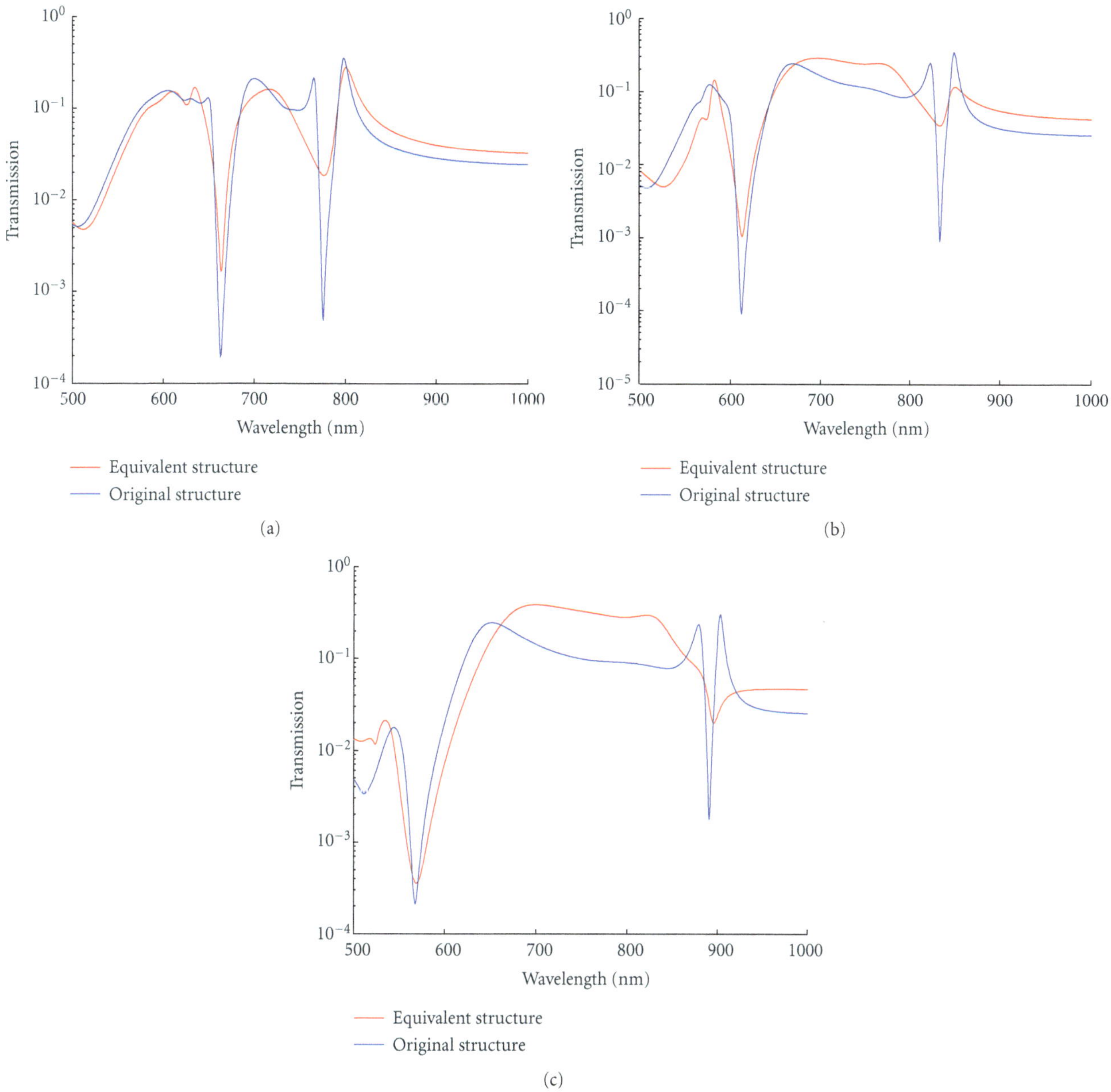

FIGURE 8: Blue lines report the transmission diagrams already presented in Figure 5 in logarithmic scale, whereas red lines display on the same scale the results obtained by the equivalent structures with the parameters listed in Table 1. The angles of incidence of the input light for the original arrays are (a) $\theta = 5^\circ$, (b) $\theta = 10^\circ$, and (c) $\theta = 15^\circ$.

TABLE 2: Comparison of the resonant wavelengths obtained by the analysis of the original structure with different angles of incidence and its equivalent counterparts.

	Original structure		Equivalent structure	
θ	$\lambda^{+}_{res,1}$ (nm)	$\lambda^{-}_{res,1}$ (nm)	$\lambda^{+}_{res,1}$ (nm)	$\lambda^{-}_{res,1}$ (nm)
5°	775.72	663.08	776.15	663.51
10°	833.61	611.23	832.78	612.36
15°	891.29	565.12	895.42	569.08

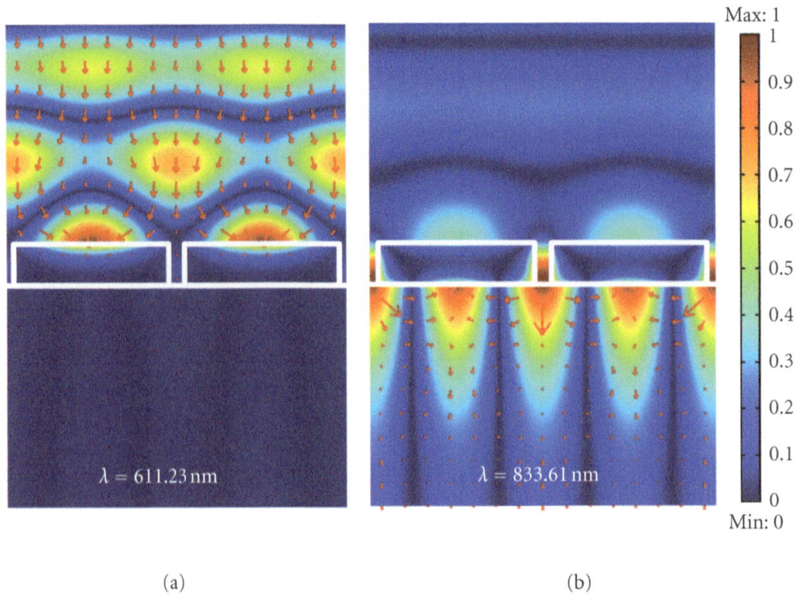

FIGURE 9: Absolute value of the magnetic field for the equivalent system applied to the case of $\theta = 10^\circ$ ($p_{eq} = 583.88$ nm and $\varepsilon_{eq} = 1.9071$). Computations are performed at (a) $\lambda = 611.23$ nm and (b) $\lambda = 833.61$ nm. Red arrows indicate the Poynting vector direction.

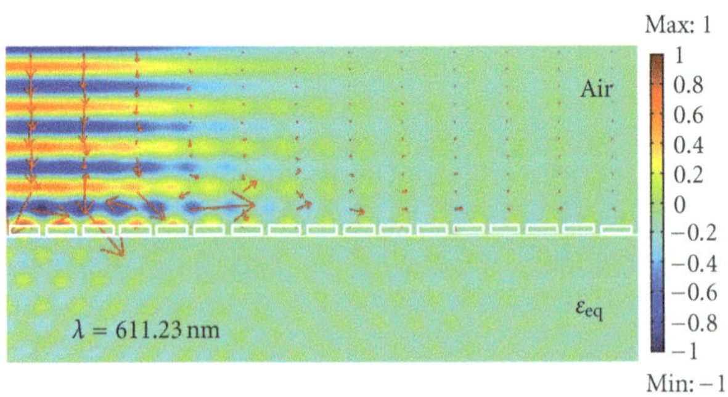

FIGURE 10: Magnetic field component H_z at $\lambda = 611.23$ nm of the equivalent system with the parameters in Figure 9 applied to the case of $\theta = 10^\circ$. Exploiting the even symmetry of the structure, the left side of the magnetic field is the mirrored replica of this plot. Red arrows indicate the direction of the Poynting vector. In this case the source has a Gaussian-shaped spatial profile.

is due to radiation phenomena owing to the presence of the sub-wavelength apertures.

The equivalence of structures allows an easy comprehension of the actual bandgap formation in slit arrays subjected to an out-of-plane excitation, although the effective incidence of a source field over a metal grating would only generate a dip of transmission following the momentum conservation in (3).

With reference to the experimental demonstration in [4], a measurement setup able to excite SPP modes over periodic arrangements of metal elements for sensing purposes makes use of microscope objectives to focus a white light over the grating. As a consequence, the impinging light is in a cone more than 10° wide. This produces a measured effect of an enlargement of the complete-reflection band. Since the proposed equivalence of structures is still valid for two-dimensional arrays of metal square elements, as the two polarizations act independently, experiencing the same periodicity, the simple original structure can be seen as a weighted superposition of results coming out by the analysis of several equivalent systems. Exploiting the mathematical formalism given by (6) and (7), we find that the equivalent structures for the air-metal interface have periods ranging in an interval of about 100 nm, with a maximum shift of the dielectric constant equal to 0.87. It means that the original system with practical normal excitation can be seen as made of many different structures with a perfectly orthogonal excitation. In this way the broadening of the stop band for input light becomes clear, although the physical meaning resides in the excitation of several SPP up and down modes over the grating. Thus, we can consider the effective opening of a finite bandgap for light impinging orthogonally on actual systems.

Moreover, this equivalence is also helpful to understand the capability of these arrangements of metal elements in solar cells designed to improve the energy absorption. In fact, a single system excited by sunlight with random incident angles behaves as a set of structures, each one with its equivalent period and substrate, with effects in the broadening of the bandgap of transmitted energy, very suitable for back reflectors.

4. Conclusions

In this paper we have demonstrated through numerical experiments that a non-zero angle of incidence induces the formation of an effective finite bandgap in practical systems. From a physical point of view, out-of-plane incidences over metal gratings made of periodic arrangements of slits on metal layers generate two SPP modes whenever the momentum conservation is fulfilled. Such modes propagate over the metal grating in opposite directions and degenerate in a standing wave if light impinges orthogonally with an infinite wavefront. By exploiting the independence of surface modes over the two interfaces of the metal film, we define an equivalence of structures able to convert the counter-propagating SPP modes, due to the non-zero incident angle of the source field, in two spatially separated SPP modes, generated by an orthogonal source field and propagating at the same wavelength of those of the original system, but on different interfaces. The equivalence of structures has been applied to the case of a realistic measurement setup, where the input light is in a cone at least 10° wide. In this case the behaviour of the actual structure can be described by a set of equivalent systems ideally illuminated by a perfectly orthogonal excitation with effective periods ranging in an interval of about 100 nm, with a maximum shift of the dielectric constant of the equivalent substrate equal to 0.87. In this way we prove that realistic systems show effective bandgaps owing to the excitation of several SPP modes.

Acknowledgments

The authors acknowledge the CINECA Award for the proposed project *SPMAPS—Surface Plasmon Management in Active and Passive Systems—2011* and for the availability of high-performance computing resources and support.

References

[1] R. Ameling, L. Langguth, M. Hentschel, M. Mesch, P. V. Braun, and H. Giessen, "Cavity enhanced localized plasmon resonance sensing," *Applied Physics Letters*, vol. 97, no. 25, Article ID 253116, 2010.

[2] Y. T. Chang, Y. C. Lai, C. T. Li, C. K. Chen, and T. J. Yen, "A multi-functional plasmonic biosensor," *Optics Express*, vol. 18, no. 9, pp. 9561–9569, 2010.

[3] M. Grande, R. Marani, F. Portincasa et al., "Asymmetric plasmonic grating for optical sensing of thin layers of organic materials," *Sensors and Actuators, B*, vol. 160, no. 1, pp. 1056–1062, 2011.

[4] M. Grande, M. A. Vincenti, T. Stomeo et al., "Experimental demonstration of a novel bio-sensing platform via plasmonic band gap formation in gold nano-patch arrays," *Optics Express*, vol. 19, no. 22, pp. 21385–21395, 2011.

[5] M. Grande, M. A. Vincenti, T. Stomeo et al., "Two-dimensional plasmonic nano-structures for linear and non-linear regimes," in *Proceedings of the Micro- and nano-photonic materials and devices (MINAP '12)*, pp. 73–76, Trento, Italy, January 2012.

[6] J. Bravo-Abad, F. J. García-Vidal, and L. Martín-Moreno, "Wavelength de-multiplexing properties of a single aperture flanked by periodic arrays of indentations," *Photonics and Nanostructures—Fundamentals and Applications*, vol. 1, no. 1, pp. 55–62, 2003.

[7] R. Marani, V. Marrocco, M. Grande, G. Morea, A. D'Orazio, and V. Petruzzelli, "Enhancement of extraordinary optical transmission in a double heterostructure plasmonic bandgap cavity," *Plasmonics*, vol. 6, no. 3, pp. 469–476, 2011.

[8] D. J. Bergman and M. I. Stockman, "Surface plasmon amplification by stimulated emission of radiation: quantum generation of coherent surface plasmons in nanosystems," *Physical Review Letters*, vol. 90, no. 2, Article ID 027402, 4 pages, 2003.

[9] M. A. Noginov, G. Zhu, A. M. Belgrave et al., "Demonstration of a spaser-based nanolaser," *Nature*, vol. 460, no. 7259, pp. 1110–1112, 2009.

[10] R. Marani, A. D'Orazio, V. Petruzzelli et al., "Gain-assisted extraordinary optical transmission through periodic arrays of subwavelength apertures," *New Journal of Physics*, vol. 14, Article ID 013020, 2012.

[11] K. R. Catchpole and A. Polman, "Plasmonic solar cells," *Optics Express*, vol. 16, no. 26, pp. 21793–21800, 2008.

[12] M. Losurdo, M. M. Giangregorio, G. V. Bianco, A. Sacchetti, P. Capezzuto, and G. Bruno, "Enhanced absorption in Au nanoparticles/a-Si:H/c-Si heterojunction solar cells exploiting Au surface plasmon resonance," *Solar Energy Materials and Solar Cells*, vol. 93, no. 10, pp. 1749–1754, 2009.

[13] V. Marrocco, R. Marani, M. Grande et al., "Modification of the scattering of silver nanoparticles induced by Fabry-Pérot resonances rising from a finite Si layer," *Journal of Optics*, vol. 13, no. 1, Article ID 015004, 2011.

[14] D. Dregely, R. Taubert, J. Dorfmüller, R. Vogelgesang, K. Kern, and H. Giessen, "3D optical Yagi-Uda nanoantenna array," *Nature Communications*, vol. 2, no. 1, article 267, 2011.

[15] T. Dattoma, M. Grande, R. Marani, G. Morea, V. Marrocco, and A. D'Orazio, "Resonancewavelength dependence and mode formation in gold nanorod optical antennas with finite thickness," *Progress In Electromagnetics Research B*, no. 30, pp. 337–353, 2011.

[16] T. W. Ebbesen, H. J. Lezec, H. F. Ghaemi, T. Thio, and P. A. Wolff, "Extraordinary optical transmission through sub-wavelenght hole arrays," *Nature*, vol. 391, no. 6668, pp. 667–669, 1998.

[17] J. A. Porto, F. J. García-Vidal, and J. B. Pendry, "Transmission resonances on metallic gratings with very narrow slits," *Physical Review Letters*, vol. 83, no. 14, pp. 2845–2848, 1999.

[18] D. F. P. Pile, T. Ogawa, D. K. Gramotnev et al., "Two-dimensionally localized modes of a nanoscale gap plasmon waveguide," *Applied Physics Letters*, vol. 87, no. 26, Article ID 261114, pp. 1–3, 2005.

[19] D. De Ceglia, M. A. Vincenti, M. Scalora, N. Akozbek, and M. J. Bloemer, "Plasmonic band edge effects on the transmission properties of metal gratings," *AIP Advances*, vol. 1, no. 3, Article ID 032151, 2011.

[20] R. Marani, M. Grande, V. Marrocco et al., "Plasmonic bandgap formation in two-dimensional periodic arrangements of gold patches with subwavelength gaps," *Optics Letters*, vol. 36, no. 6, pp. 903–905, 2011.

[21] COMSOL Multiphysics, http://www.comsol.com.

[22] E. D. Palik, *Handbook of Optical Constants of Solids*, Academic Press, San Diego, Calif, USA, 1998.

[23] J. M. McMahon, J. Henzie, T. W. Odom, G. C. Schatz, and S. K. Gray, "Tailoring the sensing capabilities of nanohole arrays in gold films with Rayleigh anomaly-surface plasmon polaritons," *Optics Express*, vol. 15, no. 26, pp. 18119–18129, 2007.

[24] Y. S. Jung, J. Wuenschell, H. K. Kim, P. Kaur, and D. H. Waldeck, "Blue-shift of surface plasmon resonance in a metal nanoslit array structure," *Optics Express*, vol. 17, no. 18, pp. 16081–16091, 2009.

[25] H. Raether, *Surface Plasmons on Smooth and Rough Surfaces and on Gratings*, Springer, Berlin, Germnay, 1986.

[26] K. Sakoda, *Optical Properties of Photonic Crystals*, Springer, Berlin, Germany, 2005.

[27] J. D. Joannopoulos, S. G. Johnson, J. N. Winn, and R. D. Meade, *Photonic Crystals—Molding the Flow of*, Princeton University Press, Princeton, NJ, USA, 2008.

[28] S. I. Bozhevolnyi, J. Erland, K. Leosson, P. M. W. Skovgaard, and J. M. Hvan, "Waveguiding in surface plasmon polariton band gap structures," *Physical Review Letters*, vol. 86, no. 14, pp. 3008–3011, 2001.

5

Coupling Schemes in Terahertz Planar Metamaterials

Dibakar Roy Chowdhury,[1] Ranjan Singh,[1] Antoinette J. Taylor,[1] Hou-Tong Chen,[1] Weili Zhang,[2] and Abul K. Azad[1]

[1] *Center for Integrated Nanotechnologies, Materials Physics and Applications Division, Los Alamos National Laboratory, Los Alamos, NM 87545, USA*
[2] *School of Electrical and Computer Engineering, Oklahoma State University, Stillwater, OK 74078, USA*

Correspondence should be addressed to Dibakar Roy Chowdhury, dibakar@lanl.gov and Ranjan Singh, ranjan@lanl.gov

Academic Editor: Zhaolin Lu

We present a review of the different coupling schemes in a planar array of terahertz metamaterials. The gap-to-gap near-field capacitive coupling between split-ring resonators in a unit cell leads to either blue shift or red shift of the fundamental inductive-capacitive (*LC*) resonance, depending on the position of the split gap. The inductive coupling is enhanced by decreasing the inter resonator distance resulting in strong blue shifts of the *LC* resonance. We observe the *LC* resonance tuning only when the split-ring resonators are in close proximity of each other; otherwise, they appear to be uncoupled. Conversely, the higher-order resonances are sensitive to the smallest change in the inter particle distance or split-ring resonator orientation and undergo tremendous resonance line reshaping giving rise to a sharp subradiant resonance mode which produces hot spots useful for sensing applications. Most of the coupling schemes in a metamaterial are based on a near-field effect, though there also exists a mechanism to couple the resonators through the excitation of lowest-order lattice mode which facilitates the long-range radiative or diffractive coupling in the split-ring resonator plane leading to resonance line narrowing of the fundamental as well as the higher order resonance modes.

1. Introduction

The interaction of electromagnetic waves with unstructured bulk matter is mainly governed by the inherent chemistry of material, and its properties are usually described in terms of permittivity and permeability. Recently developed electromagnetic metamaterials allow engineered material at the unit cell level, which has enabled novel methods to manipulate the electromagnetic properties [1]. Metamaterials have attracted a tremendous amount of interest because of their exotic properties such as negative refractive index [2], perfect focusing [3], cloaking [4, 5], and resonance modulation in the active and passive modes [6–29]. The electromagnetic responses offered by metamaterials are unique and are not found in naturally occurring materials. The fundamental building blocks of metamaterials are mainly comprised of a metallic split-ring resonators (SRRs) fabricated on a die-lectric/semiconducting substrate. The operation of most metamaterial devices depends on the designs employed to control the fundamental resonances of SRRs, since the major tuning of material permittivity and permeability is enabled by these resonances [6–29]. Apart from the single-SRR approach, several research groups have explored lateral coupling between nearest neighbor SRRs and its effects on the resonances [30–42]. It turns out that one of the prominent schemes to passively tune the metamaterial resonances is achieved by changing either the intercell distance or the relative orientation of SRRs with respect to each other. These studies revealed that the laterally coupled mechanisms provide an extra degree of freedom for the manipulation of metamaterial responses. Moreover, the terahertz frequency range offers a convenient platform for these studies, since sample fabrication is straightforward, and terahertz time-domain spectroscopy (THz-TDS) provides a phase-coherent characterization technique.

2. Coupling Mechanisms in Planar Metamaterials

The short-range coupling in metamaterials occurs via the electric and magnetic fields due to the close proximity of the nearest neighbor resonators. The electric field couples through the gaps of the SRRs, while the magnetic field couples through the circumference, producing a red shift or blue shift of the resonance, depending on the nature of near-field coupling. We employed two different SRR design schemes to experimentally probe the effect of such coupling. In the first scheme, the density of the resonators is kept fixed, while the nearest neighbor interaction between two SRRs in the unit cell is probed as a function of their separation and mutual orientation. In the second type of experimental design, the effect of packing density is also taken into account. In this case, the number of SRRs excited by the incident beam varies from one sample to another as the resonators are collectively brought closer to each other. Thus, these results highlight the coupling behavior in metamaterials with low-, medium-, and high-density SRR arrays.

Recently, it has been demonstrated that near-field electric and magnetic coupling is not the only mechanism by which the SRRs in a planar arrangement remains coupled to each other [38–40]. In addition, there is also long-range radiative coupling as demonstrated with inductive-capacitive (*LC*) resonances by Singh et al. [39] and for higher-order resonance by Bitzer et al. [38]. This long-range diffractive coupling occurs when the lattice periodicity of the SRR array matches the wavelength of incident light at the fundamental *LC* or higher-order resonances, leading to a tremendous reduction in radiation damping that enhances the quality factor of the coupled resonances.

2.1. Capacitive Coupling between SRR Pairs in a Planar Array. In this section, we discuss the effect of different gap-to-gap coupling schemes on the fundamental *LC* resonances. The metamaterial samples studied were fabricated on semi-insulating GaAs substrates by photolithography methods, electron beam deposition 200 nm thick Au using 10 nm thick Ti as adhesion layer, and followed by a lift-off process. In the fabricated samples, each of the metamaterial unit cells contains two SRRs oriented with their gaps facing toward each other, as shown in Figures 1(a)–1(d) [29]. The samples were characterized in transmission using THz-TDS in a confocal geometry [43], using a bare GaAs substrate as the reference. The incident THz beam was linearly polarized with the electric field oriented parallel to the gap-bearing side, as shown in Figure 1(d).

In Figures 1(a) and 1(b), the two SRRs are individually identical, and their mutual coupling is controlled by varying their separation. As the separation between the resonators decreases from MM1 to MM2, a blue shift of metamaterial resonance is observed (Figure 2(a)). The coupling can also be controlled by asymmetry through displacing one of the gaps but keeping the other gap at the center of the gap-bearing arm (MM3 in Figure 1(c)), or by displacing both the gaps but in the opposite directions (MM4 in Figure 1(d)).

While displacing the gaps in MM3 and MM4 from their original position in MM2, the separation between SRRs within the unit cell is kept fixed at 2 μm. We observe a red shift accompanied by a reduction in the strength of the *LC* resonance for MM3 as compared to MM2 due to the asymmetry and a further red shift and strength reduction in MM4 (Figure 2(b)).

The fundamental *LC* resonance mode arises because of the electric currents oscillating around the full circumference of the SRR loop. The electric currents are excited along the SRR arms by the incident electric field due to the asymmetric arm length of the individual elements of SRR unit cells. The position of the anticipated *LC* resonance frequency can usually be described by a simple expression, $\omega_{LC} = 1/\sqrt{LC}$, where L and C are the effective inductance and capacitance of the individual SRR. In order to have better understanding of the physical mechanisms related to the resonant frequency shift from MM1 to MM4, we simulated the electric and magnetic field distributions using the numerical solver CST Microwave Studio [44]. Simulation results reveal that the magnetic field distributions for the coupled resonators remain almost similar to the uncoupled SRRs, and therefore, do not contribute to the resonance frequency shift significantly. However, the electric fields within the SRR pairs, as shown in Figure 3, clearly go through significant changes for the various coupled resonators. When the SRRs are far apart and SRR gaps are aligned, as in the case of MM1, fringing fields around the gap-bearing side contribute significantly to the overall capacitance of the individual resonators, as seen from the electric field distribution in Figure 3(a). As the separation between SRRs is reduced, these fringing fields increasingly couple the individual responses of the rings. Coulomb repulsion occurs through the fringing fields between like charges in the paired rings, thereby opposing charge accumulation near the gap resulting in a reduction of the fringing capacitance in both rings, as shown in Figure 3(b). This is responsible for the observed blue shift of the resonance frequency in MM2 as compared to MM1. When one gap is displaced completely towards the corner, as in the case of MM3, the Coulomb interaction between charges in one of the arms of the paired rings becomes attractive, thereby increasing the net electric flux density, as shown in Figure 3(c). This produces additional capacitance between the rings causing a red shift in the resonance.

For the sample MM4, both the gaps are displaced towards the corner, therefore, enhancing the Coulomb attraction between the opposite charges between the paired rings by maximum extent. This effect gives rise to a larger value of the coupling capacitance, resulting in maximum red shift of the *LC* resonance. The red shift in MM3 and MM4 is accompanied by a weakening of the resonance, as revealed by the increased resonant transmission minimum. This is because of the weak excitation of SRRs with displaced gaps by the external electric field. Therefore, two apparently independent and simultaneous effects are observed: strong capacitive coupling between SRRs causing resonance frequency shifts and symmetry breaking in the individual rings causing variation of the resonance strength. The larger the capacitive coupling between the paired rings, the larger the resonance frequency shift. Similarly, the larger the asymmetry of the rings, the weaker the *LC* resonance strength due to the weak coupling of the incident THz radiation with the SRRs.

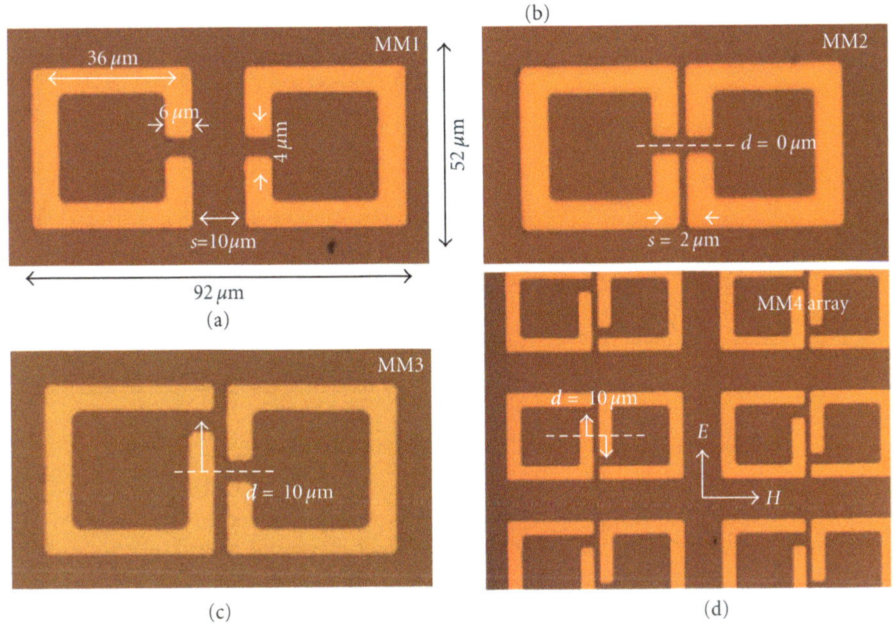

FIGURE 1: Optical microscope images of the fabricated metamaterial samples. (a) In the unit cell with symmetric gap-to-gap coupled SRRs, the separation is either (a) 10 μm (MM1) or (b) 2 μm (MM2). (c) The asymmetric SRRs with one gap displaced from the center (MM3) and (d) the SRR array where both gaps are displaced in the opposite direction from the center (MM4). The dimensions of the unit cells and their features are indicated in the figure. The unit cell periodicity for all the metamaterial samples is identical.

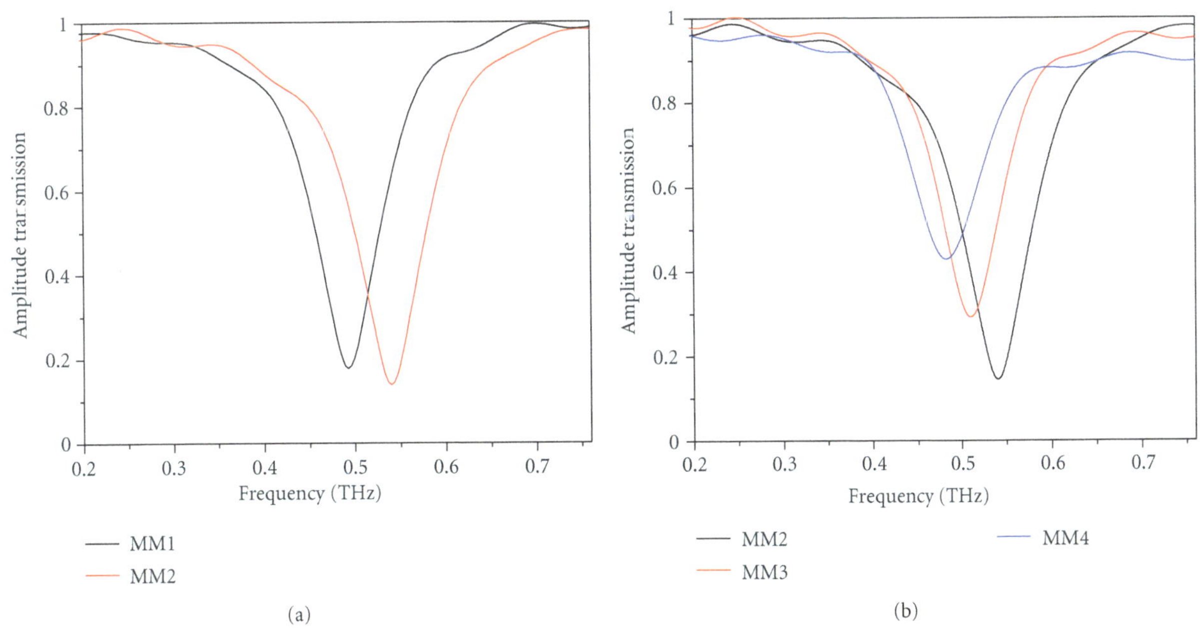

FIGURE 2: Measured THz electric field amplitude spectra when transmitting through the metamaterial samples for (a) MM1 and MM2 (b) and MM2, MM3, and MM4.

2.2. Inductive Coupling in the Metamaterial Array. The effect of strong inductive coupling was probed in terahertz SRRs by maintaining a fixed orientation and the capacitive gaps facing upwards in all samples, which minimizes the effect of the capacitive coupling [39]. Thus, the metamaterials are designed to have different periodicity from one sample to another. Nine sets of samples were prepared with periodicity of $P = 200, 150, 100, 60, 50, 40, 35, 30$, and $25\ \mu$m, keeping the geometrical dimensions of the SRR fixed. Figure 4(a) shows the metamaterial sample with $P = 25\ \mu$m. The dimensions of SRR were chosen such that its LC resonance wavelength is $\lambda \sim 200\ \mu$m (i.e., 1.5 THz) on the silicon substrate. Different periodicities in samples were designed so that the resonance wavelength to lattice constant ratio λ/P varies from a value

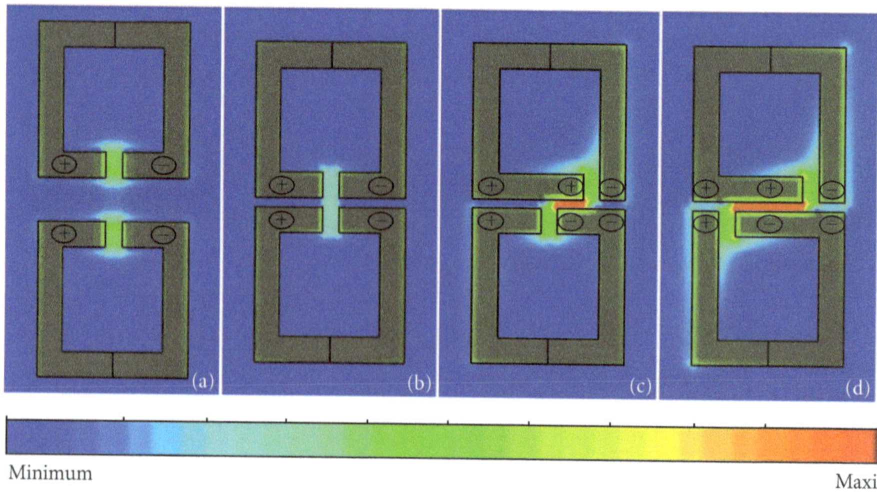

FIGURE 3: The numerically simulated electric field strength distributions for the coupled SRRs: (a) MM1 (b), MM2 (c), MM3 (d), and MM4. The electric field strength is indicated by the color map at the bottom. The polarity of the induced electronic charges on the SRR arms is indicated in the figure.

of $\lambda/P = 1$ for $P = 200\,\mu$m to $\lambda/P = 8$ for $P = 25\,\mu$m. Each metamaterial sample is excited by an incident terahertz beam with the focal diameter of 3.5 mm, with which the number of excited SRRs is gradually increased from 960 for $P = 200\,\mu$m to 61,500 for $P = 25\,\mu$m. This enables the study of the inter-SRR packing-dependent coupling and its impact on the overall collective resonant response in THz transmission.

The measured resonant transmission spectra are plotted in Figure 4(b). The evolution of the LC resonance was observed at around 1.5 THz. The resonance was fairly weak for the 200, 150, and 100 μm periodicities. A strong resonance feature is observed for the metamaterial sample with period of 60 μm, and then a gradual broadening of the resonance was seen for smaller periodicities or higher packing densities.

The dominant coupling mechanism among SRRs at the LC resonance with increasing packing density is mainly magnetic/inductive in nature. The increase in the packing density leads to spatially closer magnetic loops strongly coupled to each other. When this happens, the mutual inductance between SRRs tends to reduce the self-inductance of the SRR loop leading to a reduced effective inductance and thus an increase or blue shifting of the LC resonance frequency. Thus, the stronger is the inductive coupling between the SRRs, the higher the resonance frequency will be, since each SRR due to mutual coupling reduces the magnetic flux of its neighbors. The experimental data shows that stronger (high density) or weaker (low density) magnetic coupling leads to lowering of Q factor of SRRs as shown in Figure 4(c). Further, there seems to be an optimal packing density with periodicity of 60 μm where the SRRs are coupled to yield the sharpest LC resonance due to diffractive coupling. This coupling mechanism is explained in detail in the next section.

2.3. Long-Range Diffractive Coupling in Metamaterial Array. In this section, we describe the long-range radiative coupling encountered in the experiment described in the previous section. We highlight that the inter-SRR distance and its number density or periodicity can control the Q factor and the strength of the LC resonance of SRRs. The measured amplitude transmission in Figure 4(b) shows that the LC resonance at 60 μm periodicity yields the narrowest resonance linewidth. Coincidentally, it also corresponds to a λ/P ratio of 3.33 which is the refractive index of the silicon substrate on which the SRRs are fabricated. Thus, we observe a strong suppression of radiative losses and thus a sharp enhancement of the Q factor at a certain critical periodicity, P_c, that exactly matches the SRR resonant wavelength in the silicon substrate, that is, $P_c = \lambda_{LC}/n$, where λ_{LC} is the LC resonance wavelength in vacuum, and n is the refractive index of the substrate material. The sharpening of LC resonance is mainly due to Wood's anomaly or the diffractive lattice mode corresponding to the critical periodicity, the resonant wavelength, and the substrate index [38–40, 45]. When the periodicity is further reduced, a gradual decline in Q factor of the fundamental SRR resonance is observed in Figure 4(c).

The narrowest resonance occurs when the incident light field corresponds to diffractive grating order (i, j) change from evanescent to radiative, $P_c = \sqrt{(i^2 + j^2)}\lambda/n$; here n is the refractive index of the substrate in use, (i, j) are the diffraction grating orders, and λ is the resonance wavelength. In the experiment, P_c = 60 um corresponds to the lowest-order $(1, 0)$ diffractive lattice mode. Every time the lowest-order lattice mode matches either the lower-order fundamental LC resonance or higher-order resonances, a sudden sharpening of the resonance will occur. The narrowing of the resonance in an optimally packed metamaterial can also be best understood by the suppression of radiation damping due to the confinement or trapping of electromagnetic fields in the metamaterial plane. This mechanism is based on diffractive coupling during which the diffracted fields are trapped in the metamaterial lattice, and each SRR radiatively couples to the neighboring SRR. The lower Q factor for larger

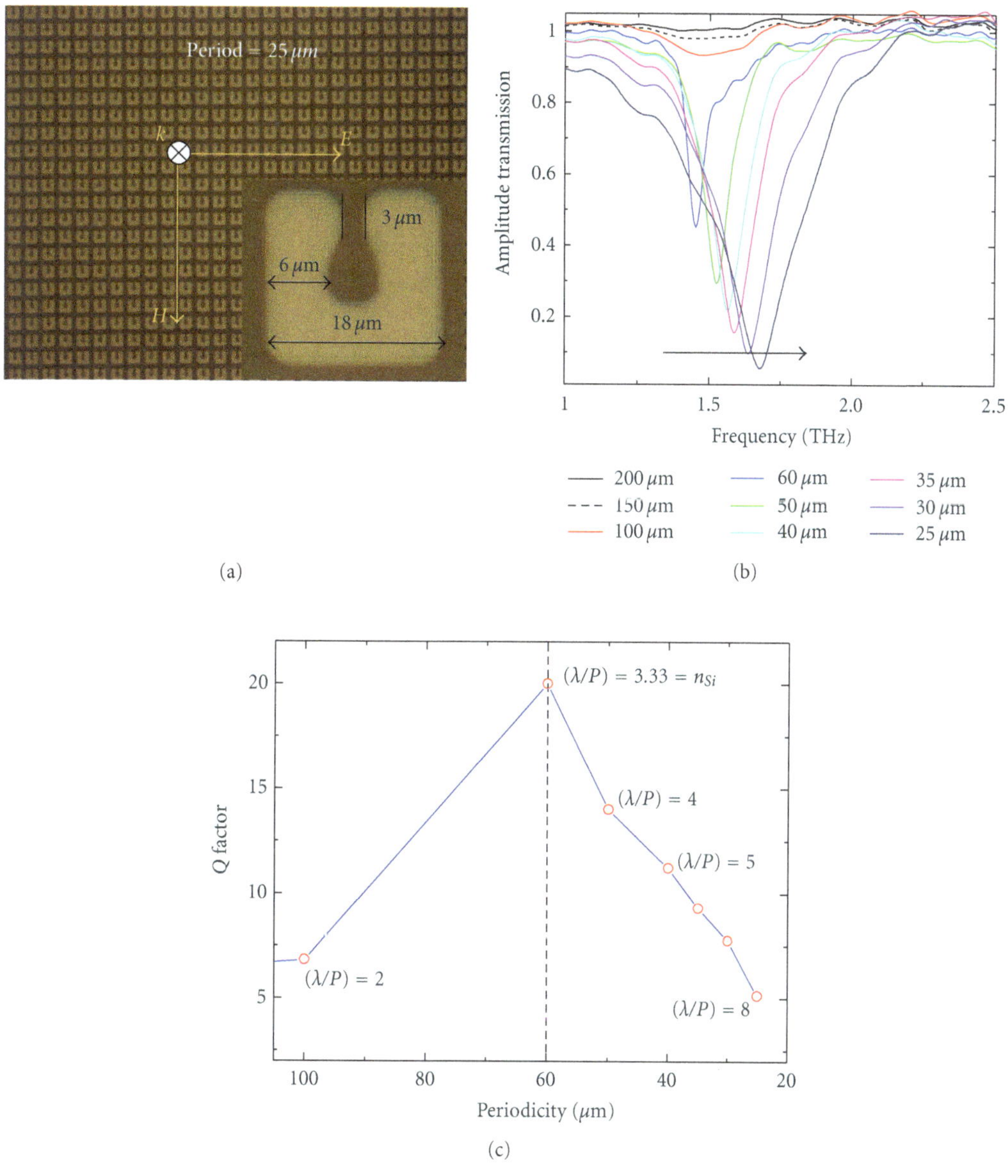

FIGURE 4: (a) Microscopic image of the highest-density metamaterial sample. The SRRs are comprised of 200 nm thick Al on 640 μm thick n-type silicon substrates. (b) Measured transmission spectra of metamaterials with different periodicities. (c) Quality factor extracted from the measured transmission for different period metamaterials.

periods is associated to an enhanced radiation damping. The lower Q factor for smaller periods is mainly due to the mismatch with the lattice mode during which the radiated fields of the coherently coupled SRRs interfere destructively in the plane of the metamaterial, leading to higher radiative losses and increase of nonradiative losses due to a capacitive and inductive coupling.

Tailoring the periodicity of metamaterials will allow the design of frequency-selective surfaces with an adjustable Q factor for metamaterials across the entire electromagnetic spectrum optimized for various applications. This finding is important for the development of metamolecule density and substrate-refractive-index-tuned THz devices and components.

2.4. Capacitive and Inductive Coupling for Different Orientation and Separation in Nearest-Neighbor SRRs. In this section, we probe scenarios where neighboring split rings have their gap either on the same or on opposite sides with the change in periodicity [37]. The lowest-order odd eigenmode and the lowest-order even eigenmodes for two different excitations are marginally affected, whereas the higher-order eigenmode experiences a significant spectral reshaping [37, 46] due to a coherent interaction.

We designed metamaterials composed of three different supercells each containing four SRRs. The geometrical arrangement of the SRRs in each supercell is shown in Figure 5. In the first sample, all SRRs have their gap on the same side. In a second sample, the SRRs are arranged

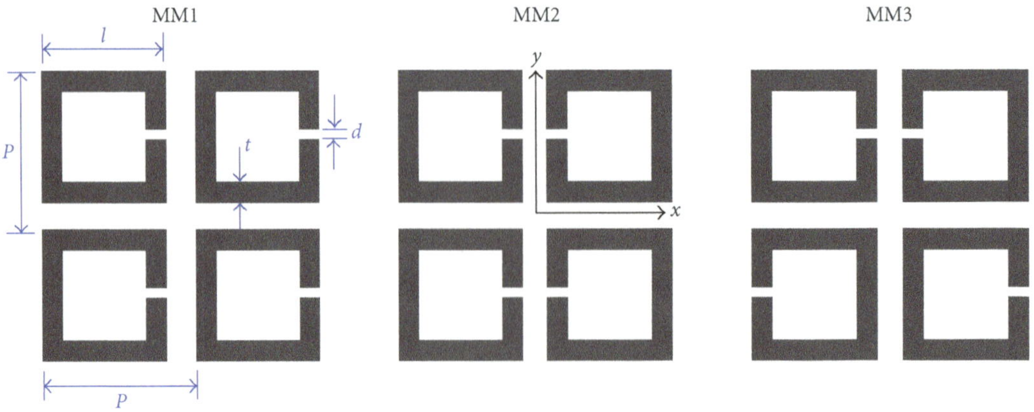

FIGURE 5: Unit cell of the three types of differently oriented split-ring resonators. The dimensions of all SRRs are $t = 6\,\mu$m, $l = 36\,\mu$m, and $d = 2\,\mu$m. In the design, the period P is varied.

with vertical mirror symmetry. In a third sample, the SRRs have the same orientation of their gaps along the diagonal directions of the supercell. In this work, the effects of the orientation on the spectral position and the width of each resonance are investigated. It was observed that control over the orientation of the SRRs and the periodicity enables extremely sharp features in the spectrum that are promising candidates for various practical applications such as sensing devices.

Simulations were done with the Fourier modal method [47] using the structure parameters shown in Figure 6 where only the periodicity parameter "P" was swept from $P = 50\,\mu$m to $P = 39\,\mu$m. Figure 6 shows the transmission for the three configurations as a function of period and frequency. In all cases, the incident E-field is polarized along the gap-bearing arms in the y-axis direction as shown in MM2. The main effects observed in the simulations are the impact of each configuration, the impact of the periodicity for each configuration on the spectral positions and the widths of the resonance, and the occurrence of coherent phenomena once SRRs are closely spaced. It can be clearly seen that the fundamental mode LC resonance at 0.5 THz does not undergo significant change in resonance position or resonance linewidth, but the higher-order dipole resonance undergoes a huge spectral change at about 1.5 THz.

Between MM1 and MM2, as the periodicity is reduced, the resonance at 1.5 THz broadens in MM2, and the same resonance in MM3 splits and exhibits a very sharp spectral feature, shown by a white-dotted ellipse. Such large spectral reshaping prompted us to carry out the experiment in all of these configurations.

Six samples of planar SRRs (denoted in the following section as MM1–MM6) were fabricated. As shown in Figure 5 above, MM1–MM3 are single SRRs with lattice constant $P = 50\,\mu$m. MM4–MM6 are single SRRs with the same symmetry as MM1–MM3 but with reduced periodicity of $P = 39\,\mu$m. The measured amplitude transmission spectra of MM1–MM3 are shown in Figure 7(a). The two significant dips in transmission are traces from the two lowest-order

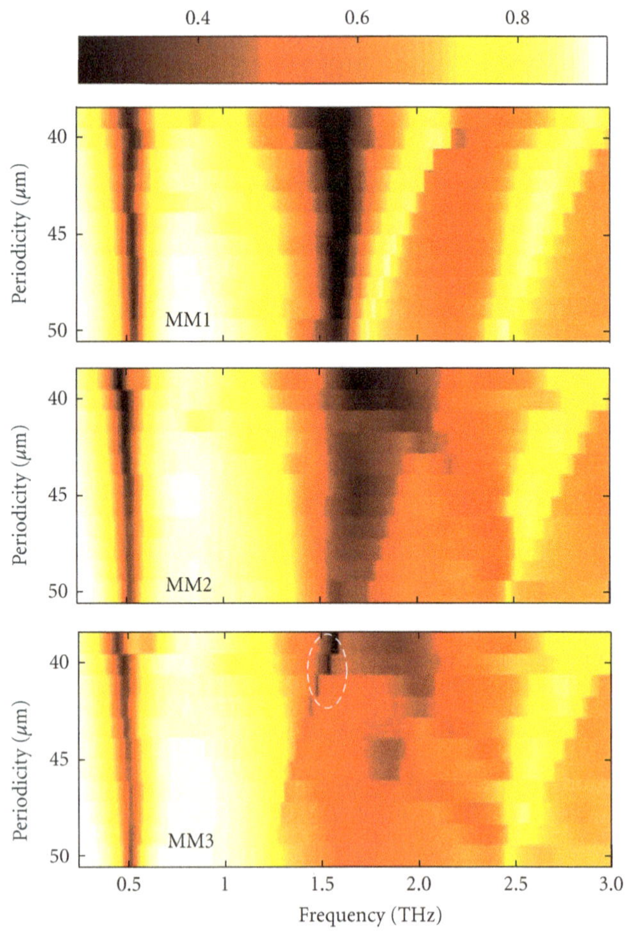

FIGURE 6: Simulated transmission spectra of the three differently oriented metamaterials with period, P, swept from $P = 50\,\mu$m to $P = 39\,\mu$m.

odd eigenmodes. Significant differences in the spectra are observed, depending on the orientation of the gap of neighboring SRRs. When comparing the resonances of MM2 with those of MM1, it can be seen that the LC resonance at 0.5 THz

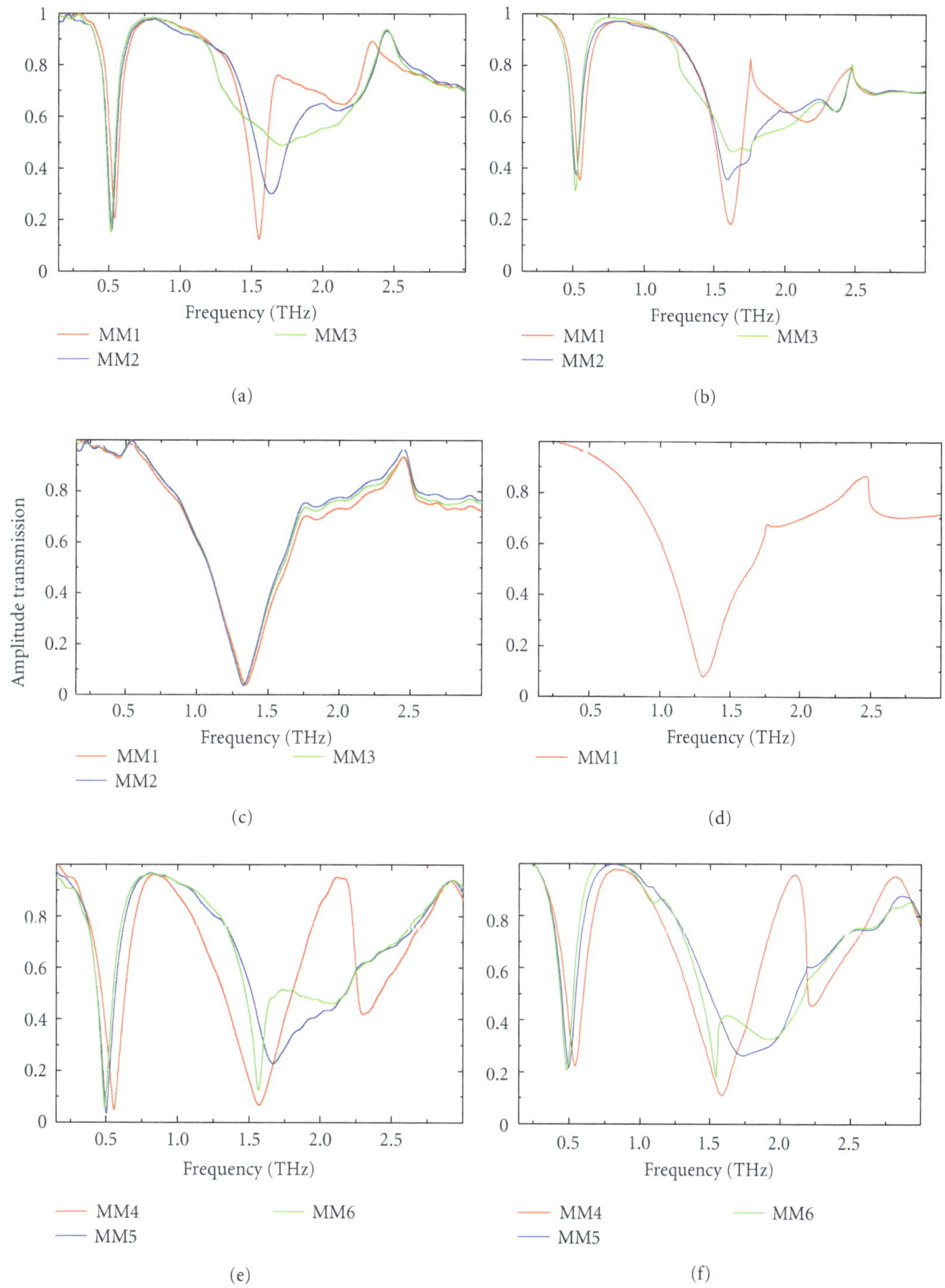

Figure 7: (a), (b) are measurement and simulations of differently oriented metamaterials, MM1, MM2, and MM3, for $P = 50\,\mu$m for excitation in x direction, (c), (d) are the same but excitation in y direction, and (e), (f) are measurement and simulations for three differently oriented unit cells with period $P = 39\,\mu$m.

is red shifted slightly. In contrast, the next higher-order odd resonance at 1.5 THz broadens, blue-shifts by 84 GHz, and the transmission increases significantly in MM2. Furthermore, the orientation of the SRRs in the supercell for MM3 causes only a marginal additional red shift of the lowest-order eigenmode by 8 GHz and nearly no change on transmission. The next higher-order resonances suffer from a severe broadening and a significant increase in transmission up to 49%. The simulated spectra in Figure 7(b) are in good agreement with the measurements except the amplitude transmission depth at the *LC* resonance due to the limited resolution of our numerical code. For an incident *E*-field perpendicular to the gap-bearing arms, the response of MM1–MM3 is shown in Figure 7(c). Independent of the

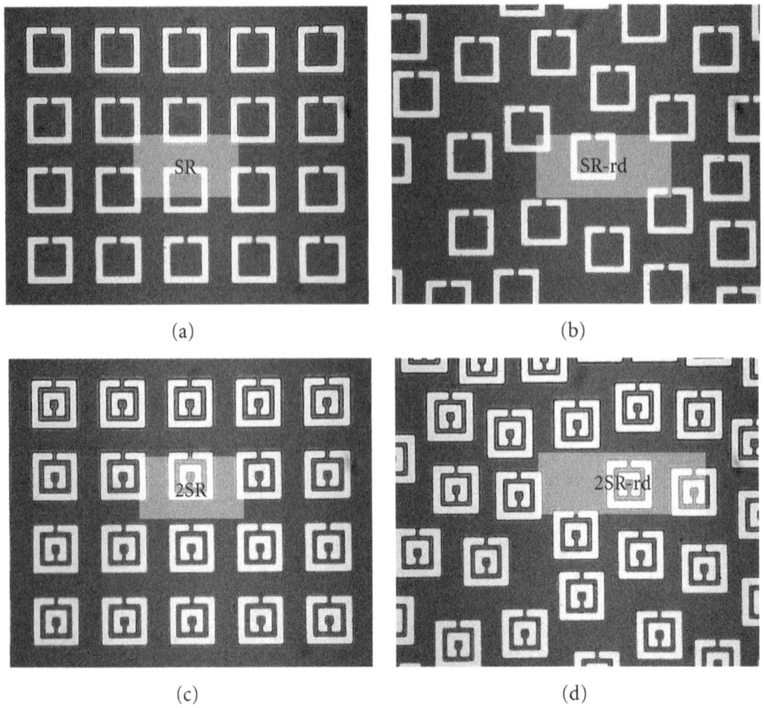

FIGURE 8: Microscopic images of the (a) periodic and (b) random single SRR array, (c) periodic and (d) random double SRR array. The dimensions of the outer ring are the same as in Figure 5 and of the inner smaller ring are the same as in Figure 4(a).

arrangement, all spectra show a strong resonance at 1.33 THz and have nearly the same spectral dependence. Exactly the same observation is made in the simulation. This resonance is the lowest-order even eigenmode.

The experimentally measured transmission spectra for MM4–MM6 are shown in Figure 7(e). The incident E-field is polarized along y-axis. Due to reduced periodicity and high density of SRRs, we observe a different spectral behavior in comparison with the samples MM1–MM3. In MM4, there are three distinct resonances, the LC resonance at 0.556 THz, the next higher-order odd eigenmode at 1.57 THz, and a third strong resonance at 2.3 THz. This third resonance, however, is a Wood's anomaly in the substrate. In MM5, the lowest-order odd eigenmode appears red-shifted and has a reduced linewidth when compared to MM4. The next higher-order odd eigenmodes are broadened and blue-shifted. In MM6, the lowest-order odd eigenmode is again marginally further red-shifted. Finally, and potentially most interesting, the next higher-order odd eigenmode appears as an extremely sharp feature with a linewidth of only 69.5 GHz, resonating approximately at the same frequency as the higher-order odd eigenmode of MM4. The resonance has a high Q-factor of 18.5. This high Q-factor is an essential property for the achievement of high sensitivity in planar THz MM sensors, as sensitivity increases with the sharpness of the spectral resonance feature. In Figure 7(f), the simulation agrees well with the measurements.

To explain the influence of the arrangement on the spectral properties of the lowest-order LC resonance mode, one may evoke analogies to transmission line theory. Placing SRRs like in MM2 and MM3 will cause an increase of the effective capacitance as the next neighbor in such situation provides a parasitic or mutual capacitance in parallel that would add up with the gap capacitance. The increase in effective capacitance causes the red shift of the LC resonance.

For the lowest-order even eigenmode when the excitation field is along x direction, placing the gap on either side of the SRR has no effect, as this mode is dominantly characterized by oscillating currents in the side arms of the SRR that have no gap. Therefore, modification of the resonance conditions for the SRR is neither observed nor expected.

For the third resonance which is the second-order odd eigenmode, the resonance shape and position depends on the chossen period and consequently on the absolute inter-SRR distance between the neighbors. By taking into account the fact that the resonance wavelength is comparable to the chosen interparticle separation, the coherent superposition of the scattered field by neighboring SRRs causes the strong modification of the transmission in this spectral domain. The formation of a *subradiant* and a *superradiant* type mode, depending on the interparticle distance and the gap orientation, causes a blue shift or a red shift of the resonance [43]. The superradiant (subradiant) mode also increases (decreases) the radiation losses of the excited eigenmode which leads to a broadening (narrowing) of the spectral resonance.

2.5. Coupling in the Randomly Distributed Metamaterial Array. In this section, we review the outcome of the introduction of positional disorder of SRRs in metamaterials on the quality factor of the fundamental LC and higher-order dipole resonances. Several existing metamaterials are

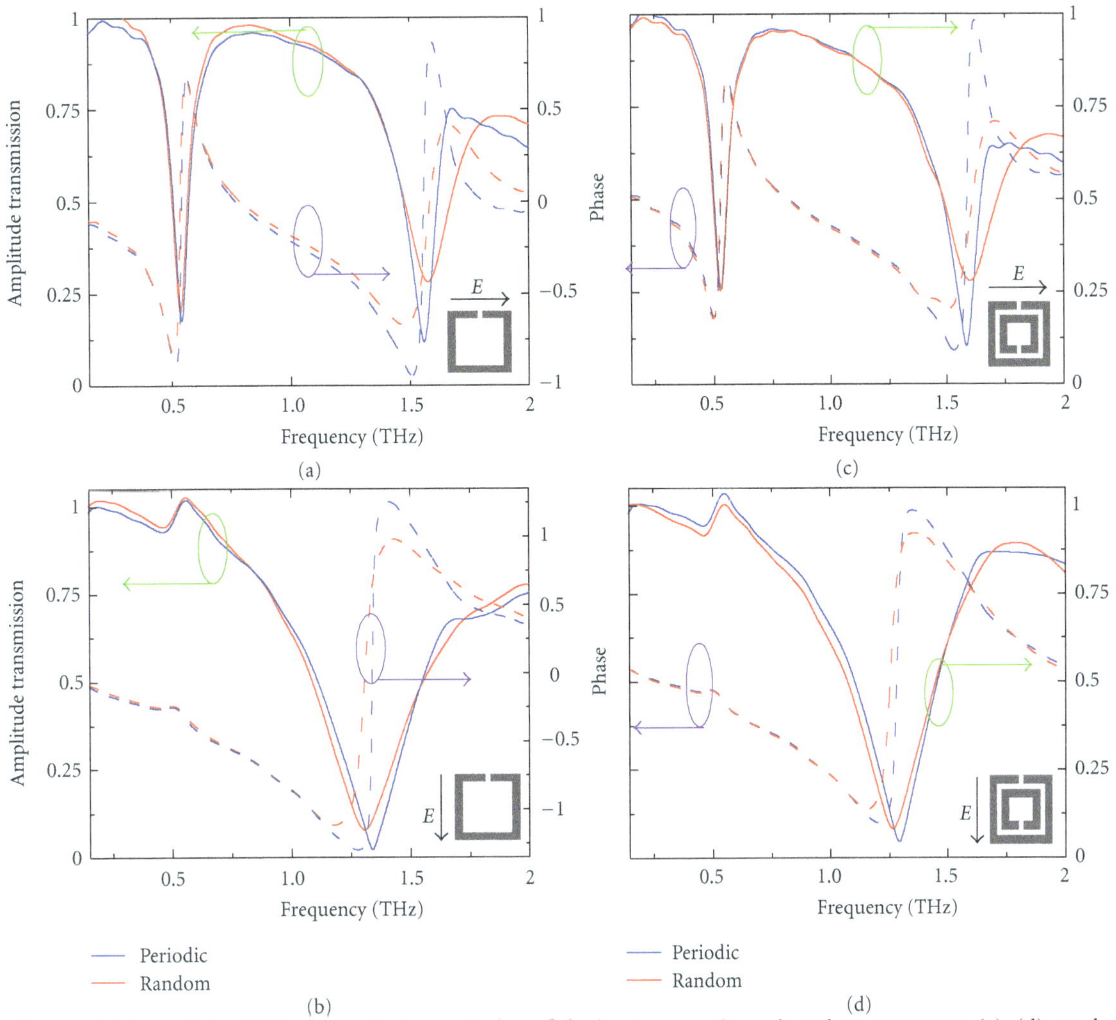

FIGURE 9: Measured (a) parallel and (b) perpendicular excitation of single-ring periodic and random resonators; (c), (d) are the same for double-ring resonators. The dash curves are the corresponding phase changes.

anisotropic in nature. However, for most practical applications across the entire electromagnetic spectrum, isotropic materials are required, and isotropic medium is sometimes based on randomly distributed cells in the volume of the host structure. Much of the previous work on metamaterials was focused on investigating the unique properties of wave propagation through periodic structures. There has been a tremendous desire to explore the collective behavior of SRRs when disorder is present in either their geometrical dimensions or periodic positioning on the host media [48–52].

The disorder has been studied in the single and the double split-ring resonators [50]. For single SRRs, only the outer ring with the same dimensions is retained. The disorder in arrangement was introduced by manually changing the lattice points of SRRs in a random fashion during the mask design. Each of the periodic SRRs was initially at a lattice position, $\vec{r}_n$, where $\vec{r} = x\vec{i} + y\vec{j}$, and individual SRRs were then displaced by $\pm\delta\vec{x}$ and $\pm\delta\vec{y}$ to introduce positioning disorder as shown in Figure 8. Thus, $\delta r = \delta x\vec{i} + \delta y\vec{j}$ is used to define the degree of randomness. In our case, $|\delta x| \leq 32\,\mu$m and $|\delta\vec{y}| \leq 32\,\mu$m, leading to a maximum disorder $|\delta\vec{r}| \leq 45.25\,\mu$m, which is equivalent to 85% of the periodicity, $P = 53\,\mu$m. The random SRRs retain the same number density as that of the counterpart periodic SRRs, giving the same metal volume filling fraction, $f = 0.25$ for the single SRRs, *SR* (periodic), and *SR-rd* (random), $f = 0.33$ for the double SRRs, *2SR* (periodic), and *2SR-rd* (random).

For the random sample *SR-rd* in the parallel orientation, as shown in Figure 9(a), no change in the *LC* resonance feature is observed compared to that of *SR*. The next higher resonance, however, blue-shifts by 16 GHz with a linewidth broadening by 80 GHz and a significant transmission modification from 12% to 28%. In the perpendicular orientation, the lowest-order eigenmode of the random structure, *SR-rd*, red-shifts by 34 GHz and broadens by 106 GHz, as shown in Figure 9(b). This behavior indicates that the dipole resonance is clearly affected by the positional disorder of SRRs.

Figures 9(c) and 9(d) illustrate the frequency-dependent amplitude transmission of the double SRR samples, *2SR* and *2SR-rd*. Similar effects are observed in the periodic and aperiodic double SRRs, as seen for the single SRRs. The *LC* resonance of *2SR* and *2SR-rd* was found identical in all respects. The second-order odd eigenresonance at 1.58 THz blue-shifted by 16 GHz and broadened by 99 GHz, and the resonance minimum decreased by 17.5%. The lowest even eigenmode red-shifts by 28.5 GHz and broadens by 49 GHz. The corresponding phase change spectra are shown in all the figures.

The character of the *LC* resonance is independent of randomness in the positioning of SRRs as its peak frequency and resonance strength remain unaffected. The response of the random SRRs at the *LC* resonance is identical to that of their periodic counterpart as long as their number density remains constant by maintaining equal volume metal filling fraction. The inter-SRR interaction between the individual elemental SRR is negligible since the collective behavior of the periodic and the random metamaterials is just a direct addition of individual SRR contributions.

Unlike the *LC* resonance, the dipole resonances in the random structures undergo a change compared to that of the periodic SRRs. This can be expected due to the different current profiles for the SRRs at *LC* and dipole resonance frequencies. The second odd and lowest even eigenmode resonances in the transmission spectra of all the SRRs are due to linear currents in the SRR side arms and have similar current density distributions. Therefore, those resonances undergo similar spectral modification when disorder is introduced in the SRRs. The incident terahertz electric field excites plasmon oscillations of conduction electrons at the surface of individual metallic SRR arms that are parallel to the illuminating field, producing a collection of oscillating dipoles with a dipole moment. The dipole resonance is due to the dipole-dipole interaction between the SRR arms along the direction of the incident electric field. The plasmon energy and the resonance linewidth strongly depend on the interparticle distance. The dipole resonance wavelength in the metamaterial structures becomes comparable to the interparticle spacing, and as the distance between the dipoles is varied, there is formation of a coupled plasmon mode which alternates between superradiative and subradiative behavior modifying the resonance spectral width and causing frequency shifts. The dipole coupling among the randomized split ring resonators is highly complicated due to their random positioning with respect to each other and is influenced by a number of other factors, hence rendering a straightforward explanation impossible. In the periodic structures, the coupling between the dipoles is stronger compared to the randomly distributed SRRs since the interaction between the oscillating particles is partially cancelled by the disorder, which leads to weaker coupling among the dipoles and higher radiative damping. As a result, broadened and weaker plasmon resonances in the randomly scattered SRR metamaterials are observed. This is also due to inhomogeneous broadening that is caused due to the randomly scattered dipoles.

3. Conclusion

In conclusion, in-plane coupling of terahertz SRRs can manipulate and tailor the fundamental as well as higher-order resonances. The near-field coupling via the SRRs electric and magnetic field causes the fundamental mode resonance to red-shift and blue-shift, respectively. The mutual orientation and the separation plays dominant roles in tuning the higher-order dipolar resonance, leading to excitation of sharp subradiant resonances that can act like hotspots for sensing applications. The effect of packing density in SRRs can be very strong, enabling the design of tunable transmission and quality factor metamaterials. Long range radiative coupling also leads to resonance sharpening and electromagnetic energy trapping in the metamaterial lattice, and these long-range interactions are mediated by the fundamental-order diffractive mode of the periodic grating. Thus, the different coupling mechanisms in planar matamaterials could be exploited for the development of terahertz devices with various functionalities.

Acknowledgments

The authors thank C. Rockstuhl, J. F. O' Hara, M. Walther, and X. Lu for their outstanding contribution and fruitful discussions in this work. Some of the work was partially supported by the U.S. National Science Foundation (NSF). This work was performed, in part, at the Center for Integrated Nanotechnologies, a U.S. Department of Energy, Office of Basic Energy Sciences user facility. Los Alamos National Laboratory, an affirmative action equal opportunity employer, is operated by Los Alamos National Security, LLC, for the National Nuclear Security Administration of the U.S. Department of Energy under Contract no. DE-AC52-06NA25396.

References

[1] J. B. Pendry, A. J. Holden, D. J. Robbins, and W. J. Stewart, "Magnetism from conductors and enhanced nonlinear phenomena," *IEEE Transactions on Microwave Theory and Techniques*, vol. 47, no. 11, pp. 2075–2084, 1999.

[2] R. A. Shelby, D. R. Smith, and S. Schultz, "Experimental verification of a negative index of refraction," *Science*, vol. 292, no. 5514, pp. 77–79, 2001.

[3] N. Fang, H. Lee, C. Sun, and X. Zhang, "Sub-diffraction-limited optical imaging with a silver superlens," *Science*, vol. 308, no. 5721, pp. 534–537, 2005.

[4] J. B. Pendry, D. Schurig, and D. R. Smith, "Controlling electromagnetic fields," *Science*, vol. 312, no. 5781, pp. 1780–1782, 2006.

[5] D. Schurig, J. J. Mock, B. J. Justice et al., "Metamaterial electromagnetic cloak at microwave frequencies," *Science*, vol. 314, no. 5801, pp. 977–980, 2006.

[6] W. J. Padilla, A. J. Taylor, C. Highstrete, M. Lee, and R. D. Averitt, "Dynamical electric and magnetic metamaterial response at terahertz frequencies," *Physical Review Letters*, vol. 96, no. 10, Article ID 107401, pp. 1–4, 2006.

[7] H. T. Chen, W. J. Padilla, J. M. O. Zide, A. C. Gossard, A. J. Taylor, and R. D. Averitt, "Active terahertz metamaterial devices," *Nature*, vol. 444, no. 7119, pp. 597–600, 2006.

[8] H. T. Chen, J. F. O'Hara, A. K. Azad et al., "Experimental demonstration of frequency-agile terahertz metamaterials," *Nature Photonics*, vol. 2, no. 5, pp. 295–298, 2008.

[9] H. Tao, A. C. Strikwerda, K. Fan, W. J. Padilla, X. Zhang, and R. D. Averitt, "Reconfigurable terahertz metamaterials," *Physical Review Letters*, vol. 103, no. 14, Article ID 147401, 2009.
[10] Z. Tian, R. Singh, J. Han et al., "Terahertz superconducting plasmonic hole array," *Optics Letters*, vol. 35, no. 21, pp. 3586–3588, 2010.
[11] T. Driscoll, H. T. Kim, B. G. Chae et al., "Memory metamaterials," *Science*, vol. 325, no. 5947, pp. 1518–1521, 2009.
[12] R. Singh, E. Plum, W. Zhang, and N. I. Zheludev, "Highly tunable optical activity in planar achiral terahertz metamaterials," *Optics Express*, vol. 18, no. 13, pp. 13425–13430, 2010.
[13] J. Gu, R. Singh, Z. Tian et al., "Terahertz superconductor metamaterial," *Applied Physics Letters*, vol. 97, no. 7, Article ID 071102, 2010.
[14] B. Jin, C. Zhang, S. Engelbrecht et al., "Low loss and magnetic field-tunable superconducting terahertz metamaterial," *Optics Express*, vol. 18, no. 16, pp. 17504–17509, 2010.
[15] H. T. Chen, H. Yang, R. Singh et al., "Tuning the resonance in high-temperature superconducting terahertz metamaterials," *Physical Review Letters*, vol. 105, no. 24, Article ID 247402, 2010.
[16] R. Singh, A. K. Azad, Q. X. Jia, A. J. Taylor, and H. -T. Chen, "Thermal tunability in terahertz metamaterials fabricated on strontium titanate single-crystal substrates," *Optics Letters*, vol. 36, no. 7, pp. 1230–1232, 2011.
[17] R. Singh, J. Xiong, A. K. Azad et al., "Optical tuning and ultrafast dynamics of high-temperature superconducting terahertz metamaterials," *Nanophotonics*. In press, http://arxiv.org/abs/1111.3917.
[18] S. Y. Chiam, R. Singh, W. Zhang, and A. A. Bettiol, "Controlling metamaterial resonances via dielectric and aspect ratio effects," *Applied Physics Letters*, vol. 97, no. 19, Article ID 191906, 2010.
[19] N.-H. Shen, M. Massaouti, M. Gokkavas et al., "Optically implemented broadband blueshift switch in the terahertz regime," *Physical Review Letters*, vol. 106, no. 3, Article ID 037403, 2011.
[20] D. Roy Chowdhury, R. Singh, J. F. O'Hara, H.-T. Chen, A. J. Taylor, and A. K. Azad, "Dynamically reconfigurable terahertz metamaterial through photo-doped semiconductor," *Applied Physics Letters*, vol. 99, no. 23, Article ID 231101, 2011.
[21] J. Gu, R. Singh, A. K. Azad et al., "An active hybrid plasmonic metamaterial," *Optical Materials Express*, vol. 2, Article ID 1617, 2011, http://www.opticsinfobase.org/ome/abstract.cfm?uri=ome-2-1-31.
[22] H.-T. Chen, J. F. O'Hara, A. K. Azad, and A. J. Taylor, "Manipulation of terahertz radiation using metamaterials," *Laser & Photonics Reviews*, vol. 5, no. 4, pp. 513–533, 2011.
[23] R. Singh, A. K. Azad, J. F. O'Hara, A. J. Taylor, and W. Zhang, "Effect of metal permittivity on resonant properties of terahertz metamaterials," *Optics Letters*, vol. 33, no. 13, pp. 1506–1508, 2008.
[24] D. R. Chowdhury, R. Singh, M. Reiten et al., "A broadband planar terahertz metamaterial with nested structure," *Optics Express*, vol. 19, no. 17, pp. 15817–15823, 2011.
[25] J. F. O'Hara, R. Singh, I. Brener et al., "Thin-film sensing with planar terahertz metamaterials: sensitivity and limitations," *Optics Express*, vol. 16, no. 3, pp. 1786–1795, 2008.
[26] I. A.I. Al-Naib, C. Jansen, and M. Koch, "Thin-film sensing with planar asymmetric metamaterial resonators," *Applied Physics Letters*, vol. 93, no. 8, Article ID 083507, 2008.
[27] R. Singh, E. Smirnova, A. J. Taylor, J. F. O'Hara, and W. Zhang, "Optically thin terahertz metamaterials," *Optics Express*, vol. 16, no. 9, pp. 6537–6543, 2008.
[28] W. Withayachumnankul and D. Abbott, "Metamaterials in the Terahertz Regime," *IEEE Photonics Journal*, vol. 1, p. 99, 2009, http://ieeexplore.ieee.org/xpls/abs_all.jsp?arnumber=5130235.
[29] D. R. Chowdhury, R. Singh, M. Reiten, J. Zhou, A. J. Taylor, and J. F. O'Hara, "Tailored resonator coupling for modifying the terahertz metamaterial response," *Optics Express*, vol. 19, no. 11, pp. 10679–10685, 2011.
[30] F. Hesmer, E. Tatartschuk, O. Zhuromskyy et al., "Coupling mechanisms for split ring resonators: theory and experiment," *Physica Status Solidi B*, vol. 244, no. 4, pp. 1170–1175, 2007.
[31] R. S. Penciu, K. Aydin, M. Kafesaki et al., "Multi-gap individual and coupled split-ring resonator structures," *Optics Express*, vol. 16, no. 22, pp. 18131–18144, 2008.
[32] I. Sersic, M. Frimmer, E. Verhagen, and A. F. Koenderink, "Electric and magnetic dipole coupling in near-infrared split-ring metamaterial arrays," *Physical Review Letters*, vol. 103, no. 21, Article ID 213902, 2009.
[33] R. Singh, I. A.I. Al Naib, Y. Yang et al., "Observing metamaterial induced transparency in individual Fano resonators with broken symmetry," *Applied Physics Letters*, vol. 99, no. 20, Article ID 201107, 2011.
[34] N. Liu, S. Kaiser, and H. Giessen, "Magnetoinductive and electroinductive coupling in plasmonic metamaterial molecules," *Advanced Materials*, vol. 20, no. 23, pp. 4521–4525, 2008.
[35] R. Singh, C. Rockstuhl, F. Lederer, and W. Zhang, "Coupling between a dark and a bright eigenmode in a terahertz metamaterial," *Physical Review B*, vol. 79, no. 8, Article ID 085111, 2009.
[36] N. Feth, M. König, M. Husnik et al., "Electromagnetic interaction of split-ring resonators: the role of separation and relative orientation," *Optics Express*, vol. 18, no. 7, pp. 6545–6554, 2010.
[37] R. Singh, C. Rockstuhl, F. Lederer, and W. Zhang, "The impact of nearest neighbor interaction on the resonances in terahertz metamaterials," *Applied Physics Letters*, vol. 94, no. 2, Article ID 021116, 2009.
[38] A. Bitzer, J. Wallauer, H. Helm, H. Merbold, T. Feurer, and M. Walther, "Lattice modes mediate radiative coupling in metamaterial arrays," *Optics Express*, vol. 17, no. 24, pp. 22108–22113, 2009.
[39] R. Singh, C. Rockstuhl, and W. Zhang, "Strong influence of packing density in terahertz metamaterials," *Applied Physics Letters*, vol. 97, no. 24, Article ID 241108, 2010.
[40] J. Wallauer, A. Bitzer, S. Waselikowski, and M. Walther, "Near-field signature of electromagnetic coupling in metamaterial arrays: a terahertz microscopy study," *Optics Express*, vol. 19, no. 18, pp. 17283–17292, 2011.
[41] D. A. Powell, M. Lapine, M. V. Gorkunov, I. V. Shadrivov, and Y. S. Kivshar, "Metamaterial tuning by manipulation of near-field interaction," *Physical Review B*, vol. 82, no. 15, Article ID 155128, 2010.
[42] R. Singh, I. A.I. Al-Naib, M. Koch, and W. Zhang, "Sharp Fano resonances in THz metamaterials," *Optics Express*, vol. 19, no. 7, pp. 6312–6319, 2011.
[43] D. Grischkowsky, S. Keiding, M. van Exter, and C. Fattinger, "Far-infrared time-domain spectroscopy with terahertz beams of dielectrics and semiconductors," *Journal of the Optical Society of America B*, vol. 7, p. 2006, 1990, http://www.opticsinfobase.org/josab/abstract.cfm?URI=josab-7-10-2006.
[44] *Computer Simulation Technology (CST)*, Darmstadt, Germany.
[45] V. G. Kravets, F. Schedin, and A. N. Grigorenko, "Extremely narrow plasmon resonances based on diffraction coupling

of localized plasmons in arrays of metallic nanoparticles," *Physical Review Letters*, vol. 101, no. 8, Article ID 087403, 2008.

[46] C. Rockstuhl, T. Zentgraf, C. Etrich, J. Kuhl, F. Lederer, and H. Giessen, "On the reinterpretation of resonances in split-ring-resonators at normal incidence," *Optics Express*, vol. 14, no. 19, pp. 8827–8836, 2006.

[47] L. Li, "New formulation of the Fourier modal method for crossed surface-relief gratings," *Journal of the Optical Society of America A*, vol. 14, no. 10, pp. 2758–2767, 1997.

[48] N. Papasimakis, V. A. Fedotov, Y. H. Fu, D. P. Tsai, and N. I. Zheludev, "Coherent and incoherent metamaterials and order-disorder transitions," *Physical Review B*, vol. 80, no. 4, Article ID 041102, 2009.

[49] C. Helgert, C. Rockstuhl, C. Etrich et al., "Effective properties of amorphous metamaterials," *Physical Review B*, vol. 79, Article ID 233107, 4 pages, 2009.

[50] R. Singh, X. Lu, J. Gu, Z. Tian, and W. Zhang, "Random terahertz metamaterials," *Journal of Optics A*, vol. 12, no. 1, Article ID 015101, 2010.

[51] G. Acuna, S. F. Heucke, F. Kuchler, H. T. Chen, A. J. Taylor, and R. Kersting, "Surface plasmons in terahertz metamaterials," *Optics Express*, vol. 16, no. 23, pp. 18745–18751, 2008.

[52] A. K. Azad, J. Dai, and W. Zhang, "Transmission properties of terahertz pulses through subwavelength double split-ring resonators," *Optics Letters*, vol. 31, no. 5, pp. 634–636, 2006.

6

Surface Plasmon Singularities

Gabriel Martínez-Niconoff,[1] P. Martinez-Vara,[2] G. Diaz-Gonzalez,[1] J. Silva-Barranco,[1] and A. Carbajal-Domínguez[3]

[1] *Departamento de Óptica, Instituto Nacional de Astrofísica Óptica y Electrónica (INAOE), Apdo. Postal 51 y 216, Tonantzintla, PUE, Mexico*
[2] *Facultad de Ingeniería, Benemérita Universidad Autónoma de Puebla (BUAP), Publa, PUE, Mexico*
[3] *Universidad Juárez Autónoma de Tabasco, (UJAT), Cunduacán, Tab, Mexico*

Correspondence should be addressed to Gabriel Martínez-Niconoff, gmartin@inaoep.mx

Academic Editor: Zhaolin Lu

With the purpose to compare the physical features of the electromagnetic field, we describe the synthesis of optical singularities propagating in the free space and on a metal surface. In both cases the electromagnetic field has a slit-shaped curve as a boundary condition, and the singularities correspond to a shock wave that is a consequence of the curvature of the slit curve. As prototypes, we generate singularities that correspond to fold and cusped regions. We show that singularities in free space may generate bifurcation effects while plasmon fields do not generate these kinds of effects. Experimental results for free-space propagation are presented and for surface plasmon fields, computer simulations are shown.

1. Introduction

The contemporary trends in plasmon optics consist of establishing a parallelism with the traditional optical models. The wave nature of the electromagnetic field in both cases implies that the mathematical analysis may present many similarities. However, the physical features may be completely different. An important common behavior is that the electromagnetic field is organized around the singular regions. The simplest singular region is kind fold, and the union of twofold focusing regions generates a cusped focusing region [1]. These are the only focusing regions that can be detected on a plane [1, 2], so that they are the expected singularities for surface plasmon fields. From this fact, some physical features can be analyzed. For optical fields in free space, the inverse process may occur, that is, when a cusped focusing region is split into two folds, vortex and bifurcation effects may occur [3]. For the case of plasmon fields, these features do not appear; however, the focusing region generates charge redistribution [4, 5]. This latter property offers many interesting applications; for example, it can be implemented to polarize particles, allowing the design of plasmonic tweezers.

In the present study, we describe the synthesis of singular regions propagating in free space and they are compared with the singular regions for surface plasmon fields. In both cases, the singularities geometry can be obtained from the phase function [6, 7], and it satisfies the nonlinear partial differential equation [6]

$$\frac{\partial^2 L}{\partial x^2}\frac{\partial^2 L}{\partial y^2} - \left(\frac{\partial^2 L}{\partial x \partial y}\right)^2 = 0, \qquad (1)$$

where L is the optical path length. When the boundary condition is a slit-shaped curve, (1) has a simple geometrical interpretation. The singularity is generated by means of the envelope region of a ray set, and each ray must satisfy the Fermat principle [8], and this construction is sketched in Figure 1. According to the transversal condition [9], the rays must emerge perpendicular to the curve slit. In the case of plasmon fields, this geometrical point of view is very important because all the electromagnetic fields propagating

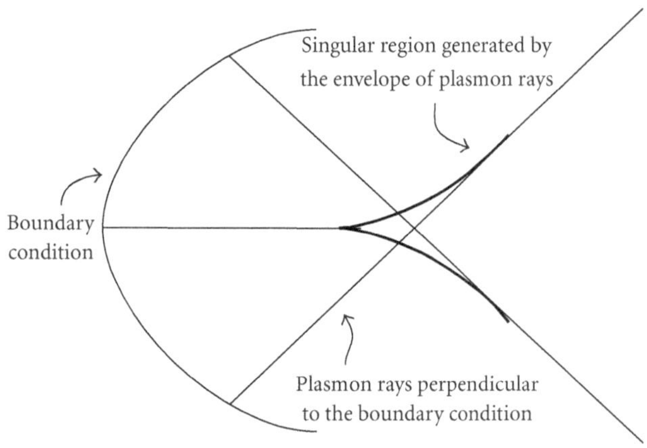

FIGURE 1: Generation of the singular region by means of a set of plasmon rays. Each ray must satisfy variational principles; this means that the rays must emerge perpendicular to the boundary condition. This construction is valid for waves in free space and surface plasmon fields.

on the metal surface can be considered as a "Geodesic Flow" [10], and in this sense, the singularities are deeply related to the stability of the flow.

2. Long-Range Surface Plasmon Modes

Surface plasmon modes are nonhomogeneous electromagnetic waves propagating on the interface of a dielectric-metal media. For a semi-infinite media, the electric field is

$$E_1(x,z) = \left(\vec{i}a_1 + \vec{k}b_1\right)\exp(-\alpha_1 z)\exp(i\beta x),$$
$$E_2(x,z) = \left(\vec{i}a_2 + \vec{k}b_2\right)\exp(-\alpha_2 z)\exp(i\beta x). \quad (2)$$

Related to this electromagnetic mode is the charge motion of the free electrons, that oscillate longitudinally and transversally to the surface. These oscillations are determined by the trajectory of the electric field, whose geometry can be obtained by establishing an analogy with the classical polarization model. It is easy to show that the trajectory of the electric field on the x-z plane is given by

$$\frac{E_x^2}{a_1^2} + \frac{E_z^2}{(\alpha a_1/|\beta|)^2} - \frac{2E_xE_z}{a_1^2\alpha/|\beta|}\cos\delta_0 = \exp(-2\alpha z)\sin^2\delta_0, \quad (3)$$

where β is the dispersion relation function that satisfies

$$\beta = \frac{w}{c}\left(\frac{\varepsilon_1\varepsilon_2}{\varepsilon_1+\varepsilon_2}\right)^{1/2}. \quad (4)$$

ε_1 is the permittivity of the dielectric media, ε_2 is the permittivity of the metal, which is a complex number, and consequently, β is a complex function. The imaginary part is responsible for surface plasmon modes propagating at very short distances, about 100 μm. This length of propagation is a serious limitation for the development of plasmon optics. The synthesis of focusing regions implies the generation of long-range surface plasmon modes modifying the dispersion relation function β. This is possible when the surface plasmon is propagating in a metallic thin film of thickness 20–80 nm. The calculus of the new dispersion relation function is described by means of the interaction between two plasmon modes, as shown in Figure 2.

The interaction between the two modes is performed by using the coupling mode theory. This implies to model the surface effects, associating an expression for the refractive index, in the neighborhood of the interface of the dielectric-metal surface, which is given by

$$n = n_1 + a\exp(-pz), \quad z \geq d, \quad (5)$$

where n_1 is the refractive index of the dielectric media, a is a small constant determined by the combination of the materials involved, d is the thickness of the film, and p is a parameter that allows us to model the surface effects. When $z \to \infty$, the proposed refractive index recovers the classical expression for the dielectric media.

For this case, the dispersion relation function acquires the form

$$\beta = \frac{\omega}{c}\left\{\left(\frac{\varepsilon_1\varepsilon_2}{\varepsilon_1+\varepsilon_2}\right)^{1/2} + 2an_1\left(\frac{\varepsilon_1\varepsilon_2}{\varepsilon_1+\varepsilon_2}\right)^{-1/2}\exp(-2\alpha d)\right\}. \quad (6)$$

The calculus details can be found in [4]. When the last expression is substituted in (3), the trajectory geometry of the electrical field on z = constant is an ellipse segment, as shown in Figure 3.

The modulus of the imaginary part given by (6) is minor, when compared with that given by (4). As a consequence the plasmon fields for a thin film propagate large distances. The geometrical interpretation is that the x-component for the electric field is minor, decreasing the amplitude oscillation of the charge along the x-axis.

The parameters implicit in (5) are sketched in Figure 2. The expression for the long-range surface plasmon mode is obtained by replacing the expression for β given by (6) in (2).

To generate the singular regions, it is necessary to describe the long-range plasmon mode on the x-y plane. The corresponding expression is obtained by rotating the z-axis, and the transformation of coordinates is $z \to z$; $x \to x\cos\theta \pm y\sin\theta$. The general expression for the plasmon mode is

$$E(x,y,z) = \left(\vec{i}a + \vec{k}b\cos\theta + \vec{j}b\sin\theta\right)$$
$$\times \exp(-\alpha_1 z)\exp(i\beta(x\cos\theta + y\sin\theta)), \quad (7)$$

with β given as that presented in (6).

3. Singular Regions in Free Space

With the purpose to analyze the singularities, we describe the generation of fold singularities propagating in free space. The transmittance function has a mathematical representation in the frequency space of the form [6]

$$T(u,v) = \delta(u - f(v)), \quad (8)$$

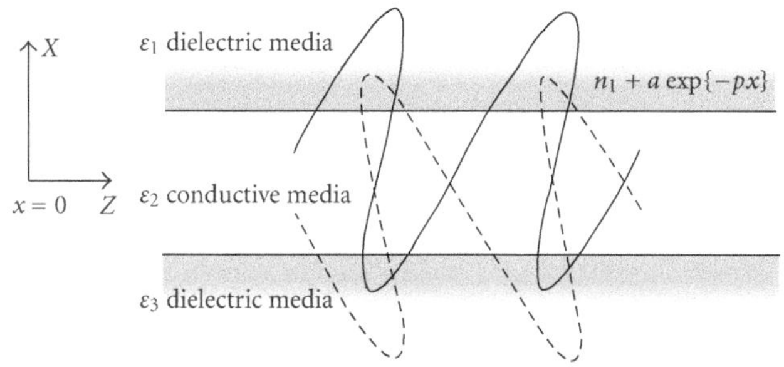

FIGURE 2: Schematic set up to describe the synthesis of long-range surface plasmon modes. The surface effects are described by assuming an exponential representation for the refractive index. The physical effects are described by assuming two surface plasmon modes propagating on each surface.

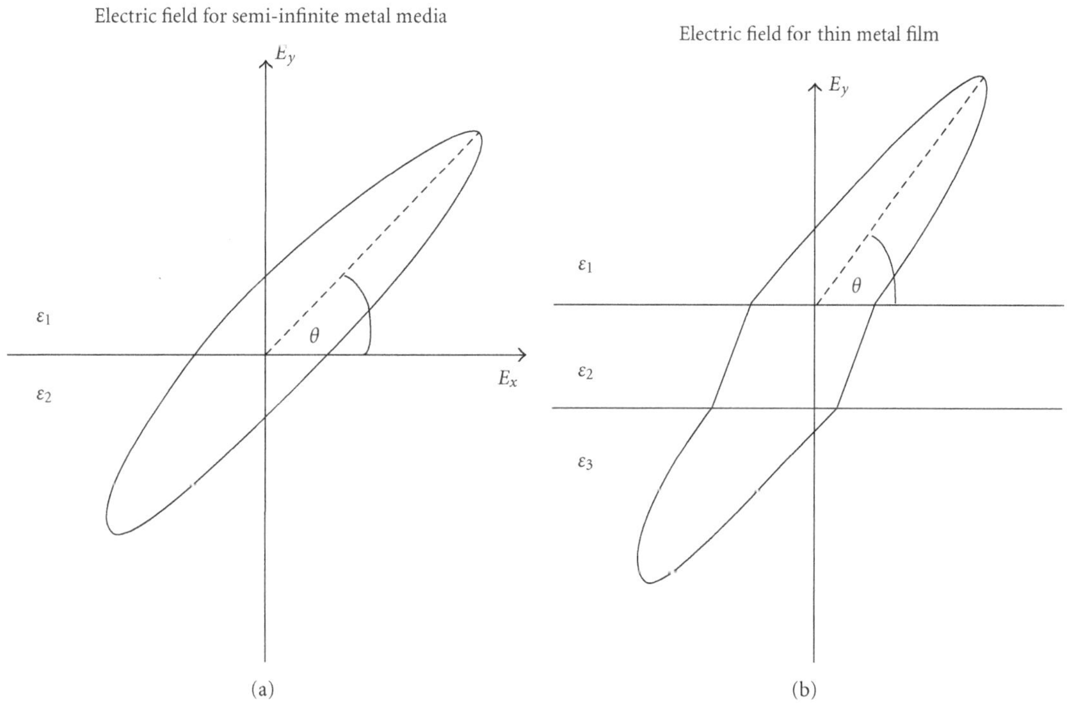

FIGURE 3: Polarization-like trajectory for surface plasmon modes. (a) Surface plasmon modes propagating in a semi-infinite media. (b) Surface plasmon modes propagating on a thin film.

where δ is the Dirac delta-function. The diffraction field, using the angular spectrum model is given by [11]

$$\varphi(x,y,z) = \iint_{-\infty}^{\infty} \delta(u - f(v)) \times \exp\{i2\pi(xu + yv + zp)\}\,du\,dv, \tag{9}$$

where $u = \cos\alpha/\lambda, v = \cos\beta/\lambda, p = \cos\gamma/\lambda$ are the spatial frequencies that satisfy $u^2 + v^2 + p^2 = 1/\lambda^2$; α, β, and γ are the angles that makes the wave vector with the (x,y,z)-axis; and λ is the wavelength. In Figure 4, the setup to generate the diffraction field associated with slit-shaped curves is shown.

As prototype, we describe the synthesis of two kinds of optical fields. The first one has a transmittance as boundary condition, whose geometry in frequency space is a cubic function given by

$$T(u,v) = \delta(u - v^3). \tag{10}$$

The optical field has the representation

$$\begin{aligned}\varphi(x,y,z) &= \iint_{-\infty}^{\infty} \delta(u - v^3)\exp\{i2\pi(xu + yv + zp)\}\,du\,dv \\ &= \int_{-\infty}^{\infty} \exp\left\{i2\pi\left(xv^3 + yv + z\left(\frac{1}{\lambda^2} - (v^6 + v^2)\right)^{1/2}\right)\right\}dv.\end{aligned} \tag{11}$$

The optical field presents bifurcation effects that are manifested during its propagation. To understand the amplitude

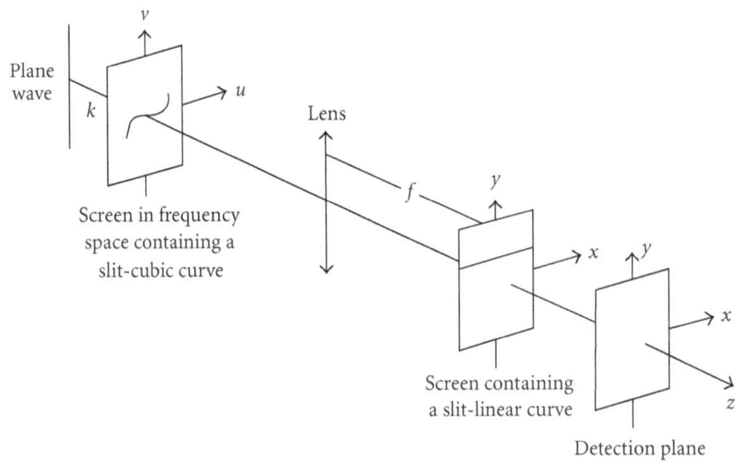

FIGURE 4: Schematic setup to generate the optical field associated with slit-shape curves.

distribution, we implement a kind of spatial filtering, which is obtained by placing a screen containing a linear slit on $z = 0$, whose representation is $t(y) = \delta(y-1)$. The amplitude distribution is

$$\begin{aligned}\Phi_1(x, y, z = 0) &= \delta(y-1)\int_{-\infty}^{\infty} \exp i2\pi(xv^3 + yv)\,dv \\ &= \delta(y-1)\int_{-\infty}^{\infty} \exp(i2\pi(xv^3 + v))\,dv.\end{aligned} \quad (12)$$

By translating the screen to the position defined by $y = -1$, the mathematical representation for the optical field is given by

$$\begin{aligned}\Phi_2(x, y, z = 0) &= \delta(y+1)\int_{-\infty}^{\infty} \exp(i2\pi(xv^3 - yv))\,dv \\ &= \delta(y+1)\int_{-\infty}^{\infty} \exp(i2\pi(xv^3 - v))\,dv.\end{aligned} \quad (13)$$

The phase function represented by (12) and (13) differs in a sign, being equals along the direction defined by $v = 0$. The bifurcations effects appear when the optical field is propagating to different planes, that is, $z \neq 0$; therefore, each optical field propagates by different paths. In Figure 5(a), the experimental result in $z = 0$ plane is shown, and in Figure 5(b) the experimental result for $z \neq 0$ where the bifurcation effects become evident are presented. From these results, it can be noted that the right branch is shifted to the bottom and the left branch is shifted to the top. The entire optical field unfolds based on the structure of the transmittance function.

4. Singular Surface Plasmon Fields

Singularities for surface plasmon fields can be obtained in a way similar to that employed for the free space case, and the corresponding surface plasmon angular spectrum model is [4, 5]

$$\begin{aligned}E_1(x, y, z) &= \iint_{-\infty}^{\infty} \left(\vec{i}A(u) + \vec{j}B(u) + \vec{k}C(u)\right) \\ &\quad \times \exp(-2\pi\alpha_1 z)\exp(i2\pi\beta(xu + yv))\,du\,dv,\end{aligned} \quad (14)$$

where the term $(\vec{i}A(u) + \vec{j}B(u) + \vec{k}C(u))$ represents the transmittance function in the frequency space. For a slit-shape curve, we have that the transmittance takes the form

$$T(u, v) = \left(\vec{i}a + \vec{j}b + \vec{k}c\right)\delta(u\tau(\lambda, d) - f(v)), \quad (15)$$

where (a,b,c) are the components of a constant vector and are necessary to satisfy the boundary conditions of the electromagnetic field. More details with respect to the value of the constants can be found in [4]. The scale term $\tau(\lambda, d)$ is proposed to describe the coupling between the diffraction field propagating in free space and surface plasmon fields. In Figure 6, the experimental setup to generate surface plasmon singularities is shown.

The singularities are generated by illuminating a slit-shape curve recorded on a thin metal film, whose thickness is in the order of 20–80 nm, and this allows generating long-range surface plasmon modes. The width d of the slit must be smaller than the illumination wavelength λ.

The expression for the surface plasmon diffraction field is given by

$$\begin{aligned}E_1(x, y, z) &= \int_{-\infty}^{\infty} \left(\vec{i}a + \vec{j}b + \vec{k}c\right)\exp(-2\pi\alpha z) \\ &\quad \times \exp(i2\pi\beta(x\tau(\lambda, d)f(v) + yv))\,du.\end{aligned} \quad (16)$$

Let us consider two examples. The first one is when the transmittance function contains a slit-cubic curve, given by

$$T(u, v) = \left(\vec{i}a + \vec{j}b + \vec{k}c\right)\delta\left(u\left(\frac{\lambda}{d} + i\frac{d}{\lambda}\right) - v^3\right), \quad (17)$$

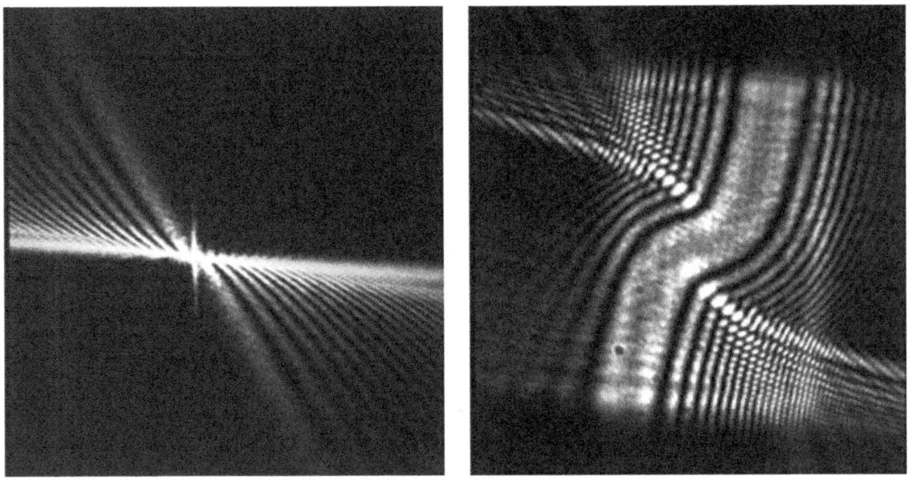

(a) (b)

Figure 5: (a) Optical field on $z = 0$ plane. This optical field is formed by two optical fields, and the spatial evolution of these two fields generates the bifurcation effects. (b) The generation of bifurcation effects. The distance between the two images along z-coordinate is approximately 1 cm; after this distance, the optical field remains practically unchanged.

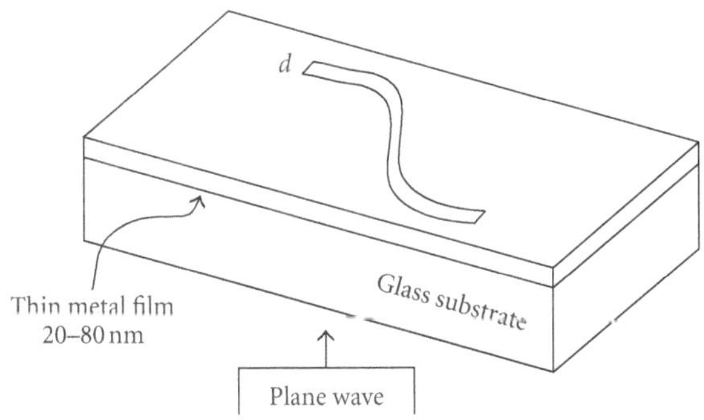

Figure 6: Experimental setup to generate surface plasmon singularities. The thin metal film contains a slit-shaped curve.

The term $\tau(\lambda, d) = (d/\lambda + i(\lambda/d))$ is proposed as an ansatz. When $d >> \lambda$, the imaginary part is very small and the optical field recovers the structure of the optical field propagating in free space. The opposite case $d << \lambda$, means that the imaginary part becomes dominant and the optical field acquires the form of evanescent waves, wich is the condition to generate surface plasmon fields. The surface plasmon diffraction field is

$$E_1(x, y, z) = \int_{-\infty}^{\infty} \left(\vec{i}a + \vec{j}b + \vec{k}c\right) \exp(-2\pi z\alpha) \times \exp\left(i2\pi\beta\left(y\tau(\lambda, d)v + x\tau(\lambda, dv)^3 f(v)^3\right)\right) dv, \tag{18}$$

having a representation of a fold caustic [2] and is similar to an Airy function [12]. The second example is when the transmittance function contains a slit-quadratic curve given by

$$T(u, v) = \left(\vec{i}a + \vec{j}b + \vec{k}c\right)\delta(v\tau(\lambda, d) - u^2), \tag{19}$$

and the surface plasmon diffraction field is

$$E_1(x, y, z) = \int_{-\infty}^{\infty} \left(\vec{i}a + \vec{j}b + \vec{k}c\right)\delta(v\tau(\lambda, d) - u^2) \times \exp(-2\pi\alpha z)\exp(i2\pi\beta(yv + xu))\,du. \tag{20}$$

The amplitude distribution must satisfy the Helmholtz equation, which means that the spatial frequencies must satisfy $\beta^2(u^2 + v^2) - \alpha^2 = 1/\lambda^2$. By substituting this expression in the integral representation and using the paraxial approximation, the amplitude function acquires the form

$$E_1(x, y, z) = \int_{-\infty}^{\infty} \left(\vec{i}a + \vec{j}b + \vec{k}c\right)\delta(v\tau(\lambda, d) - u^2)\exp(-2\pi\alpha z) \times \exp\left(i2\pi\beta\left[y\left(\frac{1}{\lambda\beta} + \frac{\lambda\beta}{2}(\alpha^2 - u^2)\right) + xu\right]\right)du,$$

$$E_1(x, y, z) = \int_{-\infty}^{\infty} \left(\vec{i}a + \vec{j}b + \vec{k}c\right)\exp(-2\pi\alpha z) \times \exp\left(i2\pi\beta\left[y\left(\frac{1}{\lambda\beta} + \frac{\lambda\beta}{2}\left(\alpha^2 - \frac{u^4}{\tau^2}\right)\right) + x\frac{u^2}{\tau}\right]\right)du. \tag{21}$$

In Figure 7, we show the computer simulation. At this point, it is convenient to compare (9) with (16). In both cases, the boundary conditions are described in the x-y plane. However the phase terms in each case are different, having a decreasing exponential function for plasmon fields and a complex dependence for free-space propagation. This means

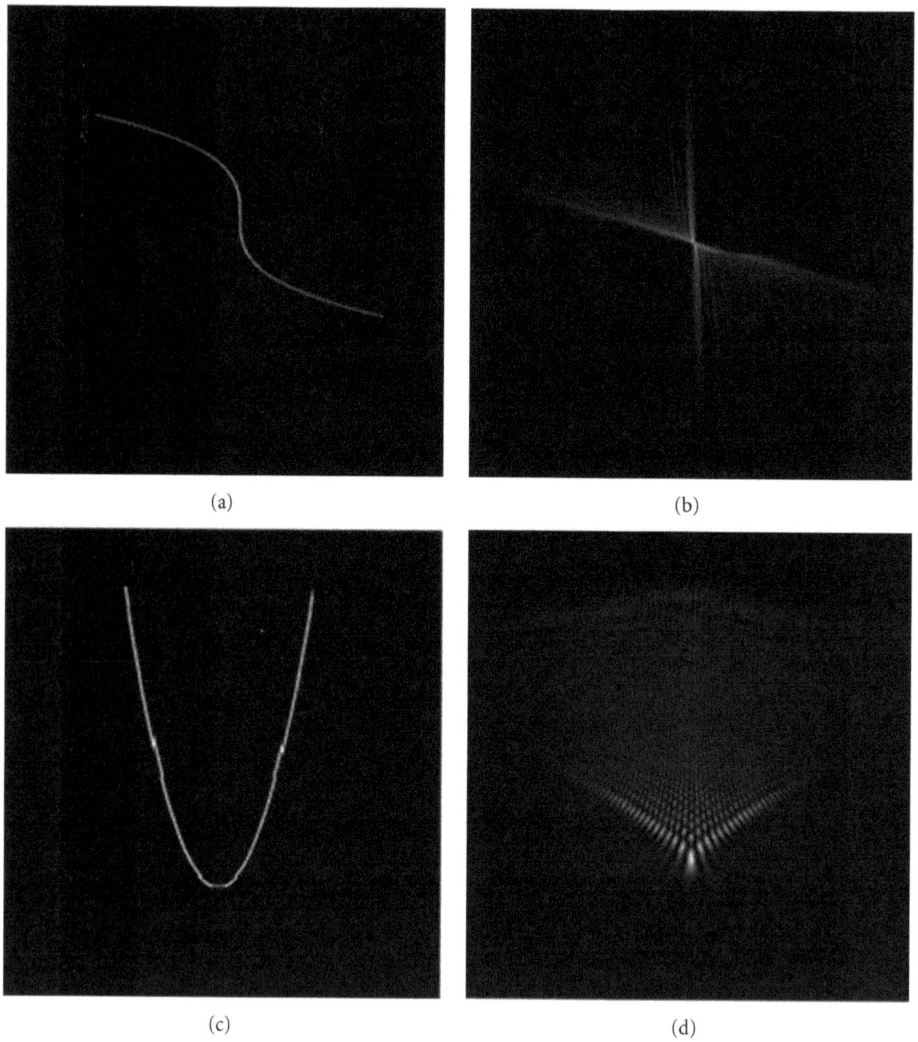

FIGURE 7: Front view of the metal surface. The illumination is incident from the back of the screen. (a) The screen containing a slit-cubic curve recorded on a thin metal film. (b) The surface plasmon diffraction field and its singular region. (c) Transmittance containing a parabolic curve and (d) singularity associated with this transmittance.

that the bifurcation effects must appear perpendicular to the plane that contains the slit-shaped curve; consequently, the plasmon singularities have a stationary character.

Finally, we remark that the phase function in (18) and (21) has the structure of a catastrophe function [2]. This means that the singularities correspond to the curvature center of the slit curve [6], and has a stationary character, generating charge redistribution given by $\nabla \cdot E_{1,2} = \rho/\varepsilon_{1,2}$, where the electrical field can be represented as

$$\begin{aligned} E(x, y, z) &= \int_{-\infty}^{\infty} \vec{\xi} \exp(-2\pi\alpha z) \\ &\quad \times \exp\left(i2\pi\beta[\text{germ function}(\nu) \right. \\ &\quad \left. + \text{perturbation function}(\nu, m, n)]d\nu\right), \end{aligned} \tag{22}$$

where m, n are parameters related with the curvature centers. The generic physical properties associated to the singularities of surface plasmon fields can be obtained from this representation.

5. Conclusions

We described the synthesis of surface plasmon singularities by analyzing the spatial evolution of the surface plasmon field emerging from a slit-shaped curve. The singularities can be easily understood by considering that they are generated by an envelope of a set of normal trajectories emerging from each point on the boundary condition. The geometry generated by the envelope of trajectories satisfies a nonlinear partial differential equation, and the surface plasmon singularity generates a charge distribution; for this reason, the singularities offer application as surface plasmon tweezers. This can be implemented by depositing micronano/particles on the metal surface, illuminating with different frequencies; we can generate a set of plasmon singularities inducing tunable resonance effects. These results and its potential applications will be presented in a forthcoming paper.

References

[1] V. I. Arnold, *Singularities of Caustics and Wave Fronts*, Kluwer Academic, 1990.

[2] M. V. Berry and C. Upstill, "Catastrophe optics: morphologies of caustics and their diffraction patterns," in *Progress in Optics*, E. Wolf, Ed., vol. 18, North-Holland, 1980.

[3] Y. S. Kivshar and E. A. Ostrovskaya, "Optical vortices: folding and twisting waves of light," *Optics and Photonics News*, vol. 12, no. 4, pp. 24–29, 2001.

[4] G. Martínez Niconoff, P. Martínez Vara, J. Munoz-Lopez, J. C. Juárez-Morales, and A. Carbajal-Dominguez, "Partially coherent surface plasmon modes," *Journal of the European Optical Society*, vol. 6, p. 52, 2011.

[5] G. Martinez Niconoff, J. A. Sanchez-Gil, H. H. Sanchez, and A. P. Leija, "Self-imaging and caustics in two-dimensional surface plasmon optics," *Optics Communications*, vol. 281, no. 8, pp. 2316–2320, 2008.

[6] G. Martinez-Niconoff, J. Carranza, and A. Cornejo-Rodriguez, "Caustics of diffraction fields," *Optics Communications*, vol. 114, no. 3-4, pp. 194–198, 1995.

[7] G. Martínez-Niconoff, J. Muñoz-Lopez, and E. Méndez-Martínez, "Description of phase singularities and their application to focusing design," *Journal of the Optical Society of America A*, vol. 18, no. 9, pp. 2089–2094, 2001.

[8] M. S. Sosskin and M. V. Vasnetsov, "Singular Optics," in *Progress in Optics*, E. Wolf, Ed., vol. 42, North-Holland, 2001.

[9] L. Elgoltz, *Differential Equations and Variational Calculus*, Mir, Moscow, Russia, 1977.

[10] F. Papoff and R. Zambrini, "Convective instability induced by nonlocality in nonlinear diffusive systems," *Physical Review Letters*, vol. 94, no. 24, Article ID 243903, 4 pages, 2005.

[11] J. P. Goodman, *Introduction to Fourier Optics*, McGraw-Hill, 2000.

[12] W. Liu, D. N. Neshev, I. V. Shadrivov, A. E. Miroshnichenko, and Y. S. Kivshar, "Plasmonic Airy beam manipulation in linear optical potentials," *Optics Letters*, vol. 36, no. 7, pp. 1164–1166, 2011.

7

Manipulating the Propagation of Solitons with Solid-Core Photonic Bandgap Fibers

O. Vanvincq, A. Kudlinski, A. Bétourné, A. Mussot, Y. Quiquempois, and G. Bouwmans

Laboratoire PhLAM, UMR CNRS 8523, IRCICA, Université Lille 1, 59655 Villeneuve d'Ascq Cedex, France

Correspondence should be addressed to A. Kudlinski, alexandre.kudlinski@univ-lille1.fr

Academic Editor: Miguel González Herráez

We review the dynamics of soliton self-frequency shift induced by Raman gain in special solid-core photonic bandgap fibers and its consequences in terms of supercontinuum generation. These photonic bandgap fibers have been designed to allow nonlinear experiments in the first bandgap without suffering from significant loss even when working close to the photonic bandgap edge. We studied experimentally, numerically, and analytically the extreme deceleration of the soliton self-frequency shift at the long-wavelength edge of the first transmission window. This phenomenon is interpreted as being due to a large variation of the group-velocity dispersion in this spectral range and has been obtained with no significant power loss. Then, we investigated experimentally and numerically the generation of supercontinuum in this kind of fibers, in both spectral and temporal domains. In particular, we demonstrated an efficient tailoring of the supercontinuum spectral extension as well as a strong noise reduction at its long-wavelength edge.

1. Introduction

Solid-core photonic bandgap (PBG) fibers are one class of microstructured optical fibers in which light is confined in a low-index solid core by the PBGs of the cladding [1]. In the case of two-dimensional PBGs, the periodic microstructured cladding is usually composed of periodic high-index inclusions embedded in a low-index background [1]. The core region then corresponds to a defect (lack of inclusion) in the center of the periodic structure. Because of the intrinsic nature of the PBG waveguidance, the fiber transmission properties are characterized by discrete spectral bands [1, 2] in which the group-velocity dispersion (GVD), attenuation, and effective mode area of the fundamental core mode are strongly wavelength dependent [3], especially near the PBG edges [1, 4]. These singular characteristics are thus of particular interest in the field of nonlinear (NL) fiber optics. The potential of solid-core PBG fibers for NL propagation experiments has already been pointed out by the report of soliton propagation and phase-matched dispersive wave generation [5, 6], even across different PBG [7]. Degenerate four-wave mixing [8], frequency doubling and tripling [9], and supercontinuum (SC) generation across adjacent PBGs have also been investigated numerically [10].

In this paper, we present an overview of our recent work showing how the particular properties of solid-core PBG fibers can influence soliton propagation and SC generation. First, we detail the linear properties of the fabricated solid-core PBG fibers under investigation through experiments and numerical modelling. Then, we demonstrate experimentally that the soliton self-frequency shift (SSFS) effect can be strongly mitigated [11]. A simple analytical model indicates that this is caused by the strong third-order dispersion experienced by solitons as they approach the PBG edge. Finally, we show that this SSFS quasicancellation allows to control the spectral extent of infrared SC in the long-pulse regime [12] and, simultaneously, leads to a significant reduction of spectral power fluctuations at the SC long-wavelength edge [13].

2. Linear Properties of the Solid-Core Photonic Bandgap Fibers

The solid-core PBG fibers used in various experiments reported in this paper are all based on the same double periodicity structure [11, 12], although their optogeometrical characteristics (summarized in Table 1) slightly vary from one fiber to another. The scanning electron microscope

TABLE 1: Table summarizing the optogeometrical properties of the three fabricated solid-core PBG fibers under investigation.

	d_{Ge} (μm)	d_{air} (μm)	Λ (μm)	ZDW (nm)	NL coefficient ($W^{-1}\cdot km^{-1}$)	PBG edge (nm)
Fiber 1	2.19	2.33	3.09	905	6 (at 1200 nm)	1535
Fiber 2	2.20	2.25	3.25	915	6.9 (at 1064 nm)	1580
Fiber 3	2.22	2.27	3.28	921	6.8 (at 1064 nm)	1590

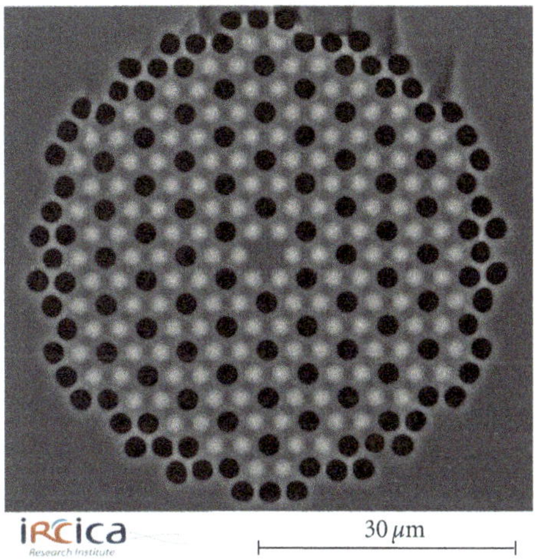

FIGURE 1: SEM image of the solid-core PBG fiber 1. Black regions are air holes, light gray regions are GeO_2-doped inclusions, and the dark gray background is pure silica.

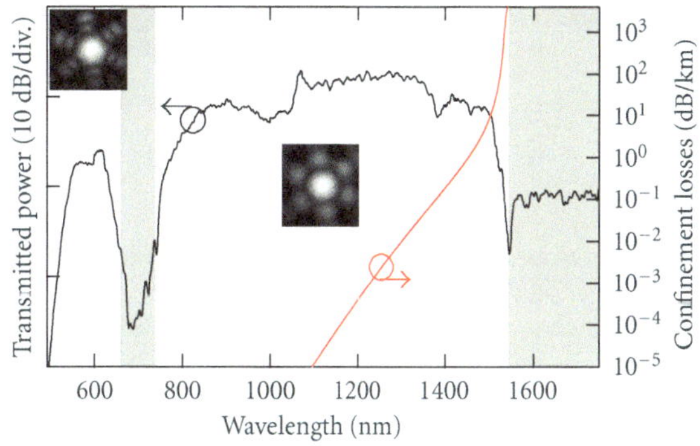

FIGURE 2: Transmission spectrum of the solid-core PBG fiber 1 (black line, left axis) measured with a SC source, and computed confinement losses across the first PBG (red line, right axis). Insets show the near-field mode profile in the first (right) and second (left) PBG. Gray areas depict spectral regions in which the PBG guidance in the core is not allowed.

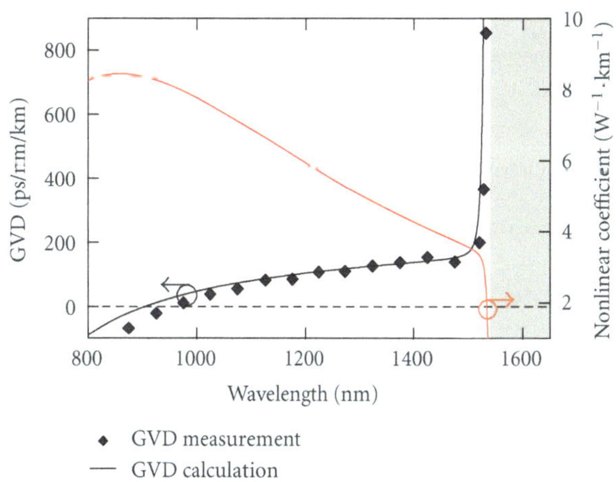

FIGURE 3: Calculated GVD curve of the solid-core PBG fiber 1 (solid black line, left axis), and corresponding measurements with a white-light interferometry setup (markers). Spectral evolution of the computed NL coefficient (red line, right axis). The gray area depicts the region outside the first PBG transmission window.

(SEM) image shown in Figure 1 illustrates such a double periodicity solid-core PBG fiber. The periodical high-index resonators (light gray regions in Figure 1) present in the cladding are responsible for the PBG guidance and are made of germanium-doped silica with a parabolic refractive index profile and a maximal index difference Δn of 32×10^{-3} (as compared to pure silica). They are embedded in a pure silica background (dark gray). Because we decided to work in the first PBG (mainly to minimize the effective area in order to maximize the NL coefficient), air holes of diameter d_{air} (depicted in black in Figure 1) were periodically added into the cladding structure for reducing the otherwise relatively high confinement losses in this PBG [14, 15]. The core region corresponds to a defect (lack of air hole) in the center of the periodic structure. The geometrical parameters of the three solid-core PBG fibers investigated here are summarized in Table 1, where the pitch Λ corresponds to the distance between two neighboring germanium-doped inclusions.

Although all fibers were characterized the same way, we only present here the linear properties of fiber 1, whose SEM image is depicted in Figure 1. Other fibers present similar characteristics, with a slight red-shift of the PBG transmission windows due to a slightly greater pitch Λ. Figure 2 shows the experimental transmission spectrum of a white light SC source through a 2 m long sample of the solid-core PBG fiber. The first PBG is very broad and extends from 740 to 1535 nm, whereas the second one is much sharper and extends from 510 to 650 nm. As for usual solid-core PBG fibers, typical resonances can be observed in the first ring of high-index resonators [4], with the number of intensity maxima indicating the PBG order (insets of Figure 2). The red curve (right axis) shows confinement losses across the first PBG calculated with a finite element mode solver. They are less than 1 dB/km across most of the PBG, and as expected they drastically increase to the dB/m range near the PBG edge. Figure 3 shows the calculated GVD curve of the solid-core PBG fiber 1 (solid black line), as well as measurements

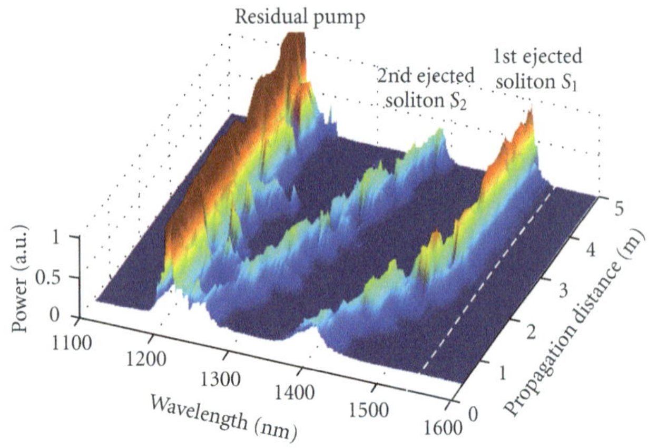

Figure 4: Evolution of the experimental output spectrum (linear scale) with fiber length (from 0.4 to 5 m) for 270 fs pump pulses at 1200 nm. The white dashed line represents the PBG edge.

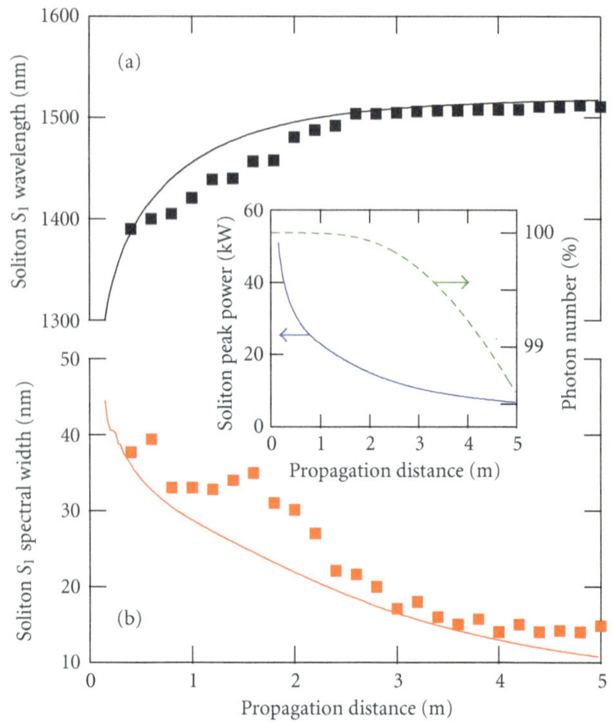

Figure 5: (a) Central wavelength of the first ejected soliton S_1 as a function of propagation distance. (b) Evolution of the soliton S_1 spectral width as a function of propagation distance. Solid lines and markers, respectively, correspond to numerical simulations and experimental measurements. Inset: numerical simulation of soliton S_1 peak power (blue solid line, left axis) and of the total photon number (green dashed line, right axis) versus propagation distance.

(markers) performed with a low coherence interferometric method [16]. The GVD slightly increases from the zero-dispersion wavelength (ZDW), located around 900 nm, to about 200 ps/nm/km at 1510 nm. For higher wavelengths, the GVD increase suddenly becomes much more important because of an anticrossing point between the fundamental core mode and cladding modes at 1535 nm, delimiting the practical PBG transmission window [12]. In this region near the PBG edge, the GVD reaches extreme values well higher than 1000 ps/nm/km: measurements (not displayed in Figure 3 for the sake of clarity) indicates a GVD value 4760 ps/nm/km at 1535 nm, in excellent agreement with the calculated curve. This leads the third-order dispersion β_3 to reach extremely high values in the order of 10^{-37} s^3/m in this spectral region. Finally, we also studied numerically the spectral dependance of the NL coefficient γ defined as $\gamma = n_2\omega/cA_{\text{eff}}$ [17], with n_2 the NL refractive index of pure silica and A_{eff} the mode effective area. Because the effective area also increases strongly near the PBG edge (due to the anticrossing phenomenon), the NL coefficient γ decreases rapidly near the PBG edge (see red curve in Figure 3) from about 6 $W^{-1}\cdot km^{-1}$ at 1200 nm to less than 1.3 $W^{-1}\cdot km^{-1}$ at 1535 nm.

In the following, we will see that the strong and unusual spectral dependencies of GVD and NL coefficient near the PBG edge can be used to tailor and manipulate the propagation of solitons approaching this spectral region through SSFS, included in the context of long-pulse SC generation.

3. Quasicancellation of the Soliton Self-Frequency Shift

3.1. Experimental Evidence. To illustrate this possible manipulation of the SSFS, NL propagation experiments were first performed by launching 270 fs pulses with a repetition rate of 1 kHz at 1200 nm into a 5 m long sample of the solid-core PBG fiber 1. This pump wavelength has been chosen from a compromise between a wavelength not too close to the long-wavelength edge of the first PBG (in order to observe a significant SSFS) and not too close to the ZDW (in order to avoid SC generation). Similarly, the input peak power was adjusted so that it was high enough to observe SSFS for several solitons while being low enough to avoid SC generation. The pump peak power effectively launched into the fiber core was estimated to about 10–20 kW. Figure 4 shows a representation of the output spectrum (in linear scale) as a function of fiber length from 0.4 to 5 m. It was obtained from a cut-back measurement in which the output spectrum was recorded every 20 cm while keeping the launch conditions unchanged. The input Gaussian pulse breaks up into two fundamental solitons (labeled S_1 and S_2) through a fission mechanism [18, 19] caused by Raman scattering and higher-order dispersion effects [19, 20]. As expected, both solitons then red-shift through Raman-induced SSFS [21, 22]. The most striking feature in the soliton dynamics is that, after 2.5 m, the first ejected soliton, S_1, becomes frequency locked at a central wavelength of about 1510 nm, that is, when its spectrum starts to overlap with the PBG located at 1535 nm (depicted in white dotted line in Figure 4). This is even more clearly illustrated in Figure 5(a) which shows a plot of the S_1 soliton central wavelength (black squares) as a function of the propagation distance. A clear slope change in the SSFS rate is observed at the fiber length of 2.5 m. Red squares in Figure 5(b) represent the evolution of soliton S_1 full width at half maximum (FWHM) spectral width versus propagation length. It is shown that, as soliton S_1 red-shifts, it initially experiences a spectral compression from about 40 nm to 15 nm within the first 3 m. For longer propagation distances, that is, when the soliton is frequency locked, its spectral width stabilizes to about 15 nm, without any significant change in its spectral power density, as observed in Figure 4. Note that the same phenomenon was observed for the soliton S_2 (at a longer fiber length) and even for more than five ejected solitons by increasing the pump power. These experiments clearly show that the SSFS is extremely decelerated (and even quasicancelled) when the solitons approach the PBG edge, this effect being accompanied by a spectral compression.

3.2. Confirmation with Numerical Simulations. In order to get further insight into this phenomenon, we performed numerical simulations by integrating the generalized NL Schrödinger equation (GNLSE) taking into account the mode profile dispersion [23]. In this case, it writes [23, 24]:

$$\frac{\partial \tilde{C}(z,\omega)}{\partial z} = i(\beta(\omega) - \beta_0 - \beta_1(\omega - \omega_0))\tilde{C}(z,\omega) - \frac{\alpha(\omega)}{2}\tilde{C}(z,\omega) + i\overline{\gamma}(\omega)\mathcal{F}\left(C(z,t)\int_{-\infty}^{\infty} R(t-t')\,|C(z,t')|^2 dt'\right), \quad (1)$$

where the NL coefficient takes the following form [24]:

$$\overline{\gamma}(\omega) = \frac{n_2 \omega n_0}{c n_{\text{eff}}(\omega)\sqrt{A_{\text{eff}}(\omega)A_{\text{eff}}(\omega_0)}} \quad (2)$$

and $\tilde{C}(z,\omega) = \tilde{A}(z,\omega)(A_{\text{eff}}(\omega_0)/A_{\text{eff}}(\omega))^{1/4}$, $\tilde{A}(z,\omega)$ being the complex spectral envelope. The coefficient β_0 corresponds to the mode propagation constant $\beta(\omega)$ at the center frequency of the input pulse spectrum ω_0, and β_1 is the first derivative of $\beta(\omega)$ (i.e., $d\beta(\omega)/d\omega$). The symbol $\mathcal{F}$ denotes the Fourier transform of the function in argument, and $\alpha(\omega)$ is the fiber linear power attenuation at ω. The silica response function is defined by $R(t) = (1 - f_R)\delta(t) + f_R h_R(t)$ where $f_R = 0.18$ and $h_R(t)$ is the Raman response function which is approximated by the analytical expression given in [25].

Figure 6 shows a propagation simulation over 5 m of a 270 fs Gaussian pulse with 15 kW peak power centered at 1200 nm, corresponding to the input experimental pulses. The frequency dependence of all parameters, that is, $\alpha(\omega)$, $\beta(\omega)$, and $A_{\text{eff}}(\omega)$, associated with the experimental fiber was taken into account. This simulation shows excellent qualitative agreement with the experimental results of Figure 4. The input pulse experiences fission leading to the ejection of two fundamental solitons, labeled S_1 to S_2, and the SSSF associated to soliton S_1 is strongly decelerated near the PBG edge. This effect is accompanied by a spectral compression and an increase of its spectral power density. Black and red solid lines in Figure 5, respectively, correspond to the simulated soliton S_1 central wavelength and FWHM spectral width. They show good quantitative agreement with experiments, especially when the solitons approach the PBG edge (for propagation distances greater than 2.5 m). The blue solid curve in the inset of Figure 5 represents the soliton S_1 peak power as a function of propagation distance. Although its spectral power density increases (as shown in Figure 6), its peak power (in the temporal domain) decreases along propagation, because the soliton temporally broadens. The variation of total photon number as a function of fiber length is displayed in the green dashed curve in the inset of Figure 5. It decreases (due to linear attenuation) by only less than 1.5% within the first 5 m, which indicates that the soliton S_1 does not significantly suffer from the high attenuation at the PBG edge. This suggests that the quasicancellation of the SSFS does not occur at the expense of significant attenuation. At this point, however, the physics at the origin of this phenomenon has not been elucidated yet.

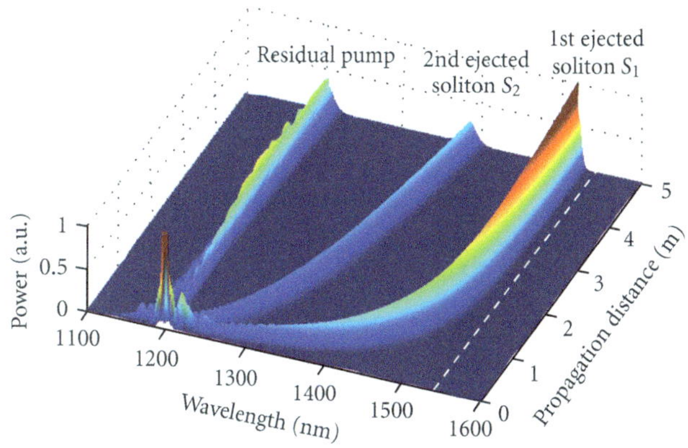

FIGURE 6: Numerical simulation of the propagation of a 270 fs Gaussian pulse with 15 kW peak power centered at 1200 nm, showing fission and ejection of two fundamental solitons S_1 and S_2. All frequency-dependant parameters of the fiber were taken into account. The white dashed line represents the PBG edge.

3.3. Discussion. First, let us recall that, as a soliton reaches the PBG edge, the values of the GVD evaluated at its central frequency strongly increase while its NL coefficient is largely reduced (see Figure 3). All these spectral evolutions are expected to be detrimental to the efficiency of the SSFS as they lead, directly or indirectly, to a significant increase of the pulse duration which is known to strongly affect the SSFS efficiency [17, 22]. In the following, we will present an analytical approach that allows to quantify the relative weight of the spectral dependence of the GVD ($d\beta_2/d\omega$) and NL coefficient ($d\overline{\gamma}/d\omega$) on the SSFS rate. For the sake of clarity, we only consider hereafter fundamental solitons as input pulses. This is justified by the fact that experimental input pulses experience fission which leads to pulse breakup and ejection of actual fundamental solitons [18–20]. The characteristics of these input pulses (hyperbolic secant pulses of 21 fs duration FWHM and 100 kW peak power) were estimated from the measurement of the soliton S_1 spectral width in Figure 5(b), assuming a soliton order $N^2 = 1$.

The starting point of our analysis is the well-known Gordon formula allowing to predict the SSFS rate, $d\langle\omega\rangle/dz$, of a fundamental soliton of mean frequency $\langle\omega\rangle$ and duration T as a function of propagation distance z [22]. It writes

$$\frac{d\langle\omega\rangle}{dz} = -\frac{f_R\,|\beta_2(z)|\,T(z)\pi}{4}I(z), \tag{3}$$

where

$$I(z) = \int \Im\left(\widetilde{h}_R(\Omega')\right)\frac{\Omega'^3}{\sinh^2(T(z)\pi\Omega'/2)}d\Omega' \tag{4}$$

and $\Im(\widetilde{h}_R)$ is the imaginary part of the Fourier transform of the Raman response, which is related to the Raman gain spectrum [22]. Then, assuming that (i) the soliton remains fundamental, (ii) confinement losses are negligible, and (iii) the photon number is conserved, it is possible to find the following evolution of the soliton duration T as a function of propagation distance z [11]:

$$T(z) = \frac{\overline{\gamma}(0)^2\,|\beta_2(z)|}{\overline{\gamma}(z)^2\,|\beta_2(0)|}T_0, \tag{5}$$

where $T_0 = T(z = 0)$ is the initial soliton temporal width. Figure 7 shows the evolution of the soliton central wavelength along propagation calculated by integrating (3) with help of (5) (red line) and by the numerical resolution of the GNLSE given by (1) (black line). A reasonable agreement is obtained between both curves, especially for the longest propagation distances (i.e., when the soliton appears to be frequency locked), the relative discrepancy on the frequency shift being less than 3% for values of z larger than 1 m. Thus, this Figure 7 shows that (i) the SSFS deceleration near the PBG edge is directly linked to the fiber properties, as claimed above and that (ii) the soliton central frequency can be predicted easily and very accurately for the longest propagation distances, without requiring the resolution of the GNLSE.

Then, in order to distinguish the effect of the spectral dependence of $\overline{\gamma}$ and the one of the spectral dependence of β_2, we now calculate the *variation* of the SSFS rate with propagation distance z. Assuming that $\beta_2 < 0$, that is, $|\beta_2| = -\beta_2$, this writes [11]

$$\frac{d^2\langle\omega\rangle}{dz^2} = \frac{f_R T(z)\pi I(z)}{4} \times \left[\frac{2\beta_2(z)}{\overline{\gamma}(z)}(4-K(z))\frac{d\overline{\gamma}}{dz} - (3-K(z))\frac{d\beta_2}{dz}\right], \tag{6}$$

where

$$K(z) = \frac{1}{I(z)}\int\left[\Im\left(\widetilde{h}_R(\Omega')\right) - \left(\frac{d\Im\left(\widetilde{h}_R\right)}{d\Omega}\right)_{\Omega'}\Omega'\right] \times \frac{\Omega'^3}{\sinh^2(T(z)\pi\Omega'/2)}d\Omega'. \tag{7}$$

From (6), it can be seen that the variation of SSFS rate during the propagation depends on the variation of β_2 and $\overline{\gamma}$ with z. Since the fiber is assumed to be invariant along its length, the variations of β_2 and $\overline{\gamma}$ with z are in fact only due to their spectral evolution as the soliton red-shifts, that is, $\beta_2\langle\omega(z)\rangle$ and $\overline{\gamma}\langle\omega(z)\rangle$. Consequently, the derivative of β_2 and $\overline{\gamma}$ with respect to z in (6) can mathematically be expressed in terms of their derivative with respect to ω. We then introduce the R ratio defined as

$$R = -\left(\frac{4-K(z)}{3-K(z)}\right)\frac{2\beta_2}{\overline{\gamma}}\frac{d\overline{\gamma}/d\omega}{d\beta_2/d\omega} \tag{8}$$

so that values of R higher than 1 indicate that the NL coefficient variation $d\overline{\gamma}/d\omega$ dominates the evolution of SSFS rate while values of R smaller than 1 mean that the dynamics of the SSFS is mainly driven by the GVD variation $d\beta_2/d\omega$, that is, by the third-order dispersion term β_3.

The blue curve of Figure 7 (right axis) shows a plot of the R ratio as a function of soliton wavelength calculated with (8). From this plot, it can be seen that the spectral variation of $\overline{\gamma}$ dominates the beginning of the SSFS dynamics, whereas for wavelengths higher than 1475 nm the effect induced by the spectral variation of β_2 becomes predominant. More importantly, the R value just below the PBG edge (located at 1535 nm) is about 0.1, which clearly means that the extreme deceleration of the SSFS is due to the large GVD slope $|\beta_3|$ near the PBG edge rather than the large NL coefficient diminution.

Finally, let us mention that, in the absence of loss, a real cancellation of the SSFS, that is, $d\langle\omega\rangle/dz = 0$, implies that β_2 and/or A_{eff} are infinite in order to get T also equal to infinite. Thus, the strong decrease of the SSFS efficiency near the PBG edge does not lead, strictly speaking, to a true suppression of the SSFS for a finite propagation distance. However, the extreme deceleration observed all along this paper is enough to consider that the soliton frequency can be considered, at least from a practical point of view, as frequency locked.

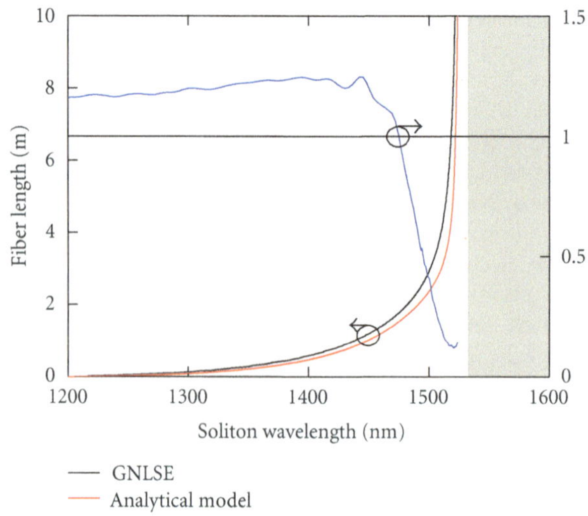

FIGURE 7: Left axis: evolution of the soliton central wavelength with fiber length calculated by integration of the GNLSE (black line) and with (5) in the case of the Gordon model (red line). Right axis: evolution of the R ratio defined by (8) with soliton wavelength.

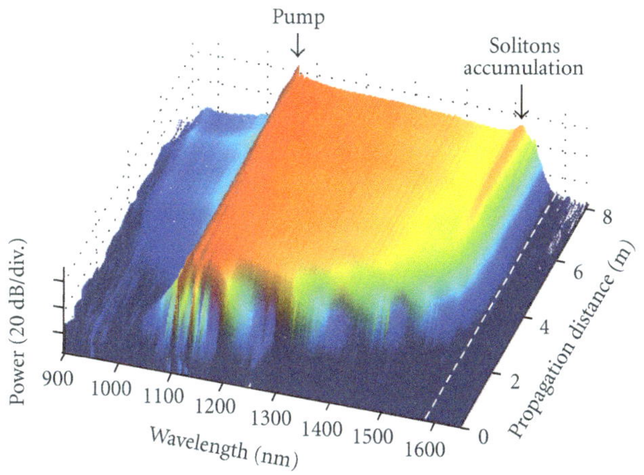

Figure 8: Experimental dynamics of the SC formation as a function of fiber length in the solid-core PBG fiber 2 for a pump power of 2.3 kW. The white dashed line represents the PBG edge.

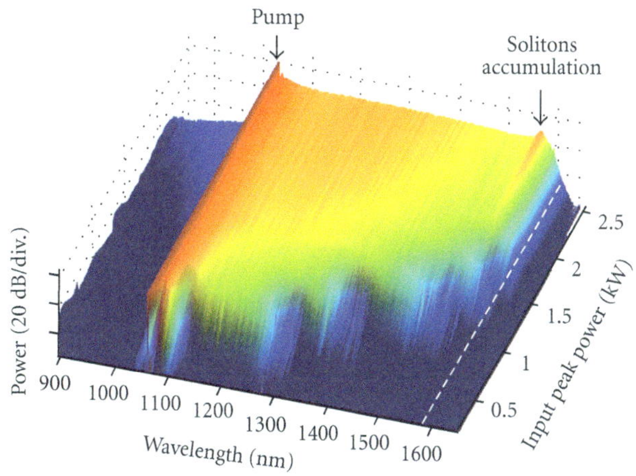

Figure 9: Experimental dynamics of the SC formation as a function of launched pump peak power. The white dashed line represents the PBG edge.

4. Tailoring the SC Extension

The SSFS is one of the main mechanisms involved in the generation of octave-spanning SC [26]. Indeed, in both the short- and long-pulse pumping regimes [27], the long-wavelength part of the SC spectrum is composed of red-shifting fundamental solitons initially created, respectively, by pulse fission and modulation instability (MI) [26]. Given that the spectral extent in the visible is intimately linked to the infrared one through a group index matching between solitons and trapped dispersive waves [28, 29], the SSFS dynamics lies at the heart of the SC generation process. There have been many attempts to optimize the spectral extension of SC sources, but, in most applications, only a small part of the SC spectrum is useful. There is thus an increasing interest in tailoring the spectral extent so that SC generation only occurs over the required spectral region. Fibers with two ZDWs have been employed in this way [30, 31]. Because the region of anomalous GVD can be suitably adjusted through a control of the microstructured cladding parameters, the SSFS can be cancelled near the long-wavelength ZDW, which results in a stop of the infrared spectral broadening. However, the SSFS cancellation is accompanied by the emission of a dispersive wave at higher wavelengths, so that the energy of frequency locked solitons exponentially decreases. In this section, we show that the cancellation of the SSFS observed in solid-core PBG fibers described above allows to control the SC spectral extent without any significant loss of energy.

4.1. Dynamics of the SC Formation. SC experiments were performed using the solid-core PBG fiber 2 of Table 1, in which long-wavelength PBG edge is located at 1580 nm. The pump laser used to generate the SC was a linearly polarized microchip laser emitting 0.6 ns pulses at 1064 nm with a repetition rate of 7 kHz. The coupling efficiency in the fibre core is estimated to be 50%. In the following, input power always refers to the power effectively launched into the fibre core.

At first, we investigated the dynamics of the SC formation with propagation distance by cutting the fiber back from 8 to 0.1 m and recording the output spectrum every 0.5 m, for a fixed pump power of 2.3 kW. Figure 8 shows the evolution of the output spectrum as a function of propagation distance. The spectrum initially presents two symmetric side lobes characteristic of MI classically observed when pumping in the anomalous GVD regime. In the temporal domain, this corresponds to the generation of a solitonic train, in which pulses strongly differ from each other in terms of peak power and duration. The spectrum then mainly broadens towards the infrared (because we pump relatively far away from the ZDW in the anomalous dispersion regime) until a fiber length of about 4 m is reached. At this propagation distance, the SC ranges from 1050 to 1580 nm. The infrared spectral broadening is due to the SSFS of each individual soliton generated from MI. Since they are characterized by very different properties, their SSFS efficiency can be very different, which leads to the SC. For further increasing fiber lengths to 8 m, the long-wavelength spectral broadening completely stops, due to the SSFS cancellation of each soliton approaching the PBG edge, following the mechanism described in Section 3. This leads to an accumulation of solitons just below the PBG edge (located at 1580 nm), which can be seen as a spectral power peak centered around 1550 nm. Consequently, the SC long-wavelength spectral extent can be tailored over a given spectral range through a suitable design of the fiber linear properties and, in particular, of the long-wavelength PBG edge.

To further illustrate this, we performed an additional experiment in which the dynamics of the SC formation was investigated as a function of pump peak power, for a fixed fiber length of 10 m. To do that, the output spectrum was recorded with 0.2 kW increments for pump peak powers ranging from 0.2 to 2.6 kW. The result of these measurements is displayed in Figure 9. Similar to previous experiments of Figure 8, the spectrum mainly broadens towards the infrared when the pump peak power is increased from 0 up to about

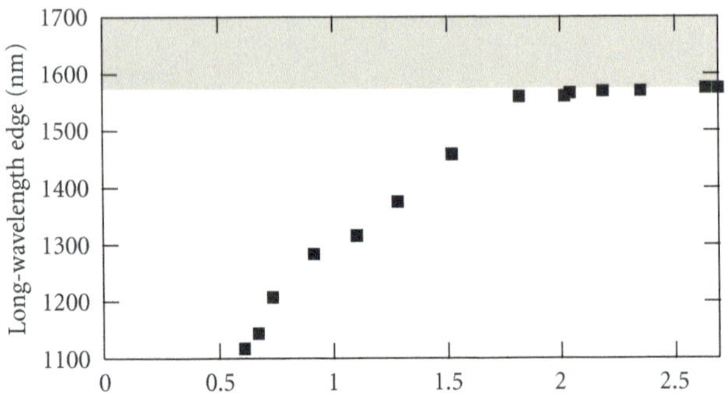

(a)

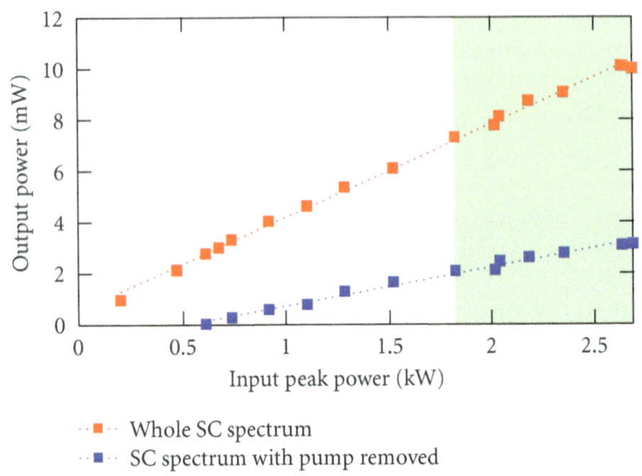

(b)

FIGURE 10: (a) Measurement of the long-wavelength edge of the SC spectrum as a function of input peak power in a 10 m-long solid-core PBG fiber sample. The gray area depicts the region outside the first PBG transmission window. (b) Measured output average power over the whole SC spectrum (red squares) and by removing the output pump peak with a 1064 nm notch filter (blue squares), as a function of input peak power. Dotted lines are linear fits. The green area corresponds to input peak powers for which the SC spectral extent stops (i.e., for input peak powers greater than 1.8 kW).

1.7 kW. For higher pump powers, the spectral broadening suddenly stops at the long-wavelength PBG edge. This is even more clearly illustrated in Figure 10(a) which represents the long-wavelength SC edge (measured for a 10 dB drop with respect to the spectral power at 1100 nm) as a function of input peak power. A clear cancellation of the spectral broadening at the PBG edge (depicted in gray) is observed for peak powers higher than 1.7 kW. As discussed above, this effect is accompanied by an accumulation of solitons which can be seen as a spectral power peak at 1550 nm, that is, just below the PBG edge located at 1580 nm.

4.2. Discussion. As shown by Figure 4 and discussed in Section 3, the SSFS cancellation occurs without any reduction of the spectral power density. This thus suggests, in the frame of SC generation, that the SC spectrum width can be limited by the PBG effect without any significant power loss. In order to check this experimentally, we plotted in Figure 10(b) the output average power measured over the whole SC spectrum (red squares) as a function of the input peak power. These measurements can be linearly fitted (red line), and the slope does not change once the SSFS cancellation has occurred and the SC spectral broadening has stopped (corresponding to the green region). In order to check that this was not due to the residual pump peak in the SC, we performed the same kind of measurements after filtering it with a notch filter at 1064 nm at the fiber output. The results are displayed with blue markers in Figure 10 and show that the output SC power (without the contribution of the 1064 nm pump) still increases with the input peak power. This confirms that the tailoring of the SC extent is not done through a simple long-wavelength filtering due to the PBG effect, which would lead to a significant output power reduction. In our case, the fact that solitons are frequency locked rather than attenuated near the PBG edge allows to keep most of the launched power into the SC and thus to optimize the pump power budget.

5. Reduction of Pulse-to-Pulse Power Fluctuations at the SC Long-Wavelength Edge

In this section, we focus our attention on pulse-to-pulse fluctuations of the SC spectral power near the long-wavelength edge of the spectrum. Indeed, in the long-pulse regime, the spectral broadening originates from the MI process which is seeded by noise and the long-wavelength SC edge is characterized by statistically rare optical rogue waves [32], originating from soliton dynamics [32–35]. Consequently, the SC is characterized by a low coherence, and pulse-to-pulse fluctuations of the spectral power at the long-wavelength SC edge are very important [32, 36]. Here, we investigate the impact of the SSFS cancellation observed in solid-core PBG fibers on pulse-to-pulse fluctuations of the long-wavelength SC edge.

5.1. Experiments. Experiments were performed using the solid-core PBG fiber 3 of Table 1, in which long-wavelength PBG edge is located at 1590 nm. Following the results of Section 4, a SC was generated using a 7 m long fiber sample and a pump power of 3 kW. Figure 11(a) shows a measurement of the spectral broadening as a function of propagation distance (made using a cutback technique). As expected, the infrared spectral broadening stops near the PBG edge, depicted with the dashed white line. Figures 11(b1) and 11(b2) shows the spectra obtained at 7 m and 2.5 m, respectively. Although the spectral extent is comparable in both cases, the shape of the spectrum is different. Indeed, while the long-wavelength edge is relatively smooth for a fiber length of 2.5 m, it becomes much sharper at 7 m, where the cancellation of the SSFS imposes a saturation of the spectral broadening. This phenomenon causes a soliton accumulation whose spectral signature is the peak located around 1560 nm, as discussed in the previous section.

Pulse-to-pulse power fluctuations of the corresponding SC obtained at 2.5 and 7 m were then studied experimentally.

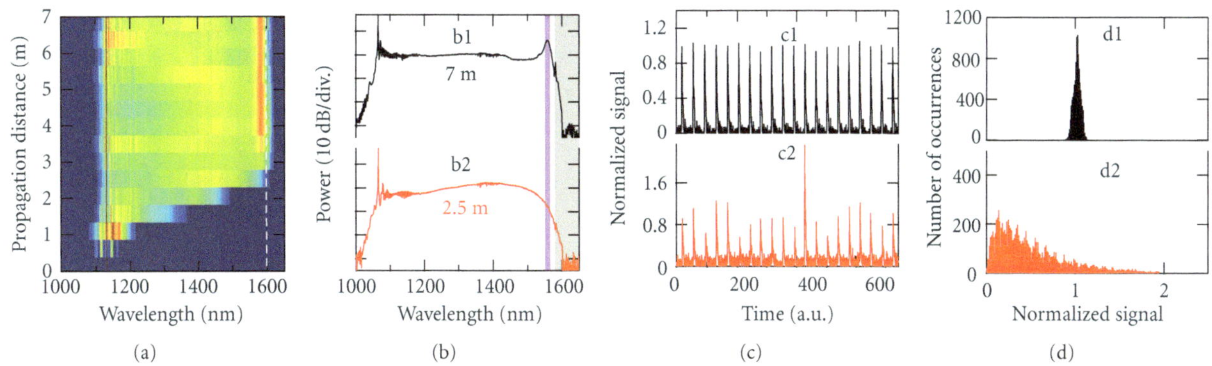

FIGURE 11: (a) Measurement of the spectral broadening as a function of propagation distance in the solid-core PBG fiber 3. The white dashed line represents the PBG edge. (b) Measured spectra for a fiber length of 7 m (b1) and 2.5 m (b2). (c1 and c2) Samples of corresponding pulse train recorded after a 10 nm bandpass spectral filtering at 1550 nm (depicted by the purple region in (b)). (d1 and d2) Corresponding statistical distribution of the signal amplitude over 10,000 records. The signal amplitude is normalized to the average over the 10,000 pulses.

To do this, the SC output was collimated and spectrally filtered with a 10 nm bandpass filter centered at 1550 nm, that is, on the soliton accumulation peak. Pulse-to-pulse fluctuations in this 10 nm wide spectral regions were measured using the method firstly proposed in [37]. The energy of spectrally filtered pulses is proportional to their peak power, so that rogue events and their associated characteristics can be captured through a simplified measurement of shot-to-shot pulse energy [37]. This was done in our experiments using a photodiode and an analog oscilloscope, with the same photodiode maximum signal amplitude and trigger level for both fiber lengths. Note that, in our experiments, this setup provides a measurement of the average energy of each pulse, but it does not allow to quantify noise-sensitive instabilities related to the actual temporal coherence of the SC, as explained in [38] for instance. The percentage of shot-to-shot variations, σ, was evaluated using $\sigma = 100 \times (V_{\max} - V_{\min})/(V_{\max} + V_{\min})$ where $V_{\max}$ and $V_{\min}$ are, respectively, the maximum and minimum photodiode signal amplitude measured at least for 10 of the 10,000 recorded pulses.

Figures 11(c1) and 11(c2) show samples of pulse trains measured for each investigated fiber length after the 10 nm bandpass filter at 1550 nm, from which pulse-to-pulse fluctuations σ can be evaluated. The pulse amplitude has been normalized to its average value over the 10,000 measurements for this plot. Pulse-to-pulse fluctuations calculated at 1550 nm are as high as 80% after 2.5 m while they drop down to 10% after 7 m. Besides, the corresponding statistical distributions at 1550 nm have very different shapes, as attested by Figures 11(d1) and 11(d2) which represent histograms calculated over 10,000 measurements for fiber lengths of 2.5 m and 7 m, respectively. The distribution is highly asymmetric after 2.5 m but becomes Gaussian-like at 7 m, that is, once solitons have reached the PBG edge. These observations thus suggest that there is a correlation between the cancellation of the SSFS occurring near the PBG edge (leading to a saturation of the SC spectrum for propagation distances longer than about 3 m), and the enhanced pulse-to-pulse stability and statistical distributions in the same spectral region.

5.2. Confirmation with Simulations. In order to get insight into these results, we performed numerical simulations using the GNLSE given by (1). At first, we simply attempted to reproduce experimental results of Figure 11. To do this, all available fiber characteristics and experimental conditions have been taken into account with no free parameter, with the exception of the pulse duration that has been reduced to 50 ps (against 600 ps in experiments). Indeed, shortening the pulse duration in simulations (while staying in the long-pulse regime though) allows to significantly reduce the computation time without significantly affecting the SC dynamics. The input pump peak power was 3 kW, and quantum noise was modeled by adding one photon per mode with random phase on each spectral discretization bin of the input field [26]. In order to study the statistical behavior of the pulse train, we performed 200 simulations with random initial noise conditions.

Simulations results are displayed in Figure 12, with the same organization as Figure 11 for easy comparison. Figures 12(a), 12(b1), and 12(b2), which show the dynamics of the SC formation, correspond to the averaged spectra over 200 simulations shots. The agreement with experiments is excellent, and the typical spectral features discussed above about the spectrum shape at its long-wavelength edge are accurately reproduced. The simulated temporal properties shown in Figures 12(c1), 12(c2), 12(d1), and 12(d2) have been calculated using a numerical gaussian filter of 10 nm (FWHM) centered at 1562 nm, which corresponds to the maximum of the spectral power peak observed near the PBG edge (see plot 12(b1)). Taking the inverse Fourier transform gives the temporal profile of the SC filtered at the corresponding wavelength, whose average power has been calculated. Samples of the modeled pulse trains corresponding to 20 simulation shots are displayed in Figures 12(c1) and 12(c2), for, respectively, fiber lengths of 7 and 2.5 m. Corresponding statistical distributions calculated from all the 200 simulations are shown in Figures 12(d1) and 12(d2) and also show excellent agreement with experiments.

All these simulation results therefore allow to reproduce not only the main spectral features observed in the

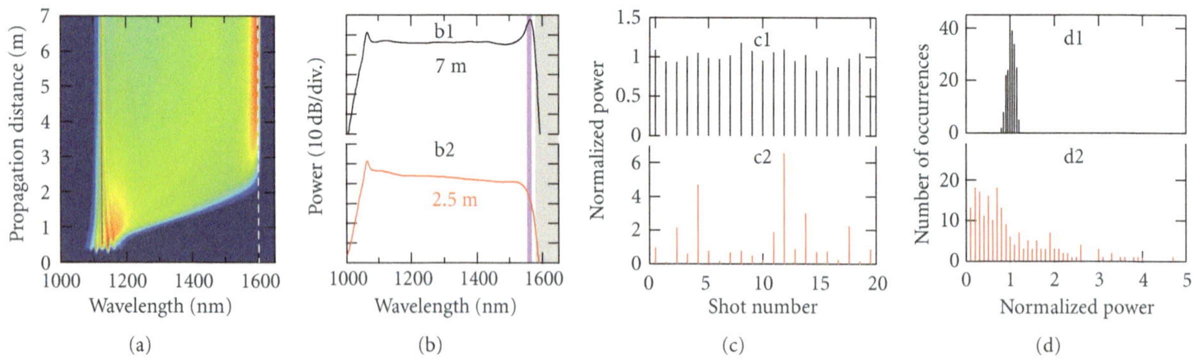

FIGURE 12: (a) Simulation of the spectral broadening as a function of propagation distance in the solid-core PBG fiber 3 (averaged over 200 simulation shots). The white dashed line represents the PBG edge. (b) Simulated averaged spectra for fiber lengths of 7 m (b1) and 2.5 m (b2). (c1 and c2) Samples of corresponding pulse train after spectral filtering at 1562 nm (depicted by the purple region in (b)). (d1 and d2) Corresponding histograms over 200 pulses at 1562 nm, where the power is normalized to its average over the 200 shots.

experiments (dynamics of the SC formation, typical spectral shape at the long-wavelength edge), but also the measured statistical features of the pulse train (pulse-to-pulse fluctuations as a function of wavelength and propagation distance, shape of the distribution). At this point, however, numerical simulations do not allow to discuss on the physical mechanisms responsible for the enhanced pulse-to-pulse stability near the PBG edge.

5.3. Discussion. Numerical simulations using the GNLSE accurately reproduce our experiments and can consequently be used to discuss the physical mechanisms causing the enhanced stability of the SC long-wavelength edge. To this end, the spectrotemporal representation has proved to be a powerful tool in studying the dynamics of SC generation [26]. Figures 13(a)–13(d) show the evolution of a simulated spectrogram for fiber lengths of 1, 2.5, 4, and 7 m.

Figure 13(a) shows how MI initially generates solitons that subsequently red-shifts through Raman-induced SSFS as can be seen from Figure 13(b). In such a long-pulse pumping regime, the presence of statistically rare temporal events in SC experiments can find two complementary physical origins. Firstly, it can be due to a single high peak power soliton generated from MI for particular initial noise conditions [32, 33]. Because of its higher peak power, it experiences a more efficient SSFS than other solitons and, consequently, becomes statistically rare at highest wavelengths due to the low probability for these particular noise conditions to happen [32]. Note that, in this case, the requirements for a soliton to be statistically rare at the long-wavelength SC edge, are only to experience a slightly more efficient SSFS than the other ones [39]. Secondly, following these early interpretations, it has been suggested that rare and brief events can arise from the collision of two or more solitons travelling with different group velocities [34, 35, 39, 40] because of the convective nature of the system [41]. Note that these two explanations are complementary for explaining long-tail statistical distributions usually observed at the SC long-wavelength edge [41].

In the case of the present experiments, however, this dynamics is strongly affected by the SSFS cancellation occurring near the PBG edge, as can be seen from Figures 13(c) and 13(d). Firstly, these figures show that, as long as solitons are frequency locked around 1562 nm, they all have very close group velocities at this location in the spectrum. Consequently, once they have reached the PBG edge, they cannot collide anymore, which prevents the formation of brief spikes [34] and leads to an enhanced pulse-to-pulse stability at the SC edge. Secondly, since all solitons whose SSFS has been cancelled near the PBG edge have very close characteristics, power fluctuations are reduced in this spectral region as compared to the usual case in which only a few solitons with slightly higher peak power are rare at the long-wavelength edge [32, 33].

As a consequence, the reduction of pulse-to-pulse fluctuations at the long-wavelength edge as a function of propagation distance is intimately linked to the cancellation of the SFSS discussed in Section 3. It is thus in fact due to the specific linear properties inherent to solid-core PBG fibers and, in particular, to the strong third-order dispersion near the PBG edge, as shown in Figure 3.

6. Conclusion

In conclusion, we showed that the SSFS and its dynamics can be strongly and easily manipulated thanks to the specific properties of the solid-core PBG fibers used in our various experimental and numerical studies. Indeed, we demonstrated that a quasicancellation of the SSFS could be obtained in this kind of fibers thanks to the large increase of β_2 observed on the long-wavelength edge of the first PBG. We established, trough experimental and numerical studies, that this phenomenon could be achieved without significant power loss. Thanks to these results, we demonstrated and clearly interpreted that these fibers could be used to tailor efficiently the spectral extension of SC at no power cost. Moreover, we revealed and explained that the long-wavelength edge of SC generated in this kind of fibers can present strong quantitative and qualitative differences

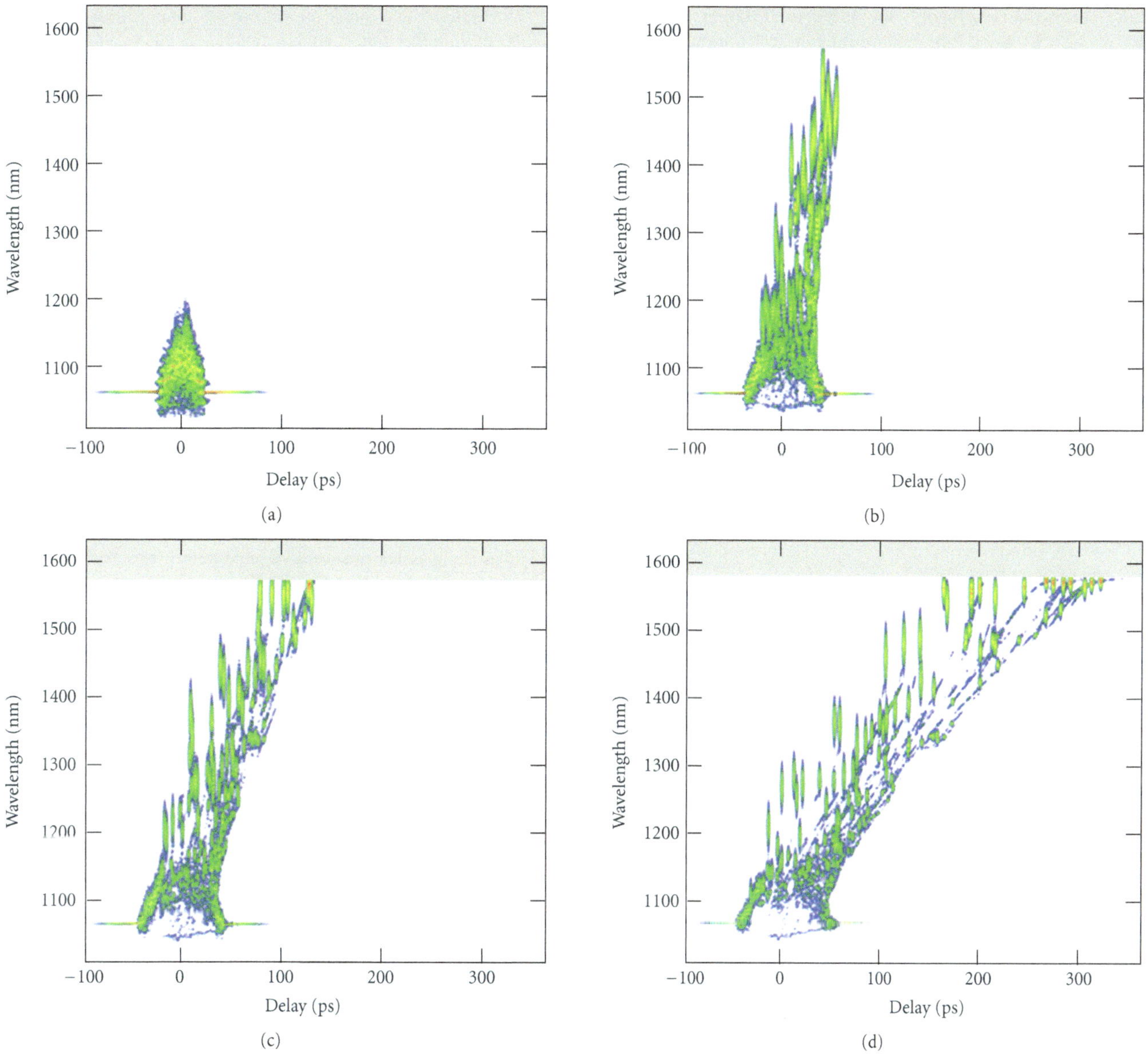

FIGURE 13: Simulated spectrograms representing the spectrotemporal dynamics of the SC generation for four propagation distances z of 1, 2.5, 4, and 7 m, respectively, from (a) to (d).

in terms of temporal stability compared to SC generated in more conventional fibres: the power fluctuations in this spectral domain are indeed strongly reduced and have a Gaussian-like statistical distribution and not the more usual L-shaped statistics associated with the presence of optical rogue waves generally observed at the long-wavelength edge of more conventional SC. We believe these results to be helpful for building SC sources for the next generation of time-resolved coherent antistokes Raman scattering (CARS) microscopes for instance [42].

Acknowledgments

The authors acknowledge Rémi Habert for experimental assistance and Karen Delplace for assistance in fiber fabrication. This work was partly supported by the French Ministry of Higher Education and Research, the Nord-Pas de Calais Regional Council, and FEDER through the "Contrat de Projets Etat Région (CPER) 2007-2013" and the "Campus Intelligence Ambiante" (CIA).

References

[1] F. Luan, A. K. George, T. D. Hedley et al., "All-solid photonic bandgap fiber," *Optics Letters*, vol. 29, no. 20, pp. 2369–2371, 2004.

[2] A. Argyros, T. A. Birks, S. G. Leon-Saval, C. M. B. Cordeiro, and P. S. J. Russell, "Guidance properties of low-contrast photonic bandgap fibres," *Optics Express*, vol. 13, no. 7, pp. 2503–2511, 2005.

[3] V. Pureur, A. Betourne, G. Bouwmans et al., "Overview on solid core photonic bandgap fibers," *Fiber and Integrated Optics*, vol. 28, no. 1, pp. 27–50, 2009.

[4] G. Bouwmans, L. Bigot, Y. Quiquempois, F. Lopez, L. Provino, and M. Douay, "Fabrication and characterization of an all-solid 2D photonic bandgap fiber with a low-loss region (< 20 dB/km) around 1550 nm," *Optics Express*, vol. 13, pp. 8452–8459, 2005.

[5] A. Fuerbach, P. Steinvurzel, J. A. Bolger, A. Nulsen, and B. J. Eggleton, "Nonlinear propagation effects in antiresonant high-index inclusion photonic crystal fibers," *Optics Letters*, vol. 30, no. 8, pp. 830–832, 2005.

[6] B. Kibler, T. Martynkien, M. Szpulak et al., "Nonlinear femtosecond pulse propagation in an all-solid photonic bandgap fiber," *Optics Express*, vol. 17, no. 12, pp. 10393–10398, 2009.

[7] A. S. Cerqueira Jr., C. M. B. Cordeiro, F. Biancalana et al., "Nonlinear interaction between two different photonic bandgaps of a hybrid photonic crystal fiber," *Optics Letters*, vol. 33, no. 18, pp. 2080–2082, 2008.

[8] P. D. Rasmussen, J. Laegsgaard, and O. Bang, "Degenerate four wave mixing in solid core photonic bandgap fibers," *Optics Express*, vol. 16, no. 6, pp. 4059–4068, 2008.

[9] A. Bétourné, Y. Quiquempois, G. Bouwmans, and M. Douay, "Design of a photonic crystal fiber for phase-matched frequency doubling or tripling," *Optics Express*, vol. 16, no. 18, pp. 14255–14262, 2008.

[10] V. Pureur and J. M. Dudley, "Nonlinear spectral broadening of femtosecond pulses in solid-core photonic bandgap fibers," *Optics Letters*, vol. 35, no. 16, pp. 2813–2815, 2010.

[11] O. Vanvincq, A. Kudlinski, A. Bétourné, Y. Quiquempois, and G. Bouwmans, "Extreme deceleration of the soliton self-frequency shift by the third-order dispersion in solid-core photonic bandgap fibers," *Journal of the Optical Society of America B*, vol. 27, no. 11, pp. 2328–2335, 2010.

[12] A. Bétourné, A. Kudlinski, G. Bouwmans, O. Vanvincq, A. Mussot, and Y. Quiquempois, "Control of supercontinuum generation and soliton self-frequency shift in solid-core photonic bandgap fibers," *Optics Letters*, vol. 34, no. 20, pp. 3083–3085, 2009.

[13] O. Vanvincq, B. Barviau, A. Mussot, G. Bouwmans, Y. Quiquempois, and A. Kudlinski, "Significant reduction of power fluctuations at the long-wavelength edge of a supercontinuum generated in solid-core photonic bandgap fibers," *Optics Express*, vol. 18, no. 23, pp. 24352–24360, 2010.

[14] A. Bétourné, G. Bouwmans, Y. Quiquempois, M. Perrin, and M. Douay, "Improvements of solid-core photonic bandgap fibers by means of interstitial air holes," *Optics Letters*, vol. 32, no. 12, pp. 1719–1721, 2007.

[15] M. Perrin, Y. Quiquempois, G. Bouwmans, and M. Douay, "Coexistence of total internal reflexion and bandgap modes in solid core photonic bandgap fibre with intersticial air holes," *Optics Express*, vol. 15, no. 21, pp. 13783–13795, 2007.

[16] M. Tateda, N. Shibata, and S. Seikai, "Interferometric method for chromatic dispersion measurement in a single-mode optical fiber," *IEEE Journal of Quantum Electronics*, vol. 17, no. 3, pp. 404–407, 1981.

[17] G. P. Agrawal, *Nonlinear Fiber Optics*, Academic Press, New York, NY, USA, 4th edition, 2007.

[18] E. M. Dianov, A. Y. Karasik, P. V. Mamyshev et al., "Stimulated-Raman conversion of multisoliton pulses in quartz optical fibers," *JETP Letters*, vol. 41, pp. 294–297, 1985.

[19] Y. Kodama and A. Hasegawa, "Nonlinear pulse propagation in a monomode dielectric guide," *IEEE Journal of Quantum Electronics*, vol. 23, no. 5, pp. 510–524, 1987.

[20] P. Beaud, W. Hodel, B. Zysset, and H. P. Weber, "Ultrashort pulse propagation, pulse breakup and fundamental soliton formation in a single-mode optical fiber," *IEEE Journal of Quantum Electronics*, vol. 23, no. 11, pp. 1938–1946, 1987.

[21] F. M. Mitschke and L. F. Mollenauer, "Discovery of the soliton self-frequency shift," *Optics Letters*, vol. 11, no. 10, pp. 659–661, 1986.

[22] J. P. Gordon, "Theory of the soliton self-frequency shift," *Optics Letters*, vol. 11, no. 10, pp. 662–664, 1986.

[23] J. Laegsgaard, "Mode profile dispersion in the generalized nonlinear Schrödinger equation," *Optics Express*, vol. 15, no. 24, pp. 16110–16123, 2007.

[24] J. C. Travers, M. H. Frosz, and J. M. Dudley, "Nonlinear fibre optics overview," in *Supercontinuum Generation in Optical Fibers*, J. M. Dudley and J. R. Taylor, Eds., chapter 3, Cambridge University Press, Cambridge, UK, 2010.

[25] D. Hollenbeck and C. D. Cantrell, "Multiple-vibrational-mode model for fiber-optic Raman gain spectrum and response function," *Journal of the Optical Society of America B*, vol. 19, no. 12, pp. 2886–2892, 2002.

[26] J. M. Dudley, G. Genty, and S. Coen, "Supercontinuum generation in photonic crystal fiber," *Reviews of Modern Physics*, vol. 78, no. 4, pp. 1135–1184, 2006.

[27] The long-pulse pumping regime refers to cases in which the pump pulse duration ΔT is much longer than the MI oscillation period ΔT_{MI}, given by $\Delta T_{\text{MI}} = 2\pi\sqrt{|\beta_2|/(2\gamma P)}$, with β_2 the second-order dispersion coefficient, γ the NL coefficient and P the pump peak power. Cases in which ΔT is in the order or less than ΔT_{MI} correspond to the short-pulse pumping regime.

[28] A. V. Gorbach and D. V. Skryabin, "Light trapping in gravity-like potentials and expansion of supercontinuum spectra in photonic-crystal fibres," *Nature Photonics*, vol. 1, no. 11, pp. 653–657, 2007.

[29] J. M. Stone and J. C. Knight, "Visibly "white" light generation in uniform photonic crystal fiber using a microchip laser," *Optics Express*, vol. 16, no. 4, pp. 2670–2675, 2008.

[30] A. Kudlinski, G. Bouwmans, Y. Quiquempois, and A. Mussot, "Experimental demonstration of multiwatt continuous-wave supercontinuum tailoring in photonic crystal fibers," *Applied Physics Letters*, vol. 92, no. 14, Article ID 141103, 2008.

[31] A. Kudlinski, G. Bouwmans, M. Douay, M. Taki, and A. Mussot, "Dispersion-engineered photonic crystal fibers for CW-pumped supercontinuum sources," *Journal of Lightwave Technology*, vol. 27, no. 11, pp. 1556–1564, 2009.

[32] D. R. Solli, C. Ropers, P. Koonath, and B. Jalali, "Optical rogue waves," *Nature*, vol. 450, no. 7172, pp. 1054–1057, 2007.

[33] J. M. Dudley, G. Genty, and B. J. Eggleton, "Harnessing and control of optical rogue waves in supercontinuum generation," *Optics Express*, vol. 16, no. 6, pp. 3644–3651, 2008.

[34] A. Mussot, A. Kudlinski, M. Kolobov, E. Louvergneaux, M. Douay, and M. Taki, "Observation of extreme temporal events in CW-pumped supercontinuum," *Optics Express*, vol. 17, no. 19, pp. 17010–17015, 2009.

[35] G. Genty, C. M. de Sterke, O. Bang, F. Dias, N. Akhmediev, and J. M. Dudley, "Collisions and turbulence in optical rogue wave formation," *Physics Letters, Section A*, vol. 374, no. 7, pp. 989–996, 2010.

[36] F. Vanholsbeeck, S. Martin-Lopez, M. González-Herráez, and S. Coen, "The role of pump incoherence in continuous-wave supercontinuum generation," *Optics Express*, vol. 13, no. 17, pp. 6615–6625, 2005.

[37] C. Lafargue, J. Bolger, G. Genty, F. Dias, J. M. Dudley, and B. J. Eggleton, "Direct detection of optical rogue wave energy

statistics in supercontinuum generation," *Electronics Letters*, vol. 45, no. 4, pp. 217–219, 2009.

[38] H. Kubota, K. R. Tamura, and M. Nakazawa, "Analyses of coherence-maintained ultrashort optical pulse trains and supercontinuum generation in the presence of soliton-amplified spontaneous-emission interaction," *Journal of the Optical Society of America B*, vol. 16, no. 12, pp. 2223–2232, 1999.

[39] M. Erkintalo, G. Genty, and J. M. Dudley, "On the statistical interpretation of optical rogue waves," *European Physical Journal*, vol. 185, no. 1, pp. 135–144, 2010.

[40] K. Hammani, B. Kibler, C. Finot, and A. Picozzi, "Emergence of rogue waves from optical turbulence," *Physics Letters, Section A*, vol. 374, no. 34, pp. 3585–3589, 2010.

[41] M. Taki, A. Mussot, A. Kudlinski, E. Louvergneaux, M. Kolobov, and M. Douay, "Third-order dispersion for generating optical rogue solitons," *Physics Letters, Section A*, vol. 374, no. 4, pp. 691–695, 2010.

[42] D. A. Sidorov-Biryukov, E. E. Serebryannikov, and A. M. Zheltikov, "Time-resolved coherent anti-Stokes Raman scattering with a femtosecond soliton output of a photonic-crystal fiber," *Optics Letters*, vol. 31, no. 15, pp. 2323–2325, 2006.

8

Nonlinear Pulse Shaping in Fibres for Pulse Generation and Optical Processing

Sonia Boscolo[1] and Christophe Finot[2]

[1] *Photonics Research Group, School of Engineering and Applied Science, Aston University, Birmingham B4 7ET, UK*
[2] *Laboratoire Interdisciplinaire Carnot de Bourgogne, UMR 6303 CNRS, Université de Bourgogne, 21078 Dijon, France*

Correspondence should be addressed to Sonia Boscolo, s.a.boscolo@aston.ac.uk

Academic Editor: Juan Diego Ania Castañón

The development of new all-optical technologies for data processing and signal manipulation is a field of growing importance with a strong potential for numerous applications in diverse areas of modern science. Nonlinear phenomena occurring in optical fibres have many attractive features and great, but not yet fully explored, potential in signal processing. Here, we review recent progress on the use of fibre nonlinearities for the generation and shaping of optical pulses and on the applications of advanced pulse shapes in all-optical signal processing. Amongst other topics, we will discuss ultrahigh repetition rate pulse sources, the generation of parabolic shaped pulses in active and passive fibres, the generation of pulses with triangular temporal profiles, and coherent supercontinuum sources. The signal processing applications will span optical regeneration, linear distortion compensation, optical decision at the receiver in optical communication systems, spectral and temporal signal doubling, and frequency conversion.

1. Introduction

The use of photonic technologies for data processing in the all-optical domain has a strong potential for a variety of interesting applications in such diverse areas as optical telecommunications, metrology, optical sensing, microwave engineering, advanced microscopy image processing, optical computing, and many others. Advantages of processing the information in the all-optical domain include the large available bandwidth and the (potential) parallelism intrinsic to the optical approach, which translate into high-processing speeds. However, today electronic techniques of signal manipulation are advanced compared to all-optical processing devices, which are still at the research stage rather than under commercial development. A key feature of electronics, enabling many different applications, is the capability of generating electrical waveforms with arbitrary temporal profiles by use of simple integrated circuits. This capability is so far unmatched in the optical frequency range, where bulky and complex devices are required to shape the light fields. Hence, considerable knowledge still has to be accumulated and new methodologies need to be explored before a true breakthrough can be achieved in this field, allowing the range of functions and operations currently accomplished electronically, to be performed in the optical domain. Furthermore, by using optical rather than electronic processing additional functionality may be possible.

In order to realize such all-optical processing, nonlinear photonics is seen as a key technology. Optical fibre materials exhibit a nonlinear response to strong electric fields, such those of optical signals confined within the small fibre core. In high-speed optical communications such nonlinear effects generally degrade the integrity of the transmitted signal but the same effects can be used to realize a variety of optical functions that have practical applications in the field of lightwave technology. In this paper, we provide a snapshot of recent results and advances in the use of nonlinear effects in optical fibres for optical waveform generation and pulse shaping and in the applications of advanced pulse profiles in all-optical signal processing. We would like to note that the main attention in this paper will be focused on results obtained in our groups, and that it is not our intention here to comprehensively cover all the possible examples of fibre-based pulse shaping and signal processing.

2. Fibre Nonlinearities

The third-order $\chi^{(3)}$ optical nonlinearity in silica-based single-mode (SM) fibres is one of the most important effects that can be used for all-optical signal processing. This happens not only because the third-order nonlinearity provides ultra-fast response times in the femtosecond range, but also because it is responsible for a wide range of phenomena such as third-harmonic generation, nonlinear refraction (Kerr nonlinearity), and stimulated Raman and Brillouin scattering [1]. The rapid recent developments of microstructured fibres with extremely small effective core areas and exhibiting enhanced nonlinear characteristics [2] and of fibres using materials with refractive indexes higher than that of the silica glass [3] have enabled dramatic reduction of the required fibre lengths for nonlinear processing compared to conventional fibres. Such progress has paved the way for the integration of fibre-based nonlinear processing functions on photonic chips.

The intensity dependence of the refractive index of a fibre gives rise to different effects depending on the shape of the input signal; the most widely studied effects are self-phase modulation (SPM) and cross-phase modulation (XPM). Considering the localized optical pulse evolution in a fibre medium and neglecting higher order effects in the fibre, pulse propagation can be described by the nonlinear Schrödinger (NLS) equation [4]:

$$\psi_z = -i\frac{\beta_2}{2}\psi_{tt} + i\gamma|\psi|^2\psi, \quad (1)$$

where $\psi(z,t)$ is the slowly varying pulse envelope in the comoving system of coordinates, β_2 and $\gamma = 2\pi n_2/(\lambda A_{\text{eff}})$ are the respective group-velocity (second-order) dispersion (GVD) and Kerr nonlinearity coefficients of the fibre, n_2 is the nonlinear-index coefficient, λ the central wavelength of the pulse, and A_{eff} the effective core area. As a result of the nonlinear term in (1), upon propagation in the fibre the pulse acquires an intensity-dependent nonlinear phase shift as $\phi(z,t) = \gamma|\psi(0,t)|^2 z$, namely, the frequency chirp $-\phi_t$. When the length scale associated with the pulse where the GVD effects take place: $L_{\text{D}} = T_0^2/|\beta_2|$ (with T_0 being some temporal characteristic value of the initial pulse) is much larger compared with both the fibre length and the nonlinear length $L_{\text{NL}} = 1/(\gamma P_0)$ (with P_0 being the peak power of the initial pulse), the dispersive term in (1) can be neglected. In this case, the presence of a chirp causes a nonlinear broadening of the pulse spectrum. When two (or more) pulses copropagate inside a fibre, one still obtains an NLS equation for each pulse but these equations are coupled through XPM [4]. In this case, the nonlinear phase evolution for each pulse depends also on the power of the other pulses according to the expression $\phi_i(z,t) = \gamma_i(|\psi_i(0,t)|^2 + 2\sum_{j\neq i}|\psi_j(0,t)|^2)z$. XPM occurs only when the pulses overlap. Similar to the SPM case, XPM leads to additional nonlinear chirping and spectral shaping.

When the effects of chromatic dispersion are considered in combination with the Kerr nonlinearity, rich pulse dynamics arise from the interplay between dispersive and nonlinear effects depending on the sign of the dispersion and the relative magnitudes of the associated length scales. A well-known and fascinating example is the formation of optical solitons [5, 6] in the anomalous dispersion regime of a fibre ($\beta_2 < 0$) as a result of a cooperation between GVD and SPM. On the contrary, in the regime of normal dispersion pulse dynamics are highly affected by the phenomenon of optical wave breaking [7–9].

In the next Section we will discuss some newly emerged qualitative features of the optical pulse evolution in a fibre under the combined action of dispersion and nonlinearity.

3. Optical Pulse Generation and Shaping Using Fibre Nonlinearities

Techniques for generating, controlling, manipulating, and measuring ultrashort optical pulses and specialized waveforms have become increasingly important in many scientific areas including, amongst others, ultrahigh-speed optical communications [10], optical signal processing, and biophotonics. To date, conventional picosecond and femtosecond pulse shaping in the optical domain have been implemented using devices, such as liquid crystal spatial light modulators [11, 12], wavelength-to-time mapping [13, 14], acoustooptic modulators, and electrooptical phase arrays, that impart user-specified spectral amplitude and phase to the pulse. Though powerful and flexible, this approach has the drawback of requiring a rather bulky and expensive apparatus, entailing a level of cost and complexity that is not often commensurate to the used laser system and target application. In order to fulfill telecommunication requirements, more compact techniques for pulse shaping have been developed, which include the use of super-structured fibre Bragg gratings [15], long-period fibre grating filters [16], and arrayed waveguide gratings that are well suited for line by line processing of high-repetition periodic pulse shapes [17, 18]. However, in all these pulse shapers using linear techniques, the bandwidth of the input pulse determines the maximum bandwidth of the output. Indeed, a linear manipulation cannot increase the pulse bandwidth, and so to create shorter pulses nonlinear effects must be used. This limitation of the linear approaches becomes very stringent when synthesizing advanced pulse shapes such as rectangular pulses, for which a large optical bandwidth is required to reproduce accurately the spectral side lobes linked to the temporal compactness of the pulse. Moreover, in those applications that require spectral narrowing, linear filtering introduces a power penalty that is at least proportional to the ratio of the target spectrum to the input one.

The combination of third-order nonlinear processes and chromatic dispersion in optical fibres can provide efficient new solutions to overcome the aforementioned limitations. The pulse-shaping examples discussed below highlight some of the advantages offered by a passive, all-optical nonlinear fibre-based approach to pulse shaping.

3.1. Ultrahigh Repetition Rate Pulse Train Generation. As a first illustration of the possible nonlinear pulse shaping mechanisms in optical fibres, we review here the generation of ultrashort pulse trains with very high repetition rates resulting from the passive evolution of optical pulses in an anomalously dispersive fibre in the presence of Kerr nonlinearity. As mentioned in the previous Section, in the anomalous GVD regime of a fibre, the NLS equation (1) admits families of soliton solutions. While fundamental solitons are characterized by an exact balance between the GVD and SPM effects ($N = \sqrt{L_D/L_{NL}} = 1$), in the case of higher order solitons ($N > 1$) SPM dominates initially but GVD soon catches up and leads to pulse contraction. This stage of temporal compression can be used for generating pulses with ultrashort durations [19–21]. However, the compressed pulses exhibit a low-amplitude structure outside the central lobe. An alternative technique that prevents sidelobe formation relies on the progressive evolution of the peak power and temporal duration of a fundamental soliton in a distributed amplifier or a dispersion decreasing (DD) fibre [22, 23]. Step-like or comb-like [24] dispersion profiled structures based on carefully chosen sets of fibres have been successfully used to emulate continuous DD fibre profiles.

The generation of high repetition rate pulse sources has largely benefited from the nonlinear compression effect taking place in the anomalous dispersion region, and several works have demonstrated the nonlinear reshaping of a sinusoidal beat-signal into well-separated pulses, the pulse repetition rate being simply determined by the frequency separation between the two continuous-wave (CW) laser sources. This technique has been successfully demonstrated with various experimental setups including DD fibres, adiabatic Raman compression in standard fibres, and step-like and comb-like dispersion profiled fibres. More recently, this nonlinear compression effect has been observed through a multiple four-wave mixing (FWM) process in a single anomalously dispersive fibre, and repetition rates ranging from a few tens gigahertz to 1 terahertz have been achieved [25]. In the frequency domain, such nonlinear compression can be viewed as the result of a modulation instability [26] experienced by the initial beat signal during propagation in the fibre, with the generation of new sidebands via FWM: two pump photons at ω_1 and ω_2 interact to generate two first-order sideband photons at $\omega_1 - \Omega$ and $\omega_2 + \Omega$, $\Omega = \omega_2 - \omega_1$ being the pump frequency detuning. High-order sidebands are then generated due to multiple FWM interactions. We note that a powerful alternative technique for realizing high-repetition-rate short-pulse sources relies on harmonic passive mode locking of fibre lasers [27–30]. The advantage of laser cavities as such compared to nonlinear pulse shaping in one-stage fibre systems is the possibility of achieving lower pulse train duty cycles. However, laser cavities have the drawbacks that they require mode locking, they cannot be referenced to an external clock directly and are not very suitable for the generation of pulses with ultrahigh repetition rates above a hundred gigahertz.

A typical experimental setup for the generation of high repetition rate pulse trains at telecommunication wavelengths is sketched in Figure 1(a). The initial beat signal is easily obtained via temporal superposition of two CWs with slightly different wavelengths delivered by external cavity lasers or a single CW directly modulated by an intensity modulator. A phase modulator driven at a frequency of a few hundreds of megahertz is used to avoid the deleterious effects of Brillouin back scattering. The beat signal is then amplified up to the Watt level by an erbium-doped fibre amplifier (EDFA) and propagated through a suitable optical fibre with anomalous dispersion to realize the nonlinear reshaping. According to the pulse source parameters that are chosen (pulse repetition rate, initial peak power, nonlinearity and dispersion coefficients of the compression fibre, and fibre length), various output pulse shapes can be obtained. Intensity and phase measurements based on the frequency resolved optical gating (FROG) technique [32] show that high-quality compressed pulses with a Gaussian intensity profile and nearly constant phase can be achieved at a typical duty cycle of 1/5 [25, 33] (Figure 1(b)). Lower duty cycles can still be obtained with a single fibre, but at the expenses of increased sidelobes in the wings of the pulses [34]. Such sidelobes can be partly avoided by using specially designed

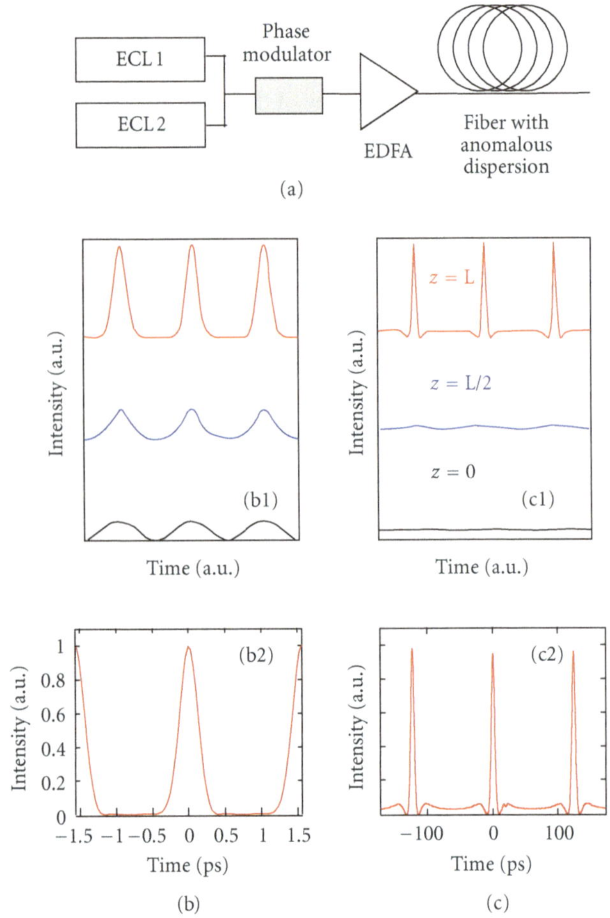

FIGURE 1: (a) Experimental setup for high repetition rate pulse train generation. (b, c) Results obtained for the optimum output shape and for the best compression, respectively. The temporal intensity profiles at different stages as predicted by numerical simulations (subplots 1) are compared to the experimental results (subplots 2) obtained by the FROG technique (b2—results adapted from [25]) or by an optical sampling oscilloscope (c2—results adapted from [31]).

arrangements of segments of fibres [35–37] or other comb-like structures.

We would like to note that the reshaping of a sinusoidal beat signal into ultrashort structures can be understood in terms of the nonlinear dynamics of "Akhmediev breathers" [38, 39]. In this context, the extreme spatiotemporal localization that can be observed (see, e.g., Figure 1(c)) is explained as a result of the formation of Peregrine solitons [31, 40], mathematical solutions of the standard NLS equation that have been first derived in 1983 in the hydrodynamics context [41] but have never been demonstrated experimentally before 2010, and using experiments in optical fibres. These observations offer a promising way to better understand the problem of the occurrence of rogue waves in optics and hydrodynamics [42, 43].

3.2. Generation of Parabolic Pulses in Active and Passive Fibres. Strong pulse reshaping can also occur in the normal regime of dispersion of a fibre. An important example that has generated a great deal of attention in the last decade is the occurrence of self-similar dynamical effects in the nonlinear propagation of ultrashort pulses in normally dispersive (ND) optical fibre amplifying media [44]. Results have demonstrated a fundamentally new operating regime where the interaction of Kerr nonlinearity, dispersion, and optical gain is exploited to generate a particular class of pulses with a parabolic intensity profile and a strictly linear frequency chirp. These pulses then propagate in a self-similar manner, holding certain relations (scaling) between pulse power, duration, and chirp parameter. The results for the self-similar parabolic pulses are obtained by assuming that a fibre amplifier can be described by the NLS equation with gain

$$\psi_z = -i\frac{\beta_2}{2}\psi_{tt} + i\gamma|\psi|^2\psi + \frac{g}{2}\psi, \tag{2}$$

where g is the distributed gain coefficient. Using a self-similar ansatz and standard amplitude/phase decomposition, one can find self-similar parabolic pulse solutions in the quasiclassical limit of (2) ($\beta_2|(|\psi|)_{tt}|/(2\gamma|\psi|^3) \ll 1$), whose dynamic evolution is summarized as [45, 46]

$$\begin{aligned} |\psi(z,t)| &= a(z)\sqrt{1 - t^2/\tau(z)^2}\,\theta(\tau(z) - |t|), \\ \arg\psi(z,t) &= b(z)t^2 + \phi_0(z), \qquad a(z) = a_0 e^{gz/3}, \\ \tau(z) &= \tau_0 e^{gz/3}, \qquad b(z) = -\frac{g}{6\beta_2}, \end{aligned} \tag{3}$$

in the case of constant gain. Here, $\theta(x)$ is the Heaviside function. The amplitude term indeed reveals a parabolic intensity profile, and the phase term a parabolic phase modulation or linear chirp with chirp rate $b/2$. The peak amplitude and width of the pulse scale from their initial values, and the initial values depend only on the initial energy and amplifier parameters. These solutions possess the remarkable property of being a global attractor to the system for arbitrary initial conditions [47, 48]. By analogy with the well-known stable dynamics of solitary waves or solitons, these self-similar parabolic pulses have come to be known as "similaritons". In contrast to solitons, similaritons can tolerate strong nonlinearity without wave breaking. The normal GVD effectively linearizes the accumulated phase of the pulse allowing for the spectral bandwidth to increase without destabilizing the pulse [49]. The unique properties of similariton pulses have motivated many theoretical and experimental studies (see, e.g., [46, 48, 50–53]).

Experimental demonstrations of similariton generation relying on amplification from either rare-earth doping (with ytterbium [45, 56–59] or erbium [51] dopants) or Raman scattering [50, 54, 60] have been achieved, confirming the potential of this method especially when dealing with the generation of high-power ultrashort pulses. Thanks to the high level of linearity of the chirp developed during parabolic amplification and as long as the impact of third-order fibre dispersion is moderate [52, 61, 62], it is possible to efficiently compensate for the chirp slope, so that compressed pulses with very low substructures and temporal durations much shorter than the initial seed pulse can be obtained [45, 51, 56–59]. As an illustration, Figures 2(a) and 2(b) show the setup and results of an experiment where similaritons are generated through Raman amplification at telecommunication wavelengths [54]. Moreover, recent fibre lasers that use self-similar pulse shaping in the normal dispersion regime have been demonstrated to achieve high-energy pulses [63–66].

In addition to fibre amplifiers and lasers, stable similariton pulses can be generated in passive fibres provided a suitable longitudinal variation of the dispersion is introduced [55, 67–69] (Figure 2(c)). This approach is based on the observation that a longitudinal decrease of the normal dispersion is formally equivalent to linear gain. Recently, a simple approach to the generation of parabolic pulses that uses progressive nonlinear pulse reshaping in a ND fibre with fixed dispersion has been demonstrated [70]. However, in contrast with the asymptotic similariton solutions obtained in fibre amplifiers, the generated parabolic waveforms represent transient states of the nonlinear pulse evolution in the passive fibre medium [71]. As such, they have a finite life distance that depends sensibly on the initial conditions (pulse shape, energy, and chirp profile). Nevertheless, stabilization of the parabolic features is possible by use of a second propagation stage in a fibre with specially adjusted nonlinear and dispersive characteristics relative to the first fibre [70, 71].

3.3. Generation of Triangular Pulses. Parabolic shapes are not the only pulse waveforms that can be generated in a passive ND fibre. Recently, the combination of pulse prechirping and nonlinear propagation in a section of ND fibre has been introduced as a method for passive nonlinear pulse shaping, which provides a simple way of generating various advanced field distributions, including flat-top- and triangular-profiled pulses with a linear chirp [71]. In this scheme, Kerr nonlinearity and GVD lead to various reshaping processes of an initial conventional pulse (e.g., a Gaussian pulse) according to the chirping value and power level at the entrance of the fibre. In particular,

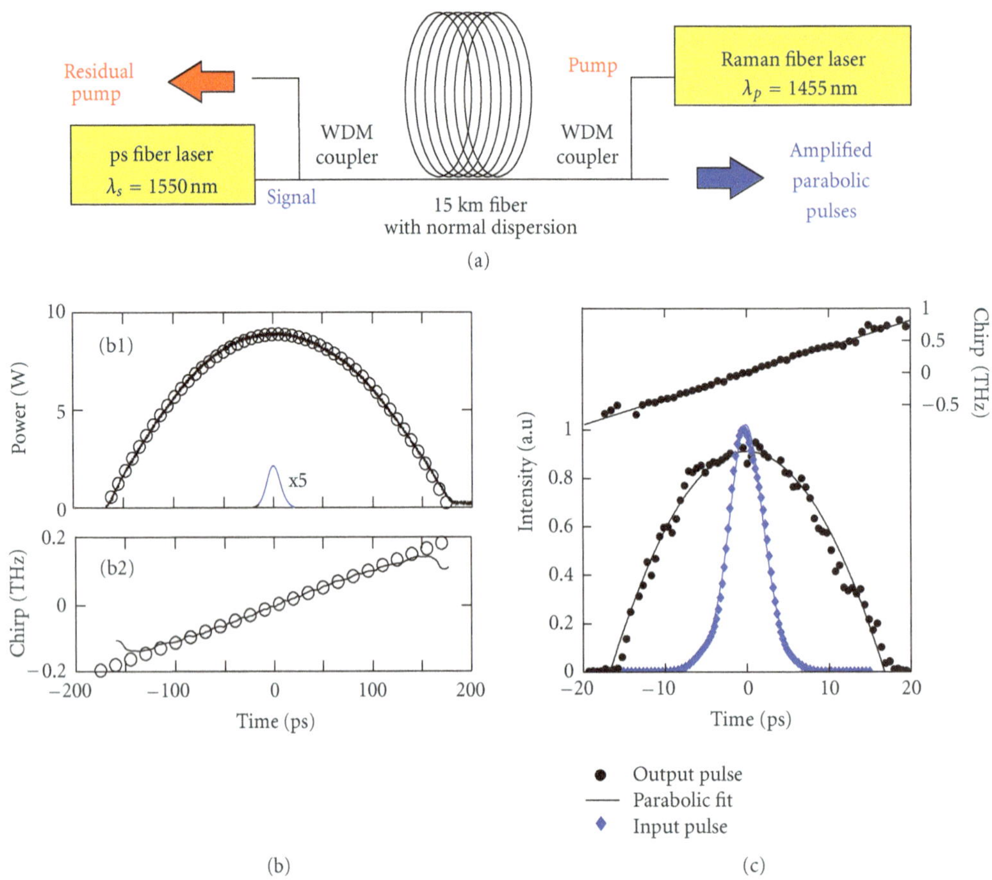

Figure 2: (a) Experimental setup for parabolic pulse generation through Raman amplification at telecommunication wavelengths. (b) Results obtained in the Raman amplifier (adapted from [54]): temporal intensity profile (subplot b1) and chirp profile (subplot b2) of the output pulse from the amplifier. The experimental results (solid black lines) are compared to a parabolic and a linear fit (open black circles), respectively. The intensity profile of the initial pulse is also shown (solid blue line). (c) Results obtained in a ND decreasing fibre (adapted from [55]): experimental temporal intensity and chirp profiles at the fibre output (black circles) compared to a parabolic and a linear fit (solid black lines), respectively. The initial intensity profile is also shown (blue diamonds).

triangular pulses can be generated for a positive initial chirp parameter (using the definition ibt^2 for the phase profile) and sufficiently high energies.

These theoretical results have been confirmed experimentally by intensity and phase measurements of the generated pulses [72] (Figure 3(b)). In the experimental setup used (Figure 3(a)), the control of the pulse prechirping value was realized by propagation through different lengths of standard SM fibre with anomalous GVD, which imposed a positive chirp parameter on the pulse. The prechirped pulses were amplified to different power levels using an EDFA and then propagated through a ND fibre to realize the pulse reshaping.

Different nonlinear dynamics may lead to the generation of triangular pulses in a passive ND fibre. Indeed, it has been recently shown qualitatively and numerically that temporal triangular intensity profiles can result from the progressive reshaping of initially parabolic pulses driven by the fourth-order dispersion (FOD) of the fibre [53]. The overall temporal effect of FOD on parabolic pulse propagation is to stretch and enhance the power reduction in the pulse wings, leading to a triangular profile. Furthermore, the possibility of triangular pulse shaping in mode-locked ring-cavity fibre lasers has been first reported in [73]. It has been numerically demonstrated that for normal net dispersion, formation of two distinct steady-state solutions of stable single pulses can be obtained in a laser cavity in different regions of the system parameter space: the previously known similariton [63] and a triangular-profiled pulse with a linear frequency chirp.

3.4. Coherent Continuums for Optical Telecommunications. One of the most well-known examples of pulse reshaping in the spectral domain is in supercontinuum (SC) generation. SC generation in optical fibres, and more specifically, in photonic crystal fibres with enhanced nonlinearity, is an established technique for producing broadband light sources in a wide range of research fields including metrology, biophotonics, and optical telecommunications [74, 75]. In telecommunications, the broadened spectrum can be spectrally sliced to generate wavelength multiplexed pulse trains [76–78]. The principle of operation of a multiwavelength source is illustrated in Figure 4(a). An initial high-power

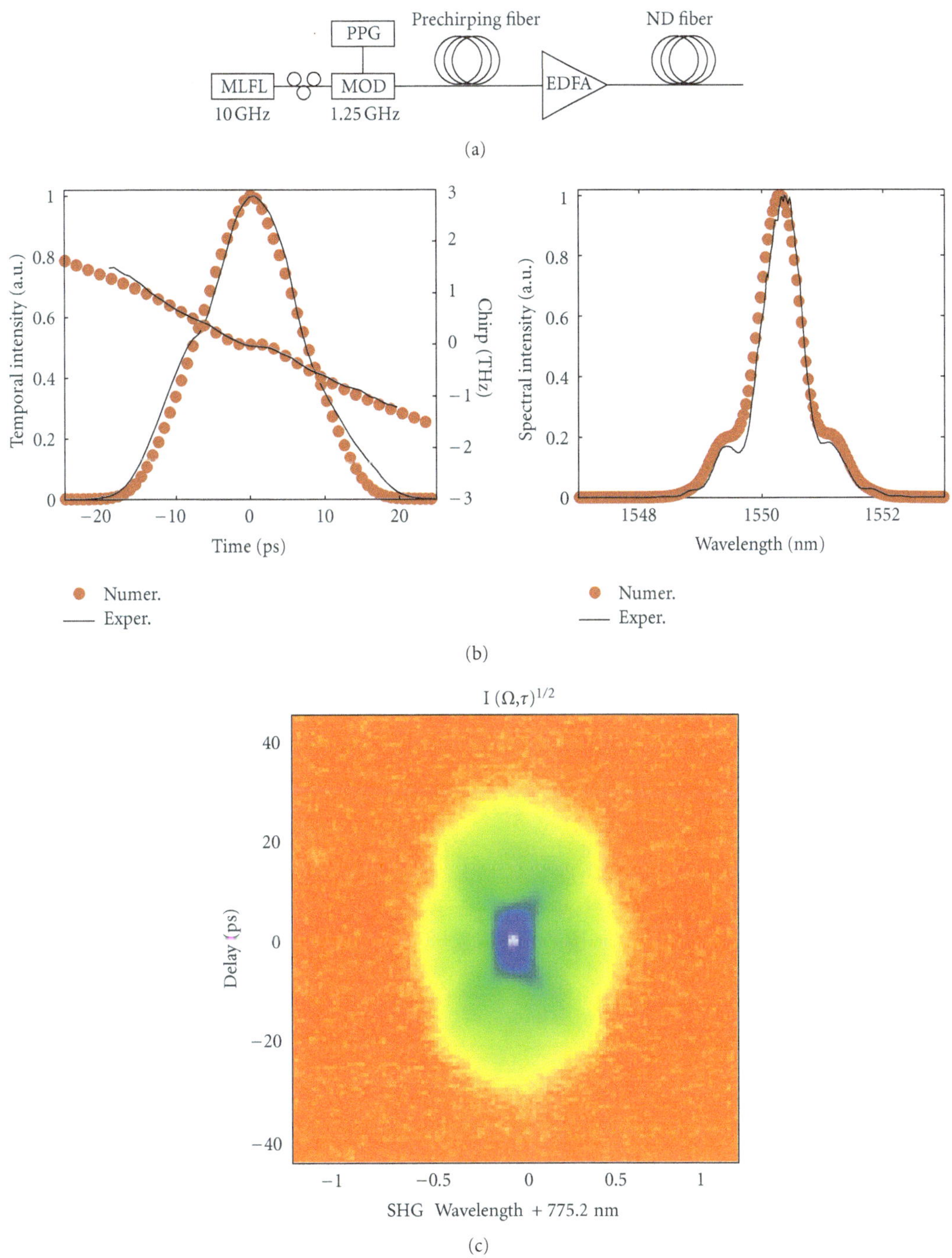

FIGURE 3: (a) Experimental setup for triangular pulse generation. (b) Experimental and numerical temporal intensity and chirp (ϕ_t) profiles retrieved from second-harmonic generation (SHG) FROG (left) and optical spectra (right) of the generated pulses for an example length of prechirping fibre. (c) SHG-FROG spectrogram of the pulse in (b). Results adapted from [72].

mode-locked picosecond or femtosecond pulse train with a repetition rate of several gigahertz is spectrally broadened through SPM in a kilometer-long highly nonlinear (HNL) fibre. At the fibre output, the spectrum is sliced by adequate bandpass filters (BPFs) to obtain pulse trains that have the same repetition rate as the initial train but new central frequencies.

Typically the broadest spectra are generated in anomalously dispersive fibres, where the broadening mechanism is strongly influenced by soliton dynamics [80]. However, solitonic effects and modulation instability lead to decreased temporal coherence and spectral flatness of the continuum generated in the anomalous dispersion regime [81]. By contrast, the nonlinear pulse evolution in ND fibres leads

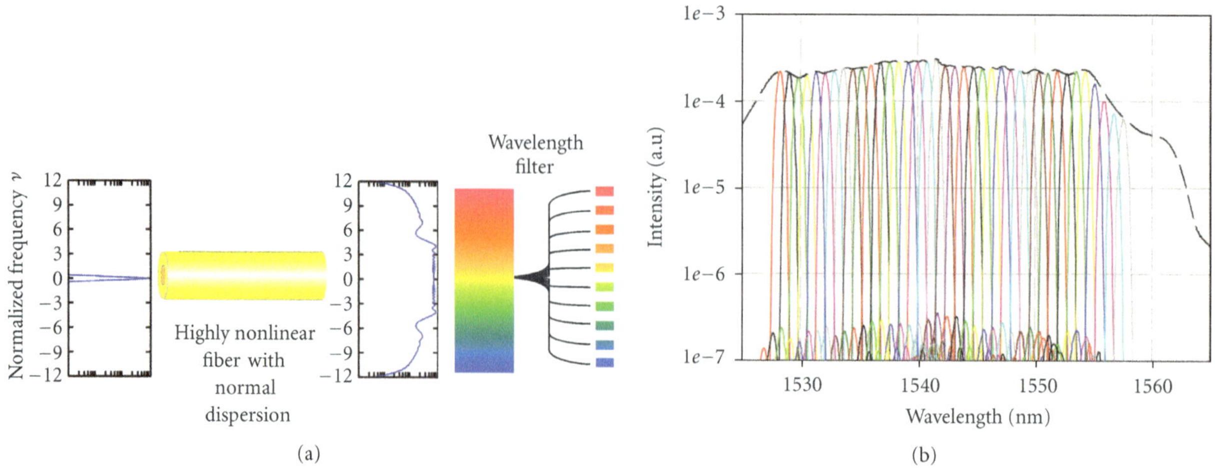

Figure 4: (a) Principle of operation of a multiwavelength picosecond source. (b) Experimental spectrum of a parabolic-shaped pulse after propagation in a ND-HNL fibre (black dashed line) and superposition of the measured spectra of 38 sliced channels. Results adapted from [79].

to flatter spectra and improved temporal coherence, but at the expenses of relatively narrower spectra [82, 83]. Hence frequency broadening using a ND fibre is an attractive choice whenever a moderate amount of broadening alongside good spectral and temporal stability of the output pulses are targeted. However, the main limits to spectral pulse quality in the normal dispersion regime are the spectral ripple arising from SPM of conventional laser pulses, and the effects of optical wave breaking which may lead to significant change in the temporal pulse shape and severe energy transfer into the wings of the spectrum. These effects can in principle be avoided by using preshaped input pulses with a parabolic temporal intensity profile, which would preserve their shape whilst propagating within the fibre and, thus, result in spectrally flat, highly coherent pulses. Continuum with low-spectral ripple and high-energy density in the central part is indeed achievable through parabolic pulse propagation in ND-HNL fibre [79], and low-noise multiwavelength picosecond sources covering the whole C-band of optical telecommunications have been demonstrated based on this technique [12, 16, 79] (Figure 4(b)).

4. Applications of Parabolic and Triangular Pulses in Signal Processing

Applications of parabolic pulses have not been limited to ultrashort high-power pulse generation and highly coherent continuum sources, but several optical processing techniques have taken advantage of their specific features. These include optical regeneration [84, 85] including pulse retiming [86], prereceiver nonlinear processing in the optical domain [87, 88], and mitigation of linear waveform distortions [89, 90]. The simple intensity profile of triangular pulses is also highly desired for a range of processing applications, including time domain add-drop multiplexing [91, 92], wavelength conversion [93–95], optical signal doubling [96], time-to-frequency mapping of multiplexed signals [97], and enhanced spectral compression [98, 99]. In this Section, we will review some of these applications.

4.1. 2R Regeneration. As a first example, here we discuss the usage of parabolic pulses in 2R (reamplification, retiming) signal regeneration in an extension of the technique proposed by Mamyshev [100]. The method relies on SPM in a nonlinear medium (fibre) and subsequent offset spectral filtering to suppress the noise in the zero-bit slots and the amplitude fluctuations in the one-bit pulses of return-to-zero (RZ) optical data streams. In the schematic of the Mamyshev regenerator depicted in Figure 5(a), the signal to be regenerated is first amplified to the optimum power level, and an optical BPF is also used to reduce the amplified spontaneous emission noise power. Then spectral broadening is achieved in a HNL fibre, and the broadened spectrum is partially sliced by a second BPF, which is shifted with respect to the input signal frequency in order to transmit SPM-generated frequencies and reject the original spectrum. The central frequency and output pulse shape of the regenerated signal are determined by the central frequency and spectral profile of the offset filter, respectively. The SPM phase shift experienced in the HNL fibre depends on the instantaneous power (cf. Section 2). Hence, a high-intensity pulse induces more SPM phase shift and spectrally broadens more than the low-intensity noise [100]. Filtering the new SPM-generated frequencies with the offset filter leads to a power transfer function for the pulses and the noise which is close to a binary one [101].

Spectral filtering of similariton pulses can improve the ability of the Mamyshev regenerator to eliminate spurious noise pulses as well as to simultaneously reduce any fluctuations in the signal one-bit level [84]. This ability relies on the specific features of self-similar amplification. Indeed, a spectrum broadened through self-similar amplification has a smooth profile (cf. Section 3.4), whereas a SPM-broadened spectrum is generally accompanied by oscillatory structures

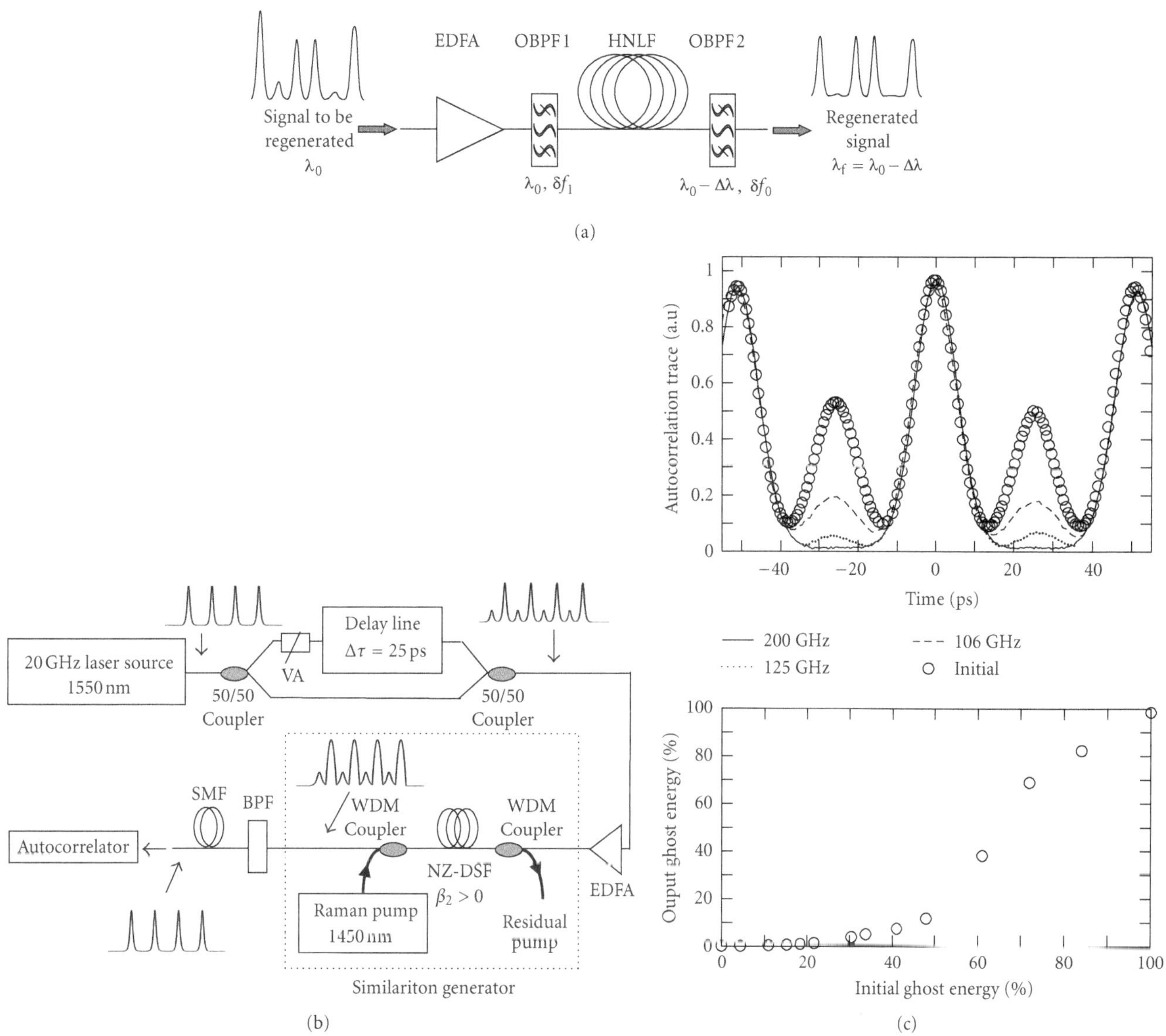

FIGURE 5: (a) Schematic of an optical regenerator based on SPM and filtering (after [100]). (b) Experimental setup for 40 Gbit/s similariton-based regeneration. (c) Top, autocorrelation traces of the pulse train at the Raman amplifier input (circles) and output for different values of the filter offset (dashed, dotted, and solid curves). Bottom, output versus input proportion of the spurious noise pulse energy with respect to the ones energy for a fixed filter offset. Results adapted from [84].

covering the entire frequency range. Moreover, the spectrum and chirp obtained by SPM both depend on the details of the initial pulse (shape, chirp, and energy), leading to some amplitude and time fluctuations of the regenerated pulses. In contrast, the similariton-based spectral broadening and similariton-induced chirp depend only on the initial pulse energy and/or amplifier parameters, respectively (cf. Section 3.2), so that both amplitude and timing jitter are reduced. In particular, the energy-dependent spectral broadening is responsible for efficient discrimination of high-energy pulses (ones) from low-energy pulses (zeros) with the offset BPF. An implementation of the similariton-based regeneration technique at 40 Gbit/s data rate where similariton pulses are generated through distributed Raman amplification in a ND fibre is illustrated in Figures 5(b) and 5(c). It is worth noting here that a distributed amplification is helpful in increasing the energy yield of the regenerator which is a highly dissipative element [102].

4.2. Linear-Distortion Compensation. As a second example of the application of parabolic pulses in signal processing, we review here the realization of a time-domain all-optical Fourier transformation (OFT) technique. Time-domain OFT in which the spectral profile is converted into a waveform in the time domain makes it possible to eliminate the waveform distortion caused by linear perturbations in optical fibres, such as higher order dispersion, jitter and polarization-mode dispersion [103–106]. Time-domain OFT is realized ideally by applying a quadratic temporal phase modulation (time lens) to a distorted pulse:

$\psi_{chirp}(t) = \psi(t)\exp(ibt^2)$ and passing the linearly chirped pulse through a dispersive medium such as a short piece of fibre. When the accumulated dispersion of the medium $D = \beta_2 L$ equates the inverse of the chirp coefficient b (i.e., when the chirp is fully compensated), the output waveform is proportional to the spectral amplitude before OFT: $\chi(t) = \sqrt{i/(2\pi D)}\exp(-ibt^2)\widetilde{\psi}(t/D)$, where $\widetilde{\psi}(\omega)$ is the Fourier transform of $\psi(t)$, and the frequency ω is converted into time t through the relationship $\omega = t/D$. Since generally the spectral envelope profile has no distortion even if its time-domain waveform is distorted by linear perturbations, this frequency-to-time mapping indicates that a distortion-free pulse waveform can be obtained in the time domain after OFT. In [103–106], this compensation technique was investigated by using electrooptic phase modulators. But this way the quadratic modulation region was restricted to around the center of the sinusoidal modulation imparted by the phase modulator. Thus, any parts of the pulse extending beyond the quadratic region could not be properly transformed, and only the section of the target pulse sitting within the quadratic region could be compensated (Figure 6).

The use of XPM with parabolic pulses as a time lens has been recently proposed as a promising way to overcome the aforementioned limitation [89, 90]. Indeed, the chirp induced by XPM on the target pulse is proportional to the gradient of the intensity profile of the parabolic control pulse: $-\phi_t = -2bt$, $b = \mp 2\gamma a^2 L/\tau^2$, so it is exactly linear and thus ideal OFT can be achieved by complete chirp compensation. Here, a and τ are the respective peak amplitude and characteristic width of the parabolic pulse (cf. Section 3.2), and the sign $\mp$ is for either a standard "bright" or a dark parabolic pulse [17]. The parabolic XPM-based OFT technique has been successfully experimentally demonstrated to achieve cancellation of the third-order dispersion-induced distortion of short-optical pulses [89, 90].

4.3. Optical Decision at the Receiver. In this section, we review a possible application of parabolic/flat-top pulses to the improvement of the signal bit-error rate (BER) in an optical communication system by using optical decision before the receiver. An attractive approach to the design of the optical receiver for high-speed fibre communication systems using RZ data formats is based on employing an additional all-optical decision element (ODE) just before the conventional receiver [87, 107–109]. In this way, the first decision is carried out in the optical domain, thus improving the quality of the received signal. The attractive possibility of improving the BER by using optical devices has been studied in some recent works [88, 110, 111]. The conceptual basis of the BER improvement by optical devices comes from the information theory: an intrinsic BER improvement is possible whenever the decision performed by the device in the optical domain is based on information that is different from the information used by the decision circuit of the (suboptimal) receiver in the electrical domain. In [88] an application of this general concept was demonstrated, based on using an advanced optical receiver enhanced by a nonlinear ODE. The example of an ODE scheme used in [88] exploits the nonlinear mechanism that leads to the formation of parabolic pulses and the resultant flattening of the signal waveform (cf. Section 3.2). The method is based on the nonlinear reshaping of a pulse towards a broad and flat temporal waveform that can occur upon propagation in a ND fibre [9] and subsequent slicing of the pulse temporal profile by an optical temporal gate (e.g., an amplitude modulator with a sharp nonlinear transfer function or a nonlinear optical loop mirror provided with a clock) [85] or XPM-based techniques.

The basic mechanism that is responsible for BER improvement in this scheme can be understood in the following way. The pulse temporal broadening and flattening in the ND fibre permits to recover the center of the bit slot for those pulses that are significantly shifted from the bit slot center in the input pulse train because of timing jitter (Figure 7(a)). Such strongly time-shifted pulses, once they are converted into electrical pulses and arrive at the decision point, might be missed by the threshold detector in those conventional receivers that use a relatively small fraction of the bit period. In contrast, the bit slot recovery in the ND fibre enables such pulses to be recognized as marks in the ODE-enhanced receiver. The slicing of the central portion of the broadened pulse temporal profile by the temporal gate enables efficient suppression of the timing jitter of the pulses which, in turn, improves the receiver sensitivity [87]. This approach is illustrated in Figure 7(b), which shows that BER improvement over the conventional scheme, indeed, can be achieved using nonlinear signal processing in the optical domain before the conventional receiver. It is worth noting that the same processing technique based on temporal slicing of nonlinearly flattened pulse waveforms can also be used in all-optical 3R (2R + re-timing) signal regeneration [85].

4.4. Optical Signal Doubling and Frequency Conversion. Here we discuss the use of triangular pulses in the context of optical signal doubling and frequency conversion. A technique of copying optical pulses in both the frequency and time domains based on a combination of either XPM with a triangular pump pulse or SPM of a triangular pulse in a nonlinear Kerr medium and subsequent propagation in a dispersive medium was introduced in [96] and experimentally demonstrated in [95]. In this scheme, spectral doubling of an optical signal is made possible by the XPM or SPM phase shift generated by a temporal intensity profile with a linear gradient: $\phi(t) = \phi_0(1 - |t/\tau|)$, $\phi_0 \propto a^2\gamma L$ (a and τ being the peak amplitude and characteristic width of a triangular pulse, resp.), which translates into a constant and distinct (opposite sign) frequency chirp induced onto the leading (downshifted) and trailing (upshifted) edges of the signal pulse. As a result of this constant and distinct chirp, under certain conditions the pulse spectrum develops a structure consisting of two equal peaks. The separation between the two spectral peaks increases with the maximum nonlinear phase shift ϕ_0, while their form is almost preserved. Under subsequent propagation of the modulated signal pulse in a linear dispersive medium, the produced spectral separation of the pulse components leads to their subsequent separation in time, thus allowing temporal replication of the pulse. In the case of normal dispersion, the pulse develops into a

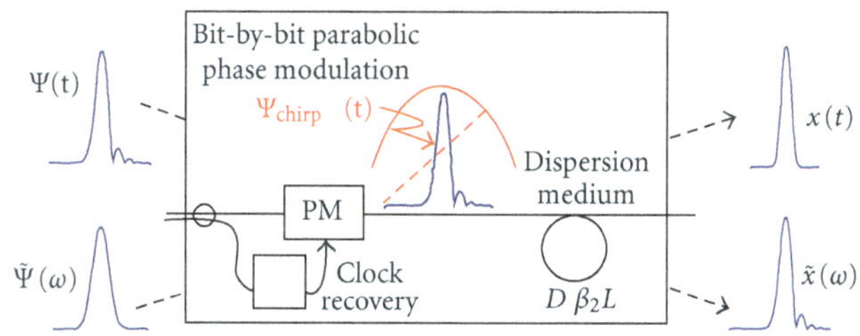

FIGURE 6: Schematic diagram of time-domain OFT (after [90]).

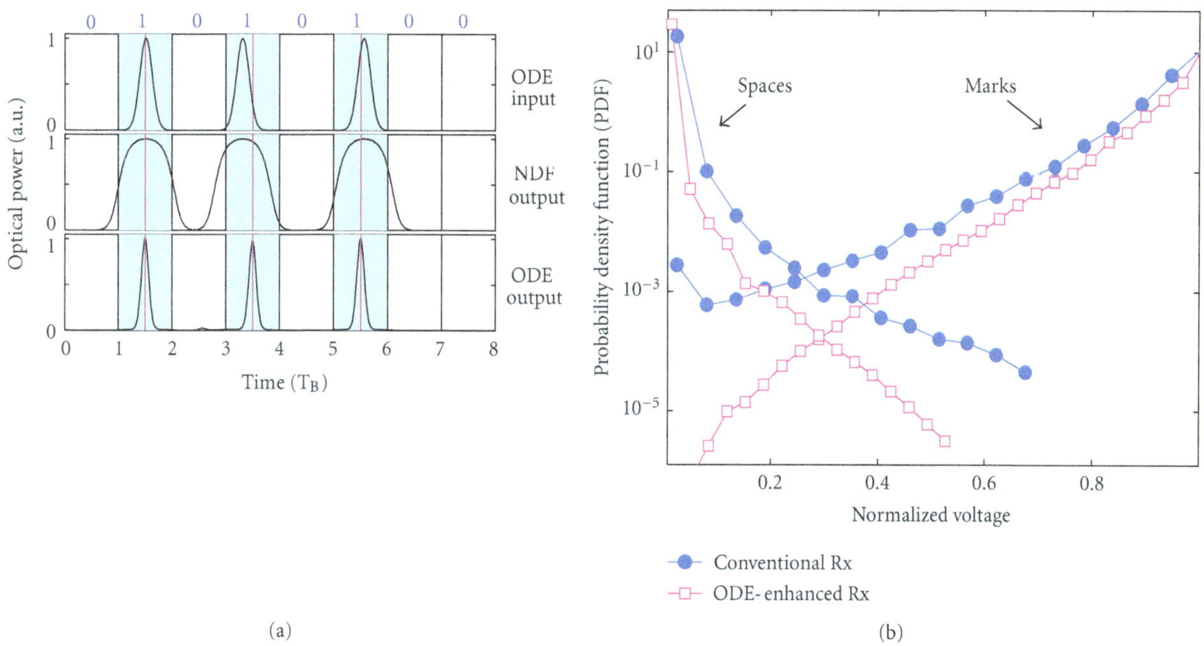

FIGURE 7: (a) Schematic illustration of the optical decision at an ODE enhanced receiver. (b) Numerically calculated probability density functions for the voltage of the marks and spaces in the conventional and ODE-enhanced receivers for a 40 Gbit/s RZ input signal with a random jitter of the pulse temporal position. Results adapted from [88].

structure consisting of two identical pulses where the original pulse splits down the middle. The two pulse parts separate in time while experiencing dispersive broadening. The relative velocity of separation is proportional to the separation between the two peaks of the nonlinearly modulated pulse spectrum. Pulse splitting and separation occur also in the case of anomalous dispersion after an initial stage of pulse compression. Exact analytic expressions for the nonlinearly modulated pulse spectrum and the pulse temporal intensity distribution clearly show the features described above [96]. Applying spectral filtering to the XPM/SPM modulated signal is also possible to realize efficient and clean frequency conversion [93–95].

Figure 8 shows results of the experiment reported in [95]. Similarly to [72], triangular pulses were generated by passive nonlinear reshaping in a ND fibre. Spectral pulse doubling was realized through the SPM experienced by the triangular pulses over their life distance in a HNL fibre. The output SPM-shaped spectrum from the HNL fibre was sent into either a linear standard SM fibre segment to realize temporal pulse doubling or a wavelength demultiplexer filter for frequency conversion processing.

5. Conclusion

We have provided an overview of several recent examples of the use of nonlinear phenomena in optical fibres for the generation and shaping of optical pulses. We have discussed the generation of ultrashort pulse trains with very high-repetition rates in the anomalous dispersion regime of a fibre, the generation of parabolic-shaped pulses in active and passive ND fibres, the generation of pulses with triangular intensity profiles in passive ND fibres, and the generation of coherent continuums for optical telecommunications in ND fibres. Examples of new techniques or improvements in existing techniques of all-optical nonlinear fibre-based signal processing enabled by specialized pulse waveforms have been reviewed. Specifically, we have presented applications of parabolic pulses in optical signal regeneration, compensation of linear waveform distortions, and signal postprocessing in

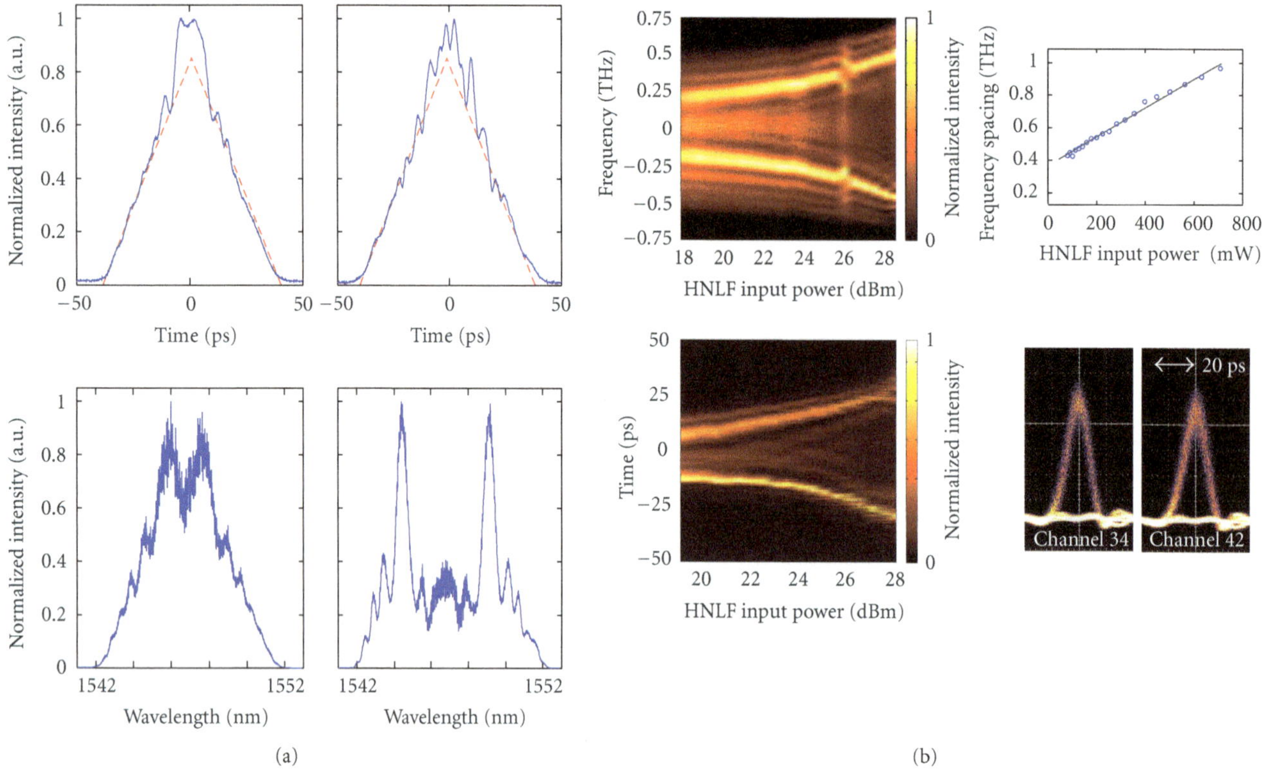

FIGURE 8: (a) Temporal (top) and spectral (bottom) intensity profiles after the ND reshaping fibre (left) and after the HNL fibre (right). (b) Top, evolution of the output spectrum from the HNL fibre with the input power (left) and dependence of the frequency spacing between the two spectral peaks on the input power (right). Bottom, evolution of the temporal intensity profile at the output of the linear SM fibre with the input power to the HNL fibre (left) and eye diagrams of a 10 Gbit/s pulse train obtained after spectral filtering for the ITU channels 34 and 42. Results adapted from [95].

optical communication systems. Applications of triangular pulses in optical signal doubling and frequency conversion have also been discussed.

Acknowledgments

The authors would like to acknowledge and thank our colleagues, namely, Prof. Sergei K. Turitsyn, Dr. Paul Harper, Dr. Anton I. Latkin, Prof. Keith J. Blow, Dr. Julien Fatome, Dr. Kamal Hammani, Dr. Bertrand Kibler, Prof. Guy Millot, and Prof. John M. Dudley and the Optoelectronics Research Center (Southampton, UK). The authors also acknowledge support from the British Council (Alliance Grant 10.002), the Ministère des Affaires Étrangères et Européennes (Alliance Grant 22836ZJ), the Agence Nationale de la Recherche, and the Conseil Regional de Bourgogne (PARI Photcom).

References

[1] Y. R. Shen, *Principles of Nonlinear Optics*, John Wiley & Sons, New York, NY, USA, 1984.

[2] J. C. Knight, "Photonic crystal fibres," *Nature*, vol. 424, no. 6950, pp. 847–851, 2003.

[3] B. J. Eggleton, B. Luther-Davies, and K. Richardson, "Chalcogenide photonics," *Nature Photonics*, vol. 5, no. 3, pp. 141–148, 2011.

[4] G. P. Agrawal, *Nonlinear Fiber Optics*, Academic Press, San Diego, Calif, USA, 4th edition, 2006.

[5] A. Hasegawa and F. Tappert, "Transmission of stationary nonlinear optical pulses in dispersive dielectric fibers I. Anomalous dispersion," *Applied Physics Letters*, vol. 23, no. 3, pp. 142–144, 1973.

[6] L. F. Mollenauer and J. P. Gordon, *Solitons in Optical Fibers: Fundamentals and Applications*, Academic Press, San Diego, Calif, USA, 2006.

[7] W. J. Tomlinson, R. H. Stolen, and A. M. Johnson, "Optical wave breaking of pulses in nonlinear optical fibers," *Optics Letters*, vol. 10, no. 9, pp. 457–459, 1985.

[8] D. Anderson, M. Desaix, M. Lisak, and M. L. Quiroga-Teixeiro, "Wave breaking in nonlinear-optical fibers," *Journal of the Optical Society of America B*, vol. 9, no. 8, pp. 1358–1361, 1992.

[9] C. Finot, B. Kibler, L. Provost, and S. Wabnitz, "Beneficial impact of wave-breaking for coherent continuum formation in normally dispersive nonlinear fibers," *Journal of the Optical Society of America B*, vol. 25, no. 11, pp. 1938–1948, 2008.

[10] J. Azaña, L. K. Oxenløwe, E. Palushani et al., "In-fiber subpicosecond pulse shaping for nonlinear optical telecommunication data processing at 640 Gbit/s," *International Journal of Optics*, vol. 2012, Article ID 895281, 16 pages, 2012.

[11] A. M. Weiner, "Femtosecond pulse shaping using spatial light modulators," *Review of Scientific Instruments*, vol. 71, no. 5, pp. 1929–1960, 2000.

[12] A. M. Clarke, D. G. Williams, M. A. F. Roelens, and B. J. Eggleton, "Reconfigurable optical pulse generator employing a Fourier-domain programmable optical processor," *Journal of Lightwave Technology*, vol. 28, no. 1, pp. 97–103, 2010.

[13] D. Nguyen, M. U. Piracha, D. Mandridis, and P. J. Delfyett, "Dynamic parabolic pulse generation using temporal shaping of wavelength to time mapped pulses," *Optics Express*, vol. 19, no. 13, pp. 12305–12311, 2011.

[14] J. Ye, L. S. Yan, W. Pan et al., "Photonic generation of triangular-shaped pulses based on frequency-to-time conversion," *Optics Letters*, vol. 36, no. 8, pp. 1458–1460, 2011.

[15] P. Petropoulos, M. Ibsen, A. D. Ellis, and D. J. Richardson, "Rectangular pulse generation based on pulse reshaping using a superstructured fiber Bragg grating," *Journal of Lightwave Technology*, vol. 19, no. 5, pp. 746–752, 2001.

[16] D. Krčmařík, R. Slavík, Y. G. Park, and J. Azaña, "Nonlinear pulse compression of picosecond parabolic-like pulses synthesized with a long period fiber grating filter," *Optics Express*, vol. 17, no. 9, pp. 7074–7087, 2009.

[17] T. Hirooka, M. Nakazawa, and K. Okamoto, "Bright and dark 40 GHz parabolic pulse generation using a picosecond optical pulse train and an arrayed waveguide grating," *Optics Letters*, vol. 33, no. 10, pp. 1102–1104, 2008.

[18] K. Kashiwagi, H. Ishizu, and T. Kurokawa, "Fiber transmission characteristics of parabolic pulses generated by optical pulse synthesizer," *Japanese Journal of Applied Physics*, vol. 50, no. 9, Article ID 092501, 2011.

[19] B. Kibler, R. Fischer, P. A. Lacourt et al., "Optimised one-step compression of femtosecond fibre laser soliton pulses around 1550 nm to below 30 fs in highly nonlinear fibre," *Electronics Letters*, vol. 43, no. 17, pp. 915–916, 2007.

[20] A. A. Amorim, M. V. Tognetti, P. Oliveira et al., "Sub-two-cycle pulses by soliton self-compression in highly nonlinear photonic crystal fibers," *Optics Letters*, vol. 34, no. 24, pp. 3851–3853, 2009.

[21] L. F. Mollenauer, R. H. Stolen, and J. P. Gordon, "Experimental observation of picosecond pulse narrowing and solitons in optical fibers," *Physical Review Letters*, vol. 45, no. 13, pp. 1095–1098, 1980.

[22] S. V. Chernikov and P. V. Mamyshev, "Femtosecond soliton propagation in fibers with slowly decreasing dispersion," *Journal of the Optical Society of America B*, vol. 8, no. 8, pp. 1633–1641, 1991.

[23] S. V. Chernikov, D. J. Richardson, E. M. Dianov, and D. N. Payne, "Picosecond soliton pulse compressor based on dispersion decreasing fibre," *Electronics Letters*, vol. 28, no. 19, pp. 1842–1844, 1992.

[24] S. V. Chernikov, J. R. Taylor, and R. Kashyap, "Comblike dispersion-profiled fiber for soliton pulse train generation," *Optics Letters*, vol. 19, no. 8, pp. 539–541, 1994.

[25] J. Fatome, S. Pitois, and G. Millot, "20-GHz-to-1-THz repetition rate pulse sources based on multiple four-wave mixing in optical fibers," *IEEE Journal of Quantum Electronics*, vol. 42, no. 10, Article ID 01703695, pp. 1038–1046, 2006.

[26] K. Tai, A. Hasegawa, and A. Tomita, "Observation of modulational instability in optical fibers," *Physical Review Letters*, vol. 56, no. 2, pp. 135–138, 1986.

[27] B. Ortaç, A. Hideur, G. Martel, and M. Brunel, "2-GHz passive harmonically mode-locked Yb-doped double-clad fiber laser," *Applied Physics B*, vol. 81, no. 4, pp. 507–509, 2005.

[28] A. Komarov, H. Leblond, and F. Sanchez, "Passive harmonic mode-locking in a fiber laser with nonlinear polarization rotation," *Optics Communications*, vol. 267, no. 1, pp. 162–169, 2006.

[29] A. Komarov, A. Haboucha, and F. Sanchez, "Ultrahigh-repetition-rate bound-soliton harmonic passive mode-locked fiber lasers," *Optics Letters*, vol. 33, no. 19, pp. 2254–2256, 2008.

[30] J. Schröder, S. Coen, F. Vanholsbeeck, and T. Sylvestre, "Passively mode-locked Raman fiber laser with 100 GHz repetition rate," *Optics Letters*, vol. 31, no. 23, pp. 3489–3491, 2006.

[31] K. Hammani, B. Kibler, C. Finot et al., "Peregrine soliton generation and breakup in standard telecommunications fiber," *Optics Letters*, vol. 36, no. 2, pp. 112–114, 2011.

[32] R. Trebino, *Frequency-Resolved Optical Gating: The Measurement of Ultrashort Laser Pulses*, Kluwer Academic, Norwell, Mass, USA, 2000.

[33] J. Fatome, S. Pitois, C. Fortier et al., "Multiple four-wave mixing in optical fibers: 1.5-3.4-THz femtosecond pulse sources and real-time monitoring of a 20-GHz picosecond source," *Optics Communications*, vol. 283, no. 11, pp. 2425–2429, 2010.

[34] S. Pitois, C. Finot, J. Fatome, B. Sinardet, and G. Millot, "Generation of 20-GHz picosecond pulse trains in the normal and anomalous dispersion regimes of optical fibers," *Optics Communications*, vol. 260, no. 1, pp. 301–306, 2006.

[35] C. Finot, J. Fatome, S. Pitois, and G. Millot, "All-fibered high-quality low duty-cycle 20-GHz and 40-GHz picosecond pulse sources," *IEEE Photonics Technology Letters*, vol. 19, no. 21, pp. 1711–1713, 2007.

[36] C. Fortier, B. Kibler, J. Fatome, C. Finot, S. Pitois, and G. Millot, "All-fibered high-quality low duty-cycle 160-GHz femtosecond pulse source," *Laser Physics Letters*, vol. 5, no. 11, pp. 817–820, 2008.

[37] I. E. Mansouri, J. Fatome, C. Finot, S. Pitois, and M. Lintz, "All-fibered high-quality stable 20- and 40-GHz picosecond pulse generators for 160-Gb/s OTDM applications," *IEEE Photonics Technology Letters*, vol. 23, no. 20, pp. 1487–1489, 2011.

[38] N. Akhmediev and V. I. Korneev, "Modulation instability and periodic solutions of the nonlinear Schrödinger equation," *Theoretical and Mathematical Physics*, vol. 69, no. 2, pp. 1089–1093, 1986.

[39] K. Hammani, B. Wetzel, B. Kibler et al., "Spectral dynamics of modulation instability described using Akhmediev breather theory," *Optics Letters*, vol. 36, no. 11, pp. 2140–2142, 2011.

[40] B. Kibler, J. Fatome, C. Finot et al., "The Peregrine soliton in nonlinear fibre optics," *Nature Physics*, vol. 6, no. 10, pp. 790–795, 2010.

[41] D. H. Peregrine, "Water waves, nonlinear Schrödinger equations and their solutions," *The Journal of the Australian Mathematical Society B*, vol. 25, no. 1, pp. 16–43, 1983.

[42] J. M. Dudley, C. Finot, G. Millot et al., "Extreme events in optics: challenges of the MANUREVA project," *European Physical Journal*, vol. 185, no. 1, pp. 125–133, 2010.

[43] D. R. Solli, C. Ropers, P. Koonath, and B. Jalali, "Optical rogue waves," *Nature*, vol. 450, no. 7172, pp. 1054–1057, 2007.

[44] J. M. Dudley, C. Finot, G. Millot, and D. J. Richardson, "Self-similarity in ultrafast nonlinear optics," *Nature Physics*, vol. 3, no. 9, pp. 597–603, 2007.

[45] M. E. Fermann, V. I. Kruglov, B. C. Thomsen, J. M. Dudley, and J. D. Harvey, "Self-similar propagation and amplification of parabolic pulses in optical fibers," *Physical Review Letters*, vol. 84, no. 26, pp. 6010–6013, 2000.

[46] S. Boscolo, S. K. Turitsyn, V. Y. Novokshenov, and J. H. B. Nijhof, "Self-similar parabolic optical solitary waves,"

Theoretical and Mathematical Physics, vol. 133, no. 3, pp. 1647–1656, 2002.

[47] V. L. Kruglov and J. D. Harvey, "Asymptotically exact parabolic solutions of the generalized nonlinear Schrödinger equation with varying parameters," *Journal of the Optical Society of America B*, vol. 23, no. 12, pp. 2541–2550, 2006.

[48] C. Finot and L. Wu, "Influence of the initial phase profile on the asymptotic self-similar parabolic dynamics," *Journal of Nonlinear Optical Physics and Materials*, vol. 18, no. 4, pp. 709–721, 2009.

[49] D. Anderson, M. Desaix, M. Karlsson, M. Lisak, and M. L. Quiroka-Teixeiro, "Wave-breaking-free pulses in nonlinear-optical fibers," *Journal of the Optical Society of America B*, vol. 10, no. 7, pp. 1185–1190, 1993.

[50] C. Finot, G. Millot, and J. M. Dudley, "Asymptotic characteristics of parabolic similariton pulses in optical fiber amplifiers," *Optics Letters*, vol. 29, no. 21, pp. 2533–2535, 2004.

[51] C. Billet, J. M. Dudley, N. Joly, and J. C. Knight, "Intermediate asymptotic evolution and photonic bandgap fiber compression of optical similaritons around 1550 nm," *Optics Express*, vol. 13, no. 9, pp. 3236–3241, 2005.

[52] B. G. Bale and S. Boscolo, "Impact of third-order fibre dispersion on the evolution of parabolic optical pulses," *Journal of Optics A*, vol. 12, no. 1, Article ID 015202, 2010.

[53] B. G. Bale, S. Boscolo, K. Hammani, and C. Finot, "Effects of fourth-order fiber dispersion on ultrashort parabolic optical pulses in the normal dispersion regime," *Journal of the Optical Society of America B*, vol. 28, no. 9, pp. 2059–2065, 2011.

[54] K. Hammani, C. Finot, S. Pitois, J. Fatome, and G. Millot, "Real-time measurement of long parabolic optical similaritons," *Electronics Letters*, vol. 44, no. 21, pp. 1239–1240, 2008.

[55] C. Finot, B. Barviau, G. Millot, A. Guryanov, A. Sysoliatin, and S. Wabnitz, "Parabolic pulse generation with active or passive dispersion decreasing optical fibers," *Optics Express*, vol. 15, no. 24, pp. 15824–15835, 2007.

[56] J. P. Limpert, T. Schreiber, T. Clausnitzer et al., "High-power femtosecond Yb-doped fiber amplifier," *Optics Express*, vol. 10, no. 14, pp. 628–638, 2002.

[57] A. Malinowski, A. Piper, J. H. V. Price et al., "Ultrashort-pulse Yb^{3+}-fiber-based laser and amplifier system producing > 25-W average power," *Optics Letters*, vol. 29, no. 17, pp. 2073–2075, 2004.

[58] P. Dupriez, C. Finot, A. Malinowski et al., "High-power, high repetition rate picosecond and femtosecond sources based on Yb-doped fiber amplification of VECSELs," *Optics Express*, vol. 14, no. 21, pp. 9611–9616, 2006.

[59] D. N. Papadopoulos, Y. Zaouter, M. Hanna et al., "Generation of 63 fs 4.1 MW peak power pulses from a parabolic fiber amplifier operated beyond the gain bandwidth limit," *Optics Letters*, vol. 32, no. 17, pp. 2520–2522, 2007.

[60] C. Finot, G. Millot, S. Pitois, C. Billet, and J. M. Dudley, "Numerical and experimental study of parabolic pulses generated via Raman amplification in standard optical fibers," *IEEE Journal on Selected Topics in Quantum Electronics*, vol. 10, no. 5, pp. 1211–1218, 2004.

[61] S. Zhang, C. Jin, Y. Meng, X. Wang, and H. Li, "Propagation of high-power parabolic pulses in cubicon fiber amplifiers," *Journal of the Optical Society of America B*, vol. 27, no. 6, pp. 1272–1278, 2010.

[62] Vladimir I. Kruglov, Claude Aguergaray, and John D. Harvey, "Propagation and breakup of pulses in fiber amplifiers and dispersion-decreasing fibers with third-order dispersion," *Physical Review A*, vol. 84, no. 2, Article ID 023823, 14 pages, 2011.

[63] F. Ö. Ilday, J. R. Buckley, W. G. Clark, and F. W. Wise, "Self-similar evolution of parabolic pulses in a laser," *Physical Review Letters*, vol. 92, no. 21, Article ID 213902, 4 pages, 2004.

[64] F. W. Wise, A. Chong, and W. H. Renninger, "High-energy femtosecond fiber lasers based on pulse propagation at normal dispersion," *Laser and Photonics Reviews*, vol. 2, no. 1-2, pp. 58–73, 2008.

[65] C. Aguergaray, D. Méchin, V. I. Kruglov, and J. D. Harvey, "Experimental realization of a mode-locked parabolic Raman fiber oscillator," *Optics Express*, vol. 18, no. 8, pp. 8680–8687, 2010.

[66] B. Oktem, C. Ülgüdür, and F. O. Ilday, "Soliton-similariton fibre laser," *Nature Photonics*, vol. 4, no. 5, pp. 307–311, 2010.

[67] T. Hirooka and M. Nakazawa, "Parabolic pulse generation by use of a dispersion-decreasing fiber with normal group-velocity dispersion," *Optics Letters*, vol. 29, no. 5, pp. 498–500, 2004.

[68] B. Kibler, C. Billet, P. A. Lacourt, R. Ferriere, L. Larger, and J. M. Dudley, "Parabolic pulse generation in comb-like profiled dispersion decreasing fibre," *Electronics Letters*, vol. 42, no. 17, pp. 965–966, 2006.

[69] A. Plocky, A. A. Sysoliatin, A. I. Latkin et al., "Experiments on the generation of parabolic pulses in waveguides with length-varying normal chromatic dispersion," *JETP Letters*, vol. 85, pp. 319–322, 2007.

[70] C. Finot, L. Provost, P. Petropoulos, and D. J. Richardson, "Parabolic pulse generation through passive nonlinear pulse reshaping in a normally dispersive two segment fiber device," *Optics Express*, vol. 15, no. 3, pp. 852–864, 2007.

[71] S. Boscolo, A. I. Latkin, and S. K. Turitsyn, "Passive nonlinear pulse shaping in normally dispersive fiber systems," *IEEE Journal of Quantum Electronics*, vol. 44, no. 12, pp. 1196–1203, 2008.

[72] H. Wang, A. I. Latkin, S. Boscolo, P. Harper, and S. K. Turitsyn, "Generation of triangular-shaped optical pulses in normally dispersive fibre," *Journal of Optics*, vol. 12, no. 3, Article ID 035205, 2010.

[73] S. Boscolo and S. K. Turitsyn, "Pulse shaping in mode-locked ring-cavity fibre lasers," in *Proceedings of Lasers and Electro-Optics Europe and 12th European Quantum Electronics Conference (CLEO EUROPE/EQEC '11)*, Munich, German, May 2011, paper EH4.6.

[74] R. R. Alfano, *The Supercontinuum Laser Source*, Springer, New York, NY, USA, 2006.

[75] J. M. Dudley and J. R. Taylor, *Supercontinuum Generation in Optical Fibers*, Cambridge University Press, New York, NY, USA, 2010.

[76] L. Boivin and B. C. Collings, "Spectrum slicing of coherent sources in optical communications," *Optical Fiber Technology*, vol. 7, no. 1, pp. 1–20, 2001.

[77] Y. Takushima and K. Kikuchi, "10-GHz, over 20-channel multiwavelength pulse source by slicing super-continuum spectrum generated in normal-dispersion fiber," *IEEE Photonics Technology Letters*, vol. 11, no. 3, pp. 322–324, 1999.

[78] Z. Yusoff, P. Petropoulos, K. Furusawa, T. M. Monro, and D. J. Richardson, "A 36-Channel x 10-GHz spectrally sliced pulse source based on supercontinuum generation in normally dispersive highly nonlinear holey fiber," *IEEE Photonics Technology Letters*, vol. 15, no. 12, pp. 1689–1691, 2003.

[79] F. Parmigiani, C. Finot, K. Mukasa et al., "Ultra-flat SPM-broadened spectra in a highly nonlinear fiber using parabolic pulses formed in a fiber Bragg grating," *Optics Express*, vol. 14, no. 17, pp. 7617–7622, 2006.

[80] J. M. Dudley, G. Genty, and S. Coen, "Supercontinuum generation in photonic crystal fiber," *Reviews of Modern Physics*, vol. 78, no. 4, pp. 1135–1184, 2006.
[81] M. Nakazawa, K. Tamura, H. Kubota, and E. Yoshida, "Coherence degradation in the process of supercontinuum generation in an optical fiber," *Optical Fiber Technology*, vol. 4, no. 2, pp. 215–223, 1998.
[82] A. M. Heidt, "Pulse preserving flat-top supercontinuum generation in all-normal dispersion photonic crystal fibers," *Journal of the Optical Society of America B*, vol. 27, no. 3, pp. 550–559, 2010.
[83] L. E. Hooper, P. J. Mosley, A. C. Muir, W. J. Wadsworth, and J. C. Knight, "Coherent supercontinuum generation in photonic crystal fiber with all-normal group velocity dispersion," *Optics Express*, vol. 19, no. 6, pp. 4902–4907, 2011.
[84] C. Finot, S. Pitois, and G. Millot, "Regenerative 40 Gbit/s wavelength converter based on similariton generation," *Optics Letters*, vol. 30, no. 14, pp. 1776–1778, 2005.
[85] S. Boscolo and S. K. Turitsyn, "All-optical signal regeneration by temporal slicing of nonlinearly flattened optical waveform," *IEEE Photonics Technology Letters*, vol. 17, no. 6, pp. 1235–1237, 2005.
[86] F. Parmigiani, P. Petropoulos, M. Ibsen, and D. J. Richardson, "Pulse retiming based on XPM using parabolic pulses formed in a fiber Bragg grating," *IEEE Photonics Technology Letters*, vol. 18, no. 7, pp. 829–831, 2006.
[87] S. Boscolo, S. K. Turitsyn, and K. J. Blow, "Time domain all-optical signal processing at a RZ optical receiver," *Optics Express*, vol. 13, no. 16, pp. 6217–6227, 2005.
[88] I. O. Nasieva, S. Boscolo, and S. K. Turitsyn, "Bit error rate improvement by nonlinear optical decision element," *Optics Letters*, vol. 31, no. 9, pp. 1205–1207, 2006.
[89] T. T. Ng, F. Parmigiani, M. Ibsen, Z. Zhang, P. Petropoulos, and D. J. Richardson, "Compensation of linear distortions by using XPM with parabolic pulses as a time lens," *IEEE Photonics Technology Letters*, vol. 20, no. 13, pp. 1097–1099, 2008.
[90] T. Hirooka and M. Nakazawa, "All-optical 40-GHz time-domain Fourier transformation using XPM with a dark parabolic pulse," *IEEE Photonics Technology Letters*, vol. 20, no. 22, pp. 1869–1871, 2008.
[91] J. Li, B. E. Olsson, M. Karlsson, and P. A. Andrekson, "OTDM add-drop multiplexer based on XPM-induced wavelength shifting in highly nonlinear fiber," *Journal of Lightwave Technology*, vol. 23, no. 9, pp. 2654–2661, 2005.
[92] F. Parmigiani, P. Petropoulos, M. Ibsen, P. J. Almeida, T. T. Ng, and D. J. Richardson, "Time domain add-drop multiplexing scheme enhanced using a saw-tooth pulse shaper," *Optics Express*, vol. 17, no. 10, pp. 8362–8369, 2009.
[93] F. Parmigiani, M. Ibsen, T. T. Ng, L. Provost, P. Petropoulos, and D. J. Richardson, "An efficient wavelength converter exploiting a grating-based saw-tooth pulse shaper," *IEEE Photonics Technology Letters*, vol. 20, no. 17, pp. 1461–1463, 2008.
[94] F. Parmigiani, M. Ibsen, P. Petropoulos, and D. J. Richardson, "Efficient all-optical wavelength-conversion scheme based on a saw-tooth pulse shaper," *IEEE Photonics Technology Letters*, vol. 21, no. 24, Article ID 5291729, pp. 1837–1839, 2009.
[95] N. Verscheure and C. Finot, "Pulse doubling and wavelength conversion through triangular nonlinear pulse reshaping," *Electronics Letters*, vol. 47, no. 21, pp. 1194–1196, 2011.
[96] A. I. Latkin, S. Boscolo, R. S. Bhamber, and S. K. Turitsyn, "Doubling of optical signals using triangular pulses," *Journal of the Optical Society of America B*, vol. 26, no. 8, pp. 1492–1496, 2009.
[97] R. S. Bhamber, S. Boscolo, A. I. Latkin, and S. K. Turitsyn, "All-optical TDM to WDM signal conversion and partial regeneration using XPM with triangular pulses," in *Proceedings of the 34th European Conference on Optical Communication (ECOC '08)*, pp. 1–2, Brussels, Belgium, 2008, paper Th.1.B.2.
[98] E. R. Andresen, J. M. Dudley, D. Oron, C. Finot, and H. Rigneault, "Transform-limited spectral compression by self-phase modulation of amplitude-shaped pulses with negative chirp," *Optics Letters*, vol. 36, no. 5, pp. 707–709, 2011.
[99] C. Finot, F. Parmigiani, P. Petropoulos, and D. J. Richardson, "Parabolic pulse evolution in normally dispersive fiber amplifiers preceding the similariton formation regime," *Optics Express*, vol. 14, no. 8, pp. 3161–3170, 2006.
[100] P. V. Mamyshev, "All-optical data regeneration based on self-phase modulation effect," in *Proceedings of the 24th European Conference on Optical Communication (ECOC '98)*, pp. 475–476, Madrid, Spain, September 1998.
[101] L. Provost, C. Finot, P. Petropoulos, K. Mukasa, and D. J. Richardson, "Design scaling rules for 2R-optical self-phase modulation-based regenerators," *Optics Express*, vol. 15, no. 8, pp. 5100–5113, 2007.
[102] C. Finot, J. Fatome, S. Pitois, G. Millot, and E. Pincemin, "Active Mamyshev regenerator," *Optical Review*, vol. 18, no. 3, pp. 257–263, 2011.
[103] M. Romagnoli, P. Franco, R. Corsini, A. Schiffini, and M. Midrio, "Time-domain Fourier optics for polarization-mode dispersion compensation," *Optics Letters*, vol. 24, no. 17, pp. 1197–1199, 1999.
[104] L. A. Jiang, M. E. Grein, H. A. Haus, E. P. Ippen, and H. Yokoyama, "Timing jitter eater for optical pulse trains," *Optics Letters*, vol. 28, no. 2, pp. 78–80, 2003.
[105] M. Nakazawa and T. Hirooka, "Distortion-free optical transmission using time-domain optical Fourier transformation and transform-limited optical pulses," *Journal of the Optical Society of America B*, vol. 22, no. 9, pp. 1842–1855, 2005.
[106] T. Hirooka and M. Nakazawa, "Optical adaptive equalization of high-speed signals using time-domain optical Fourier transformation," *Journal of Lightwave Technology*, vol. 24, no. 7, pp. 2530–2540, 2006.
[107] N. Yoshikane, I. Morita, T. Tsuritani, A. Agata, N. Edagawa, and S. Akiba, "Benefit of SPMbased all-optical reshaper for long-haul DWDM transmission systems," *Journal of Selected Topics in Quantum Electronics*, vol. 10, pp. 412–420, 2004.
[108] M. Matsuura, N. Kishi, and T. Miki, "Performance improvement of optical RZ-receiver by utilizing an all-optical waveform converter," *Optics Express*, vol. 13, no. 13, pp. 5074–5079, 2005.
[109] P. Ghelfi, M. Secondini, M. Scaffardi, F. Fresi, A. Bogoni, and L. Poti, "Impact of an additional all-optical decision element in band-limited receivers for RZ systems," *Journal of Lightwave Technology*, vol. 25, no. 7, pp. 1728–1734, 2007.
[110] M. Rochette, J. N. Kutz, J. L. Blows, D. Moss, J. T. Mok, and B. J. Eggleton, "Bit-error-ratio improvement with 2R optical regenerators," *IEEE Photonics Technology Letters*, vol. 17, no. 4, pp. 908–910, 2005.
[111] M. Rochette, J. L. Blows, and B. J. Eggleton, "3R optical regeneration: an all-optical solution with BER improvement," *Optics Express*, vol. 14, no. 14, pp. 6414–6427, 2006.

Nonlinear Coherent Directional Coupler: Coupled Mode Theory and BPM Simulation

Dharmadas Kumbhakar

Department of Electronics and Communication Engineering, Asansol Engineering College, Asansol, West Bengal 713305, India

Correspondence should be addressed to Dharmadas Kumbhakar, dkbbcasn@gmail.com

Academic Editor: Vittorio M. N. Passaro

Finite difference beam propagation method is an accurate numerical procedure, used here to explore the switching dynamics of a nonlinear coherent directional coupler. The coupling lengths derived from this simulation are compared with coupled mode theories. BPM results for the critical power follow the trend of the coupled mode theories, but it lies in between two coupled mode theories. Coupled mode theory is sensitive to numerical approximations whereas BPM results practically do not depend on grid size and longitudinal step size. Effect of coupling-region-width and core-width variations on critical power and coupling length is studied using BPM to look at the aspects of optical power-switch design.

1. Introduction

Beam propagation method [1] is a versatile tool to investigate or model various optical phenomena in photonic devices. Earlier, the method was popular to find the guided modes in dielectric waveguide structures [2]; even in microstructure fibers like holey fibers it can be used to find the modes [3]. With suitable boundary conditions, it gave proper estimates of the propagation constant in leaky structures like arrow waveguides [4]. The method was used in linear directional couplers to simulate switching of a CW signal and filtering TE/TM mode signals [5]. With growing interest in nonlinear fibers, it was the only tool to model the pulse propagation [6] in conjugation with split-step Fourier transform. It has been also used to model various other nonlinear phenomena like second harmonic generation [7], leakage loss in buried silicon substrate [8], and so forth. Beam propagation method can be incorporated either in the finite element or finite difference framework. In cases where the interfaces are parallel to the axes, the two methods are almost the same in terms of accuracy, but the latter is sometimes preferred due to its simplicity in implementation.

Nonlinear directional coupler (NLDC) is a very useful device in photonic circuits; it can be used as all optical modulator, switches, logic gates, and so forth. Jenson [9] first showed that total power exchange in a linear coherent coupler is lost in a nonlinear coherent coupler above a critical input power in one guide. This fact was used to design power-controlled all optical switches [10, 11]. The device can also be used as phase-controlled switches [12] and optical modulator [13]. So, the knowledge of the dependence of the critical power and phase variations on the NLDC parameters is important. There are numerous studies on this device; analytic methods include coupled mode theory [9, 14], variational method [15], and power flow portrait [16]. Numerical studies are mainly based on finite element [17, 18] or finite difference beam propagation methods [19–21]. There are very few experimental results reported so far (see [22] and references therein).

In some works, explicit FDBPM is used to model some nonlinear phenomena [18, 20], we preferred the implicit formulation, and it has been proved to be robust and fast. In this paper, we have simulated 2D nonlinear coherent directional coupler with implicit finite difference beam propagation method (IFDBPM). The results are compared with coupled

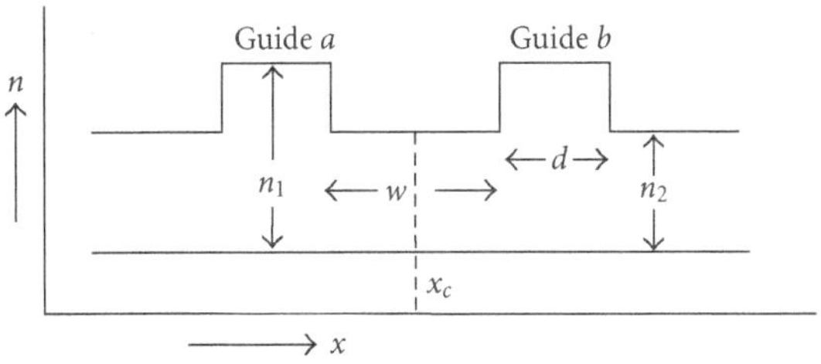

FIGURE 1: Structure of a coherent directional coupler.

mode theory. Some aspects of power-switching design are also investigated.

2. Coupled Mode Theory

Figure 1 shows the structure of a 2D symmetric or coherent directional coupler. The refractive index is given by

$$n_{dc}^2(x) = n_1^2 \quad \text{for } \left| x - x_c \pm \left(\frac{w}{2} + \frac{d}{2}\right) \right| \le \frac{d}{2}$$
$$= n_2^2 \quad \text{elsewhere.} \tag{1}$$

We assume that the coupling is small so that the power flow in the two waveguide components can be obtained from the individual waveguide modes. The individual waveguide structure is obtained removing the second core region. The index distribution in individual guides is given by
guide a

$$n_w^2(x) = n_1^2 \quad \text{for } \left| x - x_c + \left(\frac{w}{2} + \frac{d}{2}\right) \right| \le \frac{d}{2}$$
$$= n_2^2 \quad \text{elsewhere,} \tag{2a}$$

guide b

$$n_w^2(x) = n_1^2 \quad \text{for } \left| x - x_c - \left(\frac{w}{2} + \frac{d}{2}\right) \right| \le \frac{d}{2}$$
$$= n_2^2 \quad \text{elsewhere.} \tag{2b}$$

The field evolution along the length of the directional coupler can be expressed as

$$\Psi(x,z) = \left[A(z)\psi_a(x) + B(z)\psi_b(x)\right]e^{-i\beta z}. \tag{3}$$

$\psi_a(x)$ and $\psi_b(x)$ are the normalized mode fields of individual waveguides a and b, respectively. $A(z)$ and $B(z)$ are the amplitudes of the fields in waveguide a and b, respectively. Using coupled mode theory, one gets

$$\frac{dA}{dz} = i\left[K_0A + K_1B + \left(K_2|A|^2 + 2K_3|B|^2\right)A\right], \tag{4a}$$

$$\frac{dB}{dz} = i\left[K_0B + K_1A + \left(K_2|B|^2 + 2K_3|A|^2\right)B\right], \tag{4b}$$

where

$$K_0 = \frac{k_0^2}{2\beta}\int_b \Delta n^2 \left|\psi_a(x)\right|^2 dx, \tag{5a}$$

$$K_1 = \frac{k_0^2}{2\beta}\int_a \psi_a(x)\Delta n^2 \psi_b(x) dx, \tag{5b}$$

$$K_2 = \frac{k_0^2}{2\beta}\int \alpha\left|\psi_a(x)\right|^4 dx, \tag{5c}$$

$$K_3 = \frac{k_0^2}{2\beta}\int \alpha\left|\psi_a(x)\right|^2\left|\psi_b(x)\right|^2 dx, \tag{5d}$$

$k_0 = 2\pi/\lambda$, λ is the free space wavelength and β is the propagation constant in the coupler. $\alpha = \varepsilon_0 c_0 n^2 n_2'$, ε_0 is the free space electric permittivity, and c_0 is the velocity of light in free space; n_2' is the nonlinear coefficient of the nonlinear medium either core or cladding or both as the case may be and n is the linear refractive index of that medium. Δn^2 is the perturbing refractive index which will be explained in Section 4.

The index a or b in the integrals indicates integration within the core region of waveguide a or b. K_0 is the contribution of the evanescent field of one waveguide in the other; it just modifies the propagation constant β [23]. K_1 is the overlap of the evanescent field with the field into the other; it has the pivotal role in the exchange of power in a linear directional coupler. K_2 is the contribution due to self phase modulation and K_3 is due to cross-phase modulation.

For weak coupling, one can neglect the small contributions K_0 and K_3; this is justified with the evaluation of these parameters in Section 4.

If P_0 is the initial power fed into the waveguide a, $P_0 = FA(0)^2$, where $F = (1/2)\varepsilon_0 c_0 n_e$, n_e is the effective index of the waveguide. Then the power $P(z)$ in that guide at a distance z from the input end can be expressed as

$$\left(\frac{dP}{dz}\right)^2 + 4K_1P(P_0 - P) - K_2'P^2(P_0 - P)^2 = 0, \tag{6}$$

where $K_2' = K_2/F$. The above equation is obtained from (4a) and (4b) and using $P_0 = F(|A|^2 + |B|^2) =$ constant, assuming the coupler to be loss-free.

Equation (6) can be easily converted to Jacobi elliptic equation

$$\left(\frac{dP'}{dz'}\right) = \left(1 - P'^2\right)\left(1 - m + m\,P'^2\right), \tag{7}$$

where $P' = (P - P_0/2)/P_0/2$, $z' = 2K_1z$, and $m = (P_0K_2'/4K_1)^2$.
For $m \le 1$, the solution of (7) is

$$P = \frac{1}{2}P_0\left[1 + cn(2K_1z \mid m)\right]. \tag{8}$$

cn is the relevant Jacobi elliptic function. The value of the input power P_0 for which $m = 1$ is called the critical power. So, critical power $P_c = 4K_1/K_2'$.

The coupling length L is the minimum value of z at which power in the waveguide falls from initial value P_0 to zero. So,

$$cn(2K_1L \mid m) = -1. \quad (9)$$

For linear directional coupler, $m = 0$, the coupling length L_0 is given by

$$\cos(2K_1L_0) = -1 \qquad \text{or } L_0 = \frac{\pi}{2K_1}. \quad (10)$$

Above the critical input power ($m > 1$), (7) can be written as

$$\left(\frac{dP'}{d(mz')}\right) = \left(P'^2 - 1\right)\left(1 - m^{-1} - P'^2\right). \quad (11)$$

Its solution is

$$P = \frac{1}{2}P_0[1 + dn(2K_1mz \mid m^{-1})]. \quad (12)$$

3. Beam Propagation Method

In the linear regime, the paraxial scalar wave equation in finite difference form results in a linear matrix equation; evolution of the field in the z-direction is then obtained successively solving the linear equation through small steps. With Kerr or non-Kerr-type nonlinearity, the equation is not linear. However, the nonlinear contribution in the paraxial wave equation, which contains the electric field value, can be used to modify the refractive index profile. The electric field can then be obtained as in the linear case. The modal field and propagation constant of a three-layer or multilayer waveguide can then be obtained in this process with finite difference inverse iteration method or shifted power method or imaginary distance beam propagation method. But when field evolution is required as in a directional coupler, to keep or develop the proper phase of the fields, the successive solution of the original paraxial wave equation with imaginary quantities is essential.

In z-invarinat or slow z-variant structures, the light propagation is given by the paraxial wave equation

$$2i\beta\frac{\partial\Psi}{\partial z} = \frac{\partial^2\Psi}{\partial x^2} + k_0^2\left(n_{dc}^2(x) + \alpha|\Psi|^2 - \beta^2\right)\Psi. \quad (13)$$

If $\psi_n(x)$ is the field distribution in the directional coupler at the nth step, in the $(n+1)$th step, the index distribution can be taken to be effectively

$$n^2(x) = n_{dc}^2(x) + \alpha\left|\psi_n(x)\right|^2. \quad (14)$$

The field distribution at the $(n+1)$th step can be expressed as

$$\psi_{n+1} = \left(1 + \frac{G\Delta z}{2i\beta}\right)(\xi\psi_{n+1} + (1-\xi)\psi_n). \quad (15)$$

In finite difference form, x is discretised into N segments; ψ_n is an N-dimensional column matrix and G is an $N\times N$ matrix operator. n^2 is revised at each step with the latest ψ_n using (14) and then ψ_n is updated through (15). ξ is a parameter required for numerical convergence. In Crank-Nicholson scheme, it is 0.5. $\xi < 0.5$ is called explicit formulation and $\xi \geq 0.5$ is called implicit formulation. In fully explicit form ($\xi = 0$), ψ_{n+1} can be obtained directly with (15). This requires very small steps of Δz for convergence and consequently long runtime. In implicit formulation, a matrix inversion is necessary at each step; however, it is very stable and large Δz can be used [24]. Matrix inversion is very easy and fast using the sparse matrix techniques in Matlab; for that reason we have used implicit formulation for fast and stable computation.

4. Coupling Length Variations with Power

Coupled mode theory [9] considered only the linear perturbation contribution in Δn^2

$$\Delta n^2 = \left(n_{dc}^2 - n_w^2\right), \quad (16)$$

and the individual waveguide modes were also taken to be the solution of Helmholtz equation without the nonlinear term:

$$\frac{d^2\psi}{dx^2} + k_0^2\{n_w^2(x) - n_e^2\}\psi = 0. \quad (17)$$

n_e is the effective index of the waveguide.

Power flow in the coherent nonlinear directional coupler was derived from coupled mode equations.

Later improved coupled mode theory [14] derived the power flow equations from Lorentz reciprocity theorem and used the nonlinear perturbation in Δn^2

$$\Delta n^2 = \left(n_{dc}^2 - n_w^2\right) + \alpha\left|\psi_b\right|^2_{\text{actual}}. \quad (18)$$

Individual waveguide modes were obtained using the nonlinear Helmholtz equation

$$\frac{d^2\psi}{dx^2} + k_0^2\left\{n_w^2(x) - n_e^2 + \alpha\left|\psi\right|^2\right\}\psi = 0. \quad (19)$$

Actual field $|\psi|_{\text{actual}}$ and normalized ψ are related as

$$|\psi|_{\text{actual}} = \psi\sqrt{\frac{P}{\varepsilon_0 c_0 n_e}}. \quad (20)$$

Two couplers with different nonlinearity configurations are used in this paper to show the versatility of our BPM simulation model.

The mode fields are determined with highly accurate finite difference inverse iteration method which we have developed and coded in Matlab. The transverse grid size used in this evaluation is 0.001 μm and it takes only a few seconds. These fields are used to find the coupling constants K_1 and K_2.

K_0, K_3 are found to be two orders smaller in magnitude than K_1, K_2, respectively. One can solve the coupled equations (4a) and (4b) numerically; it shows that these terms practically have no effect on coupling length and critical power within the range of power level we consider.

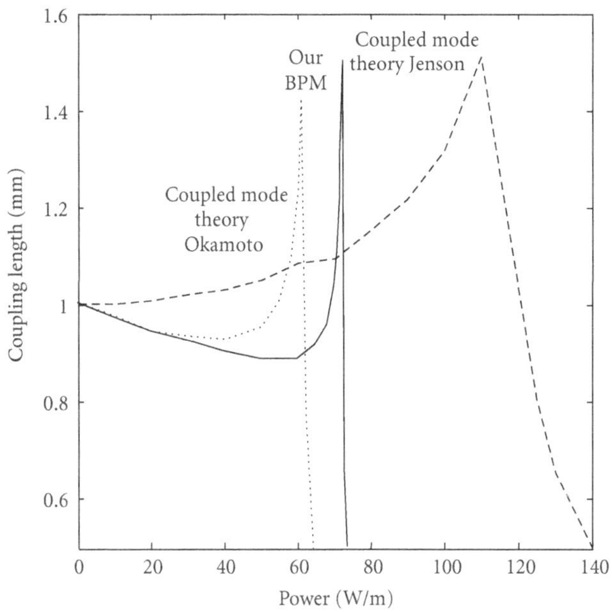

FIGURE 2: Comparison of coupling length variation with power: coupled mode theory 1 (Jenson): dashed curve, coupled mode theory 2 (Okamoto): dotted curve, our BPM simulation: solid curve.

A smaller transverse grid size like 0.04 μm is used to determine critical power with BPM simulation. A longitudinal step size of 1 μm is found to be enough. Refining transverse grid size to 0.01 and longitudinal step size to 0.1 μm does not produce any significant change in critical power estimate.

4.1. Nonlinear Coupling Region. The 2D directional coupler parameters are taken as in [14]: $n_1 = 1.57$; $n_2 = 1.55$; $\lambda = 1.064\,\mu$m; $d = 2\,\mu$m; $w = 2.4\,\mu$m; $\alpha = 6.3786\,\text{m}^2/\text{V}$. Central cladding layer is taken to be nonlinear.

Figure 2 shows the variation of coupling length in this coupler as a function of input power as obtained from two coupled mode theories and our BPM simulation. The results of two coupled mode theories are widely different; the BPM simulation results follow the trend of improved coupled mode theory [14]; the critical power is slightly higher in BPM.

4.2. Nonlinear Core Regions. We have also studied a directional coupler with nonlinear core regions. The coupler parameters are $n_1 = 1.56$, $n_2 = 1.55$, $d = 3\,\mu$m, $W = 5\,\mu$m, $\lambda = 1.3\,\mu$m, and $\alpha = 0.15\,\text{m}^2/\text{V}$. The variation of coupling length with power for this coupler is shown in Figure 3.

The critical power in our BPM simulation results lies in between the results of two coupled mode theories. For high nonlinearity, BPM simulation curve for the variation of coupling length with power shifts towards improved coupled mode theory and, for low nonlinearity, it shifts towards the other coupled mode theory.

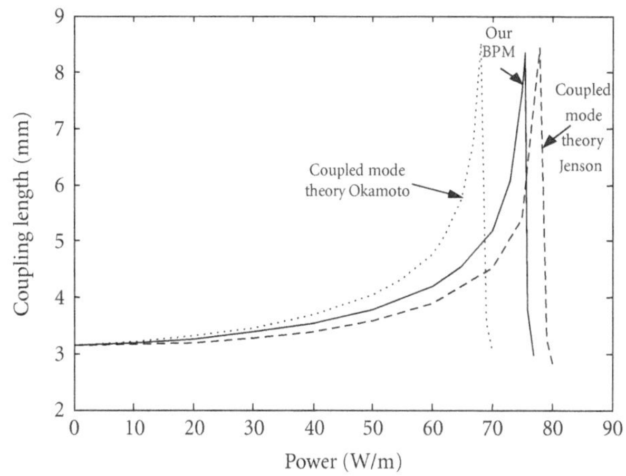

FIGURE 3: Power versus coupling length. Line properties are same as in Figure 2.

Although BPM algorithm is quite different from coupled mode theories, the functional form of the power flow in the coupler as obtained with both the approaches is very similar, apart from some shift as input power is increased. To justify this, we compare BPM simulation and coupled mode theory with the coupler structure of Section 4.2. Figure 4(a) shows the BPM simulation picture of power evolution in the coupler for low input power (0.1 W/m). In this low power regime, only one coupling constant K_1 describes power flow:

$$K_1 = \frac{\pi}{2L_0}, \tag{21}$$

where L_0 is the low power coupling length. Figure 4(b) shows the plot of the function $P = P_0[1 + \cos(2K_1z)]$ and the computed total power in the guide a from the field evolution obtained with BPM simulation shown in Figure 4(a). As revealed in Figure 4(b), agreement is excellent.

Figure 4(c) shows the field evolution with BPM for an input power 70 W/m, near but below the critical power (75 W/m). The nature of the power flow in BPM is different from cosine function as the look of the field evolution (Figure 4(c)) tells. The power flow pattern according to coupled mode theory is (8), that is, $P = (1/2)P_0(1 + cn(2K_1z \mid m))$. We check whether power flow obeys this relation. We take a slightly different (lower) value of K_1, than given by (21); find K_2, m from K_1 and critical power determined from BPM. Putting these values in (8), plot of P-z clearly matches the BPM power flow, shown in Figure 4(d). It clearly shows that power flow happens to be of cn nature below critical power.

Figure 4(e) shows the field evolution at a power 100 W/m which is above the critical power. The power is almost contained in the guide a; almost 90% on the average. Only about 10% power is exchanged between the two arms of the coupler. The match of the BPM power flow curve with the coupled mode theory equation (12) in Figure 4(f) verifies the fact that the undulations in the peak power are of dn nature according to (12).

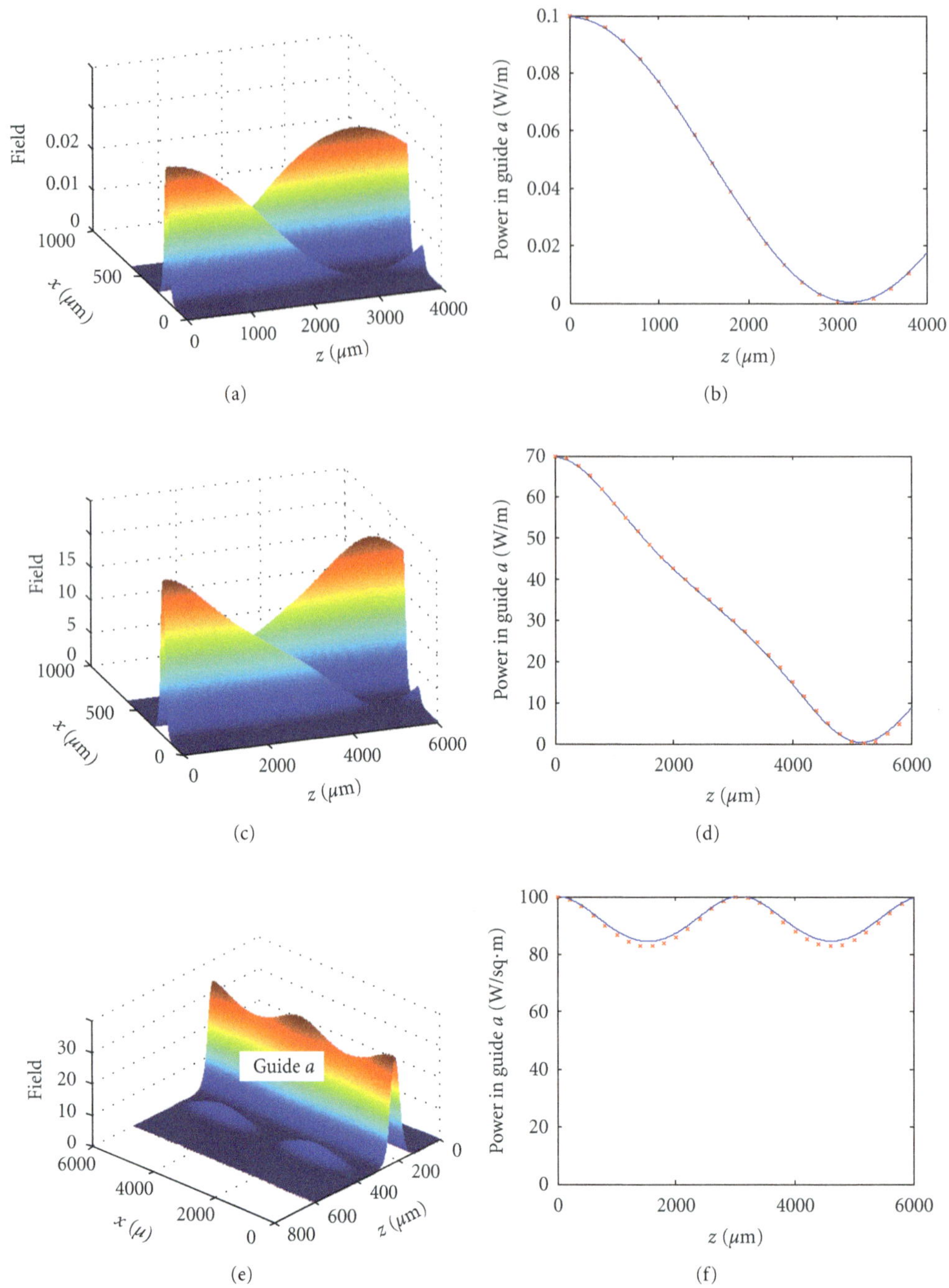

FIGURE 4: (a) Field evolution at 0.1 W/m input power: BPM. (b) Power flow pattern, (line curve) computed from BPM shown left, (crossed points) coupled mode theory. (c) Field evolution at 70 W/m power: BPM. (d) Power flow pattern (line curve), computed from BPM shown left, (crossed points) coupled mode theory fitted. (e) Field evolution at 100 W/m power: BPM. (f) Power flow pattern, (line curve) computed from BPM shown left, (crossed points) coupled mode theory fitted.

In the experiments [22] on the nonlinear switching of directional couplers, GaAs/AlGaAs multiple quantum well structures were used and operated at wavelengths near the band edge to tap sufficient Kerr nonlinearity due to exitonic resonances. However, these structures exhibit large losses. To reduce the losses, later, only the coupling region was made nonlinear. Even with such improved structures, the loss is of the order of $10\,\text{cm}^{-1}$. With high initial powers, although straight-through condition is achieved, some power is coupled to the other guide due to power loss in the lossy nonlinear medium. The experimental data reveals that power flow is different from Jenson's theory but it is insufficient to differentiate further.

In coupled mode theory, the coupling constants $K_0 - K_3$ are determined with the solution of the Helmholtz equation in individual wave guide ψ_a and ψ_b at a fixed power. However,

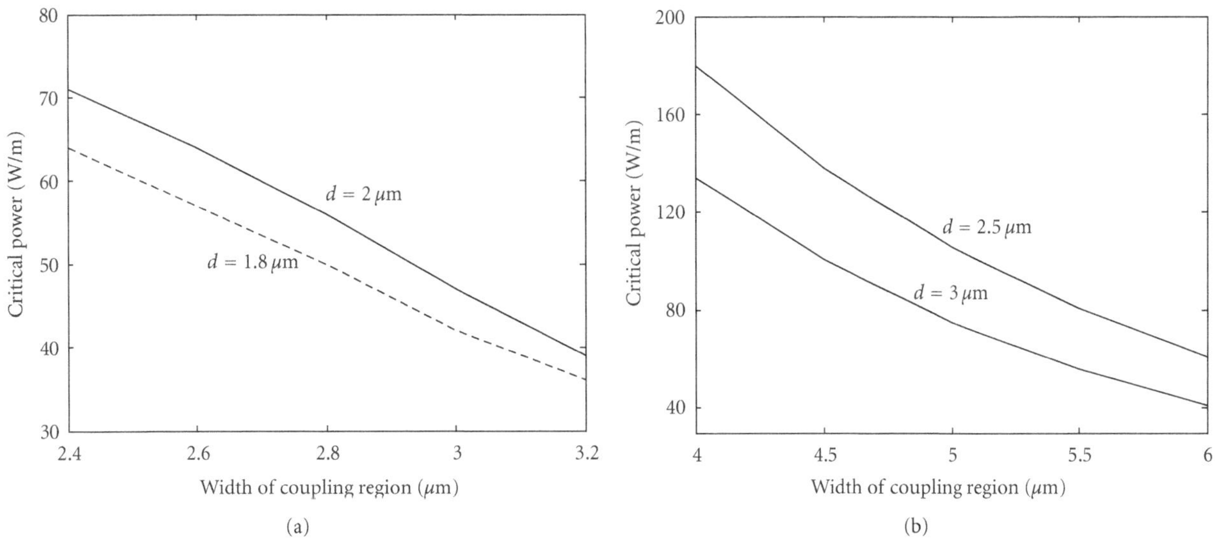

FIGURE 5: (a) Variation of critical power with core separation: nonlinear coupling region. (b) Variation of critical power with core separation: nonlinear core regions.

the power in each waveguide gets changing all along the length due to power transfer. Coupled mode theory did not incorporate the phase and field overlap changes continuously. But in the BPM, at any step, the field is computed with the instantaneous power. So, it incorporates the power and phase change continuously. Power transfer is likely to be more complicated than that predicted by existing theoretical frameworks. Some mechanism to get proper coupling constants in the coupled mode theory may be one step forward.

5. Effect of Core and Coupling Width Variations

A nonlinear directional coupler can find application as a power-controlled switch. If power is input at the waveguide a, with proper choice of the length of the directional coupler, at low powers the optical power is switched out through the other guide b; above a critical power, the optical power is retained and out through the guide a itself.

Obviously, the nonlinear medium—either the coupling region or core regions—must have high Kerr-type nonlinearity to operate the coupler as an optical switch at moderately low critical power. The coupling length of a coupler is also an important factor to optimize the size of the device. So, one must look at the coupler dimensions which govern these two quantities—critical power and coupling length. We have studied the variation of these two quantities with the change of the width of the coupling region and the width of the core regions for both the types of nonlinear couplers studied in Section 4. Figure 5(a) shows the variation of critical power with the change in w (width of the coupling region), core width d as a parameter—for the coupler with nonlinear coupling region (Section 4.1). Figure 5(b) shows those variations for the nonlinear core coupler (Section 4.2). As a general rule, the critical power falls with increase in the separation between the guides. If d is kept fixed, K_2 remains unchanged as the mode profile ψ_a (or ψ_b) does not change (see (5c)); however, as w is increased, the value of K_1 decreases due to reduced overlap of the fields (see (5b)). This is the reason that critical power ($P_c = 4K_1/K_2'$) decreases with increasing w.

But the effect of core-width d variation is different for two types of nonlinear couplers. With the decrease in d, the field inside the core decreases in the first type of coupler—where the coupling region is nonlinear. The decrease of field inside the core leads to field spreading that increases the overlap to some extent, but more is the increase in K_2 as the field increases in the nonlinear region. For the other type, just the opposite happens—field inside the core increases with decrease in d. That shrinks the fields, increasing the overlap but keeping K_2 almost the same.

Figures 6(a) and 6(b) show the variations in the low power coupling length. As the separation between the coupler arms increases, K_1 decreases due to reduction in the field overlap. Coupling length ($\pi/2K_1$) increases as a result whatever the value of d is.

To summarize, critical power decreases as the core separation is increased in both the two types of couplers we consider. But as the core separation is increased, coupling length also increases. So, to contain the size of the switching device, one has to optimize between the critical power and the coupling length.

6. Conclusion

We have developed an implicit BPM algorithm which gives the effective power flow in a nonlinear directional coupler, easily and correctly. The method is fairly insensitive to transverse and longitudinal step size. It appears to be more effective than the coupled mode theories which use some approximations. Hence IFDBPM should find wide applications in investigation of NLDC.

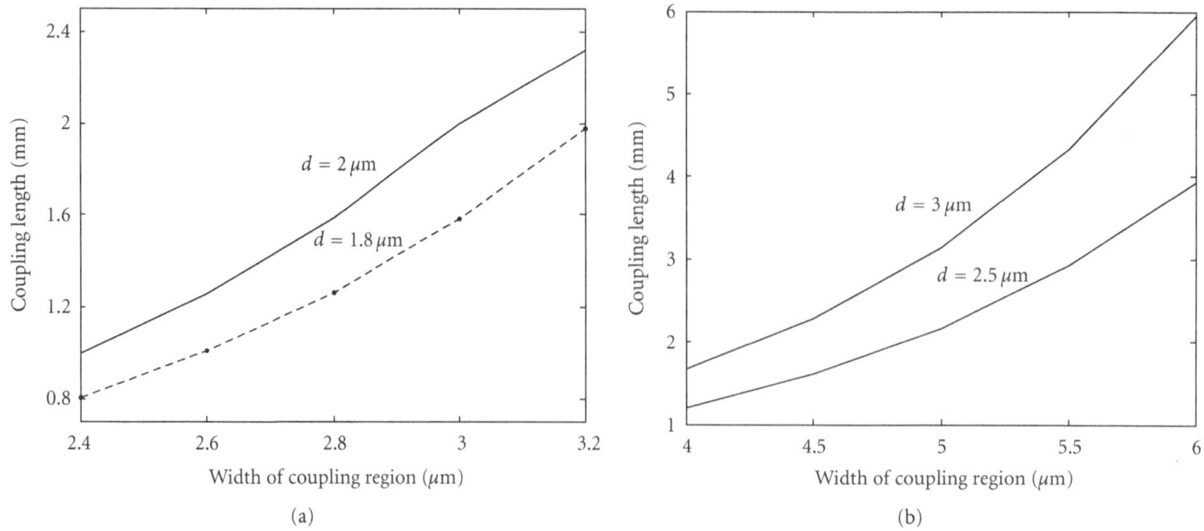

FIGURE 6: (a) Variation of coupling length with core separation: nonlinear coupling region. (b) Variation of coupling length with core separation: nonlinear core regions.

Acknowledgment

The author is very grateful to Professor S. N. Sarkar, Department of Electronic Science, Calcutta University for a critical reading of the paper and making some useful suggestions.

References

[1] D. Yevick and B. Hermansson, "New formulations of the matrix beam propagation method: application to rib waveguides," *IEEE Journal of Quantum Electronics*, vol. 25, no. 2, pp. 221–229, 1989.

[2] W. Huang, C. Xu, S. T. Chu, and S. K. Chaudhuri, "The finite-difference vector beam propagation method: analysis and assessment," *IEEE Journal of Lightwave Technology*, vol. 10, no. 3, pp. 295–305, 1992.

[3] K. Saitoh and M. Koshiba, "Full-vectorial imaginary-distance beam propagation method based on a finite element scheme: application to photonic crystal fibers," *IEEE Journal of Quantum Electronics*, vol. 38, no. 7, pp. 927–933, 2002.

[4] W. P. Huang, C. L. Xu, W. Lui, and K. Yokoyama, "The perfectly matched layer boundary condition for modal analysis of optical waveguides: leaky mode calculations," *IEEE Photonics Technology Letters*, vol. 8, no. 5, pp. 652–654, 1996.

[5] W. P. Huang, C. L. Xu, and S. K. Chaudhuri, "Application of the finite-difference vector beam propagation method to directional coupler devices," *IEEE Journal of Quantum Electronics*, vol. 28, no. 6, pp. 1527–1532, 1992.

[6] G. P. Agrawal, *Nonlinear Fiber Optics*, Academic Press, San Diego, Calif, USA, 2001.

[7] P. S. Weitzman and U. Osterberg, "Modified beam propagation method to model second harmonic generation in optical fibers," *IEEE Journal of Quantum Electronics*, vol. 29, no. 5, pp. 1437–1443, 1993.

[8] J. Yamauchi, K. Ose, J. Shibayama, and H. Nakano, "Leakage loss and phase variation of a buried directional coupler on a silicon substrate," *IEEE Photonics Technology Letters*, vol. 18, no. 17, pp. 1873–1875, 2006.

[9] S. M. Jensen, "The nonlinear coherent coupler," *IEEE Journal of Quantum Electronics*, vol. 18, no. 10, pp. 1580–1583, 1982.

[10] M. Romagnoli, S. Trillo, and S. Wabnitz, "Soliton switching in nonlinear couplers," *Optical and Quantum Electronics*, vol. 24, no. 11, pp. S1237–S1267, 1992.

[11] S. Trillo and S. Wabnitz, "Anisotropic and Nonlinear Optical Waveguides," C. S. Someda and G. Stegeman, Eds., Elsevier, Amsterdam, The Netherlands, 1992.

[12] S. Wabnitz, E. M. Wright, C. T. Seaton, and G. I. Stegeman, "Instabilities and all-optical phase-controlled switching in a nonlinear directional coherent coupler," *Applied Physics Letters*, vol. 49, no. 14, pp. 838–840, 1986.

[13] A. T. Pham and L. N. Binh, "All-optical modulation and switching using a nonlinear-optical directional coupler," *Journal of the Optical Society of America B*, vol. 8, pp. 1914–1931, 1991.

[14] X. J. Meng and N. Okamoto, "Improved coupled-mode theory for nonlinear directional couplers," *IEEE Journal of Quantum Electronics*, vol. 27, no. 5, pp. 1175–1181, 1991.

[15] I. M. Uzunov, R. Muschall, M. Gölles, Y. S. Kivshar, B. A. Malomed, and F. Lederer, "Pulse switching in nonlinear fiber directional couplers," *Physical Review E*, vol. 51, no. 3, pp. 2527–2537, 1995.

[16] A. W. Snyder, D. J. Mitchell, L. Poladian, D. R. Rowland, and Y. Chen, "Physics of nonlinear fiber couplers," *Journal of the Optical Society of America B*, vol. 8, pp. 986–992, 1991.

[17] A. D. McAulay and X. Xu, "Finite element analysis of optically controlled non-linear directional coupler switches," *Optics and Laser Technology*, vol. 26, no. 4, pp. 251–258, 1994.

[18] B. M. A. Rahman, T. Wongcharoen, and K. T. V. Grattan, "Finite element analysis of nonsynchronous directional couplers," *Fiber and Integrated Optics*, vol. 13, no. 3, pp. 331–336, 1994.

[19] K. Yasumoto, H. Maeda, and K. Nakamura, "Beam-propagation method analysis of a three-waveguide nonlinear directional coupler," *Microwave and Optical Technology Letters*, vol. 9, no. 6, pp. 319–323, 1995.

[20] Y. Chung and N. Dagli, "Analysis of Z-invariant and Z-variant semiconductor rib waveguides by explicit finite difference

beam propagation method with nonuniform mesh configuration," *IEEE Journal of Quantum Electronics*, vol. 27, no. 10, pp. 2296–2305, 1991.

[21] H. F. Chou, C. F. Lin, and G. C. Wang, "An iterative finite difference beam propagation method for modeling second-order nonlinear effects in optical waveguides," *IEEE Journal of Lightwave Technology*, vol. 16, no. 9, pp. 1686–1693, 1998.

[22] P. R. Berger, P. K. Bhattacharya, and S. Gupta, "A waveguide directional coupler with a nonlinear coupling medium," *IEEE Journal of Quantum Electronics*, vol. 27, no. 3, pp. 788–795, 1991.

[23] A. K. Ghatak and K. Thyagrajan, *Introduction to Fiber Optics*, Cambridge University Press, 1999.

[24] S. Juengling and J. C. Chen, "Study and optimization of the eigenmode calculations using the imaginary-distance beam-propagation method," *IEEE Journal of Quantum Electronics*, vol. 30, no. 9, pp. 2098–2105, 1994.

10

Mie Plasmons: Modes Volumes, Quality Factors, and Coupling Strengths (Purcell Factor) to a Dipolar Emitter

G. Colas des Francs, S. Derom, R. Vincent, A. Bouhelier, and A. Dereux

Laboratoire Interdisciplinaire Carnot de Bourgogne, UMR 6303, CNRS-Université de Bourgogne, 9 Avenue A. Savary, BP 47 870, 21078 Dijon, France

Correspondence should be addressed to G. Colas des Francs, gerard.colas-des-francs@u-bourgogne.fr

Academic Editor: Nicolas Bonod

Using either quasistatic approximation or exact Mie expansion, we characterize the localized surface plasmons supported by a metallic spherical nanoparticle. We estimate the quality factor Q_n and define the effective volume V_n of the nth mode in such a way that coupling strength with a neighbouring dipolar emitter is proportional to the ratio Q_n/V_n (Purcell factor). The role of Joule losses, far-field scattering, and mode confinement in the coupling mechanism is introduced and discussed with simple physical understanding, with particular attention paid to energy conservation.

1. Introduction

Metallic nanoparticles support localized surface plasmonpolaritons (SPP) strongly confined at the metal surface ensuring efficient electromagnetic coupling with neighbouring materials, offering a variety of applications such as surface-enhanced spectroscopies [1, 2], photochemistry [3], or optical nanoantennas [4]. This also opens the way towards control of light emission at the nanoscale [5–10]. The ratio Q/V_{eff} is generally used to quantify the coupling strength between a dipolar emitter and a cavity mode (polariton). Q and V_{eff} refers to the mode quality factor and effective volume, respectively. High ratio Q/V_{eff} may lead to strong coupling regime with characteristics Rabi oscillations revealing cycles of energy exchange between the emitter and the cavity. In the weak coupling regime, the emitter energy dissipation into the cavity mode is nonreversible and follows the Purcell factor:

$$\frac{\gamma}{n_B \gamma_0} = \frac{3}{4\pi^2}\left(\frac{\lambda}{n_B}\right)^3 \frac{Q}{V_{\text{eff}}}, \tag{1}$$

where γ is the emitter spontaneous decay rate into the cavity compared to its free-space value γ_0, n_B is the optical index inside the cavity, and λ is the emission wavelength. Spontaneous emission rate in complex systems follows the more general Fermi's golden rule that expresses γ as a function of the density of electromagnetic mode [11]. However, for describing the coupling to a cavity mode, the Purcell factor is usually preferred since it clearly introduces the cavity resonance quality factor Q and the mode extension V_{eff} on which coupling remains efficient, bringing therefore a clear physical understanding of the coupling process.

In this context, we propose to determine the quality factors and effective volumes of localized SPPs supported by a metallic nanosphere. Indeed, these quantities are useful parameters to understand and evaluate the coupling mechanisms between dipolar emitters and plasmonic nanostructures [12, 13]. This would help for achieving strong coupling regime [14, 15] or for designing plasmonic nanolasers [16–18]. We have chosen a spherical particle, a highly symmetrical system, since it is fully analytical and simple expressions can be derived with clear physical meaning [19–23]. Moreover, our results could be extended to more complex structures [24, 25].

In Section 2, we focus on the dipolar mode and present in details the derivation of its quality factor and effective volume. We extend our approach to each mode of the particle in Section 3. Finally, we discuss the coupling efficiency to one of the particle modes in the last section. For the sake of simplicity, all the analytical expressions are derived assuming a particle in air and dipolar emitter perpendicular to the particle surface. The generalisation to arbitrary emitter orientation and a background medium of optical index n_B is provided in the appendix. Exact calculations presented in the main text using Mie expansion correctly include both the background medium and dipolar orientation.

2. Dipolar Mode

We first characterize the dipolar mode of a spherical particle. For sphere radius R small compared to the excitation wavelength $\lambda = 2\pi c/\omega$, the electric field is considered uniform over the metallic particle. The metallic particle is polarized by the incident electric field $\mathbf{E_0}$ and behaves as a dipole

$$\mathbf{p}^{(1)}(\omega) = 4\pi\epsilon_0\alpha_1(\omega)\mathbf{E}_0, \tag{2}$$

$$\alpha_1(\omega) = \frac{\epsilon_m(\omega) - 1}{\epsilon_m(\omega) + 2}R^3, \tag{3}$$

where α_1 is the nanoparticle quasistatic (dipolar) polarisability and ϵ_m is the metal dielectric constant. The dipole plasmon resonance appears at ω_1 such that $\epsilon_m(\omega_1) + 2 = 0$. In case of Drude metal, the dipolar resonance is $\omega_1 = \omega_p/\sqrt{3}$ with ω_p the bulk metal plasma angular frequency. However, expression (3) does not satisfy the optical theorem (energy conservation). It is wellknown that this apparent paradox is easily overcome by taking into account the finite size of the particle and leads to define the effective polarisability [26]:

$$\alpha_1^{\text{eff}} = \left[1 - i\frac{2k^3}{3}\alpha_1\right]^{-1}\alpha_1, \quad \left(k = \frac{2\pi}{\lambda}\right). \tag{4}$$

The corrective term $(2k^3\alpha_1/3)$ is the so-called radiative reaction correction and microscopically originates from the radiation emitted by the charges oscillations induced inside the nanoparticle by the excitation field [27].

The dipolar polarisability presents a simple shape near the resonance if the metallic dielectric constant follows Drude model [20] (ω_p and Γ_{abs} refers to metal plasma frequency an Ohmic loss rate, resp.) (for the sake of clarity, we use lowercase notation γ for the emitter decay rates and uppercase Γ for the quantities associated with the metallic particle);

$$\epsilon_m = 1 - \frac{\omega_p^2}{\omega^2 + i\Gamma_{\text{abs}}\omega}; \tag{5}$$

$$\alpha_1^{\text{eff}}(\omega) \underset{\omega_1}{\sim} \frac{\omega_1}{2(\omega_1 - \omega) - i\Gamma_1}R^3, \tag{6}$$

$$\Gamma_1 = \Gamma_{\text{abs}} + \frac{2(k_1R)^3\omega_1}{3}, \quad \left(k_1 = \frac{\omega_1}{c}\right). \tag{7}$$

where Γ_1 is the decay rate of the particle dipolar mode and includes both the Joule (Γ_{abs}) and radiative [$\Gamma_1^{\text{rad}} = 2(k_1R)^3\omega_1/3$] losses rates.

2.1. Quality Factor. The dipolar response can be described by either the extinction efficiency Q_{ext}, proportional to $\text{Im}(\alpha_1)$, or scattering efficiency Q_{scatt}, proportional to $|\alpha_1|^2$, and therefore follows a Lorentzian profile centered at ω_1 and with a full width at half maximum (FWHM) $\Delta\omega_1 = \Gamma_1$:

$$\begin{aligned} Q_{\text{ext}}(\omega) &= \frac{4k}{R^2}\,\text{Im}\left[\alpha_1^{\text{eff}}(\omega)\right] \\ &\underset{\omega_1}{\propto} \frac{1}{(\omega - \omega_1)^2 + (\Gamma_1/2)^2}, \\ Q_{\text{scatt}}(\omega) &= \frac{8}{3R^2}k^4\left|\alpha_1^{\text{eff}}(\omega)\right|^2 \\ &\underset{\omega_1}{\propto} \frac{1}{(\omega - \omega_1)^2 + (\Gamma_1/2)^2}. \end{aligned} \tag{8}$$

The quality factor of this resonance is therefore

$$Q_1 = \frac{\omega_1}{\Gamma_1}. \tag{9}$$

As an example, we consider a $R = 25$ nm silver sphere in air. The Drude model parameters are $\hbar\omega_p = 9.1$ eV and $\hbar\Gamma_{\text{abs}} = 18$ meV that lead to a resonance peak at $\hbar\omega_1 = 5.2$ eV ($\omega_1 = 7.98 \times 10^{15}$ Hz) and radiative energy $\hbar\Gamma_1^{\text{rad}} = 1$ eV. The quality factor is then $Q_1 = 5$. Note that the metal optical properties are better described when including the bound electrons in the Drude model: $\epsilon_m = \epsilon_\infty - \omega_p^2/(\omega^2 + i\Gamma_{\text{abs}}\omega)$, ($\epsilon_\infty = 3.7$ for silver, see also the appendix). In that case, we obtain $\hbar\omega_1 = 3.74$ eV ($\omega_1 = 5.7 \times 10^{15}$ Hz), $\hbar\Gamma_1^{\text{rad}} = 0.13$ eV, and $Q_1 = 24$.

Although Drude model qualitatively explains the shape of the resonance, a more representative value of the quality factor can only be determined using tabulated data for the dielectric constant of the metal [28]. The extinction efficiencies associated with the dipolar resonance are represented in Figure 1. For silver particle (Figure 1(a)), the extinction efficiency closely follows a Lorentzian shape, as expected, with a quality factor $Q_1 = 21$, in good agreement with the value obtained using Drude model (with the contribution of the bound electrons included). In case of gold, the resonance profile is not well defined due to interband transitions for $\omega > 4 \cdot 10^{15}$ Hz but a quality factor of 7 can be estimated (Figure 1(b)). Figure 1(c) represents the extinction efficiency for a gold particle embedded in polymethylmethacrylate (PMMA) into which metallic particles are routinely dispersed. This leads to a small redshift of a resonance, avoiding therefore the resonance disturbance by interband absorption, and we recover partly the Lorentzian profile [29].

2.2. Effective Volume. Coupling rate of a dipolar emitter to a dipolar particle expresses for very short emitter-particle distances d ($z_0 = R + d$ is the distance to the particle center) as [19–21]

$$\frac{\gamma_1^{\perp}}{\gamma_0} \sim \frac{6}{k^3z_0^6}\,\text{Im}(\alpha_1), \tag{10}$$

for a dipole emitter orientation perpendicular to the nanoparticle surface. Using (6), we obtain, in case of a dipolar

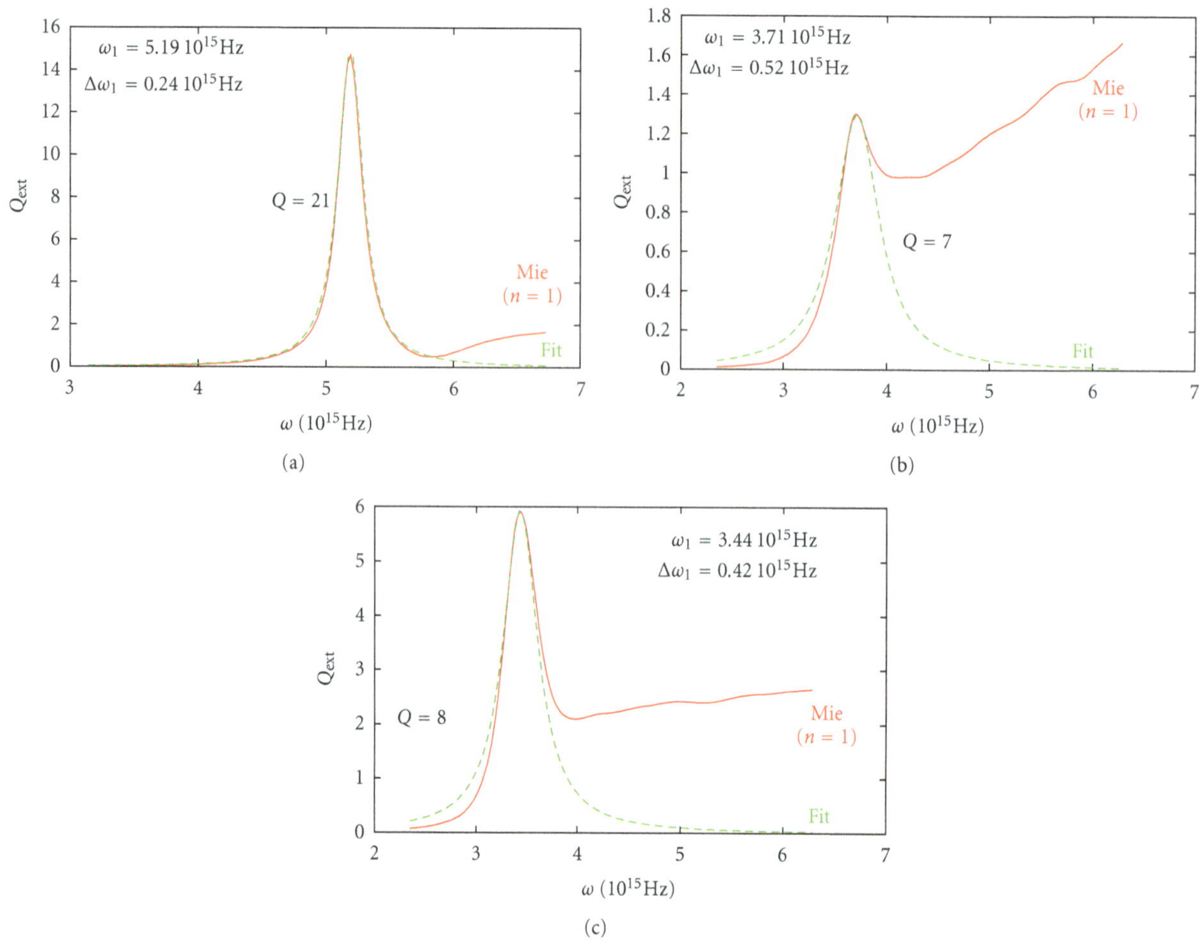

FIGURE 1: Extinction efficiency for the dipolar resonance calculated keeping only the dipolar mode ($n = 1$) in the Mie expansion and using experimental value for the metal dielectric constant [28]. (a) Silver particle in air. (b) Gold particle in air. (c) Gold particle in PMMA (optical index $n_B = 1.5$). The particle radius is $R = 25$ nm for each case. Fit refers to a Lorentzian fit using parameters indicated on the figure.

emitter emission tuned to the dipolar particle resonance ($\lambda = \lambda_1 = 2\pi c/\omega_1$)

$$\frac{\gamma_1^{\perp}}{\gamma_0} \underset{\omega_1}{\sim} \frac{6\omega_1 R^3}{k_1^3 z_0^6 \Gamma_1} \sim \frac{3}{4\pi^2}\lambda_1^3 \frac{R^3}{\pi z_0^6} Q_1. \quad (11)$$

In order to determine the dipolar mode effective volume, we now identify the coupling rate γ/γ_0 to the Purcell factor ((1), assuming $n_B = 1$), so that we obtain

$$V_1^{\perp} = \frac{\pi z_0^6}{R^3}, \quad (12)$$

in full agreement with expression recently derived by Greffet et al. by defining the optical impedance of the nanoparticle antenna [30]. For a $R = 25$ nm radius sphere in air, we estimate the mode effective volume $V_1^{\perp} = 3.7 \times 10^{-4}\,\mu\text{m}^3 = (72\,\text{nm})^3$ for an emitter 10 nm away from the particle surface ($z_0 = 35$ nm).

Unlike here, usual definition of the mode effective volume does not include the emitter position. For instance, in case of cavity quantum electrodynamics (cQED) applications, it can be expressed as $V_{\text{eff}} = \int \epsilon |\mathbf{E}(\mathbf{r})|^2 d\mathbf{r} / \max(\epsilon |\mathbf{E}(\mathbf{r})|^2)$, so that it directly characterizes the mode extension. However, in that case, Purcell factor expression (1) is only valid for an emitter located at the cavity center where the mode-emitter coupling is maximum (mode antinode). In close analogy with the definition used in cQED, we derive later another expression for the SPP effective volume, independent of the emitter position (see (22), and the discussion below).

3. Multipolar Modes

If the excitation field is generated by a dipolar emitter (fluorescent molecules, quantum dots, ...), it cannot be considered uniform anymore and the dipolar approximation fails. One needs therefore to consider the coupling strength to high-order modes (Figure 2). Here again, we discuss the mode quality factors and volume first using quasistatic approximation and then discuss their quantitative behaviour using exact Mie theory.

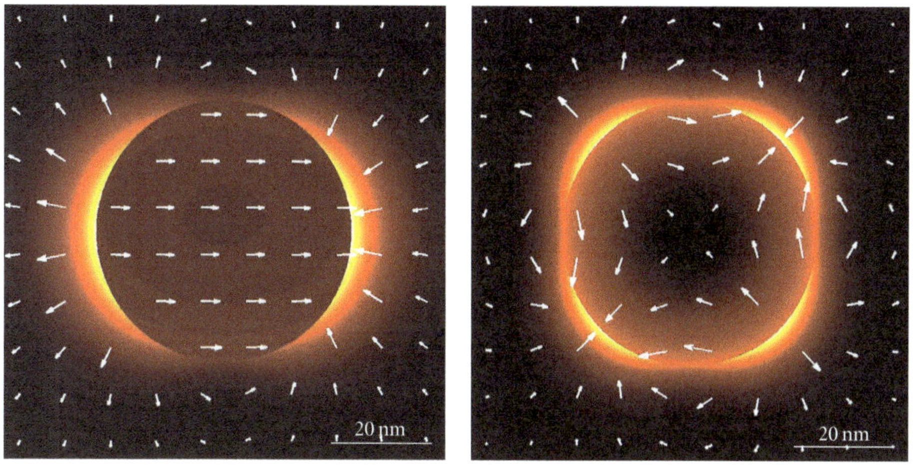

(a) (b)

FIGURE 2: (a) Dipolar and (b) quadrupolar mode profiles of a $R = 25$ nm silver sphere, embedded in PMMA calculated using exact Mie expansion. Silver dielectric constant is taken from [28]. Color and arrows refer to electric field intensity and vector respectively.

The nth multipole tensor moment of the metallic particle is given by

$$\mathbf{p}^{(n)} = \frac{4\pi\epsilon_0}{(2n-1)!!}\alpha_n \nabla^{n-1}\mathbf{E}_0, \tag{13}$$

$$\alpha_n = \frac{n(\epsilon_m - 1)}{n\epsilon_m + (n+1)} R^{(2n+1)}, \tag{14}$$

with $(2n+1)!! = 1\times3\times5\times\cdots\times(2n+1)$ and ∇ is the vector differential operator. As discussed above, the dipole moment $\mathbf{p}^{(1)} = 4\pi\epsilon_0\alpha_1\mathbf{E}_0$ is the unique mode excited in an uniform field. However, the dipolar emitter near field behaves as $1/r^3$ and strongly varies spatially so that higher modes can be excited in the particle.

For a Drude metal, the nth resonance appears at $\omega_n = \omega_p\sqrt{n/(2n+1)}$. Therefore, higher-order modes accumulate near $\omega_\infty = \omega_p/\sqrt{2}$. Moreover, as discussed for the dipolar case, quasistatic expression of the nth mode polarisability (14) does not obey energy conservation. Applying the optical theorem, we recently extend the radiative correction to all the spherical particles modes [22]. This leads to

$$\alpha_n^{\text{eff}} = \left[1 - i\frac{(n+1)k^{2n+1}}{n(2n-1)!!(2n+1)!!}\alpha_n\right]^{-1}\alpha_n. \tag{15}$$

3.1. Quality Factors. It is now a simple matter to generalize the dipolar mode analysis reported in the previous section to all the particle modes. The nth mode polarisability can be approximated near resonance. A simple expression for the polarisability is achieved considering a Drude metal:

$$\alpha_n^{\text{eff}} \underset{\omega_n}{\sim} \frac{\omega_n}{2(\omega_n - \omega) - i\Gamma_n} R^{2n+1},$$
$$\Gamma_n = \Gamma_{\text{abs}} + \Gamma_n^{\text{rad}}, \tag{16}$$
$$\Gamma_n^{\text{rad}} = \omega_n \frac{(n+1)(k_n R)^{2n+1}}{n(2n-1)!!(2n+1)!!}, \quad \left(k_n = \frac{\omega_n}{c}\right),$$

where Γ_n is the total decay rate of the nth mode, that includes both ohmic losses and radiative scattering. As expected, for a given mode n, the radiative scattering rate $\Gamma_n^{\text{rad}} \propto R^{2n+1}$ increases with the particle size since it couples more efficiently to the far field. For instance, we obtain $\hbar\Gamma_2^{\text{rad}} = 39$ meV ($\hbar\Gamma_2^{\text{rad}} = 1.8$ meV with $\epsilon_\infty = 3.7$) for the quadrupolar mode of a $R = 25$ nm silver particle in air. As expected, the radiative rate of the quadrupolar is strongly reduced compared to the dipolar mode.

The quality factor associated with the nth mode is therefore

$$Q_n = \frac{\omega_n}{\Gamma_n} = \frac{\omega_n}{\Gamma_{\text{abs}} + \Gamma_n^{\text{rad}}}. \tag{17}$$

Figure 3 details the quality factor of the two first modes of a silver sphere in PMMA using tabulated value for ϵ_m. The quadrupolar mode presents a quality factor almost 5 times higher than the dipolar mode since it has limited radiative losses. Indeed, quadrupolar mode poorly couples to the far field. Finally, similar Q values ($Q_n \approx 50$) are obtained for all the higher modes ($n \geq 3$). Here again, assuming a Drude metal and in the quasistatic approximation, we qualitatively explains this result. Actually, Q factors of high-order modes are absorption loss limited ($\Gamma_n^{\text{rad}} \ll \Gamma_{\text{abs}}$, so called dark modes) and tend to $Q_\infty = \omega_\infty/\Gamma_{\text{abs}}$. However this overestimates the mode quality factor ($Q_\infty \approx 207$) as compared to the value deduced using tabulated value for ϵ_m and exact Mie expansion.

3.2. Effective Volumes. Last, we express the *total* coupling strength of a dipolar emitter to the spherical metallic particle for very short separation distances ($kz_0 \ll 1$) [21, 22]:

$$\frac{\gamma_{\text{tot}}^{\perp}}{\gamma_0} \approx \frac{3}{2}\frac{1}{(kz_0)^3}\sum_{n=1}^{\infty}\frac{(n+1)^2}{z_0^{(2n+1)}}\operatorname{Im}\left(\alpha_n^{\text{eff}}\right), \tag{18}$$

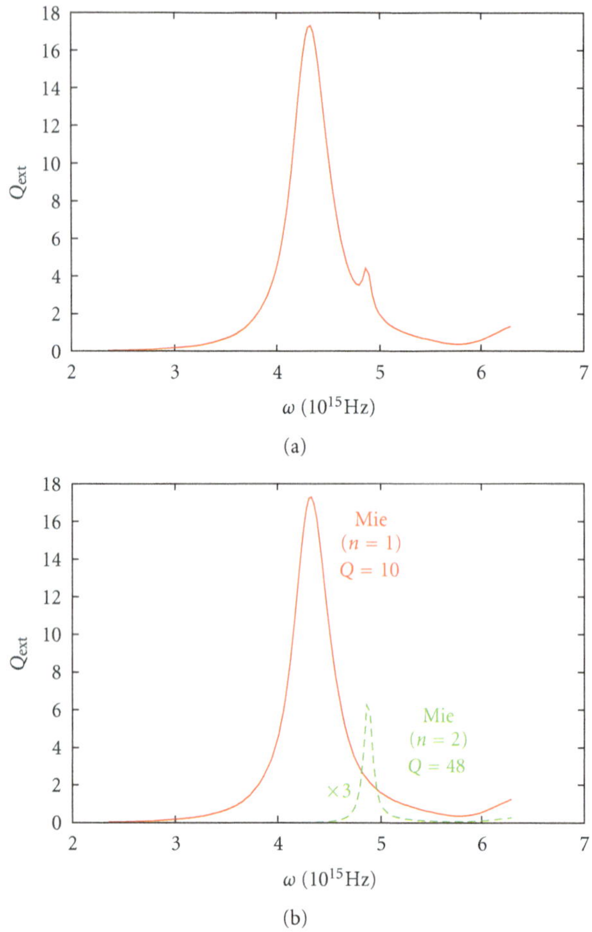

FIGURE 3: (a) Extinction efficiency of a $R = 25$ nm silver sphere embedded in PMMA, calculated using full Mie expansion. (b) Contribution of the dipolar ($n = 1$) and quadrupolar ($n = 2$) mode to the full extinction efficiency. The quality factor is indicated for each mode. Dielectric constant for silver is taken from [28].

so that the coupling strength to the nth mode is easily deduced as

$$\begin{aligned}\frac{\gamma_n^{\perp}}{\gamma_0} &\approx \frac{3}{2}\frac{1}{(kz_0)^3}\frac{(n+1)^2}{z_0^{(2n+1)}}\operatorname{Im}\left(\alpha_n^{\text{eff}}\right),\\ &\underset{\omega_n}{\sim}\frac{3}{2}\frac{R^{2n+1}}{(k_nz_0)^3}\frac{(n+1)^2}{z_0^{(2n+1)}}Q_n,\end{aligned} \tag{19}$$

where we have used approximated expression (16) for the nth polarisability. The mode effective volume is then straightforwardly deduced from comparison to the Purcell factor (1)

$$V_n^{\perp} = \frac{4\pi z_0^{2n+4}}{(n+1)^2 R^{2n+1}}. \tag{20}$$

We obtain for instance quadrupole effective volume $V_2^{\perp} = 4\pi z_0^8/(9R^5) = 3.2\times10^{-4}\,\mu\text{m}^3 = (68\,\text{nm})^3$ for an emitter 10 nm away from a 25 nm radius metallic particle in air. The calculated effective volumes of the three first modes are

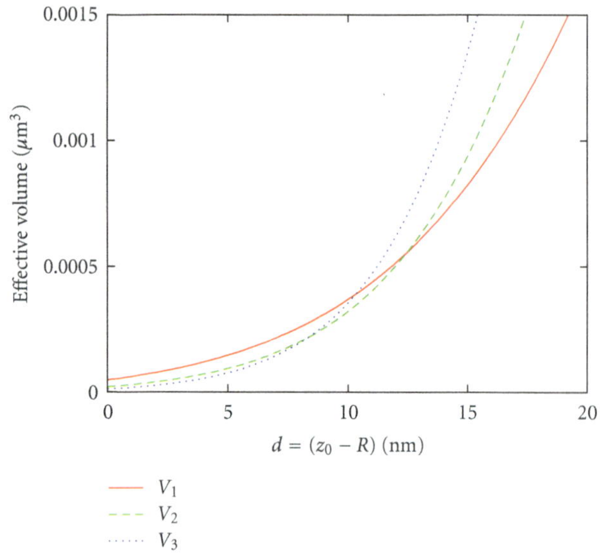

FIGURE 4: Effective volume of the dipole (V_1), quadrupole (V_2) and hexapole (V_3) modes of a $R = 25$ nm sphere as a function of the emitter-particle distance. The emitter is perpendicular to the particle surface.

shown on Figure 4. At long distance, the smaller effective volume (most efficient coupling to an emitter) is associated with the dipole mode since it presents the largest extension. When the emitter-particle distance decreases the coupling strength to the quadrupolar ($d \le 12$ nm), then hexapolar ($d \le 10$ nm) mode becomes stronger as revealed by their lower effective volume. Most efficient coupling between an emitter and one of the particle modes obviously occurs at contact since plasmons modes are confined near the particle surface.

Finally, it is worthwhile to note that mode effective volume is generally defined independently on the particle-emitter distance so that it gives an estimation of the mode extension. This is done by evaluating the maximum effective volume available and for a random emitter orientation. In case of localized SPP, this is achieved for contact ($z_0 = R$). Since the decay rate of a randomly oriented emitter expresses $\gamma = (\gamma^{\perp} + 2\gamma^{\parallel})/3$, it comes

$$\frac{1}{V_n} = \frac{1}{3V_n^{\perp}(R)} + \frac{2}{3V_n^{\parallel}(R)}, \tag{21}$$

$$V_n = \frac{9}{(2n+1)(n+1)}V_0, \tag{22}$$

where $V_0 = 4\pi R^3/3$ is the metallic sphere volume and $V_n^{\parallel}$ refers to a dipole parallel to the sphere surface (see (B.4) in Appendix B, with $\epsilon_B = 1$). Dipolar and quadrupolar mode volumes are $V_1 = 3/2V_0$ and $V_2 = 3/5V_0$, respectively. The expression (22), derived for a Drude metal, *quantifies an extremely important property of localized SPPs; their effective volume does not depend on the wavelength and is of the order of the particle volume.* Metallic nanoparticles therefore support localized modes of strongly subwavelength extension. Highest-order modes have negligible extension ($V_n \longrightarrow 0$ for $n \longrightarrow \infty$, see Figure 5). As a comparison, photonics cavity

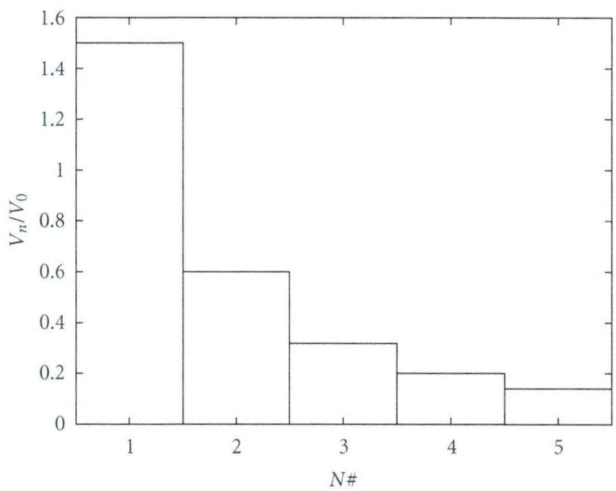

FIGURE 5: Normalized effective volume V_n/V_0 as a function of mode order.

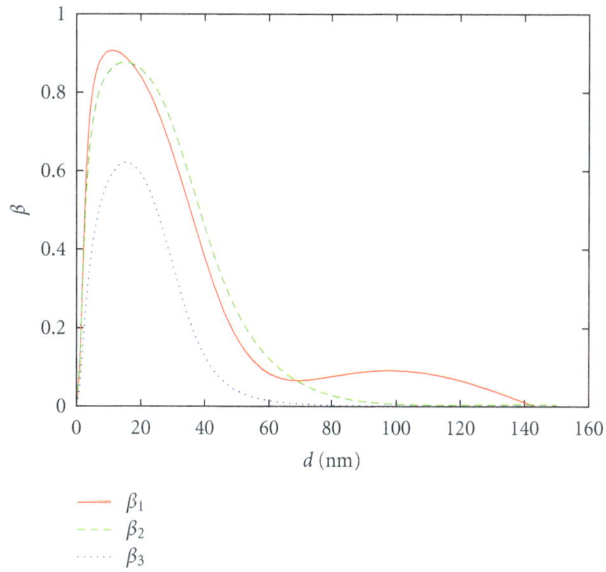

FIGURE 6: Coupling efficiency into the dipolar (β_1), quadrupolar (β_2) and hexapolar (β_3) mode of a $R = 25$ nm silver sphere in PMMA, calculated using exact Mie expansion. The emitter wavelength is assumed to match the considered mode. Therefore β_1, β_2 and β_3 are calculated at $\lambda_1 = 435$ nm (dipolar resonance), $\lambda_2 = 387$ nm (quadrupolar resonance), and $\lambda_3 = 375$ nm (hexapolar resonance), respectively.

modes are generally limited by the diffraction limit so that their effective volume is at best of the order of $(\lambda/n_B)^3$ (see [31]). Nevertheless, the extremely reduced SPP volume is achieved at the expense of the mode quality factor. It should be noticed that a low-quality factor indicates a large cavity resonance FWHM so that emitter-SPP coupling can occur on a large spectrum range.

Note that mode effective volume is generally defined by its energy confinement $V_{\text{eff}} = \int \epsilon |\mathbf{E}(\mathbf{r})|^2 d\mathbf{r} / \max(\epsilon |\mathbf{E}(\mathbf{r})|^2)$ [12]. Khurgin and Sun used this definition and obtained [32] $V_n = 6V_0/(n+1)^2$ that is in agreement with our expression for the dipolar mode volume ($V_1 = 1.5V_0$) but leads to slightly different values for other modes (e.g., $V_2 \approx 0.67V_0$ instead of $V_2 = 0.60V_0$). Recently, Koenderink showed that defining the mode volume on the basis of energy density could lead to underestimating the Purcell factor near plasmonic nanostructures [33] (however, he defined the coupling rate to the whole system rather than considering the coupling into a single mode). Oppositely, we adopt here a phenomenological approach where the mode volume is defined so that Purcell factor remains valid. Nevertheless, both methods lead to very similar results for the effective volumes of localized SPPs supported by a nanosphere. Since mode volume is generally a simple way to qualitatively characterize the mode extension, expressions derived here or by Sun et al. could be used.

4. β-Factor

Purcell factor quantifies the coupling strength between a quantum emitter and a (plasmon) mode but lacks information on the coupling efficiency as compared to all the other emitter relaxation channels. For the sake of clarity, it has to be mentioned that coupling strength to one SPP mode corresponds to the total emission decay rate induced by this coupling. It does not permit to distinguish radiative (γ_{rad}) and nonradiative (γ_{NR}) coupling into a single mode. This could be done by numerically cancelling the imaginary part of the metal dielectric function (see also [34] for similar discussion in case of coupling to delocalized SPPs). However, one can determine the coupling efficiency into a single mode as compared to all other modes. This coupling efficiency, or the so-called β-factor, is easily estimated in case of a spherical metallic particle since all the available channels are taken into account in the Mie expansion. Coupling efficiency into nth mode writes

$$\beta_n = \frac{\langle \gamma_n \rangle}{\langle \gamma_{\text{tot}} \rangle}, \tag{23}$$

where $\langle \gamma \rangle = (\gamma^{\perp} + 2\gamma^{\parallel})/3$ is the decay rate of a randomly oriented molecule. β-factor is represented on Figure 6 for the first three modes. A maximum efficiency of 90% can be achieved in the dipolar mode ($d \sim 10$ nm) and 87% into quadrupolar mode ($d \sim 15$ nm). The coupling efficiency into the hexapolar mode is lower ($\sim 60\%$ around $d \sim 15$ nm) since it has a very low extension, as indicated in Figure 5. For very short distances, all the coupling efficiencies drop down to zero since all the higher-order modes accumulate in this region, opening numerous alternative decay channels. Moreover, it is possible to efficiently couple the emitter to either the dipolar or quadrupolar mode, by matching the emitter and mode wavelengths. This is of strong importance when designing a SPASER (or plasmon laser) [5] so that the active mode can be tuned on the dipolar or quadrupolar mode. This last *spasing* mode would consist of an extremely localized and ultrafast nanosource [18].

5. Conclusion

We explicitly determined the effective volumes for all the SPP modes supported by a metallic nanosphere. Their quality factor is also approximated in the quasistatic case or calculated using exact Mie expansion. Rather low-quality factors ranging from 10 to 100 can be achieved, associated with extremely confined effective volume of nanometric dimensions. This results in high Q/V_{eff} ratios indicating efficient coupling strength with a quantum emitter. At the opposite to cavity quantum electrodynamics where the coupling strength is obtained on diffraction limited volumes thanks to ultrahigh Q factors, plasmonics structures allow efficient coupling on nanometric scales with reasonable Q factors, defining therefore high-bandwidth interaction. Finally, high coupling efficiencies (80%–90%) to dipolar or quadrupolar mode can be achieved and are of great interest for nanolasers realization.

Appendices

In this appendix, we derive the expression of mode quality factor and effective volume for a spherical particle embedded in homogeneous background of optical index $n_B = \sqrt{\epsilon_B}$. For a better description of the metal optical properties, we include the contributions of the bound electrons into the metal dielectric constant $\epsilon_m = \epsilon_\infty - \omega_p^2/(\omega^2 + i\Gamma_{\text{abs}}\omega)$.

A. Effective Polarisability Associated to *n*th SPP Mode

The nth multipole tensor moment of the metallic particle is given by

$$\mathbf{p}^{(n)} = \frac{4\pi\epsilon_0\epsilon_B}{(2n-1)!!}\alpha_n \nabla^{n-1}\mathbf{E}_0,$$
$$\alpha_n = \frac{n(\epsilon_m - \epsilon_B)}{n\epsilon_m + (n+1)\epsilon_B} R^{(2n+1)}. \tag{A.1}$$

The resonance angular frequency is then $\omega_n = \omega_p\sqrt{n/[n\epsilon_\infty + (n+1)\epsilon_B]}$ ($\omega_\infty = \omega_p/\sqrt{\epsilon_\infty + \epsilon_B}$). Finally, the effective polarisability, including finite size effects writes [22]

$$\alpha_n^{\text{eff}} = \left[1 - i\frac{(n+1)k_B^{2n+1}}{n(2n-1)!!(2n+1)!!}\alpha_n\right]^{-1}\alpha_n \tag{A.2}$$

with $k_B = n_B k$ the wavenumber in the background medium.

Considering a Drude metal, we achieve a simple approximated expression for α_n^{eff} near a resonance

$$\alpha_n^{\text{eff}} \underset{\omega_n}{\sim} \frac{(2n+1)\epsilon_B}{n\epsilon_\infty + (n+1)\epsilon_B}\frac{\omega_n}{2(\omega_n - \omega) - i\Gamma_n}R^{2n+1},$$
$$\Gamma_n = \Gamma_{\text{abs}} + \Gamma_n^{\text{rad}}, \tag{A.3}$$
$$\Gamma_n^{\text{rad}} = \frac{(2n+1)\epsilon_B}{n\epsilon_\infty + (n+1)\epsilon_B}\omega_n\frac{(n+1)(k_B R)^{2n+1}}{n(2n-1)!!(2n+1)!!},$$

so that the quality factor expression $Q_n = \omega_n/\Gamma_n$ remains valid but with the corrected expressions for resonance frequency ω_n and total dissipation rate Γ_n.

B. Effective Volumes

The total coupling strength of a dipolar emitter to the spherical metallic particle, embedded in n_B medium, expresses [21, 22]

$$\frac{\gamma_{\text{tot}}^{\perp}}{n_B\gamma_0} \approx \frac{3}{2}\frac{1}{(k_B z_0)^3}\sum_{n=1}^{\infty}\frac{(n+1)^2}{z_0^{(2n+1)}}\operatorname{Im}\left(\alpha_n^{\text{eff}}\right),$$
$$\frac{\gamma_{\text{tot}}^{\parallel}}{n_B\gamma_0} \approx \frac{3}{4}\frac{1}{(k_B z_0)^3}\sum_{n=1}^{\infty}\frac{n(n+1)}{z_0^{(2n+1)}}\operatorname{Im}\left(\alpha_n^{\text{eff}}\right). \tag{B.1}$$

The coupling strength to the nth mode is

$$\frac{\gamma_n^{\perp}}{n_B\gamma_0} \approx \frac{3}{2}\frac{1}{(k_B z_0)^3}\frac{(n+1)^2}{z_0^{(2n+1)}}\operatorname{Im}\left(\alpha_n^{\text{eff}}\right)$$
$$\underset{\omega_n}{\sim} \frac{3}{4\pi^2}\left(\frac{\lambda}{n_B}\right)^3\frac{R^{2n+1}}{4\pi z_0^3}\frac{(2n+1)\epsilon_B}{n\epsilon_\infty + (n+1)\epsilon_B}\frac{(n+1)^2}{z_0^{(2n+1)}}Q_n,$$
$$\frac{\gamma_n^{\parallel}}{n_B\gamma_0} \approx \frac{3}{4}\frac{1}{(k_B z_0)^3}\frac{n(n+1)}{z_0^{(2n+1)}}\operatorname{Im}\left(\alpha_n^{\text{eff}}\right)$$
$$\underset{\omega_n}{\sim} \frac{3}{4\pi^2}\left(\frac{\lambda}{n_B}\right)^3\frac{R^{2n+1}}{8\pi z_0^3}\frac{(2n+1)\epsilon_B}{n\epsilon_\infty + (n+1)\epsilon_B}\frac{n(n+1)}{z_0^{(2n+1)}}Q_n, \tag{B.2}$$

and the mode effective volume is deduced from comparison to the Purcell factor (1)

$$V_n^{\perp} = \frac{n\epsilon_\infty + (n+1)\epsilon_B}{(2n+1)\epsilon_B}\frac{4\pi z_0^{2n+4}}{(n+1)^2 R^{2n+1}}, \tag{B.3}$$

$$V_n^{\parallel} = \frac{n\epsilon_\infty + (n+1)\epsilon_B}{(2n+1)\epsilon_B}\frac{8\pi z_0^{2n+4}}{n(n+1)R^{2n+1}}. \tag{B.4}$$

Finally, we define the mode volume as

$$\frac{1}{V_n} = \frac{1}{3V_n^{\perp}(R)} + \frac{2}{3V_n^{\parallel}(R)},$$
$$V_n = \frac{n\epsilon_\infty + (n+1)\epsilon_B}{(2n+1)\epsilon_B}\frac{9}{(2n+1)(n+1)}V_0. \tag{B.5}$$

All the expressions obtained in this appendix reduce to the simple analytical case discussed in the main text for $\epsilon_\infty = 1$ (bound electrons contribution neglected) and $\epsilon_B = 1$ (background medium is air).

Acknowledgments

This work is supported by the Agence Nationale de la Recherche (ANR) under Grants Plastips (ANR-09-BLAN-0049-01), Fenoptiχs (ANR-09-NANO-23), and HYNNA (ANR-10-BLAN-1016). This paper was also motivated by the preparation of a lecture for the 2nd Summer School On Plasmonics organized by Nicolas Bonod in October 2011.

References

[1] M. Moskovits, "Surface-enhanced spectroscopy," *Reviews of Modern Physics*, vol. 57, no. 3, pp. 783–826, 1985.

[2] V. Giannini, A. I. Fernández-Domínguez, S. C. Heck, and S. A. Maier, "Plasmonic nanoantennas: fundamentals and their use in controlling the radiative properties of nanoemitters," *Chemical Reviews*, vol. 111, no. 6, pp. 3888–3912, 2011.

[3] C. Deeb, R. Bachelot, J. Plain et al., "Quantitative analysis of localized surface plasmons based on molecular probing," *ACS Nano*, vol. 4, no. 8, pp. 4579–4586, 2010.

[4] P. Bharadwaj, B. Deutsch, and L. Novotny, "Optical antennas," *Advances in Optics and Photonics*, vol. 1, no. 3, pp. 438–483, 2009.

[5] D. J. Bergman and M. I. Stockman, "Surface plasmon amplification by stimulated emission of radiation: quantum generation of coherent surface plasmons in nanosystems," *Physical Review Letters*, vol. 90, no. 2, Article ID 027402, 2003.

[6] S. Schietinger, M. Barth, T. Aichele, and O. Benson, "Plasmon-enhanced single photon emission from a nanoassembled metal—diamond hybrid structure at room temperature," *Nano Letters*, vol. 9, no. 4, pp. 1694–1698, 2009.

[7] C. Huang, A. Bouhelier, J. Berthelot et al., "External control of the scattering properties of a single optical nanoantenna," *Applied Physics Letters*, vol. 96, no. 14, Article ID 143116, 2010.

[8] R. Marty, A. Arbouet, V. Paillard, C. Girard, and G. Colas Des Francs, "Photon antibunching in the optical near field," *Physical Review B*, vol. 82, no. 8, Article ID 081403, 2010.

[9] R. Vincent and R. Carminati, "Magneto-optical control of Förster energy transfer," *Physical Review B*, vol. 83, no. 16, Article ID 165426, 2011.

[10] J.-H. Kang, Y.-S. No, S.-H. Kwon, and H.-G. Park, "Ultrasmall subwavelength nanorod plasmonic cavity," *Optics Letters*, vol. 36, no. 11, pp. 2011–2013, 2011.

[11] C. Girard, O. J. F. Martin, G. Lévèque, G. C. Des Francs, and A. Dereux, "Generalized bloch equations for optical interactions in confined geometries," *Chemical Physics Letters*, vol. 404, no. 1–3, pp. 44–48, 2005.

[12] S. A. Maier, "Plasmonic field enhancement and SERS in the effective mode volume picture," *Optics Express*, vol. 14, no. 5, pp. 1957–1964, 2006.

[13] V. V. Datsyuk, "Ultimate enhancement of the local density of electromagnetic states outside an absorbing sphere," *Physical Review A*, vol. 75, no. 4, Article ID 043820, 2007.

[14] A. Trügler and U. Hohenester, "Strong coupling between a metallic nanoparticle and a single molecule," *Physical Review B*, vol. 77, no. 11, Article ID 115403, 2008.

[15] S. Savasta, R. Saija, A. Ridolfo, O. Di Stefano, P. Denti, and F. Borghese, "Nanopolaritons: vacuum rabi splitting with a single quantum dot in the center of a dimer nanoantenna," *ACS Nano*, vol. 4, no. 11, pp. 6369–6376, 2010.

[16] I. E. Protsenko, A. V. Uskov, O. A. Zaimidoroga, V. N. Samoilov, and E. P. O'Reilly, "Dipole nanolaser," *Physical Review A*, vol. 71, no. 6, Article ID 063812, 7 pages, 2005.

[17] M. A. Noginov, G. Zhu, A. M. Belgrave et al., "Demonstration of a spaser-based nanolaser," *Nature*, vol. 460, no. 7259, pp. 1110–1112, 2009.

[18] M. I. Stockman, "The spaser as a nanoscale quantum generator and ultrafast amplifier," *Journal of Optics A*, vol. 12, no. 2, Article ID 024004, 13 pages, 2010.

[19] G. Colas Des Francs, C. Girard, M. Juan, and A. Dereux, "Energy transfer in near-field optics," *Journal of Chemical Physics*, vol. 123, no. 17, Article ID 174709, 7 pages, 2005.

[20] R. Carminati, J. J. Greffet, C. Henkel, and J. M. Vigoureux, "Radiative and non-radiative decay of a single molecule close to a metallic nanoparticle," *Optics Communications*, vol. 261, no. 2, pp. 368–375, 2006.

[21] G. Colas Des Francs, A. Bouhelier, E. Finot et al., "Fluorescence relaxation in the near-field of a mesoscopic metallic particle: distance dependence and role of plasmon modes," *Optics Express*, vol. 16, no. 22, pp. 17654–17666, 2008.

[22] G. C. Des Francs, "Molecule non-radiative coupling to a metallic nanosphere: an optical theorem treatment," *International Journal of Molecular Sciences*, vol. 10, no. 9, pp. 3931–3936, 2009.

[23] B. Rolly, B. Stout, S. Bidault, and N. Bonod, "Crucial role of the emitter-particle distance on the directivity of optical antennas," *Optics Letters*, vol. 36, no. 17, pp. 3368–3370, 2011.

[24] N. Bonod, A. Devilez, B. Rolly, S. Bidault, and B. Stout, "Ultracompact and unidirectional metallic antennas," *Physical Review B*, vol. 82, no. 11, Article ID 115429, 2010.

[25] G. Sun and J. B. Khurgin, "Theory of optical emission enhancement by coupled metal nanoparticles: an analytical approach," *Applied Physics Letters*, vol. 98, no. 11, 2011.

[26] A. Wokaun, J. P. Gordon, and P. F. Liao, "Radiation damping in surface-enhanced Raman scattering," *Physical Review Letters*, vol. 48, no. 14, pp. 957–960, 1982.

[27] J. Jackson, *Classical Electrodynamics*, John Wiley & Sons, Hoboken, NJ, USA, 3rd edition, 1998.

[28] P. B. Johnson and R. W. Christy, "Optical constants of the noble metals," *Physical Review B*, vol. 6, no. 12, pp. 4370–4379, 1972.

[29] C. Sönnichsen, T. Franzl, T. Wilk et al., "Drastic reduction of plasmon damping in gold nanorods," *Physical Review Letters*, vol. 88, no. 7, Article ID 077402, 4 pages, 2002.

[30] J. J. Greffet, M. Laroche, and F. Marquier, "Impedance of a nanoantenna and a single quantum emitter," *Physical Review Letters*, vol. 105, no. 11, Article ID 117701, 2010.

[31] K. J. Vahala, "Optical microcavities," *Nature*, vol. 424, no. 6950, pp. 839–846, 2003.

[32] J. B. Khurgin and G. Sun, "Enhancement of optical properties of nanoscaled objects by metal nanoparticles," *Journal of the Optical Society of America B*, vol. 26, no. 12, pp. B83–B95, 2009.

[33] A. F. Koenderink, "On the use of Purcell factors for plasmon antennas," *Optics Letters*, vol. 35, no. 24, pp. 4208–4210, 2010.

[34] J. Barthes, G. Colas Des Francs, A. Bouhelier, J.-C. Weeber, and A. Dereux, "Purcell factor for a point-like dipolar emitter coupled to a two-dimensional plasmonic waveguide," *Physical Review B*, vol. 84, no. 7, Article ID 073403, 2011.

Diffuse Optical Imaging Using Decomposition Methods

Binlin Wu,[1] M. Alrubaiee,[1] W. Cai,[1] M. Xu,[2] and S. K. Gayen[1]

[1] *Department of Physics, The City College and The Graduate Center of The City University of New York, 160 Convent Avenue, New York, NY 10031, USA*
[2] *Department of Physics, Fairfield University, 1073 North Benson Road, Fairfield, CT 06824, USA*

Correspondence should be addressed to Binlin Wu, bwu@sci.ccny.cuny.edu

Academic Editor: Baohong Yuan

Diffuse optical imaging (DOI) for detecting and locating targets in a highly scattering turbid medium is treated as a blind source separation (BSS) problem. Three matrix decomposition methods, independent component analysis (ICA), principal component analysis (PCA), and nonnegative matrix factorization (NMF) were used to study the DOI problem. The efficacy of resulting approaches was evaluated and compared using simulated and experimental data. Samples used in the experiments included Intralipid-10% or Intralipid-20% suspension in water as the medium with absorptive or scattering targets embedded.

1. Introduction

Diffuse optical imaging (DOI) for detection and retrieval of location information of targets in a highly scattering turbid medium may be treated as a blind source separation (BSS) problem [1, 2]. Various matrix decomposition methods, such as, independent component analysis (ICA) [3], principal component analysis (PCA) [4], and nonnegative matrix factorization (NMF) [5, 6] have been developed for solving the BSS problem and retrieving desired information.

Xu et al. adapted ICA of information theory to develop optical tomography using independent component analysis (OPTICA) and demonstrated its application for diffuse imaging of absorptive, scattering, and fluorescent targets [7–11]. ICA assumes the signals from different targets to be *independent* of each other and optimizes a relevant measure of independence to obtain the ICs associated with different targets. The position coordinates of targets in three dimensions are determined from the individual components separately.

PCA assumes that the PCs contributing to the signal are *uncorrelated* and explain the most variance in the signal. PCA has been widely used in various applications, such as spectroscopy [12], face recognition [13], and neuroimaging [14]. NMF seeks to factorize a matrix into two nonnegative matrices (component signals and weights) and requires the contributions to signal and the weights of the components to be *non-negative*. It does not imply any relationship between the components. NMF has also been widely used in biological analysis [15] and spectral analysis [16].

The objective of this study is to test and compare the efficacy of these algorithms when used to solve the DOI problem. Results are presented and compared using simulative data and experimental data using absorptive and scattering targets embedded in model scattering media. Our interest in solving the DOI problem derives from the need for a noninvasive modality for detecting, locating, and diagnosing breast tumors in early stages of growth.

The remainder of the paper is organized as follows. In Section 2, the formalisms of the three methods are introduced. Section 3 evaluates the resulting imaging approaches using simulated data. The approaches are further examined in Section 4 for experimental data acquired using absorptive and scattering targets embedded in model scattering media. Section 5 summarizes and discusses the results.

2. Formalism

2.1. Blind Source Separation Problem. Blind source separation (BSS), also known as blind signal separation, is a general problem in information theory that seeks to separate different individual signals from the measured signals, which are weighted mixtures of those individual signals. Assuming M individual signals, $s_j(t)$, $j = 1, \ldots M$, are linearly mixed instantaneously, the BSS problem is modeled as follows. The dimension of $s_j(t)$ is N_s, the number of sampling times. In this study, t will be replaced by spatial positions of the excitation light sources. A total of N_d detectors sense N_d different mixtures of $s_j(t)$. The mixture measured by the ith detector can be presented as $x_i(t) = \sum_{j=1}^{M} a_{ij} s_j(t)$, or $X = AS$, in a matrix notation, where $A \in R^{N_d \times M}$ is a mixing or weighting matrix, $S \in R^{M \times N_s}$, $X \in R^{N_d \times N_s}$, and $M < \min(N_s, N_d)$. The objective of BSS is to retrieve the signals $s_j(t)$ and their weights, a_{ij}. ICA, PCA, and NMF are statistical analysis methods used to solve the BSS problem.

2.2. Diffuse Optical Imaging Problem. In DOI, one measures the signal at the sample boundary, which includes a weighted mixture of contributions from embedded targets. One uses the diffusion approximation [17–19] of the radiative transfer equation [20, 21] as the forward model to describe light propagation in a highly scattering turbid medium. The perturbation in the light intensity distribution measured on the boundary of the sample due to the presence of the targets (which are localized inhomogeneities in the optical properties within the sample volume) may be written, in the first-order Born approximation, as [22, 23]

$$\begin{aligned}\Delta\phi(\mathbf{r}_d, \mathbf{r}_s) = &-\int G(\mathbf{r}_d, \mathbf{r})\delta\mu_a(\mathbf{r})cG(\mathbf{r}, \mathbf{r}_s)d^3\mathbf{r} \\ &-\int \delta D(\mathbf{r})c\nabla_{\mathbf{r}}G(\mathbf{r}_d, \mathbf{r}) \cdot \nabla_{\mathbf{r}}G(\mathbf{r}, \mathbf{r}_s)d^3\mathbf{r},\end{aligned} \tag{1}$$

where $\mathbf{r}_s$, $\mathbf{r}_d$, and $\mathbf{r}$ are the positions of a source of unit power, detector and target, respectively; $G(\mathbf{r}, \mathbf{r}_s)$ and $G(\mathbf{r}_d, \mathbf{r})$ are the Green's functions that describe light propagation from the source to the target and from the target to the detector, respectively; $\delta\mu_a$ and δD are the differences in absorption coefficient and diffusion coefficient between the targets and the background medium, respectively; and c is the light speed in the medium.

A multisource illumination and multidetector signal acquisition scheme is used to acquire light transmitted through a scattering medium. For small absorptive targets, a perturbation data matrix is constructed using $-\Delta\phi$ for all sources. The elements of the data matrix pertaining to absorptive targets represented by the first term in (1) may be written in a discrete form as

$$\begin{aligned}X_{ij} &= \sum_{m=1}^{M} G^d(\mathbf{r}_i, \mathbf{r}_m)\tau_m G^s\left(\mathbf{r}_m, \mathbf{r}_j\right) \\ &(i = 1, 2, \ldots, N_d; j = 1, 2, \ldots, N_s),\end{aligned} \tag{2}$$

where $\mathbf{r}_i$, $\mathbf{r}_j$, and $\mathbf{r}_m$ are the locations of the ith detector, jth source and mth target, respectively; N_s, N_d, and M are the numbers of sources, detectors, and targets, respectively; $\tau_m = \delta\mu_a(\mathbf{r}_m)c\delta V_m$ is the optical absorption strength of the mth target of volume δV_m; $G^s(\mathbf{r}_m, \mathbf{r}_j)$ and $G^d(\mathbf{r}_i, \mathbf{r}_m)$ are the Green's functions that describe light propagation from jth source to mth target and from mth target to ith detector, respectively. The number of targets is assumed to be less than that of sources and detectors, $M < \min(N_d, N_s)$.

The mth target may be considered to be a virtual source of strength $\tau_m G^s(\mathbf{r}_m, \mathbf{r}_j)$ excited by the real light source located at $\mathbf{r}_j$. The data matrix $X = \{X_{ij}\}$ may be considered to be a set of combinations of light signals from all virtual sources mixed by a mixing matrix $\{G^d(\mathbf{r}_i, \mathbf{r}_m)\}$. Therefore, this problem can be treated as a BSS problem.

As the second term in (1) suggests, each scattering target is represented by three colocated virtual sources of strength: $\tau_m \partial_p G^s(\mathbf{r}_m, \mathbf{r}_j)$, where $\partial_p = \partial/\partial p$, $(p = x, y, z)$, and $\tau_m = \delta D(\mathbf{r}_m)c\delta V_m$, is the optical scattering strength of the mth target [8]. The mixing matrices become $\{\partial_p G^d(\mathbf{r}_i, \mathbf{r}_m)\}$ for the three virtual sources generated by the mth target. The elements of the data matrix for scattering targets may be written as

$$X_{ij} = \sum_{m=1}^{M} \sum_{p=\{x,y,z\}} \partial_p G^d(\mathbf{r}_i, \mathbf{r}_m)\tau_m \partial_p G^s\left(\mathbf{r}_m, \mathbf{r}_j\right). \tag{3}$$

Since one absorptive target is represented by one centrosymmetric virtual source, while three virtual sources (one centrosymmetric and two dumb-bell shaped) represent one scattering target [7, 8], the number and patterns of virtual sources may be used, in favorable situations, to identify the target as absorptive or scattering in nature. In this paper, only small targets are considered since all three algorithms are suited for small targets, and early detection, when the tumors are more amenable to treatment, is of practical interest.

2.3. DOI as a BSS Problem. The data matrix for the DOI problem may be written as

$$X = AS = \sum_{m=1}^{M} A_{im} S_{mj}, \tag{4}$$

where $A \in R^{N_d \times M}$, $S \in R^{M \times N_s}$, and $X \in R^{N_d \times N_s}$. For absorptive targets,

$$A_{im} = \beta_m G^d(\mathbf{r}_i, \mathbf{r}_m), \qquad S_{mj} = \alpha_m G^s\left(\mathbf{r}_m, \mathbf{r}_j\right), \tag{5a}$$

while for scattering targets,

$$A_{im} = \beta_m \partial_p G^d(\mathbf{r}_i, \mathbf{r}_m), \qquad S_{mj} = \alpha_m \partial_p G^s\left(\mathbf{r}_m, \mathbf{r}_j\right). \tag{5b}$$

$\{S_{mj}\}$ $(j = 1, 2, \ldots, N_s)$ and $\{A_{im}\}$ $(i = 1, 2, \ldots N_d)$ are two-dimensional intensity distributions on the source and detector planes, respectively. Source and detector planes are the boundaries of the sample through which light enters and exits the sample volume, respectively. The scaling factors β_m and α_m are related to the target optical strength, $\tau_m = \alpha_m \beta_m$.

The location of the target and the scaling factors can be retrieved using a least squares fitting via

$$\underset{\alpha_m,\beta_m,\mathbf{r}_m}{\operatorname{argmin}}\left\{\sum_j\left[\alpha_m^{-1}S_{mj}-G^s\left(\mathbf{r}_m,\mathbf{r}_j\right)\right]^2 +\sum_i\left[\beta_m^{-1}A_{im}-G^d(\mathbf{r}_i,\mathbf{r}_m)\right]^2\right\},\ \text{or} \tag{6a}$$

$$\underset{\alpha_m,\beta_m,\mathbf{r}_m}{\operatorname{argmin}}\left\{\sum_p\left\{\sum_j\left[\alpha_m^{-1}S_{mj}-\partial_pG^s\left(\mathbf{r}_m,\mathbf{r}_j\right)\right]^2 +\sum_i\left[\beta_m^{-1}A_{im}-\partial_pG^d(\mathbf{r}_i,\mathbf{r}_m)\right]^2\right\}\right\}, \tag{6b}$$

for absorptive and scattering targets, respectively. However, when a scattering target is embedded deep in a turbid medium, only the $\tau_m\partial_zG^s(\mathbf{r}_m,\mathbf{r}_j)$ virtual source remains significant. So, only $p = z$ may be used for fitting in (6b) [8].

2.3.1. ICA. OPTICA assumes that the virtual sources are *independent* of each other [8]. So, they can be retrieved through an iterative process which seeks to maximize the independence among the components. In practice, the independent components are found by maximizing some measure of non-Gaussianity, such as kurtosis (the fourth-order cumulant), of the unmixed components. A Matlab program for ICA was adopted from http://sccn.ucsd.edu/eeglab/. The location of the target can be retrieved by fitting the independent component intensity distributions (ICIDs) to Green's functions or derivatives of Green's functions using (6a) and (6b).

2.3.2. PCA. PCA assumes that the virtual sources are uncorrelated so that the correlation (covariance) between them is ideally zero and minimal in practice. The covariance matrix of S, cov(S) should be diagonal. The general process of PCA is as follows. The data matrix $X = AS + \mathcal{N}$, where $\mathcal{N}$ is random noise added to the data, A and S the same as defined in (4). When S is mean centered, elements of the mean-centered matrix S' are defined as

$$S'_{mj} = S_{mj} - \frac{1}{N_s}\sum_{j=1}^{N_s} S_{mj}. \tag{7a}$$

Similarly,

$$X'_{ij} = X_{ij} - \frac{1}{N_s}\sum_{j=1}^{N_s} X_{ij}. \tag{7b}$$

PCA looks for a matrix P that decomposes X into virtual sources, $S = PX$. It also holds that $S' = PX'$, since P is just a rotation matrix which does not change the center of the data.

$$\operatorname{cov}(S) = S'S'^T = (PX')(PX')^T = PX'X'^TP^T = \Lambda, \tag{8}$$

where $\Lambda = \operatorname{diag}\{\lambda_1,\lambda_2,\ldots\}$. The eigenvalues λ_m are variances in the covariance matrix. Therefore, $X'X'^TP^T = P^T\Lambda$, where P^T is orthonormal. PCA is realized by eigenvalue decomposition (EVD) of the covariance matrix of X. The eigenvectors with leading eigenvalues (largest variances) are selected to be the PCs using the L-curve [24].

Since $X = P^TS \approx AS$, A is determined as a matrix including only PCs. S is calculated as $S \approx (A^TA)^{-1}A^TX$. Rows of S and columns of A represent principal component intensity distributions (PCIDs) on the source plane and detector plane, respectively and are proportional to the images of the virtual sources projected on the source and detector planes. The target positions are determined using (6a) and (6b).

2.3.3. NMF. NMF is a group of multivariate analysis algorithms that factorize a matrix X into A, and $S : X = AS$, A and S are nonnegative [6]. Unlike ICA and PCA, NMF does not imply any relationship between the retrieved components; instead, it just enforces non-negativity constraints on A and S. There are various algorithms developed to solve NMF, such as the multiplicative update method [5] and alternating least squares method [25, 26].

In the multiplicative update implementation of NMF, A and S can be found by minimizing the square of Euclidean distance $\|X - AS\|^2$ as the cost function, where $A \geq 0$ and $S \geq 0$, using the multiplicative update rule:

$$A_{ik} \longleftarrow A_{ik}\frac{(XS^T)_{ik}}{(ASS^T)_{ik}}, \tag{9a}$$

$$S_{kj} \longleftarrow S_{kj}\frac{(A^TX)_{kj}}{(A^TAS)_{kj}}. \tag{9b}$$

The alternating least squares implementation of NMF uses alternate least squares steps to estimate A (or S), and use that estimate to optimize S (or A), repeating the alternative steps until the desired optimization is obtained. Nonnegativity is ensured by setting any negative element of A or S equal to 0.

An NMF toolbox was obtained from http://cogsys.imm.dtu.dk/toolbox/ to perform NMF computation. A built-in command *nnmf* is also available in Matlab (R2011a).

NMF algorithm requires that the non-negativity assumption must hold in the problem. In particular, for absorptive targets, when X is constructed with $-\Delta\phi$, τ_m should be positive, that is, the targets should be more absorbing than the background. If the targets have weaker attenuation properties than the background, X needs to be constructed with $+\Delta\phi$ instead. For scattering targets, X should be treated similarly to keep its elements positive.

When NMF is applied to a scattering target, only the centrosymmetric component shows up properly, since the other two components have dumb-bell shape which includes negative values [8]. So without any prior knowledge or some other experimental means to assess if the target is absorptive or scattering, NMF may not distinguish between the two possibilities.

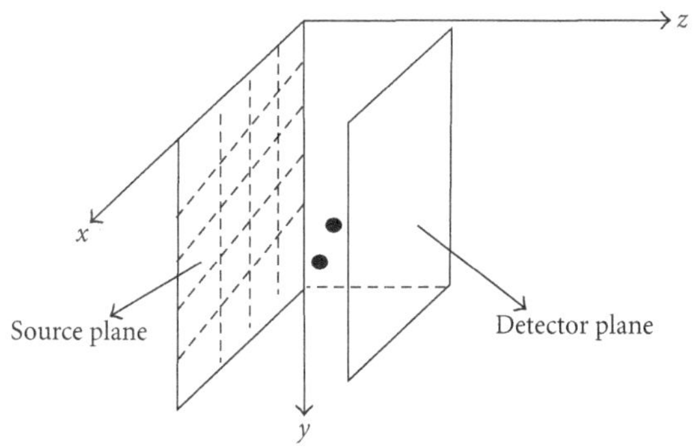

Figure 1: Light intensity distribution on the detector plane is recorded when a point source scans on the source plane.

The decomposition methods can be applied with different sample geometries such as slab and cylindrical geometries, and different measurement domains such as time-resolved domain, frequency domain, and continuous wave (CW). In this paper, Green's functions for slab geometry [23] with CW measurement were used for simulation and experiments.

3. Simulation

The sample was considered to be a 40 mm thick uniform scattering slab with lateral dimension of 80 mm × 80 mm, as shown in Figure 1. Its absorption and diffusion coefficients were taken to be $\mu_a = 0.003\,\text{mm}^{-1}$ and $D = 1/3\,\text{mm}$ (transport mean free path, $l_t = 1\,\text{mm}$), respectively, which are similar to the average value of those parameters for human breast tissue. An absorptive and a scattering point targets were placed at (50, 60, 15) mm and (30, 30, 25) mm, respectively. The index of refraction n of the medium was taken to be 1.33. The speed of light is 2.998×10^8 m/s or 299.8 mm/ns in vacuum, and 225.4 mm/ns in the medium. The absorption coefficient of the absorptive target was set to be higher than the background with $\Delta\mu_a = 0.001\,\text{mm}^{-1}$, while the diffusion coefficient was taken to be the same as that of background. The diffusion coefficient of the scattering target was set to be lower than the background (higher scattering coefficient) with $\Delta D = -0.1\,\text{mm}$ ($l_t = 0.7\,\text{mm}$), while the absorption coefficient was taken to be the same as that of the background. The volumes of both targets are set to be 8 mm^3. The optical strengths of the absorptive and scattering targets were $\Delta\mu_a c\Delta V = 1.803\,\text{mm}^3/\text{ns}$ and $\Delta D c\Delta V = -180.3\,\text{mm}^5/\text{ns}$, respectively. The incident CW beam step scanned the sample at 21 × 21 grid points covering an 80 × 80 mm^2 area, with a step size of 4 mm. Light on the opposite side was recorded at 41 × 41 grid points covering the same area. Multiplicative Gaussian noise of 5% was added to the simulated data. The data matrix X was then obtained using (2) and (3) directly and analyzed using the three different algorithms.

3.1. ICA Analysis. One independent component for the absorptive target and three independent components for the scattering target were retrieved by ICA. The independent component intensity distributions (ICIDs) on the detector plane are shown in Figures 2(a), 2(c), 2(d), and 2(e). Similar ICIDs were obtained on the source plane. Figure 2(g) shows the centrosymmetric ICID for the scattering target, and Figure 2(i) shows the ICID for the absorptive target, on the source plane.

The components in either the detector plane or the source plane can, in principle, be used to extract position and optical strength of the target(s). However, in our experimental arrangement signal is collected by a 1024 × 1024 pixels CCD camera, while the source plane is scanned in an x-y array of points, which is much smaller than the number of pixels in the CCD camera. Consequently, the resolution in the detector plane is much better, and the data set more robust than the source side. So, we used the images on the detector plane for retrieving target information using experimental data. While it would not matter in simulation, to be consistent with experimental situations, we employed detector plane images when using simulated data as well for all three algorithms. Table 1 lists the locations and strengths of the absorptive and scattering targets retrieved by fitting the spatial intensity profile of the centrosymmetric components on the detector plane to Green's functions and derivatives of Green's functions using (6a) and (6b), respectively, as shown in Figures 2(b) and 2(f). Figures 2(h) and 2(j) show the corresponding fits to the profiles on the source plane.

3.2. PCA Analysis. Eigenvalue equation of the covariance matrix of X was solved. The eigenvalues found by PCA were sorted in a descending order. Figure 3 shows a plot of leading 20 eigenvalues on a logarithmic scale.

First four leading eigenvalues were selected for PCs. The corresponding PCIDs were calculated. The PCIDs on the detector plane are shown in Figure 4. Similar images for PCIDs on the source plane were obtained. The scattering target has one centrosymmetric (Figure 4(a)) component and two dumb-bell shaped (Figures 4(c) and 4(d)) components, while the absorptive target has only one component (Figure 4(e)).

Figures 4(b) and 4(f) show fits to the spatial intensity profile of the centrosymmetric component of the scattering target and that of the absorptive target, respectively, to retrieve the locations of the two targets. The locations and optical strengths of the targets retrieved by PCA are also shown in Table 1.

3.3. NMF Analysis. The mixing matrix and virtual sources were retrieved from the data matrix X using NMF as explained in Section 2.3.3. As in the other two approaches, only one component is extracted for the absorptive target. Since NMF has a non-negativity constraint, only the centrosymmetric component for the scattering target is obtained. Nonnegative component intensity distributions (NCIDs) on detector planes are shown in Figure 5. Similar images for NCIDs on source plane were also obtained using the virtual sources in S. The results are also shown in Table 1.

3.4. Results and Discussion. The positions and optical strengths of the targets retrieved by ICA, PCA, and NMF

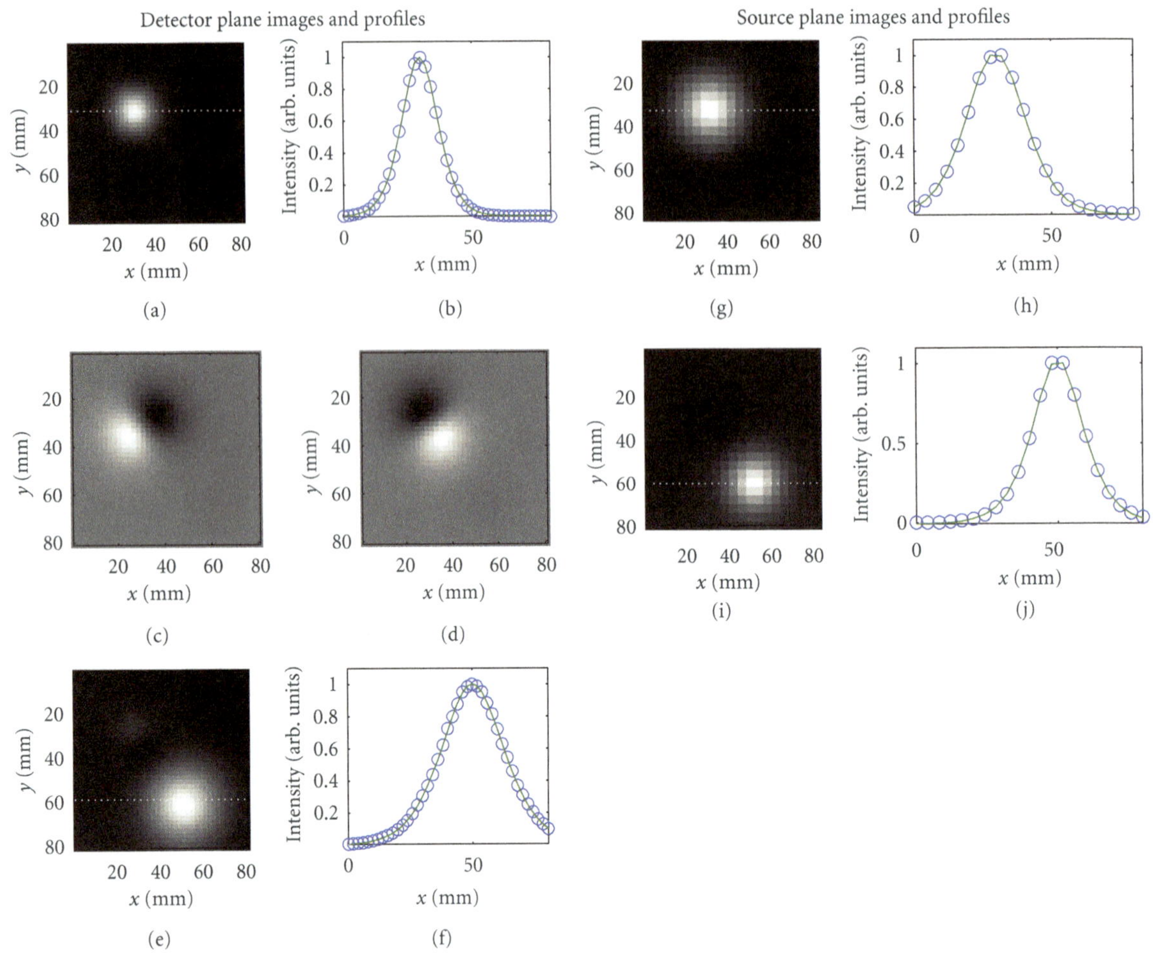

FIGURE 2: ICA-extracted two-dimensional intensity distribution on the detector plane of: (a) the centrosymmetric component; (c) and (d) dumb-bell shaped components of the scattering target; (e) the absorptive target. Similar intensity distribution on the source plane of: (g) the centrosymmetric component of the scattering target and (i) the absorptive target for comparison. Fits to the spatial intensity profile on the detector plane along the white dashed line (shown in figures) of the centrosymmetric component of the scattering target is shown in (b), and that of the absorptive target is shown in (f). Corresponding fits to spatial profiles on the source plane are displayed in (h) and (j), respectively.

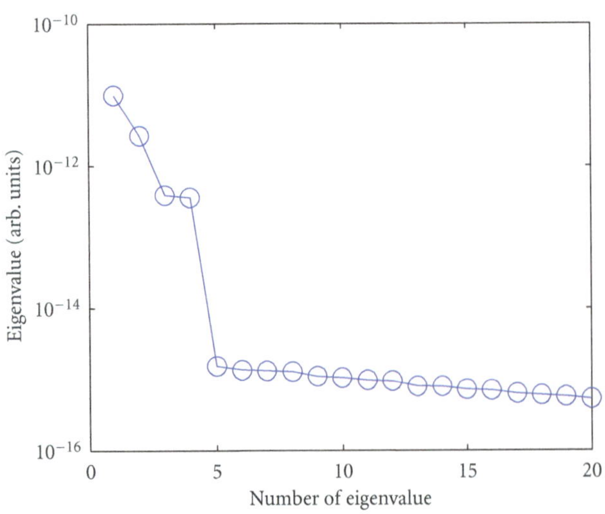

FIGURE 3: A logarithmic plot of the first 20 PCA eigenvalues.

algorithms are shown in Table 1, and compared to the known values. The retrieved results using all three algorithms from this simulated data are in excellent agreement with the known values.

4. Experiments

4.1. Experimental Materials and Methods. In this Section, the algorithms are evaluated using experimental data for absorptive and scattering targets embedded in model scattering media whose absorption and scattering properties are adjusted to mimic the average values of those parameters for human breast tissues. Two different experiments were carried out with two different samples. The first sample used a 250 mm × 250 mm × 50 mm transparent plastic container filled with Intralipid 10% suspension in water as the background medium. The concentration of Intralipid

TABLE 1: Positions and optical strengths retrieved using simulated data and ICA, PCA, and NMF algorithms.

Target	Known position (mm)	Algorithm	Fitted position (mm)	Error (mm)	Known strength*	Fitted strength*	Error (%)
		ICA	(29.9, 30.0, 25.1)	(0.1, 0, 0.1)	−180.3	−179.9	0.22
Sca.	(30, 30, 25)	PCA	(30.0, 30.0, 25.0)	(0, 0, 0)	−180.3	−180.1	0.11
		NMF	(30.0, 30.0, 25.0)	(0, 0, 0)	−180.3	−178.5	1
		ICA	(50.1, 60.2, 15.0)	(0.1, 0.2, 0)	1.803	1.826	1.28
Abs.	(50, 60, 15)	PCA	(50.1, 60.1, 14.9)	(0.1, 0.1, 0.1)	1.803	1.812	0.5
		NMF	(50.1, 60.1, 15.0)	(0.1, 0.1, 0)	1.803	1.803	0

*The unit for absorption strength of the target is mm^3/ns and for scattering strength is mm^5/ns.

10% was adjusted to provide [27, 28] an absorption coefficient of $\mu_a \sim 0.003\,mm^{-1}$, and a transport mean-free path $l_t \sim 1.43$ mm at 785 nm. The second sample used a similar container with dimension of 250 mm × 250 mm × 60 mm filled with Intralipid 20% suspension in water. The concentration of Intralipid 20% was adjusted to provide [27, 28] $\mu_a \sim 0.003\,mm^{-1}$, and $l_t \sim 1$ mm at 785 nm. These optical parameters of the medium were selected to be similar to the average values of those parameters for human breast tissue. The thickness of the samples was also comparable to that of a typical compressed female human breast.

In the first experiment, two absorptive targets were embedded in the medium. The targets were ~10-mm diameter glass spheres filled Indocyanine green (ICG) dye dissolved in Intralipid-20% suspension in water to obtain an absorption coefficient $\mu_a = 1.15\,mm^{-1}$ at 785 nm, and to match the background scattering coefficient of $1.97\,mm^{-1}$. The targets were placed at (57.2, 18.1, 20.0) mm and (19.9, 48.1, 25.0) mm, respectively.

In the second experiment, two scattering targets were embedded, which were also ~10 mm diameter glass spheres, filled with Intralipid-20% suspension in water. The transport mean free path, l_t was adjusted to be 0.25 mm, with scattering coefficient $\mu_s \approx 11\,mm^{-1}$, and absorption coefficient μ_a same as that of the background medium. The targets were placed in the middle plane ($z = 30$ mm) in the container with a lateral distance of 40 mm from each other (center to center).

The experimental setup is schematically shown in Figure 6. A 10 mW 785 nm diode laser beam was used to illuminate the first sample, while a 100 mW 785 nm diode laser beam was used for the second sample. The input surface (source plane) of the samples was scanned across the laser beam in an x-y array of grid points to realize the multi-source interrogation of the samples. The transmitted light from the exit surface (detector plane) was recorded by a 1024 pixel × 1024 pixel (pixel size = 24 μm) CCD camera (Photometrics CH350) equipped with a 60 mm focal-length camera lens. Each pixel of the CCD camera can be considered to be a detector implementing the multidetector signal acquisition arrangement. A set of 16 bit 1024 pixel × 1024 pixel images were acquired. The two samples were scanned in an array of 11 × 12 and 11 × 15 grid points, respectively, with a step size of 5 mm in both cases. The processes of scanning and data acquisition were controlled by a personal computer. At all scan positions, raw transillumination images of the samples were recorded by the computer for further analysis.

4.2. Analysis and Results. A region of interest (ROI) was cropped out from each image. Then, every 5 × 5 pixels in each cropped image were binned to one pixel to enhance signal-to-noise ratio. A background image was generated by calculating an average image for all scan positions to approximate the transillumination image without target(s) embedded.

This averaging method for generating background image is suitable for small targets used in our experiments, as the ratio of the volume of the sample to that of the target was quite high (~500 : 1). For *in vivo* imaging of tumors in early stages of growth, the breast-to-tumor volume ratio will be similarly high, and the averaging method will be applicable. Alternative approaches for generating a background image include using image of (a) a phantom that has the same average optical properties as the sample [29]; (b) the healthy contralateral breast for breast imaging [30]; (c) the sample obtained using light of wavelength for which the target(s) and the background have identical optical properties [31]. Still another approach is to compute the background using an appropriate forward model [32]. A more detailed discussion of this important issue appears in one of our earlier publications [33].

The background image was also cropped and binned corresponding to the ROI for each scan position. Perturbation in the light intensity distribution, $\Delta\phi$ due to targets in each image was found by subtracting the background image from the image. The data matrix X was then constructed using the light intensity perturbations at all scan positions. ICA, PCA, and NMF decomposition algorithms were performed on the data matrix separately. Results are shown and discussed below.

4.2.1. Absorptive Targets. The images on the detector plane obtained using the ICA, PCA, and NMF algorithms are shown in Figures 7, 8, and 9, respectively. Similar images on the source plane were also obtained using all three algorithms. The right side of each figure shows the corresponding spatial intensity profile. Locations of the targets are extracted from fits to these spatial intensity profiles, as described in Section 2.3 using (6a) and (6b). The results are presented in Table 2. In Figure 7, images on the source plane are shown in (e) and (g), and Green's function fits to their spatial profiles are shown in (f) and (h) for comparison.

It follows from the comparison of the results in Table 2 that the positions retrieved by all three algorithms are in good agreement with the known positions. The errors

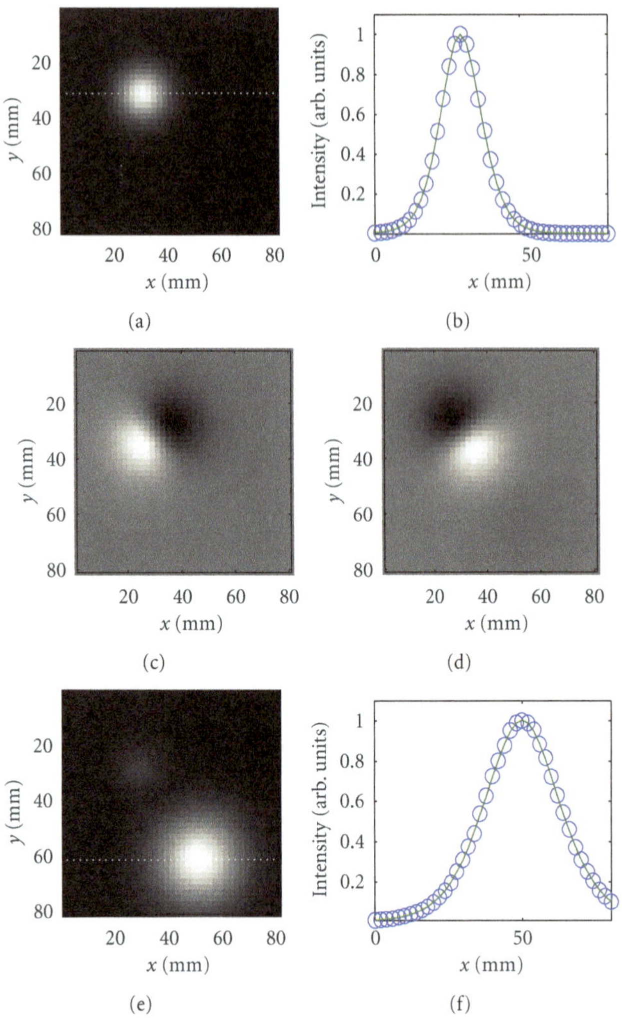

FIGURE 4: PCA-extracted two-dimensional intensity distribution on the detector plane of: (a) the centrosymmetric component; and (c) and (d) dumb-bell shaped components of the scattering target; (e) the absorptive target. Green's function fits to the spatial intensity profiles along the dashed line (shown in figures) of the (b) centrosymmetric component of the scattering target and (f) absorptive target, respectively, to retrieve the locations of the two targets.

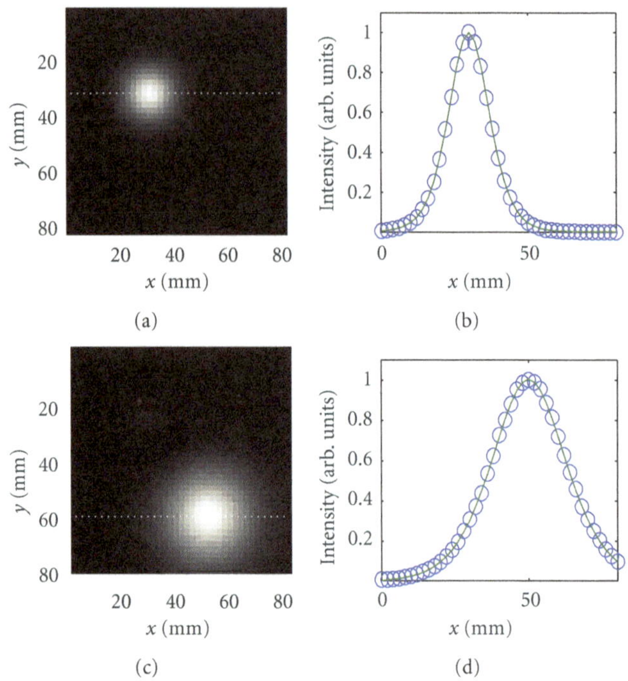

FIGURE 5: NMF-extracted two-dimensional intensity distribution on the detector plane of: (a) the centrosymmetric component of the scattering target; (c) the absorptive target. Fits to the corresponding spatial intensity profiles along the dashed line (shown in figures) are shown in (b) and (d), respectively.

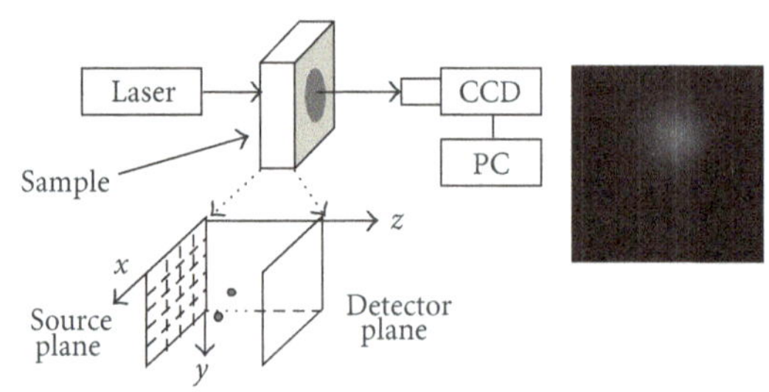

FIGURE 6: A schematic diagram of the experimental arrangement used for imaging objects embedded in a turbid medium. The inset at the bottom shows the 2D array in the input plane that was scanned across the incident laser beam; the inset to the right shows a typical raw image recorded by the CCD. (CCD: charge coupled device, and PC: personal computer).

in the retrieved locations (x, y, z) of the two targets were within 1.7 mm. The PCIDs were not totally separated. Some "residue" was observed in one PCID from the other. ICA and NMF separated two components from this dataset more clearly.

4.2.2. Scattering Targets. The "images" corresponding to the centrosymmetric components of the virtual sources (targets) on the detector plane obtained using the ICA, PCA, and NMF algorithms are shown in Figures 10, 11, and 12, respectively. Similar images on the source plane were also obtained. The right side of each figure shows the corresponding spatial intensity profile. Locations of the targets are extracted from fits to these spatial intensity profiles, as described in Section 2.3 using (6a) and (6b). The results are presented in Table 3.

Both targets were detected by all three algorithms. The target locations retrieved by three algorithms are shown in Table 3 and compared with known locations. Overall, all three algorithms detect and locate the scattering and the absorptive targets with good accuracy, the maximum deviation of any one coordinate from the known value being ~3 mm. Since the maximum difference between the known and retrieved position coordinates was larger for the scattering targets, we calculated the squared correlation coefficient γ to assess the fitting quality. NMF retrieves

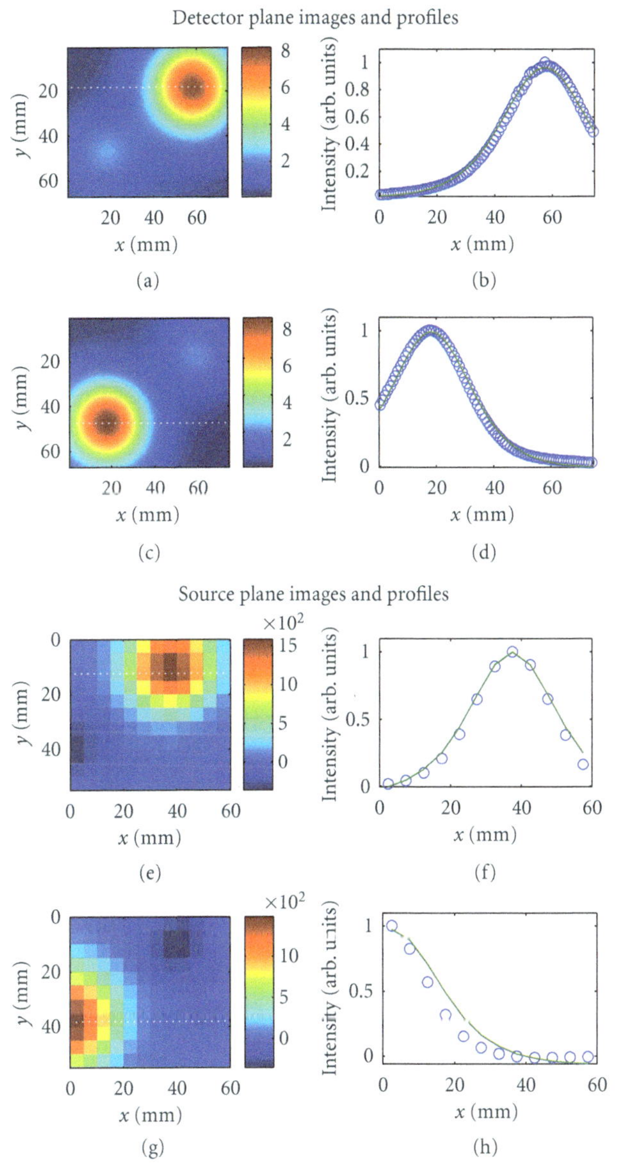

FIGURE 7: ICA-generated ICIDs on the detector plane are shown in (a) and (c); corresponding Green's function fits to the horizontal spatial profiles through the dashed lines are shown in (b) and (d). ICIDs on the source plane are shown in (e) and (g); corresponding Green's function fits to the horizontal spatial profiles through the dashed line are shown in (f) and (h).

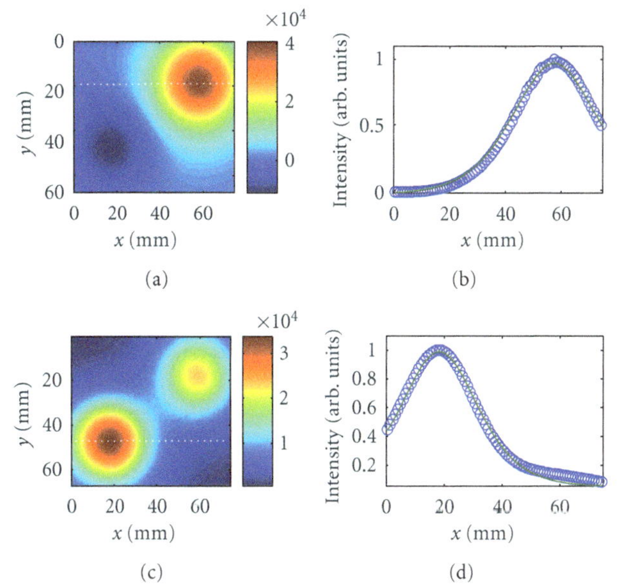

FIGURE 8: PCIDs on the detector plane are shown in (a) and (c); and corresponding Green's function fits to the horizontal spatial profiles through the dashed line are shown in (b) and (d).

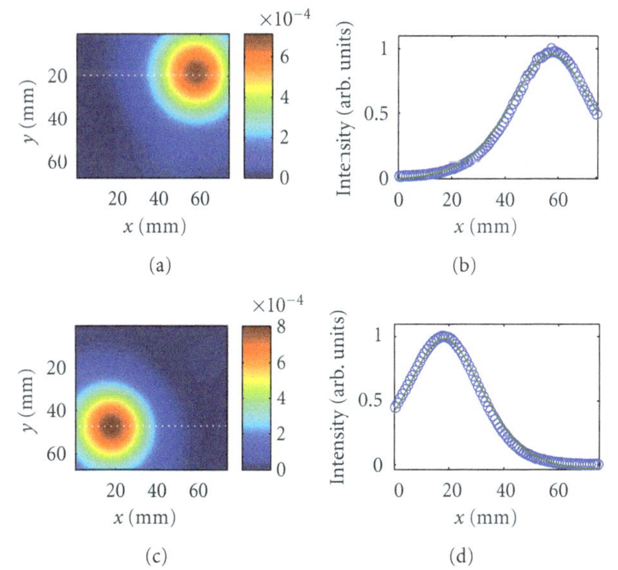

FIGURE 9: NCIDs on the detector plane are shown in (a) and (c); corresponding Green's function fits to the horizontal spatial profiles through the dashed line are shown in (b) and (d).

the position coordinates better (within 0.5 mm) for the scattering Target 2 than done by ICA and PCA (deviation from known values being between 2-3 mm). NMF retrieved the position coordinates for Target 1 with 3.0 mm error in z direction, which is not as good as that done by ICA and PCA. But γ is 0.783 and 0.778 in the fittings for ICA and PCA, respectively, as compared to 0.993 for NMF, indicating that the quality of the fitting is better for NMF. The quality of fitting is presumably affected by the efficacy of decomposition. The decomposed NCIDs by NMF were more "clean" than those decomposed by ICA and PCA. We ascribe the observed higher errors in ICA and PCA estimates of the position coordinates of the scattering Target 2 than the NMF estimates to the interference from the other virtual source (corresponding to Target 1) in ICA (Figure 10(c)) and PCA (Figure 11(c)) images. It is commonly believed that errors in locating a scattering target are higher than that for locating an absorptive target, and the results of this study conform to that notion.

Table 2: Known positions versus retrieved positions of the absorptive targets using ICA, PCA, and NMF algorithms.

Target	Known position (mm)	Algorithm	Fitted position (mm)	Error (mm)
1	(57.2, 18.1, 20)	ICA	(57.4, 18.2, 21.5)	(0.2, 0.1, 1.5)
		PCA	(57.4, 18.2, 20.6)	(0.2, 0.1, 0.6)
		NMF	(57.4, 18.2, 19.5)	(0.2, 0.1, 0.5)
2	(19.9, 48.1, 25)	ICA	(18.2, 46.7, 24.7)	(1.7, 1.4, 0.3)
		PCA	(18.2, 47.6, 25.9)	(1.7, 0.5, 0.9)
		NMF	(18.2, 47.6, 23.3)	(1.7, 0.5, 1.7)

Table 3: Known positions versus retrieved positions of the scattering targets using ICA, PCA, and NMF algorithms.

Target	Known position (mm)	Algorithm	Fitted position (mm)	Error (mm)
1	(13.0, 28.0, 30.0)	ICA	(12.6, 28.7, 29.1)	(0.4, 0.7, 0.9)
		PCA	(12.6, 28.7, 28.6)	(0.4, 0.7, 1.4)
		NMF	(12.0, 28.5, 33.0)	(1.0, 0.5, 3.0)
2	(53.3, 28.5, 30.0)	ICA	(51.0, 31.8, 26.8)	(2.3, 3.3, 3.2)
		PCA	(50.9, 31.8, 26.7)	(2.4, 3.3, 3.3)
		NMF	(53.3, 28.0, 30.3)	(0.0, 0.5, 0.3)

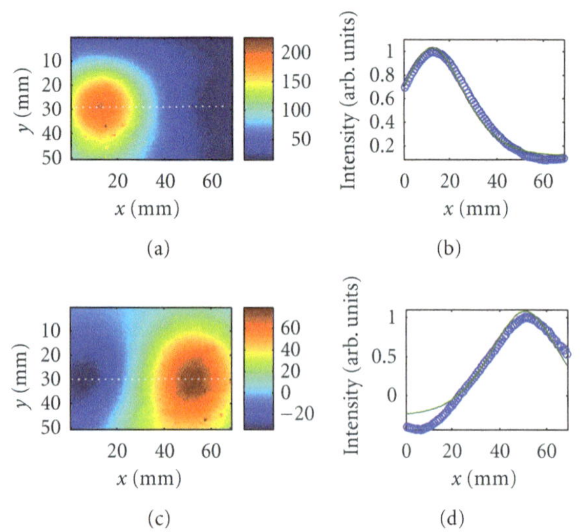

Figure 10: ICA-generated ICIDs on the detector plane are shown in (a) and (c); corresponding Green's function fits to the horizontal spatial profiles through the dashed line are shown in (b) and (d).

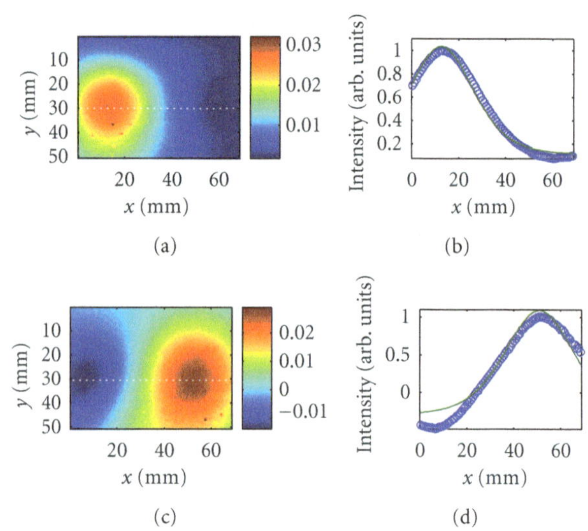

Figure 11: PCIDs on the detector plane are shown in (a) and (c); corresponding Green's function fits to the horizontal spatial profiles through the dashed line are shown in (b) and (d).

5. Summary and Discussion

Diffusive optical imaging was modeled as a BSS problem. ICA, PCA, and NMF were used to decompose the data matrix and locate the targets embedded in a highly scattering turbid medium. Only the components corresponding to the targets were extracted from a large dataset for target detection and localization.

It may be instructive to compare the objectives, scope, and computational complexity of these decomposition methods with model-based reconstruction methods. Decomposition methods obtain the 3D locations of targets (the number of targets is generally small). Based on the retrieved locations, the methods may then be further extended to retrieve size and optical property information of the targets [9]. The common practice of model-based inverse reconstruction methods is to discretize the sample volume into $N \times N \times N$ voxels and estimate absorption and/or scattering coefficient in each voxel iteratively. Voxels with significantly different optical properties than the surrounding are regions of interest and may be identified as targets. While estimating the optical properties, the forward model is solved repeatedly to calculate the intensity of the multiply scattered light on the sample boundary. The difference between the intensity of the multiply scattered light predicted by the forward model and the experimental measurements is minimized by seeking an optimal set of the optical properties of every voxel in the sample volume. The number of variables

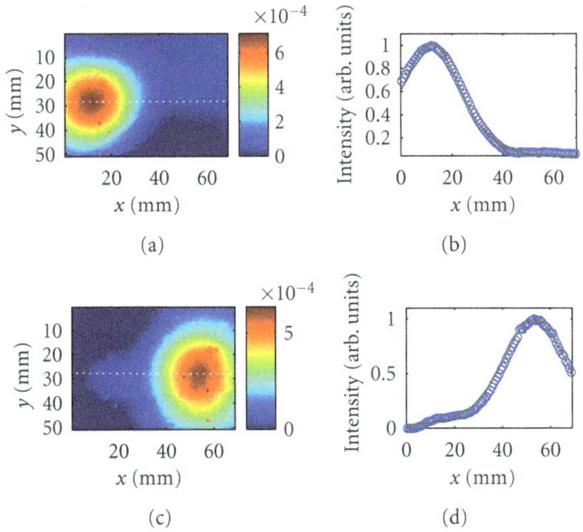

FIGURE 12: NCIDs on the detector plane are shown in (a) and (c); corresponding Green's function fits to the horizontal spatial profiles through the dashed line are shown in (b) and (d).

is thus, on the order of N^3. To determine location(s) of target(s) in three dimensions, the decomposition methods process the data matrix to retrieve the main components (A and S). Here, A and S are two-dimensional matrices with the number of unknowns on the order of N^2. The number of unknowns is, hence, reduced N times in the decomposition methods compared to the model-based approaches, which leads to a substantial saving in the computational time when N is large. No repeated solution of the forward model is involved in decomposition methods. Consequently, decomposition methods are considerably faster.

A comparison of the computational complexity of these two types of approaches may shed further light on their relative computation economy. For a model-based iterative reconstruction method, an equation of the form $b = Wx$ is solved to find the targets, where W is a weight matrix of *size* $N_dN_s \times N_v$, N_d, N_s, and N_v are the numbers of detectors, sources and voxels, respectively, b is an $N_dN_s \times 1$ vector describing the perturbation in the detected light intensity due to the presence of targets, and x is the perturbation in the optical properties from the background values with dimension of $N_v \times 1$. The computational complexity is typically $O(N_dN_sN_v^2)$ for a single iteration. For the decomposition approach, b is written as a $2D$ matrix X with dimension $N_d \times N_s$. To decompose matrix X, the computational complexity per iteration is typically of order $O(N_dN_k)$ for ICA [34], and $O(N_dN_sN_k)$ for NMF [16], where N_k is the number of components that relates to the number of targets and is usually a small number. For PCA using SVD, the complexity is $O(N_s^2N_k)$ [34]. The computational complexity of the intrinsic iterative process involved in the matrix decomposition algorithms is much lower than that in the model-based inverse reconstruction methods.

All three matrix decomposition methods presented in this manuscript can potentially be used in *in-vivo* real-time breast cancer imaging. The three algorithms have different assumptions, which may lead to different favored conditions. In this study, the algorithms were evaluated using simulative and experimental data using model scattering media and absorptive and scattering targets. The (x, y, z) positions of the targets were retrieved with good accuracy. The decomposition provided by ICA is "cleaner" than that of the PCA. PCA did not clearly separate the two absorptive targets used in the first experiment. NMF decomposition seems to provide residue-free "cleaner" images than the other two methods in this study. However, since NMF is based on nonnegativity assumption, the results might deteriorate when such a non-negativity assumption does not hold well. While continuous wave measurements were used in the work presented in this paper, the approaches could be used with frequency domain and time domain measurements as well.

The work presented here focuses on detecting and locating small targets, which derive impetus from the need to detect tumors in early stages of growth when those are more amenable to treatment. All three methods are applicable for extended targets as well and are expected to provide the "center of optical strength" as the location of the target.

All three approaches are applicable for both scattering and absorbing targets and may be used in clinical setting. The contrast between a tumor and surrounding normal tissue can be due to differences in absorption, scattering, or both absorption and scattering properties and may depend significantly on the wavelength of light used. However, *a priori* knowledge of the optical characteristics (absorptive or scattering) is not crucial. As has been shown in (2) and (3), the expression for elements of the data matrix for absorptive targets involves Green's Functions G, while that for scattering targets involves $\partial G/\partial z \approx -\kappa G$, where $\kappa = \sqrt{\mu_a/D}$ in CW [9]. This relationship with G provides basis for detection and localization of target(s), whether contrast is due to absorption, scattering, or both. We are using transillumination geometry, which is one of the approaches used by other researchers, and adequate signal for *in vivo* breast imaging is obtained [29, 35–38].

In this paper, we presented results when the approaches were used to detect and obtain three-dimensional location information of the targets. We have demonstrated, while developing OPTICA that a backprojection formalism can be further implemented to get a cross-section image of the target [11], or the retrieved target locations can be fed into other DOI methods as *a priori* information to get three-dimensional tomographic images. Since the approaches are suited for small targets, this hold promise for detecting and locating breast tumors in early stages of growth, which is crucially important for effective treatment. Further work involving *ex vivo* (model) and *in vivo* imaging of cancerous breast will be needed to establish the full potential of these approaches.

Acknowledgment

The paper is supported in part by US Army Medical Research and Materiel Command under Contract no. W81XWH-07-1-0454.

References

[1] X. R. Cao and R. W. Liu, "General approach to blind source separation," *IEEE Transactions on Signal Processing*, vol. 44, no. 3, pp. 562–571, 1996.

[2] J. F. Cardoso, "Blind signal separation: statistical principles," *Proceedings of the IEEE*, vol. 86, no. 10, pp. 2009–2025, 1998.

[3] A. Hyvärinen, J. Karhunen, and E. Oja, *Independent Component Analysis*, Wiley, New York, NY, USA, 2001.

[4] I. T. Jolliffe, *Principal Component Analysis*, Springer, New York, NY, USA, 1986.

[5] D. D. Lee and H. S. Seung, "Learning the parts of objects by non-negative matrix factorization," *Nature*, vol. 401, no. 6755, pp. 788–791, 1999.

[6] M. W. Berry, M. Browne, A. N. Langville, V. P. Pauca, and R. J. Plemmons, "Algorithms and applications for approximate nonnegative matrix factorization," *Computational Statistics and Data Analysis*, vol. 52, no. 1, pp. 155–173, 2007.

[7] M. Xu, M. Alrubaiee, S. K. Gayen, and R. R. Alfano, "Optical imaging of turbid media using independent component analysis: theory and simulation," *Journal of Biomedical Optics*, vol. 10, no. 5, Article ID 051705, 2005.

[8] M. Xu, M. Alrubaiee, S. K. Gayen, and R. R. Alfano, "Three-dimensional localization and optical imaging of objects in turbid media with independent component analysis," *Applied Optics*, vol. 44, no. 10, pp. 1889–1897, 2005.

[9] M. Xu, M. Alrubaiee, S. K. Gayen, and R. R. Alfano, "Optical diffuse imaging of an ex vivo model cancerous human breast using independent component analysis," *IEEE Journal on Selected Topics in Quantum Electronics*, vol. 14, no. 1, pp. 43–49, 2008.

[10] M. Alrubaiee, M. Xu, S. K. Gayen, M. Brito, and R. R. Alfano, "Three-dimensional optical tomographic imaging of scattering objects in tissue-simulating turbid media using independent component analysis," *Applied Physics Letters*, vol. 87, no. 19, Article ID 191112, pp. 1–3, 2005.

[11] M. Alrubaiee, M. Xu, S. K. Gayen, and R. R. Alfano, "Localization and cross section reconstruction of fluorescent targets in *ex vivo* breast tissue using independent component analysis," *Applied Physics Letters*, vol. 89, no. 13, Article ID 133902, 2006.

[12] D. L. Massart, B. G. M. Vandeginste, S. M. Deming, Y. Michotte, and L. Kaufman, *Chemometrics: A Textbook*, Elsevier, Amsterdam, The Netherlands, 1988.

[13] M. Turk and A. Pentland, "Eigenfaces for recognition," *Journal of Cognitive Neuroscience*, vol. 3, no. 1, pp. 71–86, 1991.

[14] L. K. Hansen, J. Larsen, F. A. Nielsen et al., "Generalizable patterns in neuroimaging: how many principal components?" *NeuroImage*, vol. 9, no. 5, pp. 534–544, 1999.

[15] J. P. Brunet, P. Tamayo, T. R. Golub, and J. P. Mesirov, "Metagenes and molecular pattern discovery using matrix factorization," *Proceedings of the National Academy of Sciences of the United States of America*, vol. 101, no. 12, pp. 4164–4169, 2004.

[16] V. P. Pauca, J. Piper, and R. J. Plemmons, "Nonnegative matrix factorization for spectral data analysis," *Linear Algebra and Its Applications*, vol. 416, no. 1, pp. 29–47, 2006.

[17] A. Ishimaru, "Diffusion of a pulse in densely distributed scatterers," *Journal of the Optical Society of America*, vol. 68, no. 8, pp. 1045–1050, 1978.

[18] K. Furutsu, "Diffusion equation derived from the space-time transport equation," *Journal of the Optical Society of America*, vol. 70, no. 4, pp. 360–366, 1980.

[19] M. S. Patterson, B. Chance, and B. C. Wilson, "Time resolved reflectance and transmittance for the non-invasive measurement of tissue optical properties," *Applied Optics*, vol. 28, no. 12, pp. 2331–2336, 1989.

[20] S. Chandrasekhar, *Radiative Transfer*, Clarendon Press, Oxford, UK, 1950.

[21] A. Ishimaru, *Wave Propagation and Scattering in Random Media*, vol. 1 of *Single Scattering and Transport Theory*, Academic, New York, NY, USA, 1978.

[22] S. R. Arridge and J. C. Schotland, "Optical tomography: forward and inverse problems," *Inverse Problems*, vol. 25, Article ID 123010, 2009.

[23] S. R. Arridge, "Photon-measurement density functions. Part I: analytical forms," *Applied Optics*, vol. 34, no. 31, pp. 7395–7409, 1995.

[24] P. C. Hansen, "Analysis of discrete ill-posed problems by means of the L-curve," *SIAM Review*, vol. 34, pp. 561–580, 1992.

[25] P. Paatero and U. Tapper, "Positive matrix factorization: a non-negative factor model with optimal utilization of error estimates of data values," *Environmetrics*, vol. 5, no. 2, pp. 111–126, 1994.

[26] P. Paatero, "The multilinear engine: a table-driven, least squares program for solving multilinear problems, including the n-way parallel factor analysis model," *Journal of Computational and Graphical Statistics*, vol. 8, no. 4, pp. 854–888, 1999.

[27] H. J. van Staveren, C. J. M. Moes, J. van Marle, S. A. Prahl, and M. J. C. van Gemert, "Light scattering in Intralipid-10% in the wavelength range of 400–1100 nm," *Applied Optics*, vol. 31, pp. 4507–4514, 1991.

[28] C. Bordier, C. Andraud, E. Charron, J. Lafait, M. Anastasiadou, and A. De Martino, "Illustration of a bimodal system in Intralipid-20% by polarized light scattering: experiments and modeling," *Applied Physics A*, vol. 94, no. 2, pp. 347–355, 2009.

[29] T. Nielsen, B. Brendel, R. Ziegler et al., "Linear image reconstruction for a diffuse optical mammography system in a noncompressed geometry using scattering fluid," *Applied Optics*, vol. 48, no. 10, pp. D1–D13, 2009.

[30] Y. Ardeshirpour, M. Huang, and Q. Zhu, "Effect of the chest wall on breast lesion reconstruction," *Journal of Biomedical Optics*, vol. 14, no. 4, p. 044005, 2009.

[31] B. W. Pogue, M. S. Patterson, H. Jiang, and K. D. Paulsen, "Initial assessment of a simple system for frequency domain diffuse optical tomography," *Physics in Medicine and Biology*, vol. 40, no. 10, pp. 1709–1729, 1995.

[32] A. Poellinger, J. C. Martin, S. L. Ponder et al., "Near-infrared laser computed tomography of the breast. First clinical experience," *Academic Radiology*, vol. 15, no. 12, pp. 1545–1553, 2008.

[33] B. Wu, W. Cai, M. Alrubaiee, M. Xu, and S. K. Gayen, "Time reversal optical tomography: locating targets in a highly scattering turbid medium," *Optics Express*, vol. 19, no. 22, pp. 21956–21976, 2011.

[34] T. Ristaniemi and J. Joutsensalo, "Advanced ICA-based receivers for block fading DS-CDMA channels," *Signal Processing*, vol. 82, no. 3, pp. 417–431, 2002.

[35] M. Cutler, "Transillumination as an aid in the diagnosis of breast lesions," *Surgery, Gynecology and Obstetrics*, vol. 48, pp. 721–729, 1929.

[36] R. Choe, A. Corlu, K. Lee et al., "Diffuse optical tomography of breast cancer during neoadjuvant chemotherapy: a case study with comparison to MRI," *Medical Physics*, vol. 32, no. 4, pp. 1128–1139, 2005.

[37] A. Pifferi, P. Taroni, A. Torricelli, F. Messina, R. Cubeddu, and G. Danesini, "Four-wavelength time-resolved optical

mammography in the 680-980-nm range," *Optics Letters*, vol. 28, no. 13, pp. 1138–1140, 2003.

[38] B. J. Tromberg, B. W. Pogue, K. D. Paulsen, A. G. Yodh, D. A. Boas, and A. E. Cerussi, "Assessing the future of diffuse optical imaging technologies for breast cancer management," *Medical Physics*, vol. 35, no. 6, pp. 2443–2451, 2008.

12

Zeeman Effect of Sm Atoms by High-Resolution Diode-Laser Spectroscopy

Wei-Guo Jin and Tatsuya Minowa

Department of Physics, Faculty of Science, Toho University, Miyama 2-2-1, Funabashi, Chiba 274-8510, Japan

Correspondence should be addressed to Wei-Guo Jin; jin@ph.sci.toho-u.ac.jp

Academic Editor: Gong Ru Lin

High-resolution atomic-beam diode-laser spectroscopy in Sm I has been performed. Zeeman spectra have been measured for the three optical transitions at different external magnetic fields and well resolved at the magnetic fields of stronger than 6.0 mT. Using the known precise Landé g-factors of the ground multiplet, the Landé g-factors of the upper $4f^66s6p$ $^9F_1$ and $^9D_3$ levels have been determined, and their precision has been improved compared with the reference values.

1. Introduction

Studies of the interaction of atoms with an external magnetic field, the Zeeman effect, have a long history and have been of repeated interests to spectroscopists [1–5]. The atomic Landé g-factor, a measure of the atomic magnetic moment, can be deduced from the Zeeman spectroscopy and provides a sensitive test of atomic calculation [6–8]. The Zeeman effect is also related to the research of the atomic parity nonconservation [9]. Moreover, precise information of magnetic fields for interstellar clouds, which is important for understanding the role played by magnetic fields in star formation, can be obtained from the Zeeman observation [10–12]. Study of the Zeeman effect is, therefore, of much interest not only from the point of view of atomic physics but also from the point of view of other fields such as astrophysics.

Samarium, a typical rare earth element, is suitable for measurements of the Zeeman effect because it has rich optical transitions and many stable isotopes. The Landé-g factors of the ^7F ground multiplet were precisely determined using an atomic-beam magnetic resonance [13, 14]. Some high-lying levels were measured by means of atomic-beam laser spectroscopy [15–17]. Martin et al. [18] tabulated Landé-g factors for almost all known excited levels of Sm, and errors were not given.

In this paper, we report high-resolution atomic-beam diode-laser spectroscopy in Sm I around 680 nm. Zeeman spectra at different external magnetic fields are measured and analyzed for three transitions. Using the known precise values of the Landé g-factor of the ground multiplet, the Landé g-factors of the upper levels are obtained, and results are discussed.

2. Experiment

The present experiment was performed using a diode-laser beam and an atomic beam. The experimental setup is shown in Figure 1 and is essentially identical to that used in our previous works [19, 20]. Evaporation of Sm atoms was made using a resistance heating of a molybdenum oven. The oven temperature was controlled at about 900°C in order to produce a sufficient vapor pressure for Sm. An atomic beam was formed by a 2 mm diameter aperture at a distance of 30 cm from the oven.

A laser beam was produced using a commercial tunable diode laser with an external cavity system (EOSI ECU2010-01A). The wavelength of the laser, from 665 nm to 690 nm, was measured with a high-precision wavemeter. The output power was about 8 mW and the linewidth of the laser beam was smaller than 100 kHz.

In order to reduce the Doppler broadening, the laser beam crossed the atomic beam perpendicularly. At the interaction region, the intensity of the atomic beam with a diameter of about 2 mm was estimated to be about 2×10^{12}

TABLE 1: Wavelengths of the studied transitions and properties of the lower and upper levels in Sm I.

Wavelength (nm)	Lower level			Upper level		
	Configuration	Level	Energy (cm^{-1})	Configuration	Level	Energy (cm^{-1})
672.59	$4f^6 6s^2$	$^7F_0$	0.0	$4f^6 6s6p$	$^9F_1$	14863.85
680.30	$4f^6 6s^2$	$^7F_2$	811.92	$4f^6 6s6p$	$^9D_3$	15507.35
686.09	$4f^6 6s^2$	$^7F_1$	292.58	$4f^6 6s6p$	$^9F_1$	14863.85

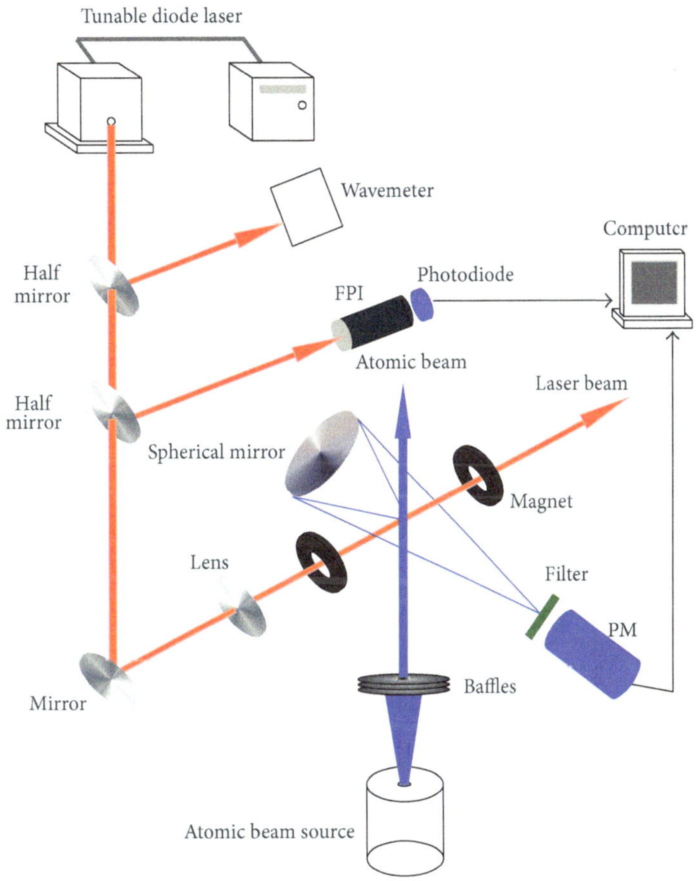

FIGURE 1: Experimental setup. FPI denotes Fabry-Perot interferometer and PM photomultiplier.

atoms/s from the oven temperature using the equation given by Ramsey [21], and the intensity of the laser beam with a diameter of about 3 mm was estimated to be about 3×10^{15} photons/s. Fluorescence from the atomic beam, induced by the laser beam, was collected with a spherical mirror and detected with a cooled photon-counting photomultiplier (Hamamatsu R2257P). Relative frequency calibration was made by measuring the spectrum of transmitted light through a confocal Fabry-Perot interferometer (FPI) with a free spectral range of 300 MHz.

The magnetic field at the interaction region between the laser beam and the atomic beam was produced using a pair of permanent magnets. The magnets had a ring shape with an inner diameter of 13–32 mm, an outer diameter of 40–60 mm, and a thickness of 4.9–8.2 mm, and the magnetic field could be changed using different magnets. The magnets were set along the laser beam and, therefore, the magnetic field was almost parallel to the laser beam, that is, perpendicular to the atomic beam.

3. Results and Analysis

Three transitions in Sm I were studied in this experiment. Wavelengths of the transitions, electronic configurations, and energies of the lower and upper levels [18, 22] are presented in Table 1. All transitions are from the levels of the ground term $4f^6 6s^2\ ^7F$ to the levels of the $4f^6 6s6p$ configuration, that is, the s^2-sp transitions. For the 672.59 nm and 686.09 nm transitions, the upper levels are of a same level of $4f^6 6s6p\ ^9F_1$.

Figure 2 shows a typically observed fluorescence spectrum for the 680.30 nm transition in Sm I. Stable Sm has five even-mass isotopes, ^{144}Sm, ^{148}Sm, ^{150}Sm, ^{152}Sm, and ^{154}Sm, and two odd-mass isotopes, ^{147}Sm, and ^{149}Sm. ^{147}Sm and ^{149}Sm, both, with the nuclear spin of $I = 7/2$, have

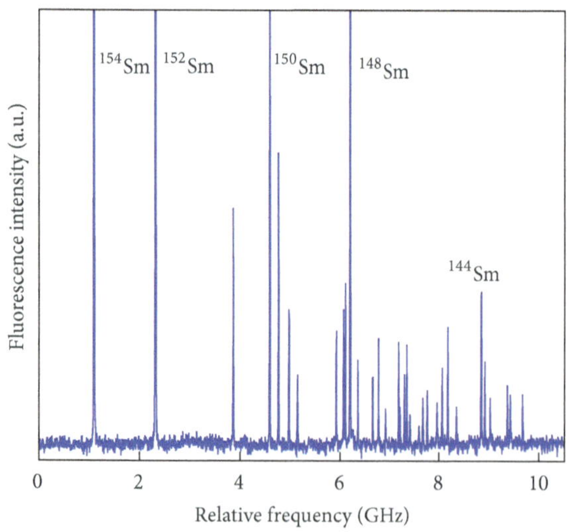

FIGURE 2: Observed fluorescence spectrum of the $4f^66s^2\ ^7F_2$–$4f^66s6p\ ^9D_3$ transition at 680.30 nm in Sm I. Peaks of the even-mass isotopes are labeled with their atomic symbol. Other peaks are hyperfine structure peaks of the two odd-mass isotopes ^{147}Sm and ^{149}Sm.

complicated hyperfine structures. No hyperfine structures exist for the even-mass isotopes because they have no nuclear spin: the even-mass isotopes are suitable for studying the Zeeman effect. It can be seen from Figure 2 that there are no other peaks around the peaks of ^{152}Sm and ^{154}Sm. Therefore, spectra of ^{152}Sm and ^{154}Sm were used for measurement of the Zeeman effect in this experiment.

Zeeman spectra were observed at three different magnetic fields. The observed Zeeman spectra of ^{152}Sm and ^{154}Sm are shown in Figure 3 for the 680.30 nm transition. Since the Zeeman spectra were observed with the permanent magnets, and the fluorescence spectra were observed without the permanent magnets, starting of the frequency scanning was different for these two spectra as shown in Figures 2 and 3. Splittings of the spectra become larger as magnetic fields become stronger. At the magnetic field of 4.1 mT, the spectra overlap considerably while the spectra become well resolved at the magnetic fields of stronger than 6.0 mT. The spectra of ^{154}Sm show a same pattern as those of ^{152}Sm. The full width at half maximum (FWHM) of the peaks is about 14 MHz. This width is considered to be due to the natural width of the upper level of the transition and the residual Doppler broadening of the atomic beam since the linewidth of the laser is smaller than 100 kHz.

Peak centers of the measured spectra were determined from a least-squares fit with a Lorentz function and calibrated with the FPI spectra. For each transition, measurements were performed more than 10 times. Thus, relative differences between different peaks were determined with the uncertainty of about 0.5 MHz, which includes the error of peak-center determination, the error of the free spectral range of the FPI (0.046 MHz), and the error of linearity correction for frequency scanning.

For atoms in an external magnetic field B, the atomic level with an electronic angular momentum J splits into $(2J + 1)$

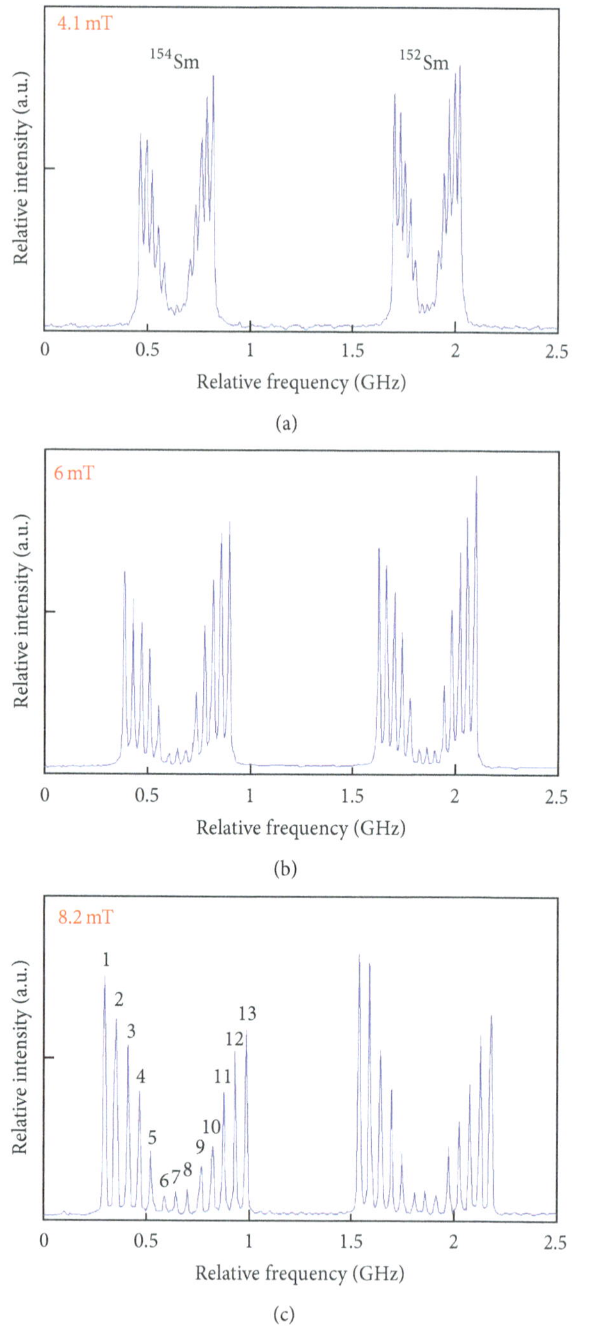

FIGURE 3: Observed Zeeman spectra of ^{152}Sm and ^{154}Sm at the 680.30 nm transition for three magnetic fields. The numbers labeled on peaks correspond to the numbers on Zeeman transitions in Figure 4.

magnetic sublevels with a magnetic quantum number m_J. The Zeeman energy ΔE from the interaction between the external weak magnetic field and the atomic magnetic moment is expressed as follows [23]:

$$\Delta E = m_J g_J \mu_B B, \quad (1)$$

where g_J is the atomic Landé g-factor and μ_B is the Bohr magneton.

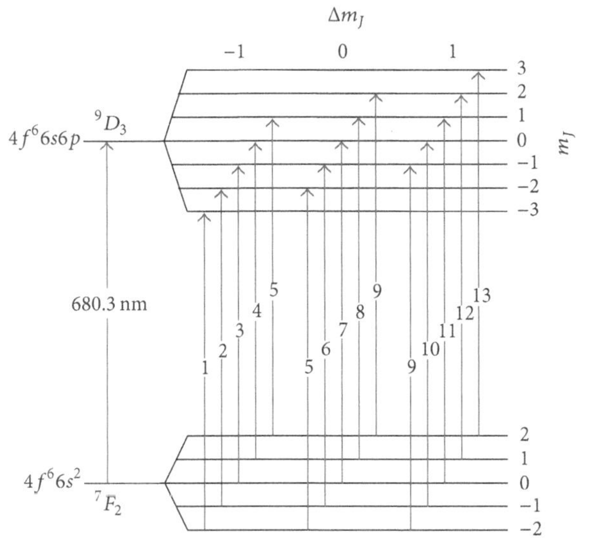

FIGURE 4: Zeeman splittings and transitions for the 680.30 nm transition. m_J is the magnetic quantum number and Zeeman transitions are divided into the three series of $\Delta m_J = -1, 0, +1$. The numbers on the Zeeman transitions correspond to the numbers labeled on peaks in Figure 3.

TABLE 2: Determined atomic Landé g-factors, g_J, for the $4f^6 6s6p\ ^9F_1$ and $^9D_3$ levels in Sm I together with the reference values.

Magnetic Field (mT)	g_J $^9F_1$ ^{154}Sm	^{152}Sm	$^9D_3$ ^{154}Sm	^{152}Sm
6.0	3.097 (3)	3.094 (3)	1.946 (3)	1.946 (3)
8.2	3.094 (3)	3.087 (3)	1.955 (3)	1.956 (3)
Average	3.093 (2)		1.951 (3)	
Reference [18]	3.10		1.965	

For the 680.30 nm transition, the lower $4f^6 6s^2\ ^7F_2$ level with $J = 2$ splits into the 5 sublevels, and the upper $4f^6 6s6p\ ^9D_3$ level with $J = 3$ splits into the 7 sublevels. The Zeeman splittings of the 680.30 nm transition are shown in Figure 4 together with the possible transitions according to the selection rule of $\Delta m_J = m_J$ (the upper $^9D_3$ level) $-m_J$ (the lower $^7F_2$ level) $= 0, \pm 1$.

The observed Zeeman spectra were identified with the Zeeman transitions as shown in Figure 4. For the 680.30 nm transition, the observed spectra shown in Figure 3 can be divided into three series of $\Delta m_J = -1, 0, +1$. The peak intensities of the $\Delta m_J = -1$ series are comparable with those of the $\Delta m_J = +1$ series. Peaks of the $\Delta m_J = 0$ series are also observed with lower intensities, and two of them are overlapped with those of the $\Delta m_J = \pm 1$ series.

From the observed Zeeman spectra, Zeeman splitting energies of both the lower and the upper levels can be obtained and are related to the external magnetic field and the g_J values of the lower and upper levels as shown in (1). Using the known precise g_J values of the $^7F_1$ and $^7F_2$ levels, the external magnetic fields B were calibrated from the measured Zeeman splitting energies of the lower levels. Since the g_J values of the lower levels were precisely measured, the uncertainty of the external magnetic field was determined by the uncertainty of the measured Zeeman splitting energies (about 0.5 MHz) to be about 0.1–0.2%. Further, the g_J values of the upper levels were determined and are presented in Table 2. The g_J values were derived from the spectra at the magnetic fields of 6.0 mT and 8.2 mT; data at the magnetic field of 4.1 mT were not used because the spectra are overlapped considerably. The uncertainty of the g_J values of the upper levels was determined by the uncertainties of the measured Zeeman splitting energies and the calibrated magnetic field and is about 0.1–0.2% as shown in Table 2.

It can be seen from Table 2 that the g_J values at the magnetic field of 6.0 mT agree with the values at the magnetic field of 8.2 mT at the margin of experimental uncertainties. The g_J values of ^{154}Sm are in good agreement with those of ^{152}Sm and this shows that the Landé g-factor is independent of the isotope. Thus, the averaged g_J values were obtained for the $4f^6 6s6p\ ^9F_1$ and $^9D_3$ levels. The g_J values tabulated by Martin et al. [18] are also listed in Table 2 for comparison. The reference g_J values are very close to the present values. Because errors are not given for the reference g_J values, it is difficult to make detailed comparison between the present values and the reference values. For the $4f^6 6s6p\ ^9F_1$ level, the precision of the present g_J value is, however, improved at least one order of magnitude compared with the reference value.

4. Summary

High-resolution laser spectroscopy in Sm I has been performed using the diode-laser beam together with the collimated atomic beam. Fluorescence spectra have been observed for the three optical transitions, and the spectra of ^{152}Sm and ^{154}Sm have been used for measurements of the Zeeman effect. The Zeeman spectra have been measured at the three external magnetic fields and well resolved at the magnetic fields of stronger than 6.0 mT. The observed spectra have been identified with the Zeeman transitions and the magnetic fields have been calibrated with the known precise g_J values of the ground multiplet. The g_J values of the upper $4f^6 6s6p\ ^9F_1$ and $^9D_3$ levels have been determined and show no isotopic dependence. The precision of the present g_J values is improved compared with the reference values. The precise g_J values of various atomic levels are useful for astrophysics.

Acknowledgments

The authors thank Mr. K. Moronaga and Dr. M. Kawamura for their help in this experiment.

References

[1] T. Jeong, J. Y. Won, and H.-R. Noh, "Line shapes in polarization spectroscopy for the rubidium D_1 line in an external magnetic field," *Optics Communications*, vol. 292, pp. 106–110, 2013.

[2] G. Sinuco-León and B. M. Garraway, "Radio-frequency dressed atoms beyond the linear Zeeman effect," *New Journal of Physics*, vol. 14, Article ID 123008, 12 pages, 2012.

[3] J.-C. Pain and F. Gilleron, "Characterization of anomalous Zeeman patterns in complex atomic spectra," *Physical Review A*, vol. 85, Article ID 033409, 14 pages, 2012.

[4] J. Li, C. Dong, P. Jönsson, and G. Gaigalas, "M_F-dependent hyperfine induced transition rates in an external magnetic field for Be-like $^{47}Ti^{18+}$," *Physics Letters A*, vol. 375, no. 5, pp. 914–917, 2011.

[5] D. Shulyak, A. Reiners, S. Wende, O. Kochukhov, N. Piskunov, and A. Seifahrt, "Modelling the molecular Zeeman-effect in M-dwarfs: methods and first results," *Astronomy and Astrophysics*, vol. 523, article A37, 12 pages, 2010.

[6] R. Soria Orts, J. R. Crespo López-Urrutia, H. Bruhns et al., "Zeeman splitting and g factor of the $1s^2 2s^2 2p\, ^2P_{3/2}$ and $^2P_{1/2}$ levels in Ar^{13+}," *Physical Review A*, vol. 76, Article ID 052501, 7 pages, 2007.

[7] Y. Zhang, J. Xu, W. Zhang et al., "Lifetime and Landé factor measurements of *5p7p* levels of Sn I by time-resolved laser spectroscopy," *Physical Review A*, vol. 78, no. 2, Article ID 022505, 6 pages, 2008.

[8] D. L. Moskovkin, V. M. Shabaev, and W. Quint, "Zeeman effect of the hyperfine-structure levels in lithiumlike ions," *Physical Review A*, vol. 77, no. 6, Article ID 063421, 14 pages, 2008.

[9] E. J. Angstmann, T. H. Dinh, and V. V. Flambaum, "Parity non-conservation in atomic Zeeman transitions," *Physical Review A*, vol. 72, no. 5, Article ID 052108, 5 pages, 2005.

[10] R. M. Crutcher, B. Wandelt, C. Heiles, E. Falgarone, and T. H. Troland, "Magnetic fields in interstellar clouds from zeeman observations: inference of total field strengths by Bayesian analysis," *The Astrophysical Journal*, vol. 725, no. 1, pp. 466–479, 2010.

[11] O. Kochukhov, "The Paschen-Back effect in the Li I 6708 Å line and the presence of lithium in cool magnetic Ap stars," *Astronomy and Astrophysics*, vol. 483, no. 2, pp. 557–566, 2008.

[12] A. López Ariste, S. Tomczyk, and R. Casini, "Hyperfine structure as a diagnostic of solar magnetic fields," *The Astrophysical Journal*, vol. 580, no. 1, pp. 519–527, 2002.

[13] F. M. J. Pichanick and G. K. Woodgate, "Zeeman effect in the ground multiplet of samarium," *Proceedings of the Royal Society A*, vol. 263, pp. 89–100, 1961.

[14] W. J. Childs and L. S. Goodman, "Reanalysis of the hyperfine structure of the $4f^6 6s^2\ ^7F$ multiplet in $^{147,149}Sm$, including measurements for the $^7F_6$ state," *Physical Review A*, vol. 6, no. 6, pp. 2011–2021, 1972.

[15] T. Kobayashi, I. Endo, A. Fukumi et al., "Measurement of hyperfine structure constants, g values and tensor polarizability of excited states of Sm I," *Zeitschrift für Physik D*, vol. 39, no. 3, pp. 209–216, 1997.

[16] A. Fukumi, I. Endo, T. Horiguchi et al., "Stark and Zeeman spectroscopies of $4f^6 6s6p\, ^7G_{1-6}$ levels in Sm I under external electric and magnetic fields," *Zeitschrift für Physik D*, vol. 42, no. 4, pp. 243–249, 1997.

[17] T. Kuwamoto, I. Endo, A. Fukumi et al., "Stark and Zeeman spectroscopies of Sm I from metastable-state atomic beam produced by electric discharge," *Journal of the Physical Society of Japan*, vol. 67, no. 4, pp. 1213–1219, 1998.

[18] W. C. Martin, R. Zalubas, and L. Hagan, *Atomic Energy Levels—The Rare-Earth Elements*, NSRDS-NBS60, US GPO, Washington, DC, USA, 1978.

[19] W. G. Jin, Y. Nemoto, H. Nakai, M. Kawamura, and T. Minowa, "Isotope shifts in Gd I and Er I by UV laser spectroscopy," *Journal of the Physical Society of Japan*, vol. 77, no. 12, Article ID 124301, 5 pages, 2008.

[20] T. Wakui, W. G. Jin, K. Hasegawa, H. Uematsu, T. Minowa, and H. Katsuragawa, "High-resolution diode-laser spectroscopy of the rare-earth elements," *Journal of the Physical Society of Japan*, vol. 72, no. 9, pp. 2219–2223, 2003.

[21] N. F. Ramsey, *Molecular Beams*, Oxford University Press, Oxford, UK, 1985.

[22] W. F. Meggers, C. H. Corliss, and B. F. Scribner, *Tables of Spectral-Line Intensities—Part I: Arranged by Elements*, NBS Monograph 145, US GPO, Washington, DC, USA, 1975.

[23] H. Haken and H. C. Wolf, *The Physics of Atoms and Quanta*, Springer, Berlin, Germany, 1996.

13

Polishing Sapphire Substrates by 355 nm Ultraviolet Laser

X. Wei, X. Z. Xie, W. Hu, and J. F. Huang

School of Electromechanical Engineering, Guangdong University of Technology, Guangzhou 510090, China

Correspondence should be addressed to X. Wei, weixin@gdut.edu.cn

Academic Editor: Jian Xu

This paper tries to investigate a novel polishing technology with high efficiency and nice surface quality for sapphire crystal that has high hardness, wear resistance, and chemical stability. A Q-switched 355 nm ultraviolet laser with nanosecond pulses was set up and used to polish sapphire substrate in different conditions in this paper. Surface roughness R_a of polished sapphire was measured with surface profiler, and the surface topography was observed with scanning electronic microscope. The effects of processing parameters as laser energy, pulse repetition rate, scanning speed, incident angle, scanning patterns, and initial surface conditions on surface roughness were analyzed.

1. Introduction

Sapphire is a kind of important photoelectronic material with excellent properties such as high hardness, wear resistance, and chemical stability. So it is applied not only as important substrate for GaN and then LED, but also as important components for solid lasers, IR window and precision bearing. All these applications require sapphire surface with high machining quality and thin damaged layer [1]. Conventional polishing techniques as abrasive polishing, chemical mechanical polishing (CMP), and chemical polishing [2] are very difficult to get high quality for sapphire crystal polishing due to machining defects like mechanical scratch, low efficiency, and yield. Laser polishing technology shows potential for sapphire crystal because of high polishing quality and efficiency. Recently, series of studies on laser polishing of diamond film, polymer, ceramics, fiber, and metal have been conducted [3–7]. However, the research on laser polishing of sapphire crystal is still lack.

Polishing sapphire substrate [8] by ultraviolet laser with picosecond pulses had been conducted in our previous research. The results indicated that better surface quality can be achieved via shorter wavelength and pulse duration. However, the high cost of picosecond ultraviolet lasers affects its further industrialization. Currently, the Q-switched ultraviolet laser with wavelength of 355 nm and nanosecond pulses is widely introduced to industry. Such type of ultraviolet lasers can be used to machine and remove sapphire material in micrometer scale. Thus, a Q-switched 355 nm ultraviolet laser with nanosecond pulses was used to polish sapphire substrate in this paper. The effects of processing parameters as laser energy, pulse repetition rate, laser scanning speed, incident angle, scanning patterns, and initial sapphire surface conditions on surface quality were analysed according to the measurement of surface roughness R_a by surface profiler and the observation of surface topography by scanning electronic microscope. The optimized processing parameters were obtained by orthogonal experiments.

2. Experimental Setup

Figure 1 shows schematically the laser polishing system. The experimental system mainly included an ultraviolet laser source, optical system, the work table, and the PC-based controller. Movements of the work table were controlled by a computer. Laser beam was delivered by the optical system and focused by f-theta focusing lens on the specimen surface. Hence, the focusing beam could ablate the surface of sapphire sample. The sapphire sample was fixed on the work table and the angle between the sapphire surface and the light beam was adjusted by inclining the work table.

The laser polishing system consisted of a ultraviolet laser operator with 355 nm wavelength and laser mode of TEM00, average power of 2 Watts, pulse width of 20 ns, pulse

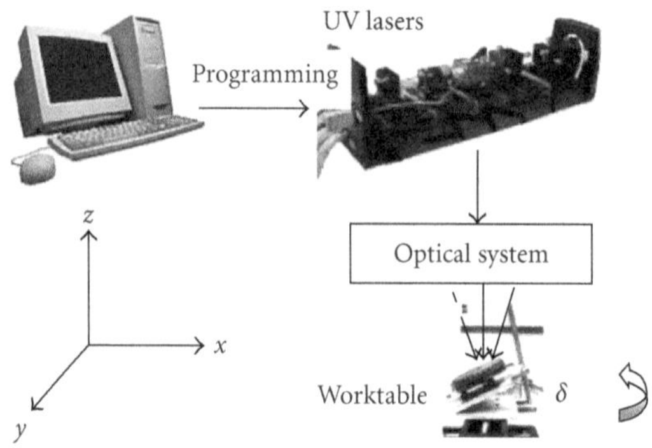

Figure 1: Schematic diagram of laser polishing system.

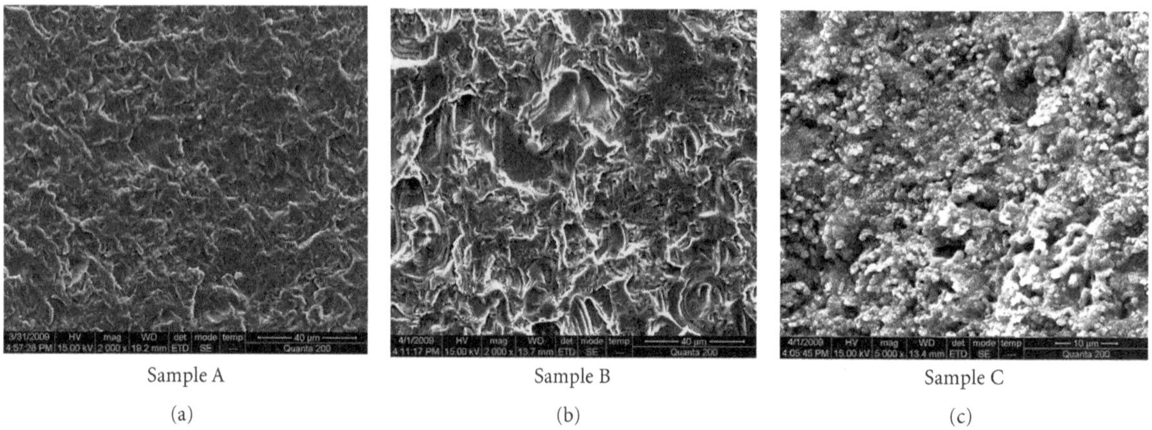

Figure 2: Samples used in this research.

Table 1: The laser parameters.

Laser energy (mJ)	0.03, 0.036, 0.044, 0.050, 0.056, 0.06, 0.064
Laser pulse repetition rate (kHz)	10, 20, 30, 40, 50, 60
Laser scanning speed (mm/s)	10, 30, 50, 70, 90, 110, 130
Incident angle (°)	0, 10, 30, 50
Scanning patterns	A, B, C, D, E

repetition rate of 100 kHz, scanning speed of 700 mm/s, and focal length of 14 μm.

Three sapphire samples (see Figure 2) were prepared as follows.

(1) Sample A: mechanical lapped surface, with dimension of 60 × 40 × 2 mm and surface roughness R_a 0.75 μm.

(2) Sample B: sawed surface, with dimension of 18 × 10 × 1.5 mm and surface roughness R_a 1.1 μm.

(3) Sample C: processed surface by 532 nm green laser, with dimension of 60 × 40 × 2 mm and surface roughness R_a 1.5 μm.

The processing parameters were selected as listed in Table 1.

For evaluating the polishing results, the surface roughness R_a of the polished sapphire substrate was measured by a surface profiler (MarSurf XR20) and the surface topography was observed by a scanning electronic microscope.

3. Results and Discussions

3.1. Effects of Laser Energy on Surface Roughness. Figure 3(a) shows the effect of laser energy on surface roughness of the polished sapphire. With the increase of laser energy, the surface roughness R_a decreases when laser energy is lower than 0.056 mJ and increases when the energy is higher than 0.056 mJ.

When the laser with pulse energy of 0.064 mJ was used, the polished sapphire surface consisted mainly of microcracks caused by thermal stress because of the rapid resolidification, as shown in Figure 3(b). If a small laser energy like 0.044 mJ was applied, the polished sapphire surface was observed with the features of submicropores or ablated pits, and micromelting and resolidification structure. The ablated pits were caused mainly by photoionization and impact ionization. However, there were few sign of microcrack in this energy condition, as shown in Figure 3(c). These results showed that the sapphire material radiated by small pulse energy laser is mainly removed by thermal and photochemical action together, and if radiated by high pulse energy laser,

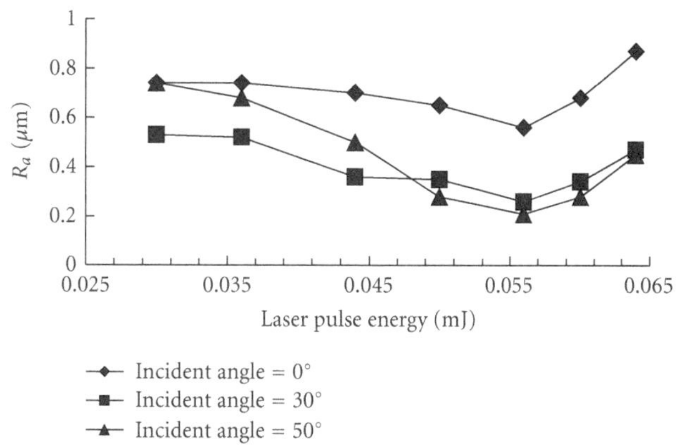

(a) Effect of laser energy on sapphire surface roughness

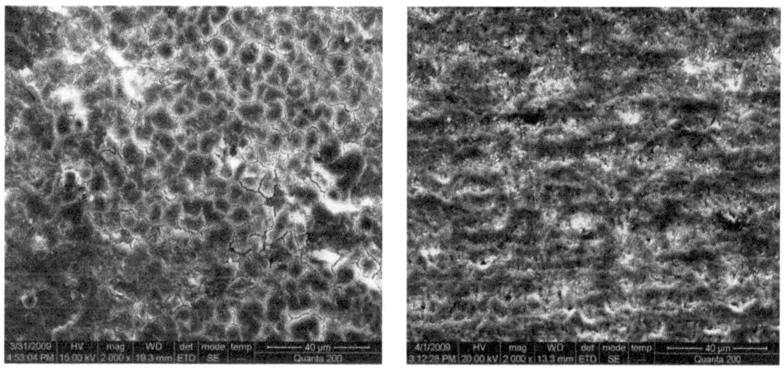

(b) Polished surface (0.064 mJ)

(c) Polished surface (0.044 mJ)

FIGURE 3: Effect of laser energy on sapphire polishing (pulse repetition rate of 30 kHz, scanning speed of 90 mm/s, single-scan polishing, Sample A, incident angle of 50° in (b), (c)).

the material is mainly removed by thermal action with lots microcracks, which causes the surface roughness to become worse.

The sapphire material is removed quite a little and the surface cannot be improved obviously if laser energy is too small. With the increase of the pulse energy continuously, the photochemical action by laser radiation increases even accompanied by thermal action of micromelting, and then the surface roughness of polished sapphire decreases prominently. When the pulse energy is higher than the transition point (e.g., 0.056 mJ in this paper), the thermal action by laser radiation becomes strong and dominant the material removal in the polishing process, which causes lots of microcracks and the surface roughness worse.

3.2. Effects of Laser Pulse Repetition Rate on Surface Roughness. The laser pulse repetition rate has obviously influences on pulse energy and then on the surface of sapphire, as shown in Figure 4.

When the pulse repetition rate was of 10 kHz, the laser energy became higher and material removal became larger comparing with those with higher pulse repetition rate. As a result, the surface roughness increased, as shown in Figures 4(a) and 4(b). Figure 4(b) also showed some features of deep furrows at the surface, which should be the ablation traces consistent with the laser scanning traces.

When the pulse repetition rate went up to 30 kHz, the smoothest surface was obtained. It can be explained that at this condition the laser energy comes to an appropriate value at which the thermal action and photochemical action collaborate favourably. When the pulse repetition rate increased up to 50 kHz or even higher, the laser energy became too small to remove the sapphire material effectively. So the surface roughness of polished sapphire was approximately the same with the original surface, as shown in Figures 4(a) and 4(c).

3.3. Effects on Surface Roughness by Laser Scanning Speed. The laser scanning speed affects the sapphire polishing by laser-matter interaction time. The faster the laser scanning speed, the shorter the interaction time between the laser and material was and the better the surface roughness became, as shown in Figure 5.

When laser scanning speed as low as 10 mm/s was used, the surface material was melted fully because of enough laser-matter interaction time and rapidly resolidified. The material was removed by thermal fracture, resulting in lots of residual microcracks and pits at the surface, as shown in Figure 5(b). So, the surface was very rough (see in Figures 5(a) and 5(b)). The better surface was achieved at the laser scanning speed of 50 mm/s, as shown in Figure 5(c). When scanning speed went up to 90 mm/s, the surface

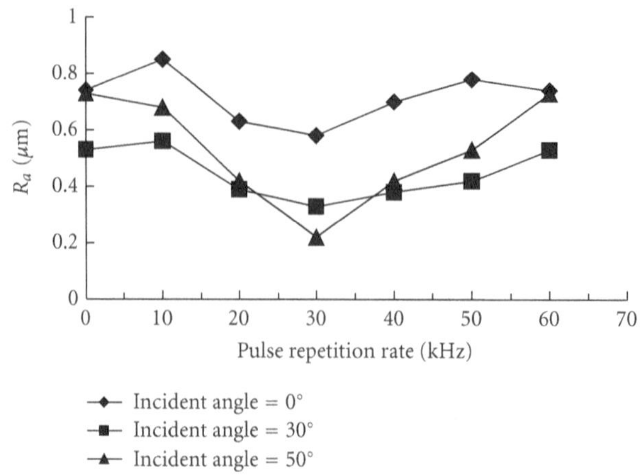

(a) Effect of laser pulse repetition rate on surface roughness

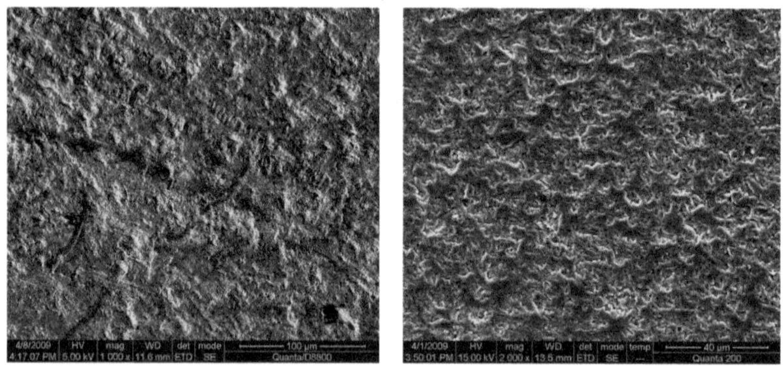

(b) Polished surface (10 kHz) (c) Polished surface (50 kHz)

FIGURE 4: Effect of laser pulse repetition rate on sapphire polishing (laser energy of 0.036 mJ, scanning speed of 90 mm/s, single-scan polishing, Sample A, incident angle of 50° in (b), (c)).

roughness decreased to the lowest value of R_a 0.27 μm and the polished surface was very smooth and dense, as shown in Figure 5(d). When the scanning speed faster than 90 mm/s was applied, the surface roughness increased again because the overlapping coefficient of focal spot on the surface became very small.

3.4. Effects of Incident Angle on Surface Roughness. The laser incident angle means the angle between the laser beam and the right direction of the sample surface. Changing incident angle affects the polishing results mainly by changing the focal spot shape of laser light on the radiated surface and then radiated energy density.

With the increase of the laser incident angle, the focal spot shape became elliptic with a little big area and the polished surface trended to became smooth (shown in Figure 6) as the slight decrease of laser energy. Actually, the convex spot of the surface is much easier to be removed under large incident angle than the concave port of the surface, which contributes to the decrease of surface roughness. However, the incident angle greater than 50 degree is difficult to be applied because of experimental set-up limitation in this study.

3.5. Effects of Laser Scanning Patterns on Surface Roughness. Laser scanning pattern has great influence on laser polishing of sapphire. Variation of scanning patterns will affect the overlapping ratio of focal spot. Five types of scanning patterns were considered in this study.

Pattern A: single-laser beam scans along one direction, as shown in Figure 7(a).

Pattern B: laser beam scans along one direction firstly, and then along the same direction after the sample rotates by θ of 90°, as shown in Figure 7(b).

Pattern C: laser beam scans along one direction firstly, and then along the same direction after the sample rotates by θ of 75°.

Pattern D: laser beam scans along one direction for twice, as shown in Figure 7(a).

Pattern E: laser beam scans firstly along one direction for twice, and then along the same direction for twice after the sample rotates by of 90°, as shown in Figure 7(b).

Figure 8 shows the effects of different laser beam scanning patterns on sapphire surface roughness. Surface

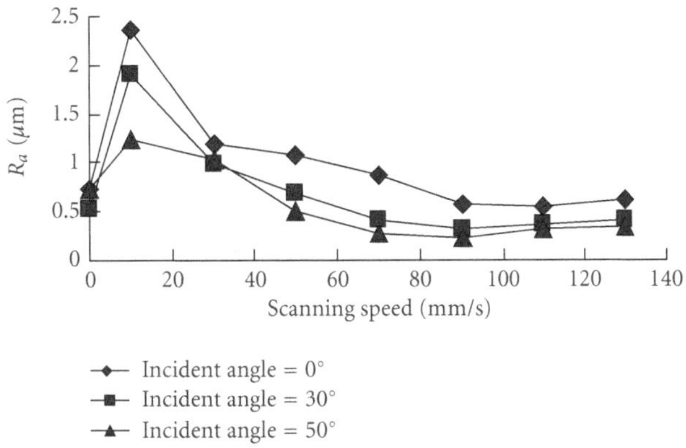

(a) Effect of laser scanning speed on sapphire surface roughness

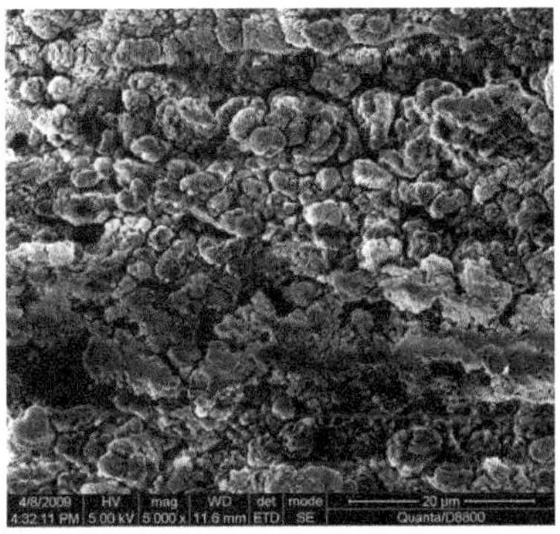

(b) Polished surface (10 mm/s)

(c) Polished surface (50 mm/s)

(d) Polished surface (90 mm/s)

FIGURE 5: Effect of scanning speed on sapphire polishing (laser energy of 0.036 mJ, pulse repetition rate of 30 kHz, single-scan polishing, Sample A, incident angle of 50° in (b), (c), and (d)).

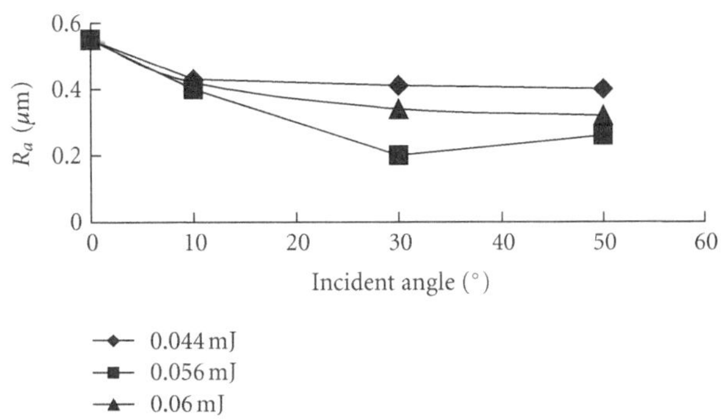

FIGURE 6: Effect of incident angle on sapphire surface roughness (pulse repetition rate of 30 kHz, scanning speed of 90 mm/s, single-scan polishing, Sample A).

roughness after polishing with scanning pattern A was lower than those polished by the other four scanning patterns. This is because single scanning can remove the convex spot sufficiently. Multiple scanning may remove both the convex and concave spots and produces deep furrow, which results in the increasing of surface roughness, as shown in Figures 8(b) and 8(c). Surface roughness after polishing with scanning pattern B was lower than that of scanning patterns C. This means that the smaller intersection angle, the lower surface roughness can obtained.

3.6. Effects of Initial Surface Conditions on Surface Roughness. Three samples with different surface conditions were prepared for this research, namely, mechanical lapped, sawed, and green laser (532 nm) polished surface, as shown in Figure 2.

After being polished by ultraviolet laser, the surface roughness of the mechanical lapped sample dropped by 50 percent, which is 36 percent higher than that of green laser polished sample and 20 percent higher than that of sawed sample. Figure 9 shows the effects of initial sapphire

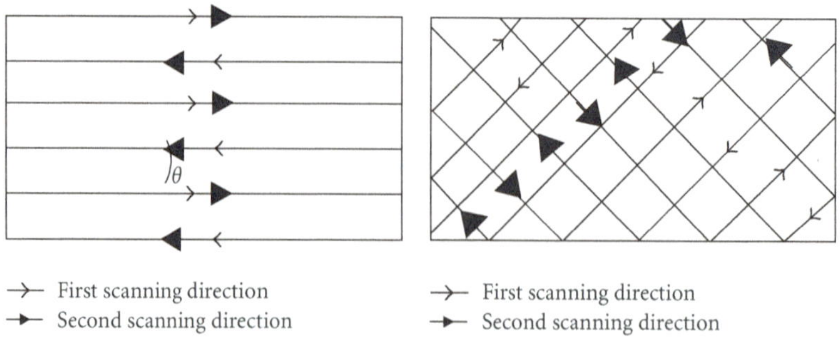

(a) Parallel scanning (b) Intersectional scanning

Figure 7: Laser beam scanning patterns.

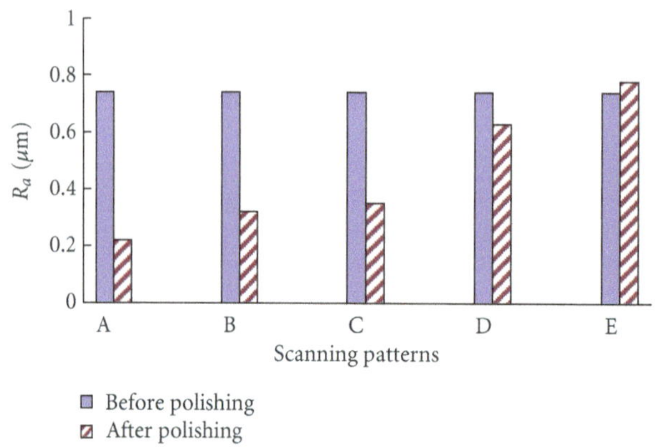

(a) Effects of scanning patterns on sapphire surface roughness

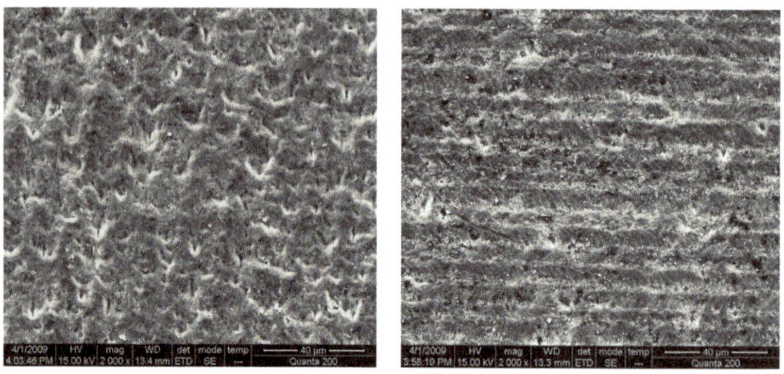

(b) Polished with Pattern C (c) polished with Pattern D

Figure 8: Effect of laser scanning patterns on sapphire polishing (laser energy of 0.056 mJ, scanning speed of 90 mm/s, pulse repetition rate of 30 kHz, incident angle of 50°, Sample A).

surface conditions on the surface roughness. The smaller roughness the initial surface of sample, the larger the reduced rate of surface roughness was obtained after ultraviolet laser polishing. It is indicated that ultraviolet laser polishing is more appropriate for precision polishing.

4. Conclusion

(1) With the increase of laser energy, the surface roughness R_a decreases when laser energy is lower than 0.056 mJ and increases again when the energy is higher than 0.056 mJ because of different laser actions.

(2) The faster the laser scanning speed, the shorter the interaction time is. The surface becomes rougher when scanning speed is as low as 10 and 30 mm/s.

(3) With the increase of the laser incident angle, the surface roughnes tends to decrease slightly.

(4) Single-scanning pattern is superior to multiple-scanning pattern and intersectional scanning pattern.

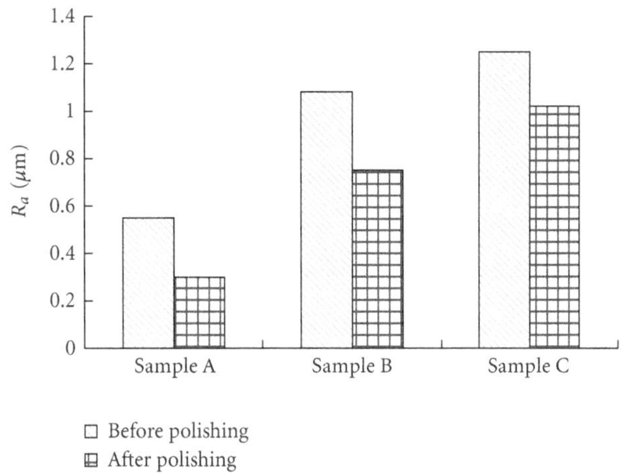

FIGURE 9: Effects of different original sapphire surface conditions on surface roughness.

The smaller intersection angle, the lowersurface roughness obtained after laser polishing.

(5) The smaller roughness of the initial surface of sample, the larger percentage of roughness decreases after ultraviolet laser polishing. It is indicated that ultraviolet laser polishing is more suitable for precision polishing.

(6) Sapphire surface roughness R_a can be decreased from 0.74 to 0.2 μm by ultraviolet laser polishing under the condition of pulse width of 20 ns, pulse energy of 0.056 mJ, pulse repetition rate of 30 kHz, scanning speed of 90 mm/s, incident angle of 50 degree and patterns A for mechanical lapped surface in this experiment.

Acknowledgments

The research was supported by Nature Science Foundation of China (nos. 50675038, 50805027), Science and Technology Plan Project of Guangdong Province (no. 2004B10201018), and Science and Technology Plan Project of Guangzhou (no. 2005J1-C0081).

References

[1] H. Park and H. M. Chan, "A novel process for the generation of pristine sapphire surfaces," *Thin Solid Films*, vol. 422, no. 1-2, pp. 135–140, 2002.

[2] X. Guo, X. Wei, and X. Xie, "Development of sapphire polishing technology," *Mechanical & Electrical Engineering Technology*, vol. 35, no. 9, pp. 76–78, 2006.

[3] Z. Li, P. Li, J. Fan, R. Fang, and D. Zhang, "Energy accumulation effect and parameter optimization for fabricating of high-uniform and large-area period surface structures induced by femtosecond pulsed laser," *Optics and Lasers in Engineering*, vol. 48, no. 1, pp. 64–68, 2010.

[4] I.-B. Sohn, Y.-C. Noh, S.-C. Choi, D.-K. Ko, J. Lee, and Y.-J. Choi, "Femtosecond laser ablation of polypropylene for breathable film," *Applied Surface Science*, vol. 254, no. 16, pp. 4919–4924, 2008.

[5] M. Folwaczny, A. Mehl, C. Haffner, and R. Hickel, "Polishing and coating of dental ceramic materials with 308 nm XeCL excimer laser radiation," *Dental Materials*, vol. 14, no. 3, pp. 186–193, 1998.

[6] M. Udrea, H. Orun, and A. Alacakir, "Laser polishing of optical fiber end surface," *Optical Engineering*, vol. 40, no. 9, pp. 2026–2030, 2001.

[7] T. M. Shao, M. Hua, H. Y. Tam, and E. H. M. Cheung, "An approach to modelling of laser polishing of metals," *Surface and Coatings Technology*, vol. 197, no. 1, pp. 77–84, 2005.

[8] X. Wei, X. Guo, and X. Xie, "Influence of the condition parameters on UV pulsed laser polishing of sapphire wafer," in *Proceedings of the 8th Pacific Rim Conference on Lasers and Electro-Optics (CLEO/Pacific Rim '09)*, September 2009.

14

Fresnel Lens with Embedded Vortices

Sunil Vyas,[1] Rakesh Kumar Singh,[1] Devinder Pal Ghai,[2] and P. Senthilkumaran[1]

[1] *Department of Physics, Indian Institute of Technology Delhi, New Delhi 110016, India*
[2] *Laser Science and Technology Centre, Metcalfe House, Delhi 110054, India*

Correspondence should be addressed to P. Senthilkumaran, psenthilk@yahoo.com

Academic Editor: Jan Masajada

Vortices of different charges are embedded in a wavefront that has quadratic phase variation, and the intensity distribution near the focal plane is studied. This method may be useful in realizing complicated beam profiles. We have experimentally demonstrated the generation of vortex arrays having integer as well as fractional topological charges that produce different intensity profiles at the focal plane. The phase variation realized on a spatial light modulator (SLM) acts as a Fresnel lens with embedded vortices.

1. Introduction

A modulo 2π operation carried out on a spherical wavefront results in concentric zones called Fresnel zones. A Fresnel lens is formed by segmenting the continuous surface of a conventional lens into a set of surfaces of the same curvature. This process allows fabrication of a large diameter lens with very short focal length thereby ensuring substantial increase in its numerical aperture. Further, there is a significant reduction in the thickness and weight of the lens. Thus, a plane wave incident on a Fresnel lens is split into multiple segments (zones), and the combination of multiple successive wavefront zones reconstructs the spherical wavefront.

An optical beam with phase singularity also called optical vortex or a screw dislocation is a helical phase ramp which focuses to a ring-like intensity pattern [1]. An optical vortex is characterized by azimuthal phase dependence of the type $\exp(im\theta)$ where m is the topological charge. The magnitude of topological charge represents the total phase accumulated by the helical wave in one complete revolution around the vortex point. The sign of topological charge represents left- or right-handed helicity. When the topological charge has a noninteger value, it results in the formation of a fractional vortex. Wavefront containing fractional charge vortex exhibits a half-cut (low-intensity radial opening in the doughnut structure), corresponding to a mixed edge-screw dislocation [2]. Such vortices, on propagation, transform into several opposite-sign screw dislocations which disappear on the beam periphery [3].

Optical vortices find applications in particle trapping [4], data encryption [5], phase contrast microscopy [6], image processing [7], collimation testing [8], vortex coronagraphy, [9] and in various other areas. Recently, optical vortex arrays which are regular nets of vortices have generated an enormous interest among many research groups [10, 11]. Optical vortex arrays, also called vortex lattices, find applications in multiple trap optical tweezers [12], spatial solitons [13], microfluidic sorting [14], metrology [15], micromechanics [16], and atom optics [17]. Vortex arrays are generated using interferometry [18–20], Talbot imaging [21], and spatial filtering [22]. In this paper, we present a simple and effective method of generating array of vortices by embedding vortex phase distribution in the Fresnel lens. Specially designed phase masks are produced which are encoded onto the SLM to achieve array of vortices near the focal plane. A vortex array of integer as well as fractional topological charge can be produced by this method. As this method uses a single lens function, there is no cross-talk and overlapping of different beams. The method offers flexibility in the selection of location, number density, and topological charge of vortices in the beam.

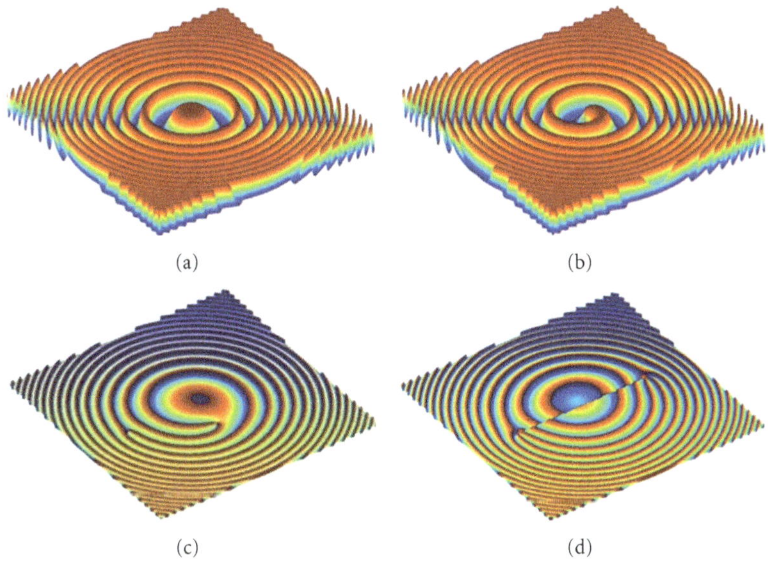

FIGURE 1: (a) Surface profile of a conventional Fresnel lens. (b) Surface profile of a Fresnel vortex lens. (c) Surface profile of a Fresnel lens with vortices embedded at off-axis locations. (d) Surface profile of a Fresnel lens with half-integer vortices embedded at off-axis locations.

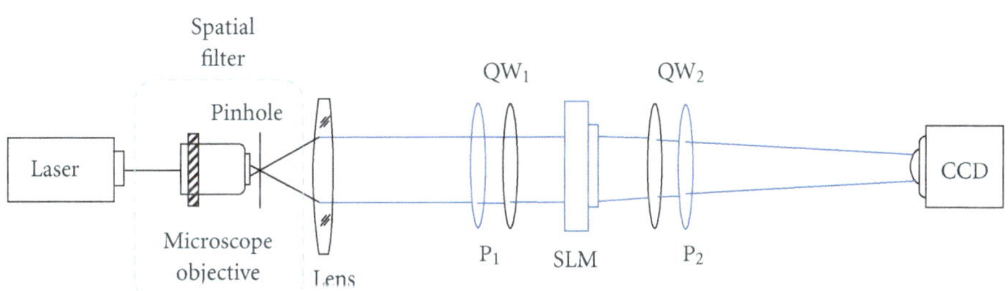

FIGURE 2: Experimental setup of Fresnel lens with embedded vortices; P_1 and P_2: Polarizers, QW_1 and QW_2: Quarter wave plates, SLM: Spatial light modulator, and CCD: Camera.

2. Theory

A convex lens brings a collimated incident beam to a focus at a distance f from the lens. The phase delay $\Delta\psi(\rho,\theta)$ suffered by the beam due to lens is

$$\Delta\psi(\rho,\theta) = \frac{-k\rho^2}{2f}, \tag{1}$$

where f is the focal length, $k = 2\pi/\lambda$, and λ is the wavelength of light used. The polar coordinates in the element plane ρ, θ are related to Cartesian coordinates by $\rho^2 = x^2 + y^2$, $\theta = \tan^{-1}(y/x)$. A thin lens can be considered equivalent to a transparency with a complex amplitude transmittance given by

$$t_1(x,y) = \exp(i\Delta\psi(\rho,\theta)). \tag{2}$$

Fresnel zone sectioning of this transmittance function leads to a Fresnel lens. The radius of the nth zone is $\rho_n \approx \sqrt{2fn\lambda}$. Figure 1(a) shows the profile of a conventional Fresnel lens.

A vortex or phase singularity in the wave field can be implanted by using a phase mask or by a spiral phase plate. The transmission function of such a phase mask is given by

$$t_2(\rho,\theta) = \begin{cases} \exp(im\theta), & \text{for } 0 \le \rho \le 1, \\ 0, & \text{otherwise}, \end{cases} \tag{3}$$

where m is the topological charge of vortex. A vortex lens [23] has the transmission function given by

$$t_3(\rho,\theta) = t_1(\rho,\theta)t_2(\rho,\theta) = \exp\left(i\left(m\theta - \frac{k}{2f}\rho^2\right)\right). \tag{4}$$

A phase mask having this transmittance function transforms incident plane wave into converging helicoidal wave $E(\rho,\theta,z)$. Presence of Fresnel lens factor in a wave field $e^{im\theta}$ gives desired curvature to the beam. When the topological charge m is a fraction, it is called fractional vortex lens [24]. This specific transmittance imparts an on-axis vortex on the incident optical field. Phase variation corresponding to

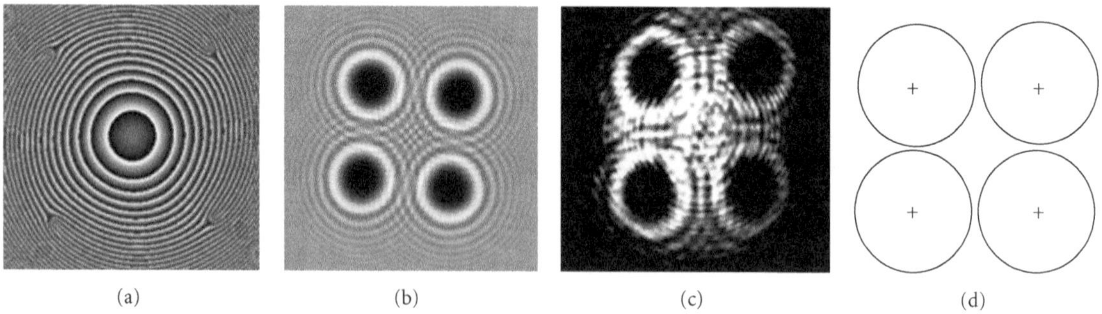

Figure 3: Fresnel lens with four vortices, each of topological charge 2 embedded at off-axis locations. (a) Phase distribution displayed on SLM. (b) Simulated intensity distribution near the focal plane. (c) Experimentally observed intensity distribution near the focal plane. (d) Distribution of topological charge in the intensity pattern.

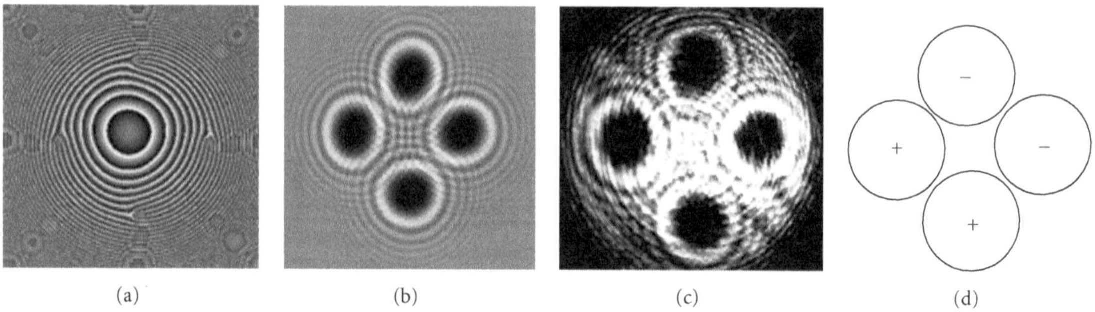

Figure 4: Fresnel lens with two positive and two negative charge vortices embedded at off-axis locations. (a) Phase distribution displayed on SLM. (b) Simulated intensity distribution near the focal plane. (c) Experimentally observed intensity distribution near the focal plane. (d) Distribution of topological charge in the intensity pattern.

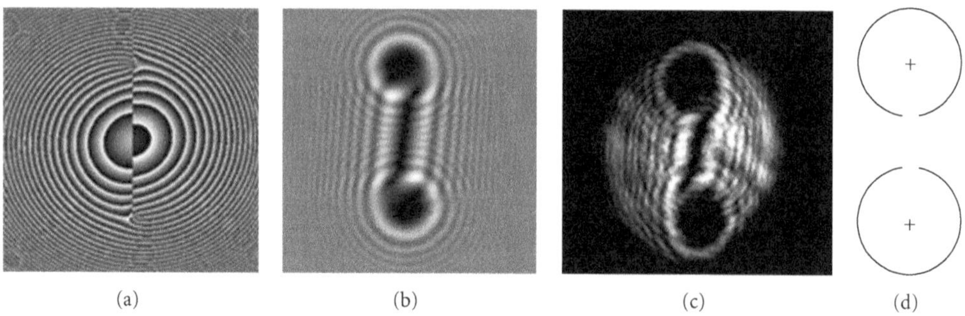

Figure 5: Fresnel lens with fractional vortices of the same charge (3.5) embedded at off-axis locations. (a) Phase distribution displayed on SLM. (b) Simulated intensity distribution near the focal plane. (c) Experimentally observed intensity distribution near the focal plane. (d) Distribution of topological charge in the intensity pattern.

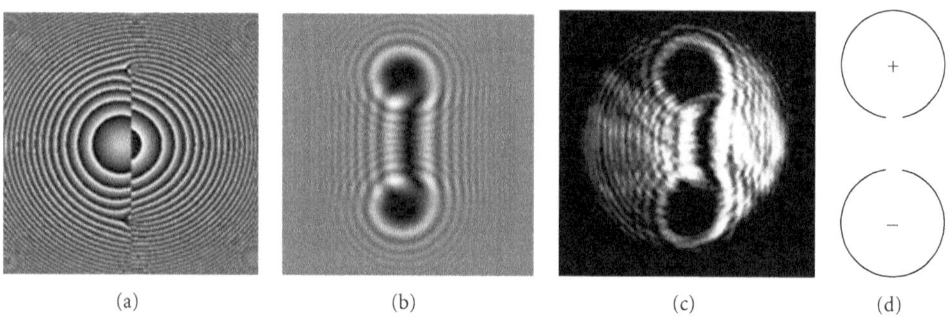

Figure 6: Fresnel lens with fractional vortices of opposite charge (3.5) embedded at off-axis locations. (a) Phase distribution displayed on SLM. (b) Simulated intensity distribution near the focal plane. (c) Experimentally observed intensity distribution near the focal plane. (d) Distribution of topological charge in the intensity pattern.

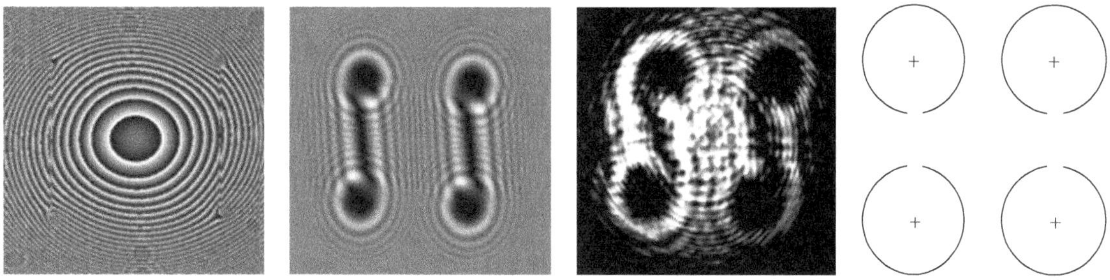

(a) (b) (c) (d)

Figure 7: Fresnel lens with 2×2 lattice of fractional vortices of the same charge $m = 3.5$ embedded at off-axis locations. (a) Phase distribution displayed on SLM. (b) Simulated intensity pattern near the focal plane. (c) Experimentally observed intensity distribution near the focal plane. (d) Distribution of topological charge in the intensity pattern.

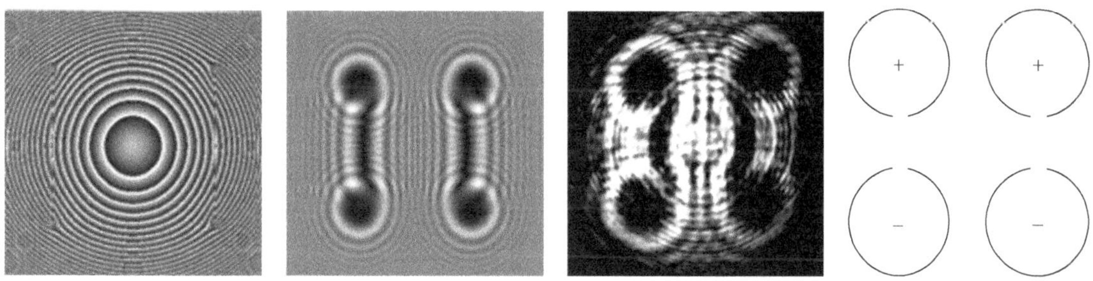

(a) (b) (c) (d)

Figure 8: Fresnel lens with 2 × 2 lattice of fractional vortices with two positive and two negative topological charge vortices, each of magnitude 3.5 embedded at off-axis locations. (a) Phase distribution displayed on SLM. (b) Simulated intensity pattern near the focal plane. (c) Experimentally observed intensity distribution near the focal plane. (d) Distribution of topological charge in the intensity pattern.

multiple vortices can be embedded into the wavefront by using the expression

$$t_4(\rho,\theta) = \exp\left[i\sum_{j=1}^{N} m_j\theta\left(\rho_j,\theta_j\right)\right], \tag{5}$$

where m_j is the topological charge of jth vortex, and θ is helical phase variation corresponding to the jth vortex. The position of the core of the jth vortex is (ρ_j, θ_j), where

$$\rho_j = \sqrt{x_j^2 + y_j^2}, \qquad \theta_j = \tan^{-1}\left(\frac{y-y_j}{x-x_j}\right), \tag{6}$$

where x_j, y_j are the Cartesian coordinates of the jth vortex core. Similar method of embedding vortices at desired locations is used in the iterative computational phase synthesis algorithms for diffractive elements, in wave optical engineering [25, 26]. This process is used to annihilate stagnation causing vortices, in the iterative algorithms.

When the extremum of a quadratic phase variation of a lens $\Delta\psi(\rho,\theta)$ coincides with the position of the core of the vortex, spiral zones are formed. Figure 1(b) shows the profile of a vortex Fresnel lens. When the position of the vortex (ρ_j,θ_j) does not coincide with the extremum of $\Delta\psi(\rho,\theta)$, branching of the zone boundary occurs as can be seen from Figure 1(c). The number of branches is equal to the magnitude of the charge of the vortex. The sign of the vortex decides the sense (clockwise or anticlockwise) in which the branching occurs. Figure 1(d) shows profile of a vortex Fresnel lens in which opposite vortices of unequal fractional charges are embedded at two different locations.

The complex amplitude at the focal plane can be evaluated by using the Fresnel-Kirchhoff diffraction integral which can be written [27] as

$$E(r,\phi,f) = A\int_0^1\int_0^{2\pi} E(\rho,\theta,z=0) \times \exp\left[-i\frac{2\pi}{\lambda f}r\rho\cos(\theta-\phi)\right]\rho d\rho d\theta, \tag{7}$$

where

$$A = \frac{1}{i\lambda f}e^{ikf}\exp\left(\frac{i\pi r^2}{\lambda f}\right). \tag{8}$$

(ρ,θ,z) and (r,ϕ,f) correspond to object and diffracted field planes expressed in cylindrical coordinates. λ is the wavelength of light used, and f is the focal length of the lens. The multiplicative factor A is ignored since the amplitude $1/\lambda f$ is only a scale factor, and the phase $(kf+(\pi r^2/\lambda f-\pi/2))$

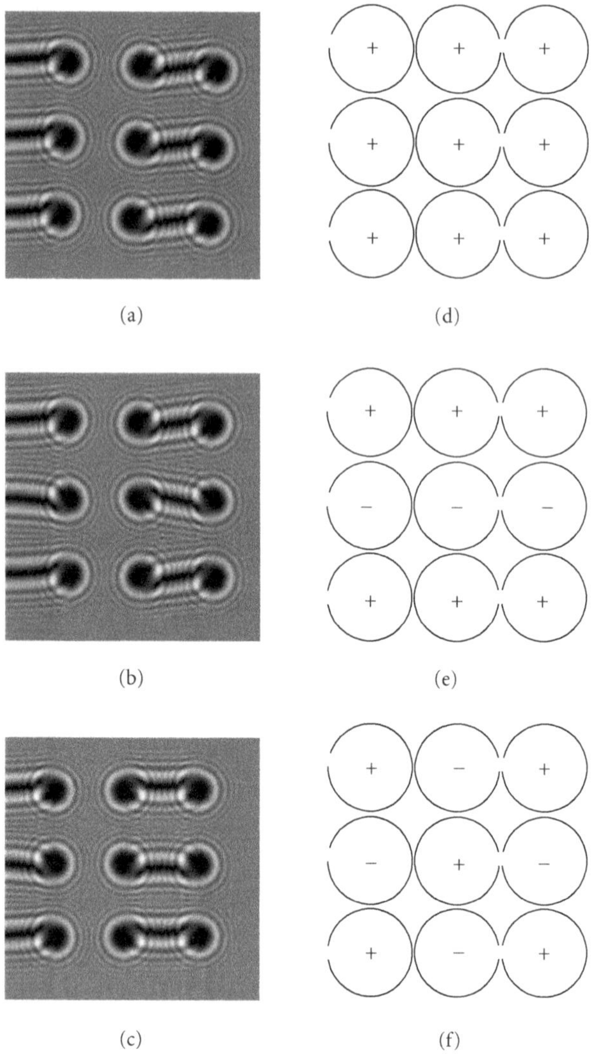

FIGURE 9: 3×3 lattices of fractional vortices each of charge ± 3.5 embedded in a Fresnel lens. Simulated intensity distributions (a–c) and the corresponding charge distributions (d–f) are shown.

has no effect on the intensity. The field at the focal plane is proportional to the Fourier transform of the transmittance function given in (7). Intensity distribution at the observation plane is then given by

$$I(r,\phi,f) = |E(r,\phi,f)|^2. \tag{9}$$

Here, the intensity distribution refers to the far-field image of the object. We have carried our simulation studies to find the distribution of intensity in the focal plane.

3. Experimental Setup

A phase-only SLM can impose an azimuthal phase term directly on the incident beam converting it into a helical beam. The method of generating optical vortices using phase mask is more versatile because it allows precise and dynamic control over vortex parameters and possibility of generating specific patterns of optical vortices. The experimental arrangement is shown in Figure 2. A linearly polarized light from a He-Ne laser ($\lambda = 632.8$ nm) is spatially filtered, expanded, and collimated to form a plane wave. The collimated beam of light is directed on the SLM (Holoeye LC-2002 with 832×624 pixels, pitch $= 32\,\mu$m). By using properly oriented linear polarizer or by generating appropriate elliptically polarized light, one can modulate the phase while keeping the amplitude constant. We have used elliptically polarized light to achieve phase-only modulation for the SLM using a combination of two polarizers and two quarter wave plates, as shown in Figure 2. This architecture consists of a combination of an input polarizer P and a quarter wave plate QW_1 in front of the SLM and an output quarter wave plates QW_2 and polarizer P_2 behind the SLM. We adjust the polarizers and wave plates such that the SLM modulates only the phase, leaving the amplitude constant. A gray-scale phase mask corresponding to the transmittance of a lens, embedded with required vortex phase distributions, is displayed on SLM, and the corresponding intensity distribution is recorded on a CCD.

4. Results and Discussion

The Airy pattern at the focal plane of a lens is replaced by a doughnut pattern in a vortex lens. In a Fresnel vortex lens, the vortex phase is embedded at the extremum of the lens phase distribution that can produce a ring focus. By embedding multiple vortices in different locations in a Fresnel lens, vortex lattices can be realized at the focal plane [28]. Figure 3(a) shows the phase distribution of the modified Fresnel lens that can generate a 2×2 vortex lattice of four vortices with positive topological charge. Figures 3(b) and 3(c) depict the simulation and experimental results of the intensity pattern in the focal plane of the Fresnel lens embedded with 2×2 vortex lattice, as described above. The position and the charge of each of the vortices can be modified. Figure 4 shows one such configuration where two positive charge and two negative charge vortices are formed. The phase distributions are similar to that of a conventional Fresnel lens with branching of zones occurring at off-axis locations [29]. For positive (negative) vortex, the branch or fork occurs in anticlockwise (clockwise) sense as we move along the zone boundary. For a diverging Fresnel lens, the sense of fork formation is opposite to that of a positive lens. Similarly, vortex phase of fractional charge can also be embedded in a Fresnel lens. Figure 5 shows two positively charge fractional vortices of magnitude 3.5, embedded at off-axis locations. In this case, apart from branching of zones at vortex points, there is an additional edge dislocation connecting the two vortices. In the intensity distribution near the focal plane, the two dark cores are connected by a curved dark line. The two dark cores and the connecting dark curved line are surrounded by bright intensity pattern. In Figure 6, oppositely charged fractional vortices are embedded in the phase distribution of Fresnel lens. The dark cores in this case are connected by the dark line that has different geometry. These intensity patterns are unique. Two fractional vortices are connected to each other, irrespective of the charge of the individual vortex. Figures 7 and 8 depict 2×2 arrays of vortices with fractional topological charges. The dark line that connects two vortices is decided by the manner, the edge dislocation connecting the two fractional vortices, appears in the phase distribution. If the edge dislocation line starting from one fractional vortex is not connected to another fractional vortex, it runs towards the boundary of the intensity pattern and terminates. Simulations are carried out to see the focal plane intensity distributions of such lenses. In all the Figures 3–8, it can be clearly seen that the experimental and simulation results match exactly. Figure 9 depicts vortex lattices of 3×3 arrays of the vortices with fractional topological charges. The charge distributions of the vortices in each of the lattices are different. Simulated intensity distributions at the focal plane corresponding to these lattices reveal that all the vortices are not connected by dark lines to form pairs. Only three pairs are connected, and in the remaining three fractional vortices, the dark lines from their cores are diverging outward.

5. Conclusion

Array of vortices of different charges are embedded in a wavefront with quadratic phase variation and the intensity distribution near the focal plane is studied. The intensity patterns, when fractional vortices are embedded in the Fresnel lens, possess unique interesting features. The intensity patterns produced by any two fractional charge vortices are connected by dark line, irrespective of the sign of the vortices. The geometry of this dark line is decided by the manner in which the edge dislocation line appears in the phase distribution of the Fresnel lens.

References

[1] J. F. Nye and M. V. Berry, "Dislocations in wave trains," *Proceedings of the Royal Society A*, vol. 336, pp. 165–190, 1974.

[2] W. M. Lee, X.-C. Yuan, and K. Dholakia, "Experimental observation of optical vortex evolution in a Gaussian beam with an embedded fractional phase step," *Optics Communications*, vol. 239, no. 1–3, pp. 129–135, 2004.

[3] M. V. Berry, "Optical vortices evolving from helicoidal integer and fractional phase steps," *Journal of Optics A*, vol. 6, no. 2, pp. 259–268, 2004.

[4] K. T. Gahagan and G. A. Swartzlander, "Optical vortex trapping of particles," *Optics Letters*, vol. 21, no. 11, pp. 827–829, 1996.

[5] A. Jesacher, S. Furhapter, S. Bernet, and M. Ritsch-Marte, "Spiral interferometery," *Optics Letters*, vol. 30, pp. 1953–1955, 2006.

[6] G. Gibson, J. Courtial, M. J. Padgett et al., "Free space information transfer using light beams carrying orbital angular momentum," *Optics Express*, vol. 12, no. 22, pp. 5448–5456, 2004.

[7] J. A. Davis, I. Moreno, and K. Crabtree, "Optical processing with vortex-producing lenses," *Applied Optics*, vol. 43, no. 6, pp. 1360–1367, 2004.

[8] P. Senthilkumaran, "Optical phase singularities in detection of laser beam collimation," *Applied Optics*, vol. 42, no. 31, pp. 6314–6320, 2003.

[9] G. A. Swartzlander Jr., "Peering into darkness with a vortex spatial filter," *Optics Letters*, vol. 26, no. 8, pp. 497–499, 2001.

[10] J. Leach, M. R. Dennis, J. Courtial, and M. J. Padgett, "Knotted threads of darkness," *Nature*, vol. 432, no. 7014, p. 165, 2004.

[11] D. P. Ghai, S. Vyas, P. Senthilkumaran, and R. S. Sirohi, "Vortex lattice generation using interferometric techniques based on lateral shearing," *Optics Communications*, vol. 282, no. 14, pp. 2692–2698, 2009.

[12] M. M. Burns, J. M. Fournier, and J. A. Golovchenko, "Optical matter: crystallization and binding in intense optical fields," *Science*, vol. 249, no. 4970, pp. 749–754, 1990.

[13] G. A. Swartzlander Jr. and C. T. Law, "Optical vortex solitons observed in Kerr nonlinear media," *Physical Review Letters*, vol. 69, no. 17, pp. 2503–2506, 1992.

[14] M. P. MacDonald, G. C. Spalding, and K. Dholakia, "Microfluidic sorting in an optical lattice," *Nature*, vol. 426, no. 6965, pp. 421–424, 2003.

[15] J. Masajada, "Small-angle rotations measurement using optical vortex interferometer," *Optics Communications*, vol. 239, no. 4–6, pp. 373–381, 2004.

[16] K. Ladavac and D. G. Grier, "Micro-opto-mechanical pumps assembled and driven by holographic optical vortex arrays," *Optics Express*, vol. 12, no. 6, pp. 1144–1149, 2004.

[17] N. Chattrapiban, E. A. Rogers, I. V. Arakelyan, R. Roy, and W. T. Hill, "Laser beams with embedded vortices: tools for atom optics," *Journal of the Optical Society of America B*, vol. 23, no. 1, pp. 94–103, 2006.

[18] J. Masajada and B. Dubik, "Optical vortex generation by three plane wave interference," *Optics Communications*, vol. 198, no. 1–3, pp. 21–27, 2001.

[19] S. Vyas and P. Senthilkumaran, "Interferometric optical vortex array generator," *Applied Optics*, vol. 46, no. 15, pp. 2893–2898, 2007.

[20] J. Masajada, A. Popiołek-Masajada, and M. Leniec, "Creation of vortex lattices by a wavefront division," *Optics Express*, vol. 15, no. 8, pp. 5196–5207, 2007.

[21] G. X. Wei, L. L. Lu, and C. S. Guo, "Generation of optical vortex array using Talbot array illuminators," *Optics Communications*, vol. 282, pp. 2665–2669, 2009.

[22] C. S. Guo, Y. Zhang, Y. J. Han, J. P. Ding, and H. T. Wang, "Generation of optical vortices with arbitrary shape and array via helical phase spatial filtering," *Optics Communications*, vol. 259, no. 2, pp. 449–454, 2006.

[23] G. A. Swartzlander Jr., "Achromatic optical vortex lens," *Optics Letters*, vol. 31, no. 13, pp. 2042–2044, 2006.

[24] S. Vyas, R. Kumar Singh, and P. Senthilkumaran, "Fractional vortex lens," *Optics and Laser Technology*, vol. 42, no. 6, pp. 1150–1159, 2010.

[25] P. Senthilkumaran and F. Wyrowski, "Phase synthesis in wave-optical engineering: mapping and diffuser type approaches," *Journal of Modern Optics*, vol. 49, no. 11, pp. 1831–1850, 2002.

[26] P. Senthilkumaran, F. Wyrowski, and H. Schimmel, "Vortex Stagnation problem in iterative Fourier transform algorithms," *Optics and Lasers in Engineering*, vol. 43, no. 1, pp. 43–56, 2005.

[27] J. W. Goodman, *Introduction to Fourier Optics*, McGraw-Hill Companies, New York, NY, USA, 2nd edition, 1996.

[28] X. Gan, J. Zhao, S. Liu, and L. Fang, "Generation and motion control of optical multi-vortex," *Chinese Optics Letters*, vol. 7, no. 12, pp. 1142–1145, 2009.

[29] F. S. Roux, "Diffractive lens with a null in the center of its focal point," *Applied Optics*, vol. 32, no. 22, pp. 4191–4192, 1993.

Generation of Bessel Surface Plasmon Polaritons in a Finite-Thickness Metal Film

S. N. Kurilkina, V. N. Belyi, and N. S. Kazak

B.I. Stepanov Institute of Physics of National Academy of Sciences of Belarus, 68 Nezavisimosti Avenue, 220072 Minsk, Belarus

Correspondence should be addressed to S. N. Kurilkina; s.kurilkina@ifanbel.bas-net.by

Academic Editor: Yu S. Kivshar

A theory of generation of low- and high-index Bessel surface plasmon polaritons and their superposition in a metal film of a finite thickness is developed. Correct analytical expressions are obtained for the field of two families of Bessel surface plasmon polariton modes formed inside and outside the metal layer. The intensity distribution near the boundary of the layer has been calculated and analyzed. A scheme for the experimental realization of a superposition of Bessel surface plasmon polaritons is suggested. Our study demonstrates that it is feasible to use the superposition of Bessel surface plasmon polaritons as a virtual tip for near-field optical microscopy with a nanoscale resolution.

1. Introduction

Surface plasmon polaritons (SPPs) are surface electromagnetic waves related to collective electron oscillations near the metal surface excited by light [1–3]. These fields arise under the resonance condition, and due to such nature of excitation, they are attractive in the context of enhancing the resolution of imaging systems substantially by increasing the amplitude of evanescent waves [4–6].

In 1987, Durnin et al. suggested a new type of waves, namely, Bessel light beams (BLBs), that kept the transverse spot size unchanged much longer than the Rayleigh range [7, 8]. Such a localized radiation mode was called the nondiffracting beam (or the diffraction-free beam). The transverse profile of the amplitude of this beam is described by a Bessel function of the first kind. In the domain of spatial frequencies, BLB is represented as a superposition of plane waves which are wrapped around a conical surface. Within the last decades, an intensive study of the scalar and vector BLBs has been made theoretically and experimentally (see, e.g., [9–18]). Bessel light beams are used in numerous applications, such as the optical manipulation of microsized particles [19], the fabrication of long polymer fiber induced by the photopolymerization [20] and microchanneling by structural modification in glass materials [21] the enhancement of energy gain in inverse-free electron lasers and inverse Cherenkov accelerators [22].

The authors of [23] pioneered in obtaining solutions of Maxwell's equations which correspond to evanescent BLBs formed in the condition of the internal total reflection in an optically less dense dielectric medium. A more detailed theory of evanescent Bessel beams is presented in [24–29]. These beams exponentially decay while moving off the surface but retain their original transversal shape. In those investigations, of particular interest was the structure of the central lobe of evanescent BLBs. It is established that its diameter can be reduced to a nanosize value. This makes it possible to use evanescent Bessel beams in optical microscopy [30].

But the evanescent BLBs investigated before [24–29] possess an essential disadvantage, namely, they are weak, which causes the necessity of application of strong laser fields for their generation. One of the ways of taking Bessel light field advantages for microscopy is the formation of Bessel surface plasmon polaritons (BSPPs). Kano et al. [31] reported the first experimental result concerning an efficient excitation of surface plasmon polaritons by using the zeroth-order BLBs. More recently, radially polarized BLBs have been demonstrated to provide the TM polarization required for an effective coupling to the SPPs, which can be used as a virtual probe for the two-photon fluorescence microscopy [32]. In paper [33], the authors have demonstrated the existence of SPP in a multilayered metal-dielectric structure that evolves without distortion within an energy attenuation length and whose in-plane profile traces a Bessel function along the transverse direction relative to the beam axis.

However, despite of intensive theoretical and experimental investigations up to now there is no detailed analysis of the problem of generation of different types of Bessel surface plasmon polaritons in the structure including a metallic absorbing layer separating a substrate and targeted (dielectric) medium. Also of great interest is the study of the intensity distribution in the structure containing a thin metal film. An exact solution of these problems is presented in this paper.

The study of the superposition of two vortex Bessel surface plasmon polaritons with opposite topological charges (J_m and J_{-m}) is undoubtedly beneficial in practical terms. Such superposed beams are promising for creating near the interface an array of diffraction-free nanoscale light needles. At the same time, we attach importance to the development of a scheme for experimental realization of a superposition of Bessel surface plasmon polaritons. And the theory developed ought to be best suited to experiment. This practical motivation necessitates a more detailed description of the field which is a superposition of Bessel surface plasmon polaritons.

The paper is structured as follows. In Section 2, the description of features of generation of different types of Bessel surface plasmon polaritons in layered metal-dielectric structure is presented. The intensity distribution in the structure is analyzed in details analytically and numerically in Section 3. The possibility of the use of a superposition of Bessel surface plasmon polaritons is considered in Section 4. A conclusion is given in Section 5.

2. Description of Bessel Surface Plasmon Polaritons Generation in a Metal-Dielectric Layered Structure

2.1. Formation of Bessel Surface Plasmon Polaritons in a Finite-Thickness Metal Film. Let us consider a vector TM-polarized high-order ($m \neq 0$) Bessel light beam J_m (BLB) incident onto the boundary of two semi-infinite media. We use the cylindrical coordinate system with the unit vectors $\vec{e}_\rho$, $\vec{e}_\varphi$, and $\vec{e}_z$ ($\vec{e}_z$ is collinear to the propagation direction). It is known (see, e.g., [10, 17, 18]) that the electric $\vec{E}(R)$ and magnetic $\vec{H}(R)$ vectors of BLB, propagating normally (in z-direction) in a medium with the refractive index n_i, can be represented as

$$\begin{aligned}\vec{E}(R) &= A\vec{E}^{\text{TM}}(\rho)\exp i(k_{zi}z + m\varphi),\\ \vec{H}(R) &= A\vec{H}^{\text{TM}}(\rho)\exp i(k_{zi}z + m\varphi),\end{aligned} \tag{1}$$

where $R = (\rho, \varphi, z)$ are the cylindrical coordinates, m is an integer, k_{zi} is the z-component of the wave vector, A is a complex constant, the common phase multiplier $\exp[i(qx - \omega t)]$ is omitted, and $q = \sqrt{k_0^2 n_i^2 - k_{zi}^2}$ is the transversal component of the wave vector.

Using the solution of Maxwell's equations for the components of the vector amplitudes $\vec{E}^{\text{TM}}(\rho)$, $\vec{H}^{\text{TM}}(\rho)$, the problem of transmission and reflection of vector Bessel beams at the boundary of two dielectrics has been solved (see [17, 18]). Following [17, 18], in the case of TM-polarized BLB with an amplitude A_{inc} incident on the interface between two media characterized by refractive indices n_i and n_j, the electric $\vec{E}_t(R)$ and magnetic $\vec{H}_t(R)$ vectors of the refracted Bessel beam (symbol "t" denotes refracted BLB) can be represented in the form

$$\begin{aligned}\vec{E}_t(R) &= \vec{E}_t^{\text{tr}}(R) + \vec{E}_t^{l}(R),\\ \vec{E}_t^{l}(R) &= A_{\text{inc}}t_{ij}^{\text{TM}}\exp i\left[m\varphi + k_{zj}z\right]\sin\gamma_j J_m(q\rho)\,\vec{e}_z,\\ \vec{E}_t^{\text{tr}}(R) &= \frac{iA_{\text{inc}}}{\sqrt{2}}\exp i\left[(m-1)\varphi + k_{zj}z\right]t_{ij}^{\text{TM}}\\ &\quad\times\cos\gamma_j\left[J_{m-1}(q\rho)\,\vec{e}_+ - J_{m+1}(q\rho)\exp(2i\varphi)\,\vec{e}_-\right],\end{aligned} \tag{2}$$

$$\begin{aligned}\vec{H}_t(R) &= \frac{n_j A_{\text{inc}}}{\sqrt{2}}\exp i\left[(m-1)\varphi + k_{zj}z\right]t_{ij}^{\text{TM}}\\ &\quad\times\left[J_{m-1}(q\rho)\,\vec{e}_+ + J_{m+1}(q\rho)\exp(2i\varphi)\,\vec{e}_-\right].\end{aligned} \tag{3}$$

Here $\cos\gamma_j = k_{zj}/k_0 n_j$, $k_0 = \omega/c$, ω is the frequency of the Bessel beam, c is the light velocity in vacuum, $J_m(q\rho)$ is the m-order Bessel function; symbols "tr," "l" denote the transversal and longitudinal component of the electric (magnetic) vector, respectively; $t_{ij}^{\text{TM}} = A_t^{\text{TM}}/A_{\text{inc}}$ is the amplitude transmission coefficient; A_t^{TM} is the amplitude factor for TM BLB; $\vec{e}_\pm = (\vec{e}_1 \pm i\vec{e}_2)/\sqrt{2}$ are the unit circular vectors which are orthogonal to the $\vec{e}_z$ vector.

In a similar manner, for the reflected field (it is marked with the symbol "r"), we obtain

$$\begin{aligned}\vec{E}_r(R) &= \vec{E}_r^{\text{tr}}(R) + \vec{E}_r^{l}(R),\\ \vec{E}_r^{l}(R) &= A_{\text{inc}}r_{ij}^{\text{TM}}\exp i\left[m\varphi - k_{zi}z\right]\sin\gamma_i J_m(q\rho)\,\vec{e}_z,\\ \vec{E}_r^{\text{tr}}(R) &= -\frac{iA_{\text{inc}}}{\sqrt{2}}\exp i\left[(m-1)\varphi - k_{zi}z\right]r_{ij}^{\text{TM}}\\ &\quad\times\cos\gamma_i\left[J_{m-1}(q\rho)\,\vec{e}_+ - J_{m+1}(q\rho)\exp(2i\varphi)\,\vec{e}_-\right],\end{aligned} \tag{4}$$

$$\begin{aligned}\vec{H}_r(R) &= \frac{n_i A_{\text{inc}}}{\sqrt{2}}\exp i\left[(m-1)\varphi - k_{zi}z\right]r_{ij}^{\text{TM}}\\ &\quad\times\left[J_{m-1}(q\rho)\,\vec{e}_+ + J_{m+1}(q\rho)\exp(2i\varphi)\,\vec{e}_-\right].\end{aligned} \tag{5}$$

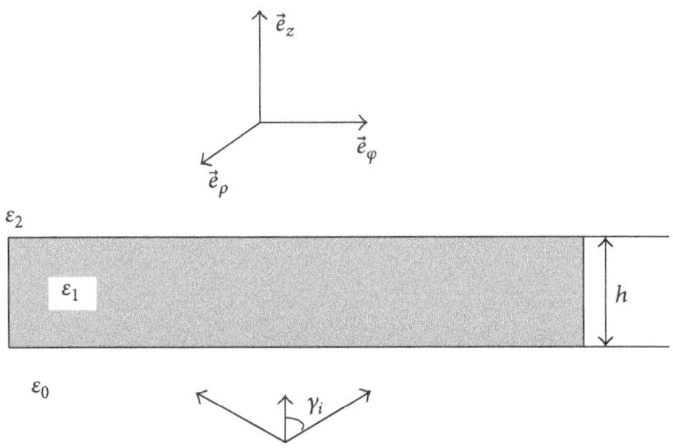

FIGURE 1: Schematic representation of the structure including a metallic layer (ε_1), a substrate (ε_0), and a targeted (ε_2) medium.

Here $r_{ij}^{TM} = A_r^{TM}/A_{inc}$ is the amplitude reflection coefficient; A_r^{TM} is the amplitude factor for TM BLB. The amplitude refraction and reflection coefficients are

$$t_{ij}^{TM} = \frac{2n_i \cos\gamma_i}{n_i\cos\gamma_j + n_j\cos\gamma_i},$$
$$r_{ij}^{TM} = \frac{n_j\cos\gamma_i - n_i\cos\gamma_j}{n_j\cos\gamma_i + n_i\cos\gamma_j}. \quad (6)$$

Equations (6) can be used for the description of the transformation of BLB at the boundary between dielectric with the real refractive index n_i and metal having the complex refractive index $n_j = n_j' + in_j''$, where $n_j' = \mathrm{Re}(n_j)$, $n_j'' = \mathrm{Im}(n_j)$.

Using (2), (3), and (6), let us consider now the formation of Bessel surface plasmon polaritons in a layered structure containing a metal film with the dielectric constant $\varepsilon_1 = n_1^2$ and thickness h between two semi-infinite dielectric media with dielectric constants $\varepsilon_0 = n_0^2$, $\varepsilon_2 = n_2^2$ (Figure 1). The origin of the coordinated system $z = 0$ is assumed to be chosen at the interface between the dielectric substrate and the layer ε_1 (Figure 1).

In view of (2), (4), the electric vector of the field inside the metal layer and outside it is expressed as

$$\vec{E}_0(R) = \vec{E}_0^{tr}(R) + \vec{E}_0^{l}(R),$$
$$\vec{E}_0^{l}(R) = A_{inc}\frac{q}{k_0 n_0}\exp i\left[m\varphi - k_{z0}z\right] J_m(q\rho)\,\vec{e}_z,$$
$$\vec{E}_0^{tr}(R) = -\frac{iA_{inc}}{\sqrt{2}}\exp i\left[(m-1)\varphi - k_{z0}z\right] \times \sqrt{1-\left(\frac{q}{k_0 n_0}\right)^2}\left[J_{m-1}(q\rho)\,\vec{e}_+ - J_{m+1}(q\rho)\exp(2i\varphi)\,\vec{e}_-\right],$$
$$\vec{E}_2(R) = \vec{E}_2^{tr}(R) + \vec{E}_2^{l}(R),$$
$$\vec{E}_2^{l}(R) = A_{inc}\frac{q}{k_0 n_2} t\exp i\left[m\varphi + k_{z2}(z-h)\right] J_m(q\rho)\,\vec{e}_z,$$
$$\vec{E}_2^{tr}(R) = \frac{iA_{inc}}{\sqrt{2}}\exp i\left[(m-1)\varphi + k_{z2}(z-h)\right] t \times \sqrt{1-\left(\frac{q}{k_0 n_2}\right)^2}\left[J_{m-1}(q\rho)\,\vec{e}_+ - J_{m+1}(q\rho)\exp(2i\varphi)\,\vec{e}_-\right], \quad (7a)$$

$$\vec{E}_1(R) = \vec{E}_1^{tr}(R) + \vec{E}_1^{l}(R),$$
$$\vec{E}_1^{l}(R) = \left(\vec{E}_1^{l}\right)^f + \left(\vec{E}_1^{l}\right)^b,$$
$$\vec{E}_1^{tr}(R) = \left(\vec{E}_1^{tr}\right)^f + \left(\vec{E}_1^{tr}\right)^b,$$
$$\left(\vec{E}_1^{l}\right)^{f,b} = A_{inc}\frac{q}{k_0 n_1} s^{f,b}\exp i\left[m\varphi \pm k_{z1}z\right] J_m(q\rho)\,\vec{e}_z,$$
$$\left(\vec{E}_1^{tr}(R)\right)^{f,b} = \pm\frac{iA_{inc}}{\sqrt{2}} s^{f,b}\exp i\left[(m-1)\varphi \pm k_{z1}z\right] \times \sqrt{1-\left(\frac{q}{k_0 n_1}\right)^2}\left[J_{m-1}(q\rho)\,\vec{e}_+ - J_{m+1}(q\rho)\exp(2i\varphi)\,\vec{e}_-\right]. \quad (7b)$$

Here, symbols "0", "1", and "2" denote the substrate, metal layer, and targeted medium, respectively. Similarly, for the magnetic vector we have

$$\vec{H}_0(R) = \frac{n_0 A_{inc}}{\sqrt{2}}\exp i\left[(m-1)\varphi - k_{z0}z\right] \times \left[J_{m-1}(q\rho)\,\vec{e}_+ + J_{m+1}(q\rho)\exp(2i\varphi)\,\vec{e}_-\right],$$

$$\vec{H}_2(R) = \frac{n_2 A_{\text{inc}}}{\sqrt{2}} \exp i\left[(m-1)\varphi + k_{z2}(z-h)\right] \times t\left[J_{m-1}(q\rho)\vec{e}_+ + J_{m+1}(q\rho)\exp(2i\varphi)\vec{e}_-\right], \tag{8a}$$

$$\vec{H}_1(R) = \left(\vec{H}_1(R)\right)^f + \left(\vec{H}_1(R)\right)^b,$$
$$\left(\vec{H}_1(R)\right)^{f,b} = \frac{n_1 A_{\text{inc}}}{\sqrt{2}} s^{f,b} \exp i\left[(m-1)\varphi \pm k_{z1}z\right] \times \left[J_{m-1}(q\rho)\vec{e}_+ + J_{m+1}(q\rho)\exp(2i\varphi)\vec{e}_-\right]. \tag{8b}$$

Here, $t = A_2/A_{\text{inc}}$, $s^f = A_f/A_{\text{inc}}$, $s^b = A_b/A_{\text{inc}}$ are the amplitude coefficients for the forward and backward Bessel fields inside the metal film, respectively; $A_{f,b}$ are the amplitude factors inside the metal layer.

Using the boundary conditions of continuity of the tangential components of the electric and magnetic fields, which have to be satisfied at the planes $z = 0$ and $z = h$, as well as (7a), (7b), (8a), (8b), and (6), we obtain the system of equations for the coefficients t, $s^{f,b}$. As a result of its solution, one can find the dispersion equation determining the existence of Bessel surface plasmon polaritons in the structure shown in Figure 1:

$$1 + r_{01}^{\text{TM}} r_{12}^{\text{TM}} \exp 2i(k_{z1}h) = 0. \tag{9}$$

As known (see [34]), the amplitude reflection coefficient of the layer is

$$r = \frac{r_{01}^{\text{TM}} + r_{12}^{\text{TM}} \exp 2i(k_{z1}h)}{1 + r_{01}^{\text{TM}} r_{12}^{\text{TM}} \exp 2i(k_{z1}h)}. \tag{10}$$

Then, (9) determines the poles of the reflection coefficient r of the metal layer. Taking into account (6), this expression can be transformed to the form

$$F = -itg(k_{z1}h)\left(1 + \frac{k_{z1}^2 \varepsilon_0 \varepsilon_2}{k_{z0} k_{z2} \varepsilon_1^2}\right) + \frac{k_{z1}\varepsilon_0}{k_{z0}\varepsilon_1} + \frac{k_{z1}\varepsilon_2}{k_{z2}\varepsilon_1} = 0. \tag{11}$$

Here, k_{zi} is the z-component of the wave vector, and the index i refers to the medium number. As seen from (11), unlike plane waves, the dispersion equation relates the frequency with the transversal component $q = \sqrt{k_0^2 n_j{}^2 - k_{zj}^2}$ of a Bessel surface plasmon polariton. Due to complexity of the metal refractive index, q is a complex quantity. Here, Re q determines the radius of rings of the excited Bessel surface plasmon polariton.

The boundary conditions allow one to determine the quantities t, s^f, and s^b entering into (7a), (7b), (8a), and (8b):

$$t = \frac{t_{01}^{\text{TM}} t_{12}^{\text{TM}} \exp(ik_{z1}h)}{1 + r_{01}^{\text{TM}} r_{12}^{\text{TM}} \exp(2ik_{z1}h)},$$
$$s^f = \frac{t_{01}^{\text{TM}}}{1 + r_{01}^{\text{TM}} r_{12}^{\text{TM}} \exp(2ik_{z1}h)}, \tag{12}$$
$$s^b = \frac{t_{01}^{\text{TM}} r_{12}^{\text{TM}} \exp(2ik_{z1}h)}{1 + r_{01}^{\text{TM}} r_{12}^{\text{TM}} \exp(2ik_{z1}h)}.$$

Here, the coefficients t_{ij}^{TM}, r_{ij}^{TM} are determined by (6).

Then, (7a), (7b), (8a), (8b), and (12) at q, which meets condition (11), describe Bessel surface plasmon polaritons formed in the structure "dielectric substrate metal layer dielectric." The field of a superposition of Bessel surface plasmon polaritons can be defined as a sum of the beams with the electric (magnetic) vector in the form (7a), (7b) ((8a), and (8b)).

It would be interesting to give a physical interpretation of the Bessel SPPs being localized and to compare them with traditional surface plasmon polaritons. It is well-known that the traditional SPPs are associated with the propagating wave process on metal/dielectric surface. In contrast to this, according to (7a), (7b), (8a), and (8b), the Bessel SPP is excited by an incident BLB, which can be considered as a superposition of p-polarized plane monochromatic waves having frequency ω and wave vectors lying on the surface of a cone.

The SPP propagating in the metal film is excited by every p-polarized plane-wave component of the incident BLB in conditions of the plasmon resonance. Here, the phase of every SPP is determined by the phase of the plane-wave component of the incident BLB. Thus, in the metal film, an array of propagating SPPs arises, wave vectors of which are being oriented in the direction of the center determining the intersection of the incident BLB axis with the surface of the metal film. As a result, there occurs the generation of pairs of two counter-propagating SPP waves with the wave vectors $\pm\vec{q}$. The generated SPPs will propagate in all the possible radial directions to form a localized SPP field. This results in a complex high-symmetric interference light structure emerging in sections parallel to the metal-dielectric surface.

For the case where the longitudinal components of the electric field $\vec{E}_1^l$ for every radially propagating SPP has the same phase, in the center of such a standing light structure a maximum appears, that is, the shape of the $\vec{E}_1^l$ component of localized Bessel SPP is described by zero-order Bessel function $J_0(q\rho)$. In a special case, each pair of counter propagating SPPs is in counter-phase and the local plasmonic field appears with a minimum in the center, that is, the so-called vortex localized Bessel SPP is formed. In this case, the electric field $\vec{E}_1^l$ of vortex Bessel SPP is described by $J_m(q\rho)$ Bessel function (see (7b)).

The problem should be studied now of attenuation of the Bessel SPP inside the metal layer. With this aim, the limited narrow Bessel light beam of radius r_0 in the

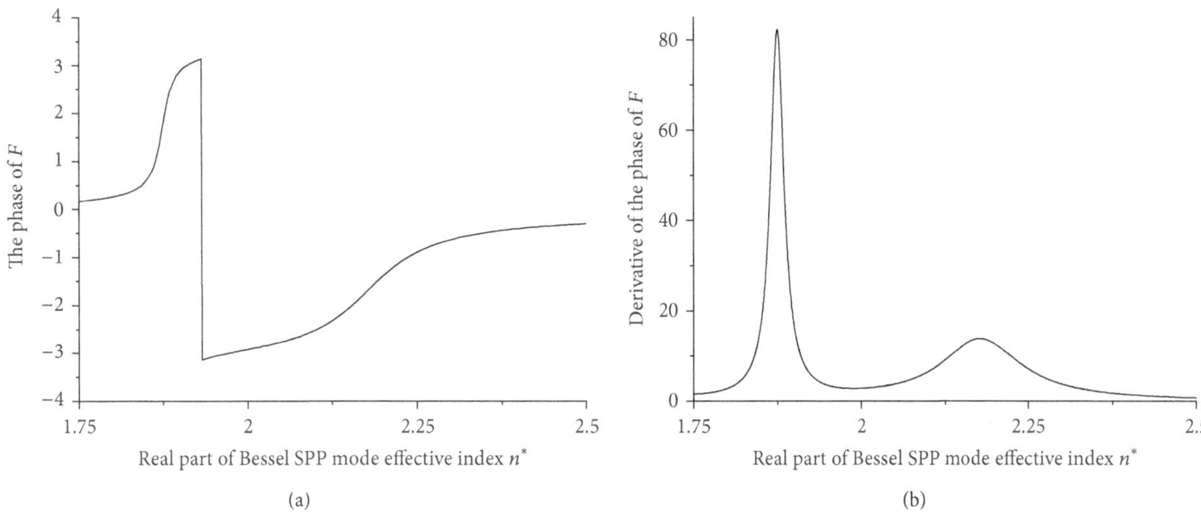

FIGURE 2: The phase of F function (a) and the derivative of the phase of F function (b) versus the real part of the Bessel surface plasmon polariton mode effective index.

transversal section should be considered. Such an apertured beam excites Bessel SSP in a cylindrical region limited by r_0. As follows from (7a), (7b), the transversal distribution of the longitudinal component of the electric field of Bessel SPP is determined by $J_m(q\rho)$ Bessel function which can be represented as a sum of the cylindrical Hankel functions of the first $H_m^{(1)}(q\rho)$ (outgoing) and second $H_m^{(2)}(q\rho)$ (incoming) kinds [35]:

$$J_m(q\rho) = \frac{\left(H_m^{(1)}(q\rho) + H_m^{(2)}(q\rho)\right)}{2}. \tag{13}$$

According to (13), in the region $\rho < r_0$, the Bessel SPPs are formed by converging and diverging conical beams described by appropriate Hankel functions. Outside the area $\rho > r_0$ there exists only a diverging conical beam, and that is why in the decomposition (13) it is supposed that $H_m^{(2)} = 0$. Using the asymptotic approximation of Hankel function [36], we obtain that the intensity of the longitudinal component of Bessel SPP electric vector $|\vec{E}^l|^2 \sim |J_m^2(q\rho)|^2$ is determined by the expression

$$\left|\vec{E}^l\right|^2 \sim \frac{1}{|q|^2 \rho^2} \exp\left[-2(\operatorname{Im} q)\rho\right]. \tag{14}$$

Thus, from (14), it follows that beyond the boundary of the exciting source the Bessel SPP decays exponentially in the radial direction. The $1/e$ energy-attenuation radius R_{BSSP} is determined by the imaginary part of the SPP wave number and given by the expression

$$R_{\text{BSSP}} = \frac{1}{2 \operatorname{Im} q}. \tag{15}$$

Thus, the energy-attenuation radius of Bessel SPP, generated by limited narrow Bessel beam, coincides with the decay length L_{SP} of propagating surface plasmon. From the physical point of view, it is explained by the fact that the Bessel SPP (outside the area of excitation) can be represented by the superposition of radially propagating plane wave SPPs.

2.2. Formation of Bessel Surface Plasmon Polaritons in a Symmetric Metal-Dielectric Layered Structure. Finding the roots of the complex equation (11) is numerically challenging. The reflection pole method (RPM) provides the phase of F as a function of the real part of the Bessel surface plasmon polariton mode effective index $n^* = q/k_0$. The Bessel surface plasmon polaritons correspond to the rapid changes of the phase of F function. According to the Bode plot theory [37], a peak of the derivative of the phase of F function corresponds to the real part of the Bessel surface plasmon polariton mode effective index, and the full width at the half maximum of the phase derivative curve is the imaginary part of the Bessel surface plasmon polariton mode effective index.

Using the reflection pole method, we searched the Bessel surface plasmon polaritons for a gold (Au) metal film ($\varepsilon_1 = -11.679 + 1.181i$ at the 0.633 micron wavelength [38]) of 50 nm thickness embedded in SF10 glass ($\varepsilon_0 = \varepsilon_2 = (1.723)^2$). Figures 2(a) and 2(b) are the plots of the phase of F function and its derivative versus the real part of the Bessel surface plasmon polariton mode effective index. From Figures 2(a) and 2(b), we see two rapid changes of the phase of F which correspond to two Bessel surface plasmon polariton modes. The complex Bessel surface plasmon polariton mode effective indices we found are $n_1^* = 1.874 + 0.012i$ and $n_2^* = 2.179 + 0.073i$.

The first (so-called low-index Bessel surface plasmon polariton (LIBSPP)) and the second (high-index Bessel surface plasmon polariton (HIBSPP)) BSPPs are characterized by the first ring radius $R_1 \approx 2.4/(k_0 \operatorname{Re} n^*)$ equal to 129 nm and 110 nm, respectively.

It is interesting to consider the symmetry of the longitudinal ($\vec{E}_1^l = E_1^l \vec{e}_z$) and transversal, for example, radial (E_1^ρ)

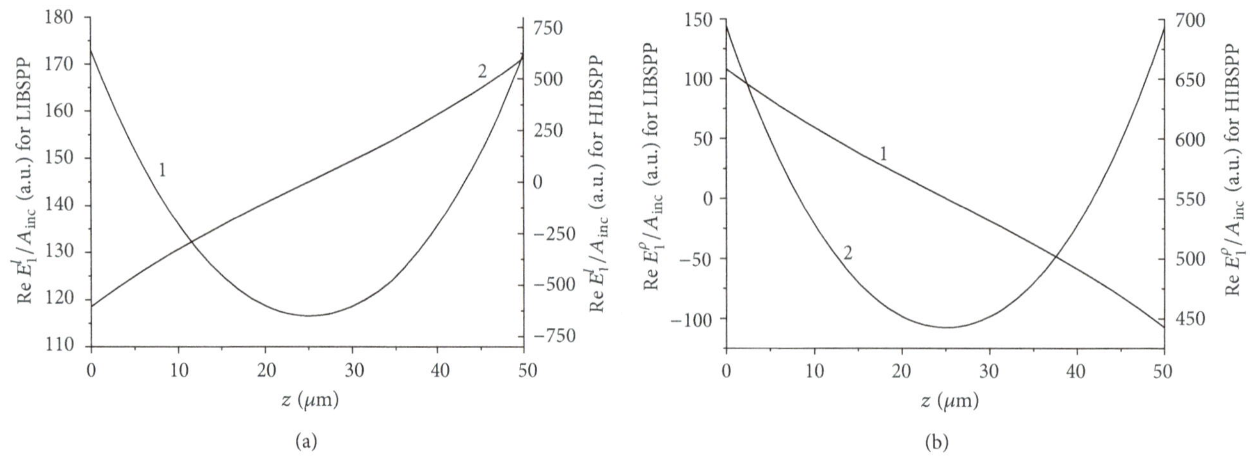

FIGURE 3: The normalized real part of the longitudinal E_1^l (a) and radial E_1^ρ (b) components of the electric vector for low-index (1) and high-index (2) TM-polarized zero-order Bessel surface plasmon polaritons versus the distance inside the gold film of 50 nm thickness embedded in SF10 glass. The distance from the Bessel surface plasmon polariton axis is $\rho = 0.05\,\mu$m, and the wavelength is $\lambda = 0.633\,\mu$m.

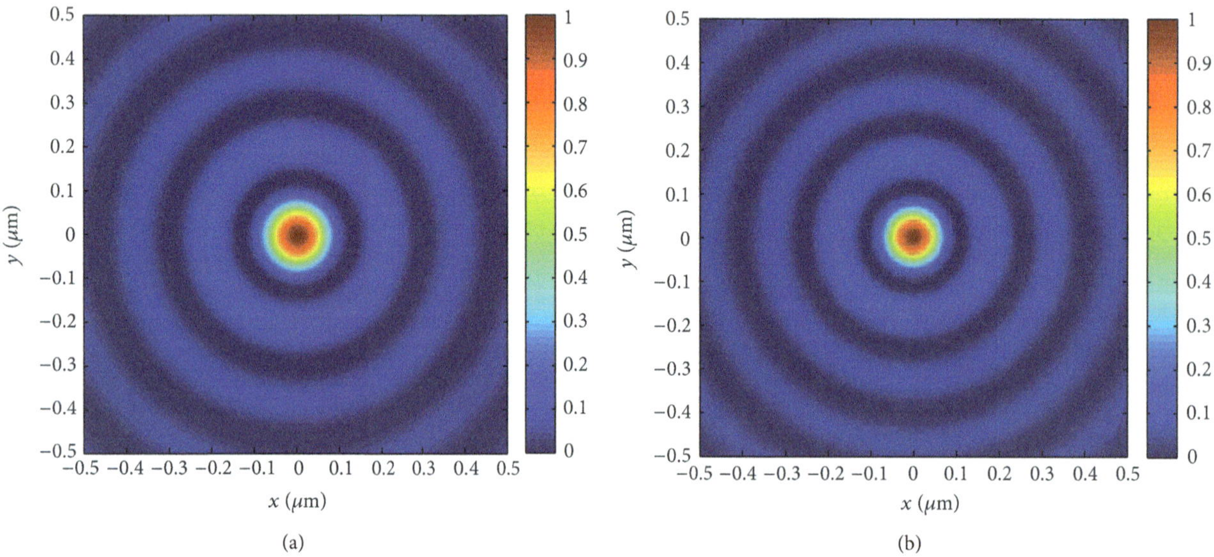

FIGURE 4: 2D pattern of normalized intensity distribution $|E_2^l|^2/|E_2^l|^2_{max}$ for low- (a) and high-index (b) zero-order Bessel surface plasmon polaritons outside the gold layer of 50 nm thickness embedded in SF10 glass. BSPPs are generated by a purely radially polarized Bessel beam having a wavelength of 633 nm. The distance from the entrance surface of the metal layer is $z = \lambda/3$.

components of the BSPP electric vector inside the metal layer. Note that the equation for E_1^ρ can be obtained from (7b). As calculation shows, the symmetry of real and imaginary parts of longitudinal (as well as transversal) components of electric vector of Bessel SPP inside the metal layer is the same. Figure 3 illustrates the dependence of the function Re E_1^l normalized to the amplitude factor A_{inc} (at the center $\rho = 0$ of Bessel surface plasmon polariton) on the distance z in a 50 nm thickness gold film surrounded by dielectric media for low- and high-index zero-order BSPPs. It is seen that for the low-index Bessel SPP with the topological charge $m = 0$, the distribution of E_1^l (E_1^ρ) turns out to be symmetrical (antisymmetrical) (Figure 3) with respect to the middle plane of the metal film. For the high-index Bessel SPP, it is antisymmetrical (symmetrical).

Figure 4 demonstrates a 2D-pattern of the normalized intensity distribution $|E_2^l|^2/|E_2^l|^2_{max}$ for low- and high-index Bessel surface plasmon polaritons formed outside the gold layer of 50 nm thickness embedded in SF10 glass. This is the most interesting case where BSPPs are generated by a purely radially polarized Bessel beam. The results of calculations presented in Figure 4 prove that the widths of the central maximum in the transverse distribution, which can be estimated approximately by the ratio $D_1 = 2R_1 = 4.8/(k_0\,\mathrm{Re}(n^*))$, for LIBSPP as well as for HIBSPP, turn out to be less than the wavelength. It should be noted that as

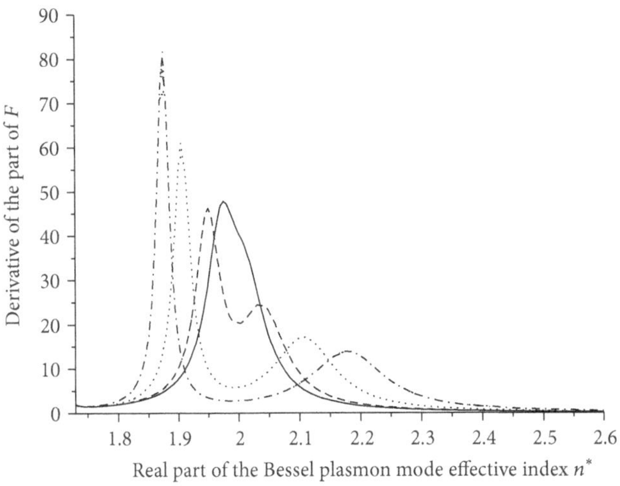

FIGURE 5: The derivative of the phase of F function versus the real part of Bessel surface plasmon polariton mode effective index for gold film of 100 nm (—), 80 nm (- - -), 60 nm (···), and 50 nm (·-·-·-) embedded in SF10 glass.

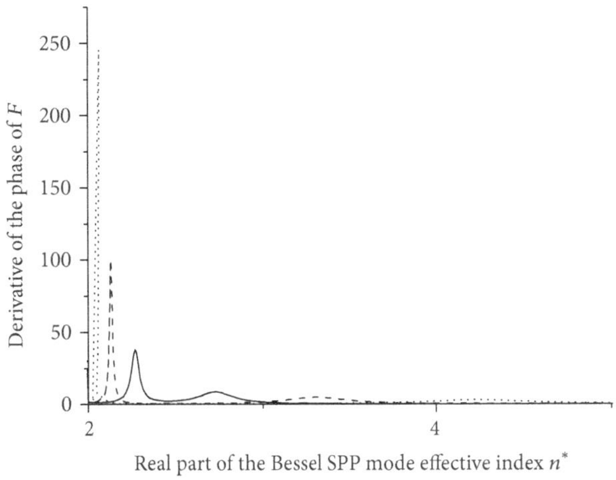

FIGURE 6: The derivative of the phase of F function versus the real part of Bessel surface plasmon polariton mode effective index for the gold film of 50 nm (—), 30 nm (- - -), and 20 nm (···) embedded in ZrO_2.

follows from (7a) while increasing the distance from the surface of the metal layer, the central peak of BSPP, generated in the structure, decays as $\exp(-\operatorname{Im} k_{zi} z)$ $(i = 1, 2)$, but its transversal size D_1 does not change practically.

Let us investigate now how the conditions of excitation of Bessel surface plasmon polaritons depend on the thickness of the metal layer. Figure 5 shows the dependence of the derivative of the phase of F function versus the real part of the Bessel surface plasmon polariton mode effective index at various thicknesses of the gold film. It is seen that at an increase in the metal layer thickness two maxima of the curve merge into one, which testifies the degeneration of Bessel surface plasmon polariton modes. Here, the Bessel surface plasmon polariton mode effective index approaches the n^*_{free} for the case of BSPP excitation at the boundary of two semi-infinite media—dielectric and metal [38]:

$$n^*_{\text{free}} = \sqrt{\varepsilon_2 \frac{\varepsilon'_1 (\varepsilon'_1 + \varepsilon_2) + \varepsilon''^2_1 + i\left[\varepsilon''_1 (\varepsilon'_1 + \varepsilon_2) - \varepsilon'_1 \varepsilon''_1\right]}{(\varepsilon'_1 + \varepsilon_2)^2 + \varepsilon''^2_1}}, \quad (16)$$

where $\varepsilon'_1 = n'^2_1 - n''^2_1$, $\varepsilon''_1 = 2n'_1 n''_1$. Equation (16) can be obtained from dispersion equation (11) at $h = 0$.

At a decrease in thickness of the metal film, the difference between the Bessel surface plasmon polariton mode effective indices increases. So, for example, in the structure containing gold (Au) metal film of 20 nm thickness embedded in SF10 glass, the effective indices take on the values $n^*_1 = 1.759 + 0.0018i$ and $n^*_2 = 3.212 + 0.234i$. At a metal film thickness of 15 nm, we have $n^*_1 = 1.744 + 0.0009i$ and $n^*_2 = 3.940 + 0.331i$. As is seen, in the first case the generated high-index Bessel surface plasmon polariton has the first ring radius R_1 equal to 75 nm. In the second case, while decreasing the thickness of metal film by 15 nm, it is possible to obtain HIBSPP with $R_1 \approx 61$ nm.

Taking into account that ε''_1 is significantly less than $|\varepsilon'_1|$, (16) enables one to get the expressions for the real $(n^{*\prime}_{\text{free}})$ and imaginary $(n^{*\prime}_{\text{free}})$ parts of the effective index n^*_{free} for Bessel surface plasmon polariton mode exited at the boundary of two semi-infinite media-dielectric and metal:

$$n^{*\prime}_{\text{free}} = \sqrt{\frac{\varepsilon'_1 \varepsilon_2}{\varepsilon'_1 + \varepsilon_2}}, \qquad n^{*\prime\prime}_{\text{free}} = \sqrt{\left(\frac{\varepsilon'_1 \varepsilon_2}{\varepsilon'_1 + \varepsilon_2}\right)^3 \frac{\varepsilon''_1}{2\varepsilon'^2_1}}. \quad (17)$$

The calculation, made according to (17), shows that for Bessel surface plasmon polariton, formed at the boundary of two semi-infinite media "gold-SF10 glass," $n^{*\prime\prime}_{\text{free}} = 0.034$. As established above, for the structure containing the 50 nm thickness gold film embedded in SF10 glass, $\operatorname{Im}(n^*_1) = 0.012$. Thus, in a thin metal film immerged in a dielectric, there may be the generation of a low-index Bessel surface plasmon polariton, for which the imaging part of the Bessel surface plasmon polariton mode effective index appears to be lower than that for BSPP excited at the boundary of semi-infinite media "metal-dielectric."

As it is mentioned above, in the z-direction, the electric field of the Bessel surface plasmon polariton decays as $\propto \exp(-\operatorname{Im} k_{zi} z)$ $(i = 0, 1, 2)$. The penetration depth of the field in the medium is determined by the expression $d = 1/\operatorname{Im} k_{zi}$. As the calculation shows, for the LIBSPP penetration depth d_{LIBSPP} into the targeted medium turns out to be larger than d_{free} for the case of the plasmon formation at the "metal-dielectric" boundary. So, in the case under study $d_{\text{LIBSPP}} = 137$ nm, $d_{\text{free}} = 75$ nm.

Now using the reflection pole method, we consider the Bessel surface plasmon polaritons generation for a gold film of 50 nm thickness embedded in material for which the difference $|\varepsilon_1| - \varepsilon_0$ is lower than that in the case studied earlier. Such parameters belong, for example, to ZrO_2 ($\varepsilon_0 = \varepsilon_2 = 4$). Figure 6 is the derivative of the phase of F function versus the real part of the Bessel surface plasmon polariton mode effective index for this case. We see that there are two fast changes of the phase of F corresponding to two Bessel surface plasmon polariton modes with the effective indices

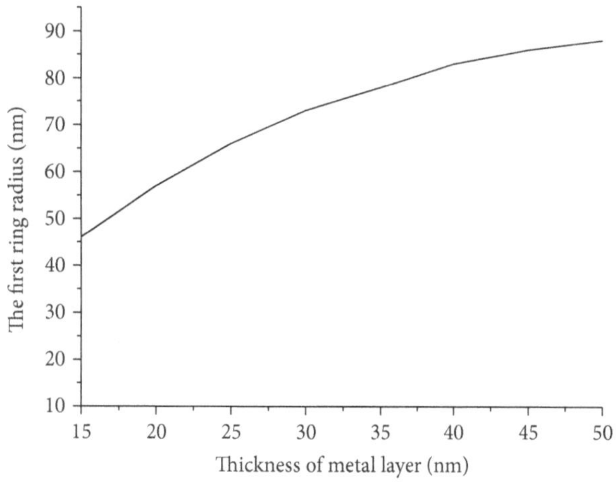

FIGURE 7: Dependence of the first ring radius of Bessel surface plasmon polariton on the thickness of Au layer embedded in ZrO_2 for high-index BSPP.

dependent on the thickness h of the metal layer. As it follows from Figure 7, for $h = 50$ nm, we have $n_1^* = 2.270 + 0.027i$ and $n_2^* = 2.732 + 0.120i$; for $h = 30$ nm, we obtain $n_1^* = 2.1320 + 0.0097i$ and $n_2^* = 3.312 + 0.217i$; for $h = 20$ nm, we find $n_1^* = 2.065 + 0.004i$ and $n_2^* = 4.235 + 0.351i$. It is seen that if the difference between the dielectric constants of a targeted medium (substrate) and metal decreases, the Bessel surface plasmon polariton mode effective indices increase. Moreover, if h decreases, the difference $\Delta n^* = n_2^* - n_1^*$ increases. Thus, in a thin metal layer with a symmetric dielectric surrounding, the generation of a Bessel surface plasmon polariton with the first ring radius of several tens of nanometers is possible (see Figure 7).

2.3. Formation of Bessel Surface Plasmon Polaritons in Metal Film with Asymmetric Dielectric Surrounding. The asymmetric configuration leads to more interesting phenomena. Using the reflection pole method for solving dispersion equation (11), we searched the Bessel surface plasmon polaritons for a gold metal film of 50 nm thickness separating SF10 glass and air. The complex Bessel surface plasmon polariton mode effective indices are found to be $n_1^* = 1.0480 + 0.0001i$ and $n_2^* = 2.022 + 0.043i$. Here, the profile of $E_1^l(z)$ changes essentially inside the metal film for both generated types of TM_0 Bessel surface plasmon polaritons (Figure 8), that is, it losses symmetry with respect to the middle plane of the metal layer. Moreover, Figure 8 shows that there is a certain value z_e on the BSPP axis ($\rho = 0$), for which the longitudinal component of electrical vector $\vec{E}_1^l$ of high-index BSPP is equal to zero, in this concrete case $z_e = 45$ nm.

As a result of the calculation performed with the help of the reflection pole method for solving dispersion equation (11), it is revealed that at an increase in the refraction index of the targeted medium Bessel surface plasmon polariton mode effective indices increase too, but their difference is determined, mainly, by the thickness of the metal film. And the value z_e decreases.

3. Intensity Distribution of Superposition of Bessel Surface Plasmon Polaritons in Layered Metal-Dielectric Structure

Let us consider now the intensity distribution of the superposition of two TM_m Bessel surface plasmon polaritons with different topological charges ($m_1 \neq m_2$) and phase changes ($\Delta_1 \neq \Delta_2$) inside the targeted medium:

$$I \sim \left|\vec{E}^l(R)\right|^2,$$

$$\begin{aligned}\vec{E}^l(R) = {} & \frac{A_{\text{inc}}}{k_0 n_2} \\ & \times \left[q_1 t_1 \exp i\right. \\ & \times \left[m_1\varphi + k_{z2}^{(1)}(z - h_1) + \Delta_1\right] J_{m_1}(q_1\rho) \\ & + q_2 t_2 \exp i\left[m_2\varphi + k_{z2}^{(2)}(z - h_1) + \Delta_2\right] \\ & \left.\times J_{m_2}(q_2\rho)\right] \vec{e}_z.\end{aligned} \tag{18}$$

Here, parameters t_1, t_2 determine the amplitudes of interfering plasmons and can be calculated from (12), $k_{z2}^{(1,2)} = \sqrt{k_0^2 n_2^2 - q_{1,2}^2}$. We will consider in more detail the case of practical interest where Bessel surface plasmon polaritons are generated in a Au film separating SF10 glass and air. This case is described in Section 2.3. Two different variants are realizable: case 1 where BSPPs with different n^* (and hence, q) interfere; case 2 when BSPPs with equal n^* (and hence, q) but different topological charges and/or phase changes interfere. Now consider one more important case where interfering Bessel surface plasmon polaritons with the same mode effective indices have topological charges equal in value but different in sign. In this case, the superposition of two BSPPs J_m and J_{-m}, leads to the excitation of Bessel surface multiplasmon polariton for which, as it follows from (18), the azimuthal modulation of intensity is brought about. Owing to this, the interference structure being a set of annular maxima is separated into substructures at a nanoscale transverse size. Figure 9 illustrates the 2D distribution of normalized intensity in the air at the distance $z = \lambda/3\,\mu$m from the boundary with metal layer. Here, the plasmon polariton field is a superposition of J_3 and J_{-3} low-index (a) and high-index (b) BSPPs. Figures 9(c) and 9(d) show that the field formed by two interfering high-index Bessel surface plasmon polaritons is characterized by more narrow near-axial maxima and essential suppression of lateral maxima while moving off the field axis to its periphery. The above-mentioned interference patterns are rather stable when moving off the interface of metal film. In Figure 10, as an example, the 2D normalized intensity distribution of the field formed by J_3 and J_{-3} low-index BSPPs is given at various distances from the "metal-air" boundary.

Thus, the superposition of two oppositely charged Bessel surface plasmon polaritons allows one to form near the interface of metal an array of quasi-nondiffracting nanoscale spots.

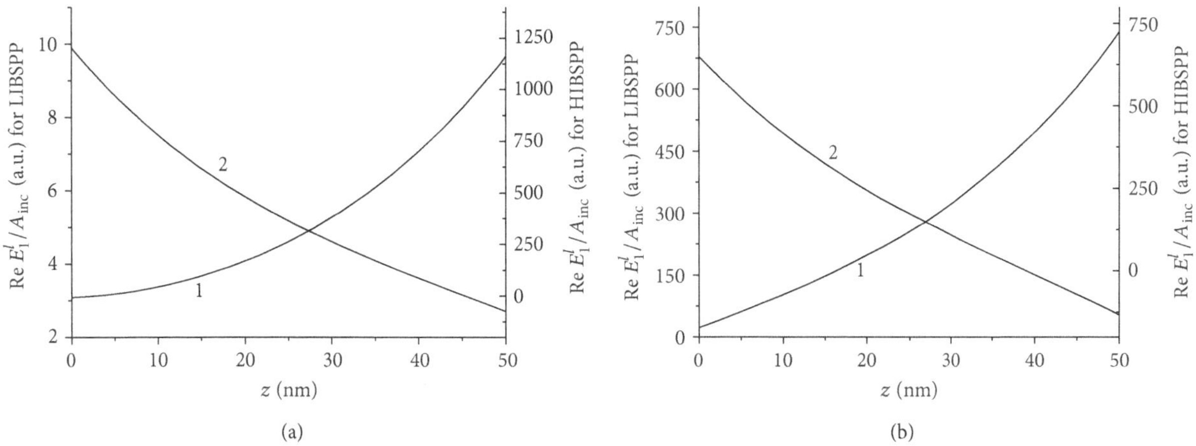

FIGURE 8: The normalized real part of the longitudinal component E_1^l of electric vector of low-index (1) and high-index (2) TM0 Bessel surface plasmon polariton versus the distance inside the 50 nm gold film separating the SF10 glass and air (a) or fluid with refractive index 1.33 (b). The distance from the Bessel surface plasmon polariton axis is $\rho = 0$, and the wavelength is $\lambda = 0.633\ \mu$m.

4. Possible Applications of a Superposition of Bessel Surface Plasmon Polaritons

Thus, the established property of a superposition of Bessel surface plasmon polaritons is applicable to the creation of a virtual multitip for near-field optics. A conceptual sketch of formation of such a multitip, the so-called virtual Bessel surface multiplasmon polariton tip, is illustrated in Figures 11 and 12.

As is seen from Figure 11, the lateral surface of conical lens 2 is illuminated by two conical beams (their extreme rays 1 and 1′ are presented in Figure 11) having the same half-cone angles. These beams are necessary for further exiting BSPPs. One of the pointed conical beams is obtained from the TM_m Bessel mode, and another conical beam is formed from the TM_{-m} BLB using a scheme represented in Figure 12 [39] and composed of a reflective conical lens (RCL) and conical mirror (CM). Both conical beams incident on the conical lens 2 perturb Bessel surface plasmon polaritons in metallic film 3. In the region of overlapping of conical beams (shaded area in Figure 11), a superposition of two Bessel surface plasmon polaritons is generated. In turn, in the air near the base surface of the conical lens covered by metal layer 3, a Bessel surface multiplasmon polariton is formed. If the tested surface is placed at the distance $z \leq \lambda$ from the metal layer, the forming exponentially decaying field penetrates into the surface. Thus, the superposition of Bessel surface plasmon polaritons can be used for testing the surfaces of various specimens.

The first crucial advantage of the near-field technique based on a superposition of Bessel surface plasmon polaritons is related to the nondiffractional nature of these fields. Namely, the virtual Bessel surface multiplasmon polariton tip is characterized by the constant size of a probing beam while increasing the probe-to-sample distance which can vary from zero up to $\sim\ \lambda$. Thus, the use of the virtual Bessel surface multiplasmon polariton tip allows one to extend potentially the working distance between an object and apparatus.

The important feasible application of beams under consideration is also creating multiple optical traps when using two interfering high-order Bessel surface plasmon polaritons.

5. Conclusion

In this paper, a theory is developed of the generation of single Bessel SPPs and their superposition in the structure including a metal layer separating two semi-infinite dielectrics.

The comparison was made of Bessel surface plasmon polaritons investigated in this paper and traditional surface plasmon polaritons. The traditional SPP is a propagating wave process on the metal/dielectric surface. Unlike it, Bessel SPP is a superposition of counter propagating SPPs in all the possible radial directions. It is a complex high-symmetric interference light structure in sections parallel to the metal/dielectric interface. It should be noted as opposed to the propagating surface plasmon polariton, the Bessel one is a standing light structure.

The problem is studied of attenuation of the Bessel SPP excited by a limited narrow Bessel beam. It is shown that outside the region of the exciting source the Bessel SPP decays exponentially in the radial direction. It is established that the $1/e$ energy-attenuation radius of Bessel SPP, generated by the limited narrow Bessel beam, coincides with the decay length of the propagating surface plasmon.

We have analyzed the cases of symmetric and asymmetric "dielectric-metal-dielectric" structures. A dispersion equation has been derived and numerically analyzed for these cases. The possibility is shown of excitation of two types (low- and high-index) of Bessel surface plasmon polaritons characterized by various profiles of longitudinal component $E^l(z)$ of electric vector inside the metal film. It has been

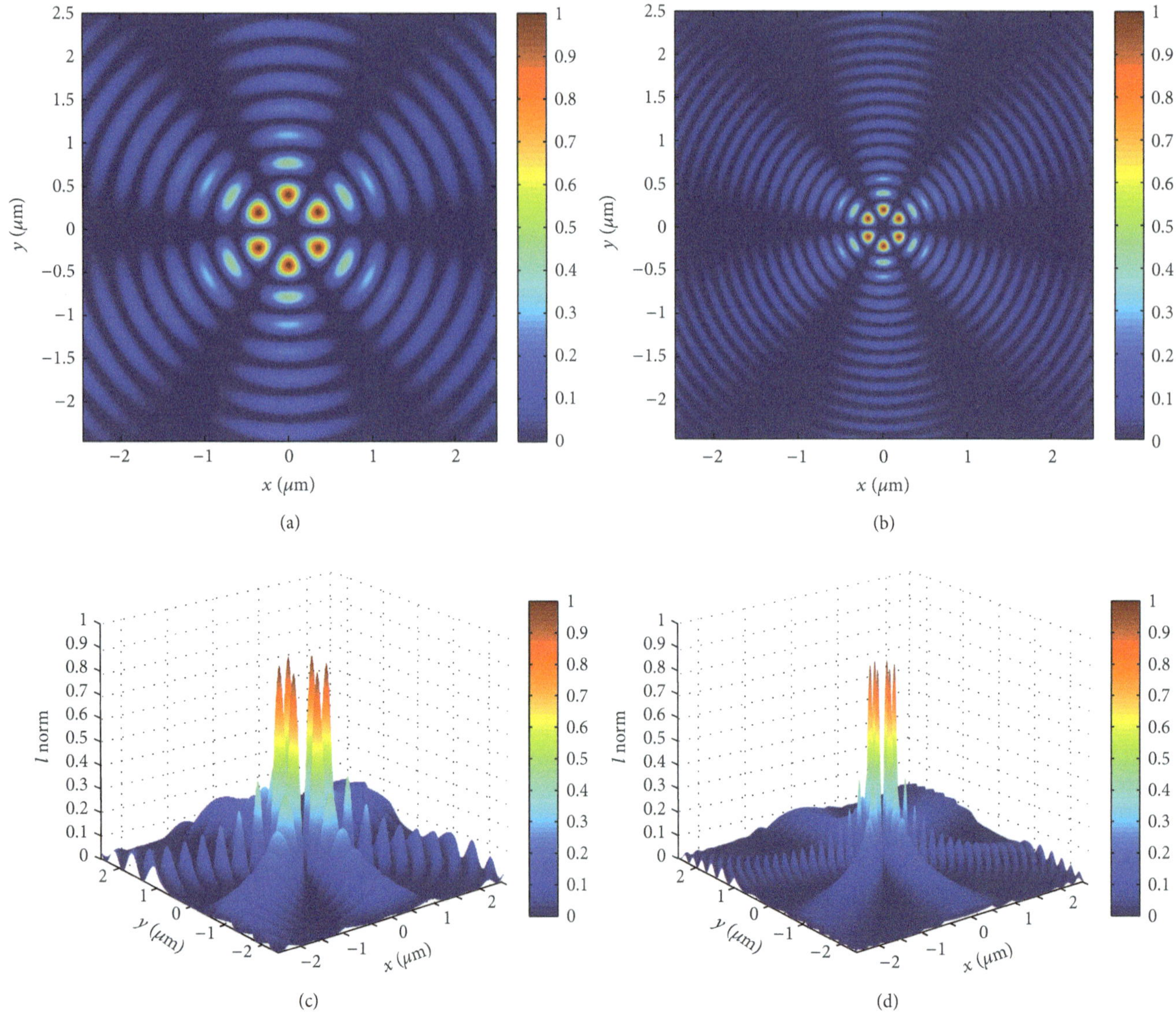

FIGURE 9: 2D and 3D normalized intensity distribution $|E_2^l|^2/|E_2^l|^2_{\max}$ of Bessel multiplasmon in the air at the distance $z = \lambda/3\ \mu$m from the interface of 50 nm thickness gold film of "SF10 glass—Au layer—air" structure. Bessel surface multiplasmon polariton is a superposition of J_3 and J_{-3} low-index (a, c) and high-index (b, d) BSPPs with wavelength $\lambda = 0.633\ \mu$m.

established that in a symmetric "dielectric-metal-dielectric" structure, the profile of $E^l(z)$ for low-index Bessel surface plasmon polaritons turns out to be a symmetric relatively medium plane of the metal film. For high-index Bessel surface plasmon polariton mode it is antisymmetric. In an asymmetric "dielectric-metal-dielectric" structure, the distribution of $E^l(z)$ inside metal film losses symmetry with respect to the medium plane of the metal layer for both generated plasmon polaritons. Moreover, there is a definite value z_e of the longitudinal coordinate inside the layer, for which the longitudinal component of the electric vector $\vec{E}^l$ for high-index Bessel SPP is equal to zero. It is shown that if the difference between the refraction indices of adjoining dielectrics and a metal layer decreases, the position z_e shifts toward the medium plane of metal film.

The dependence has been investigated of conditions of excitation of Bessel surface plasmon polaritons on the metal layer thickness h. It has been shown that if h increases, the degeneration of Bessel plasmon modes occur. Here, the Bessel surface plasmon polariton mode effective index approaches the one for the case of excitation of Bessel SPP at the boundary of two semi-infinite media, that is, dielectric and metal.

The intensity distribution of low- and high-index Bessel surface plasmon polaritons inside and outside the metal layer have been calculated and analyzed. A two- and three-dimensional pattern of the intensity for the single Bessel SPPs as well as for a superposition of two Bessel SPPs with opposite

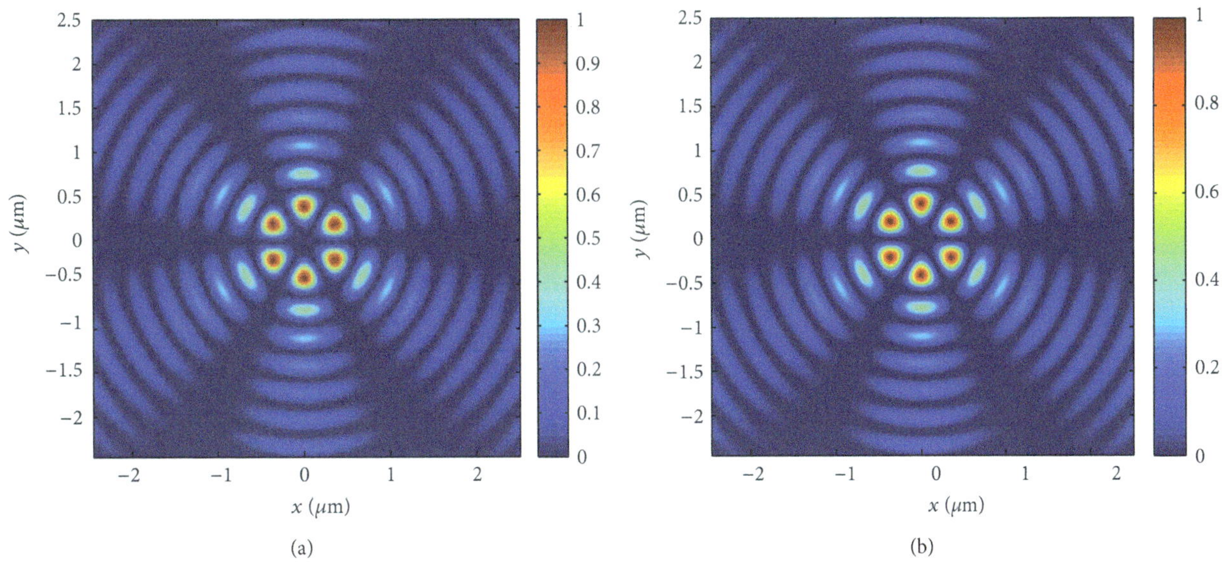

FIGURE 10: 2D normalized intensity distribution $|E_2^l|^2/|E_2^l|_{max}^2$ of Bessel surface multiplasmon polariton in the air at the distance $z = \lambda/3\,\mu$m (a), $z = \lambda/2\,\mu$m (b), and $z = \lambda\,\mu$m (c) from the interface of 50 nm thickness gold film of "SF10 glass—Au layer—air" structure. Bessel surface multiplasmon polariton is a superposition of J_3 and J_{-3} low-index BSPPs with wavelength $\lambda = 0.633\,\mu$m.

topological charges has been calculated. It is shown that while increasing the distance from the surface of the metal layer, the central peak of Bessel SPP generated in the structure decays exponentially, but its transversal size does not change.

It is established that the plasmon polariton field, excited by the superposition of two Bessel beams with opposite topological charges, is characterized by the azimuthal modulation of intensity. Owing to it, the intensity pattern is separated into substructures of a nanoscale transverse size. It is shown that the above-mentioned substructures are rather stable when moving off the interface. Thus, using the superposition of two opposite charged vortex Bessel SPPs, which keeps their transversal size at any distance ranging from zero up to about the wavelength, one can form near the interface an array of diffraction-free nanoscale light "needles." This field is associated with creating a Bessel surface plasmon polariton multitip which is perspective for probing the surface simultaneously through several channels. An experimental set-up is proposed for producing such a Bessel surface plasmon polariton multitip.

The results obtained can be used for the development and optimization of techniques and devices for testing the quality of the surface of various substrates by Bessel surface plasmon polaritons; for creating Bessel surface plasmon polariton-based optical tweezing systems; and for the capturing and manipulating of nanoparticles; applications are expected to be found in near-field optical microscopy with a nanoscale resolution.

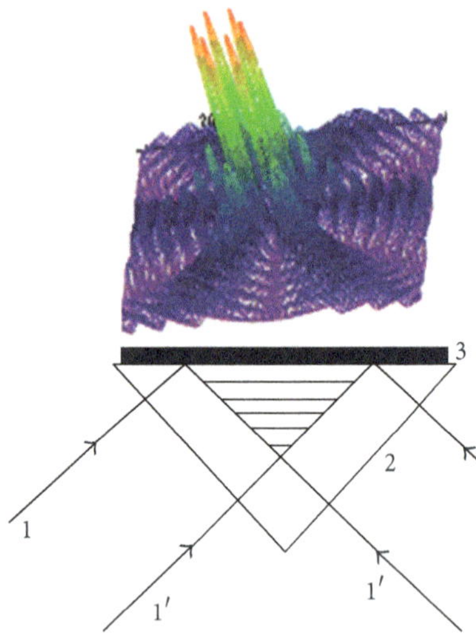

FIGURE 11: Schematic set-up for excitation of Bessel surface multi-plasmon polariton. 1 and 1′ are extreme rays of conical beams excited Bessel surface plasmon polaritons, and 2 is conical lens on the basic surface of which metal film 3 is plotted.

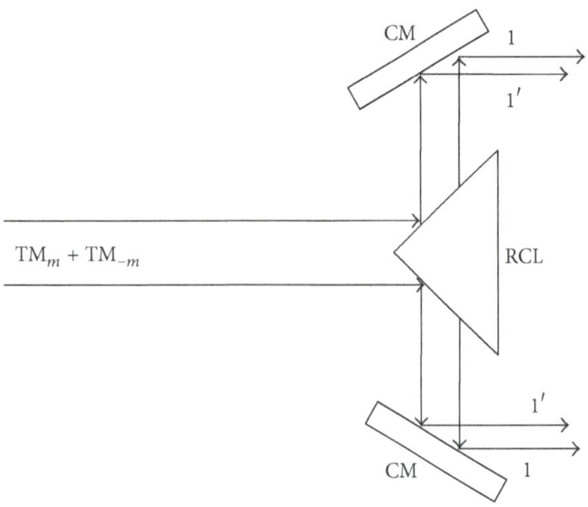

FIGURE 12: Basic scheme of obtaining of conical beams. RCL is a reflective conical lens, CM is a conical mirror, 1 and 1′ are extreme rays of the formed conical beam.

References

[1] H. Raether, *Surface Plasmons on Smooth and Rough Surfaces and on Gratings*, Springer, Berlin, Germany, 1988.

[2] E. N. Economou, "Surface plasmons in thin films," *Physical Review*, vol. 182, pp. 539–554, 1969.

[3] J. J. Burke, G. I. Stegeman, and T. Tamir, "Surface-polariton-like waves guided by thin, lossy metal films," *Physical Review B*, vol. 33, no. 8, pp. 5186–5201, 1986.

[4] W. L. Barnes, A. Dereux, and T. W. Ebbesen, "Surface plasmon subwavelength optics," *Nature*, vol. 424, no. 6950, pp. 824–830, 2003.

[5] A. V. Zayats, I. I. Smolyaninov, and A. A. Maradudin, "Nano-optics of surface plas-mon polaritons," *Physics Reports*, vol. 408, pp. 131–134, 2005.

[6] J. A. Sánchez-Gil and A. A. Maradudin, "Near-field and far-field scattering of surface plasmon polaritons by one-dimensional surface defects," *Physical Review B*, vol. 60, no. 11, pp. 8359–8367, 1999.

[7] J. Durnin, "Exact solutions for nondiffracting beams. I. The scalar theory," *Journal of the Optical Society of America A*, vol. 4, pp. 651–654, 1987.

[8] J. Durnin, J. J. Muceli, and J. H. Eberly, "Diffraction-free beams," *Physical Review Letters*, vol. 58, pp. 1499–1501, 1987.

[9] P. Sprangle and B. Hafizi, "Comment on nondiffracting beams," *Physical Review Letters*, vol. 66, pp. 837–839, 1991.

[10] Z. Bouchal, J. Wagner, and M. Chlup, "Self-reconstruction of a distorted non-diffracting beam," *Optics Communications*, vol. 151, pp. 207–211, 1998.

[11] Y. Lin, W. Seka, J. H. Eberly et al., "Experimental investigation of Bessel beam characteristics," *Applied Optics*, vol. 31, pp. 2708–2713, 1992.

[12] D. McGloin and K. Dholakia, "Bessel beams: diffraction in a new light," *Contemporary Physics*, vol. 46, pp. 15–28, 2005.

[13] J. Turunen and A. T. Friberg, "Self-imaging and propagation-invariance in electromagnetic fields," *Pure and Applied Optics*, vol. 2, pp. 51–60, 1993.

[14] R. Horak, Z. Bouchal, and J. Bajer, "Nondiffracting stationary electromagnetic field," *Optics Communications*, vol. 133, pp. 315–327, 1997.

[15] T. A. Fadeyeva, C. N. Alexeyev, P. M. Anischenko, and A. V. Volyar, "Engineering of the space-variant linear polarization of vortex-beams in biaxially induced crystals," *Applied Optics*, vol. 51, pp. C224–C230, 2012.

[16] I. A. Litvin and A. Forbes, "Bessel Gauss resonator with internal amplitude filter," *Optics Communications*, vol. 281, pp. 2385–2390, 2008.

[17] V. N. Belyi, N. S. Kazak, S. N. Kurilkina, and N. A. Khilo, "Generation of TE- and TH-polarized Bessel beams using one-dimensional photonic crystal," *Optics Communications*, vol. 282, no. 10, pp. 1998–2008, 2009.

[18] S. N. Kurilkina, V. N. Belyi, and N. S. Kazak, "Transformation of high-order Bessel vortices in one-dimensional photonic crystals," *Journal of Optics A*, vol. 12, no. 1, Article ID 015704, 12 pages, 2010.

[19] V. Garcés-Chávez, D. McGloin, H. Melville et al., "Simultaneous micromanipulation in multiple planes using a self-reconstructing light beam," *Nature*, vol. 419, pp. 145–147, 2002.

[20] J. Jerel, T. Cizmar, V. Nedela, and P. Zemanek, "Formation of long and thin polymer fiber using nondiffracting beam," *Optics Express*, vol. 14, pp. 8506–8515, 2006.

[21] M. K. Bhuyan, F. Courvoisier, P. A. Lacourt et al., "High aspect ratio taper-free microchannel fabrication using femtosecond Bessel beams," *Optics Express*, vol. 18, no. 2, pp. 566–574, 2010.

[22] B. Hafizi, E. Esarey, and P. Sprangle, "Laser-driven acceleration with Bessel beams," *Physical Review E*, vol. 55, no. 3, pp. 3539–3545, 1997.

[23] S. Rushin and A. Leizer, "Evanescent Bessel beams," *Journal of the Optical Society of America A*, vol. 15, pp. 1139–1143, 1998.

[24] S. N. Kurilkina, V. N. Belyi, and N. S. Kazak, "Features of evanescent Bessel light beams formed in structures containing a dielectric layer," *Optics Communications*, vol. 283, no. 20, pp. 3860–3868, 2010.

[25] A. M. Goncharenko, N. A. Khilo, and E. S. Petrova, "Evanescent Bessel light beams," in *Lightmetry: Metrology, Spectroscopy, and*

Testing Techniques Using Light, vol. 4517 of *Proceedings of SPIE*, pp. 95–99, 2001.

[26] Q. Zhan, "Evanescent Bessel beam generation via surface plasmon resonance excitation by a radially polarized beam," *Optics Letters*, vol. 31, no. 11, pp. 1726–1728, 2006.

[27] J. Xi, Q. Li, and J. Wang, "Numerical simulation of evanescent Bessel beams and apodization of evanescent field in near-field optical virtual probe," in *Nanophotonics, Nanostructure, and Nanometrology*, Proceedings of SPIE, pp. 42–51, 2005.

[28] A. V. Novitsky and L. M. Barkovsky, "Total internal reflection of vector Bessel beams: imbert–Fedorov shift and intensity transformation," *Journal of Optics A*, vol. 10, Article ID 075006, 7 pages, 2008.

[29] M. K. Al-Muhanna, S. N. Kurilkina, V. N. Belyi, and N. S. Kazak, "Energy flow patterns in an optical field formed by a superposition of evanescent Bessel light beams," *Journal of Optics*, vol. 13, no. 10, Article ID 105703, 2011.

[30] T. Grosjean, D. Courjon, and D. Van Labeke, "Bessel beams as virtual tips for near-field optics," *Journal of Microscopy*, vol. 210, no. 3, pp. 319–323, 2003.

[31] H. Kano, D. Nomura, and H. Shibuya, "Excitation of surface-plasmon polaritons by use of a zeroth-order Bessel beam," *Applied Optics*, vol. 43, no. 12, pp. 2409–2411, 2004.

[32] K. J. Moh, X. C. Yuan, J. Bu, S. W. Zhu, and B. Z. Gao, "Radial polarization induced surface plasmon virtual probe for two-photon fluorescence microscopy," *Optics Letters*, vol. 34, no. 7, pp. 971–973, 2009.

[33] C. J. Zapata-Rodriguez, S. Vuković, M. R. Belić, D. Pastor, and J. J. Miret, "Nondiffracting Bessel plasmons," *Optics Express*, vol. 19, pp. 19572–19581, 2011.

[34] M. Born and E. Wolf, *Principles of Optics*, Cambridge University Press, Cambridge, UK, 1999.

[35] G. N. Watson, *A Threatise on the Theory of Bessel Functions*, Cambridge University Press, Cambridge, UK, 1966.

[36] G. B. Artken, H. J. Weber, and F. E. Harris, *Mathematical Methods for Physicists*, Academic Press, Orlando, Fla, USA, 1985.

[37] A. Papoulis, *Circuits and Systems*, Holt, Rinehart and Winston, New York, NY, USA, 1980.

[38] M. Yamamoto, "Surface plasmon resonance. Theory: tutorial," *Review of Polarography*, vol. 48, pp. 209–237, 2002.

[39] V. N. Belyi, N. A. Khilo, N. S. Kazak, A. A. Ryzhevich, and A. Forbes, "Propagation of high-order circularly polarized Bessel beams and vortex generation in uniaxial crystals," *Optical Engineering*, vol. 50, Article ID 059001, 9 pages, 2011.

16

Subwavelength Plasmonic Waveguides and Plasmonic Materials

Ruoxi Yang and Zhaolin Lu

Microsystems Engineering, Kate Gleason College of Engineering, Rochester Institute of Technology, Rochester, NY 14623, USA

Correspondence should be addressed to Zhaolin Lu, zxleen@rit.edu

Academic Editor: Xiaoyue Huang

With the fast development of microfabrication technology and advanced computational tools, nanophotonics has been widely studied for high-speed data transmission, sensitive optical detection, manipulation of ultrasmall objects, and visualization of nanoscale patterns. As an important branch of nanophotonics, plasmonics has enabled light-matter interactions at a deep subwavelength length scale. Plasmonics, or surface plasmon based photonics, focus on how to exploit the optical property of metals with abundant free electrons and hence negative permittivity. The oscillation of free electrons, when properly driven by electromagnetic waves, would form plasmon-polaritons in the vicinity of metal surfaces and potentially result in extreme light confinement. The objective of this article is to review the progress of subwavelength or deep subwavelength plasmonic waveguides, and fabrication techniques of plasmonic materials.

1. Introduction

Plasmonics, a name coined for surface-plasmon-(SP-) related photonics, has inspired booming research interests since late 1990s. Actually, the history of surface plasmon research dates back to the investigation of surface plasmon resonance (SPR) on metallic thin films [1], or even light scattering from nanoscale metallic particles as early as 1970s [2]. Research focus has then shifted to the integration of plasmonic components into subsystems for optical communications and information exchange, when many novel plasmonic devices have been developed for communications in recent years [3]. Accordingly, plasmonic waveguides, couplers, and modulators become the core topics of the new plasmonics era. Two factors have catalyzed and contributed to this transition. Firstly, along with the advanced numerical tools backed up by computational electromagnetics, powerful computers and computing clusters have been more affordable and accessible. More importantly, with the fast development of nano- and microfabrication technology, researchers have been empowered to fabricate complex structures and manipulate wide selection of materials. In this paper, we will review the development and progress of plasmonic waveguides, especially the applications of metal-insulator-metal waveguides, and various plasmonic materials based on their application spectrum and fabrication techniques.

2. Motivation

Light has emerged as one of the most important carriers for large volume data, as well as a reliable and powerful aide for perception and manipulation of small particles. Compared to its classical form featured with bulky objects and diffraction-limited beams, in recent years the science of optics has utilized much finer elements where the light-matter interactions are strengthened inside a much smaller length scale compared to the lightwave's wavelength. Microoptics and nanophotonics, despite the miniature size of their elements of interest, have shown remarkable impact on all kinds of applications in communications, data storage, sensing, and imaging, and so forth. However, the mode size of a dielectric waveguide is still ruled by the diffraction-limit (close to λ/n, n being the guiding area's refractive index, and λ being the vacuum wavelength of the incident wave), and further shrinking the dimension of dielectric waveguides will inevitably lead to cutoff. This limitation leads to explorations of novel guiding mechanism and materials. Waveguides based on surface plasmons and metals, as found out later, can support propagation mode tightly bounded to the metallic surfaces and possibly confine the guiding wave in deep subwavelength scale. Accordingly, plasmonics has received tremendous attention for its scope of overcoming diffraction limit.

The current trend of technology evolution is marked by the demand of small and compact devices and high-volume integration of multiple modules on the same chip. As Moore's law in Microelectronics enforces more computational power in unit area, the telecommunications has harbored the fiber-to-the-home (FTTH) conception [4] which put high emphasis on reducing the size and cost of individual optical modules. This poses significant challenges for installation of next-generation optical interconnects featuring both compact integration and faster operation, which can be addressed possibly by resorting to plasmonics.

Moreover, light has been anticipated as an indispensable tool to print ultrafine patterns and to manipulate and observe miniscule items. For instance, optical lithography has been pivotal for the prosperity of semiconductor technology [5], but its ability to pace the roadmap of very-large-scale-integration (VLSI) technology has been slowed down because of the diffraction limit of most optical imaging systems. As another example, with complex systems such as DNA sequencing facility [6] designed and developed to incorporate ultrafine optical subsystems, the cost, and quality of optical consumables interfacing with small amount of biological or chemical substances have proved to be decisive. It has also been predicted that plasmonic nanofocusing can revolutionize the traditional magnetic recording by optically heating the media stack in a highly confined fashion, according to the technology called heat-assisted magnetic recording (HAMR) or thermal assisted magnetic recording (TAMR) [7]. Plasmonics, in this sense, shall play a much larger role for optical imaging and sensing subsystem development.

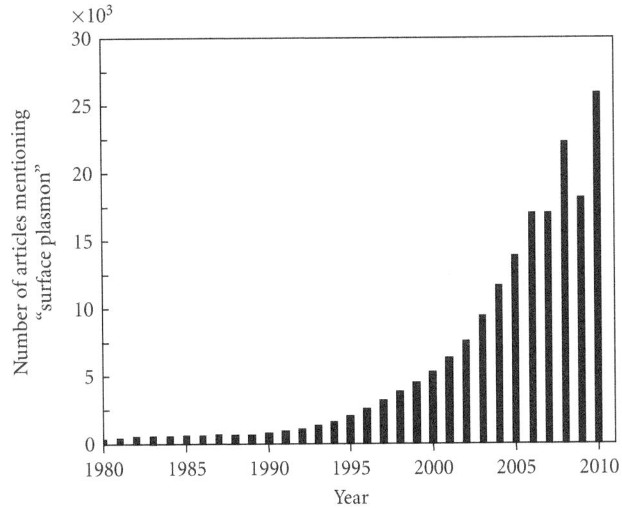

FIGURE 1: The growth of SP-related research in number of articles from 1980 to 2010, data acquired from Google Scholar (http://scholar.google.com/, accessed on Feb 10, 2012).

TABLE 1: Plasmonic frequencies and damping factors for Al, Au, Cu, and Al in NIR spectrum [12].

Material	f_p (THz)	γ_τ (THz)
Aluminum	3570	19.79
Gold	2183	6.46
Copper	1914	8.34
Silver	2180	4.353

3. Nanoplasmonic Waveguide

Research in plasmonics or other SP-related topics has gone through very fast progress in the last decade, while significant breakthroughs were reported by the end of last century (Figure 1) [8]. The study of localized SP remains active, which still finds applications from sensing, detecting, lithography, and imaging to optical storage. Since 1990, the study of plasmonic waveguides and plasmonic-enhanced (extraordinary) transmission [9] has greatly boosted the exposure of the subject. In this section, the fundamentals of plasmonics and the theory of plasmonic waveguides will be reviewed in detail.

3.1. Metal Optics with Drude Model. Metallic films are typically regarded as reflectors or cladding layers in guided-wave micro-optics for visible and near-infrared (NIR) spectrum. While metals' role in optics seems limited, the history of plasma optics is surprisingly long, although it in general lacked detailed descriptions from a guided wave perspective. Plasma optics assumes the rationality of effective dielectric constant of conduction electrons or electron gas, which determines the behavior of plasma oscillations in response to external optical driving field [10]. A plasmon is a quantized plasma oscillation, and the motion of the collective oscillation can rightly support electromagnetic (EM) waves at optical frequencies. With a frequency-dependent conductivity being prescribed, a classical Lorentz equation of motion can be used to derive the effective permittivity [11], providing the frequency independent plasma frequency $\omega_p = \sqrt{Ne^2/\varepsilon_0 m}$, while N describes the electron concentration, e the unit electric charge, ε_0 the free space permittivity, and m the electron mass.

The volume plasma has this characteristic plasma frequency ω_p, out of which the Drude model can be used to determine the dielectric constant of metals (ε_0 the free space permittivity):

$$\varepsilon_e(\omega) = \varepsilon_0 + \frac{\sigma_0}{j\omega(1 + j\omega\tau)} \approx \varepsilon_0 - \frac{\sigma_0}{\omega^2\tau} = \varepsilon_0\left(1 - \frac{\omega_p^2}{\omega^2}\right). \quad (1)$$

In this paper, we mostly consider gold and silver for applications engaging NIR and visible light. The plasma frequencies for them are well above 1000 THz, and are in general higher than the frequency of incident optical beams. The plasma frequencies given by $f_p = \omega_p/2\pi$, and the damping factors by $\gamma_\tau = 1/\tau$ for NIR spectrum of aluminium, gold, copper, and silver are shown in Table 1, adapted from [12]. The dielectric constants of these four metals are then plotted in Figure 2 by referring to [13].

By comparing the plasma frequency with the damping factor and considering the optical application regime, we conclude that external driving field from light will in general

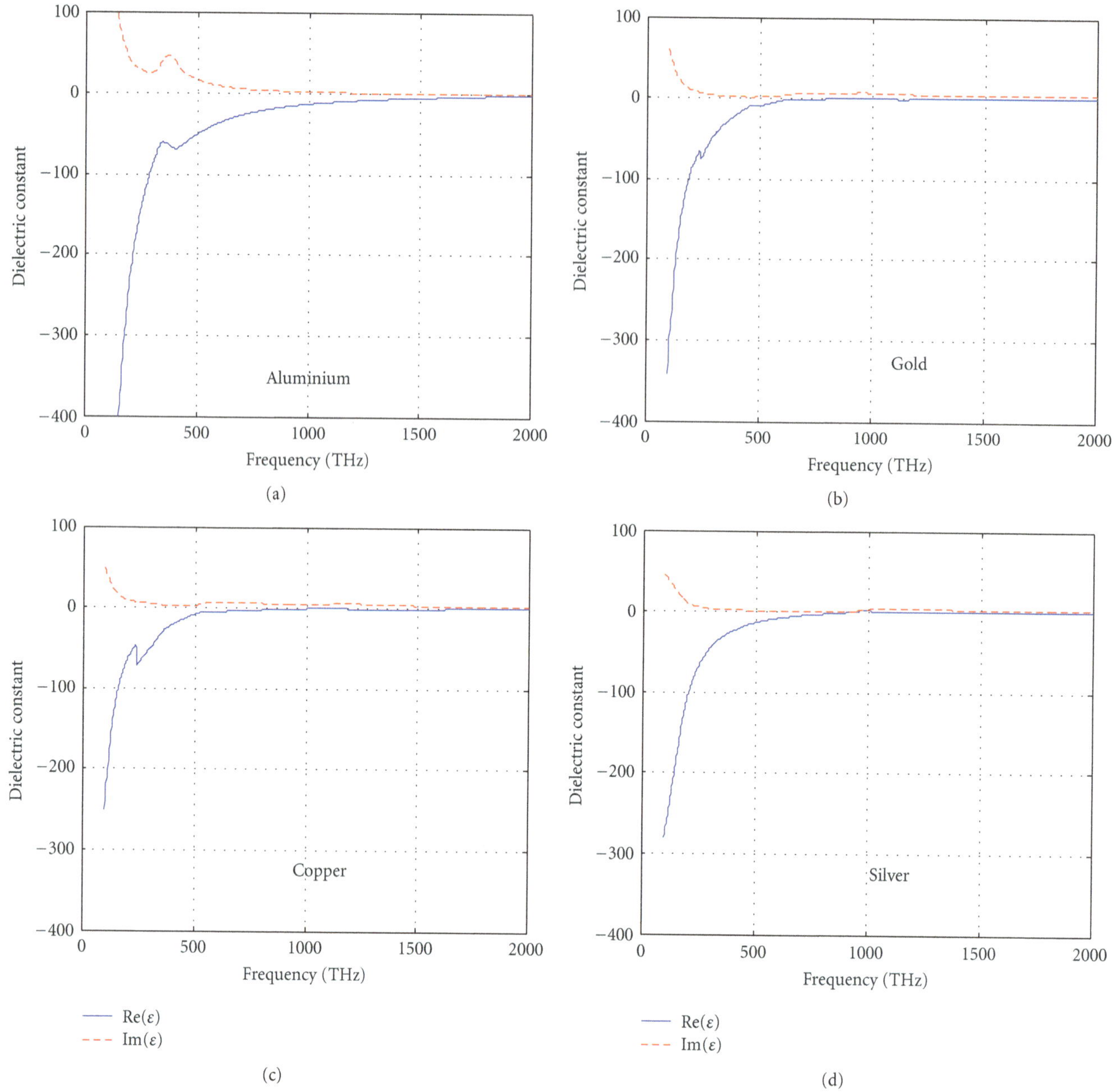

FIGURE 2: Dielectric constants of (a) aluminum, (b) gold, (c) copper, and (d) silver [13].

meet the prerequisite of $\omega \gg 1/\tau$ for the Drude model and the effective permittivity formalism cited above to hold. Therefore, EM waves inside plasmon environment using the effective permittivity formalism (no additional free current or conductivity term) can be characterized by the classical Helmholtz equation derived from Maxwell's equation

$$(\nabla^2 + \varepsilon_e k_0^2)\vec{E} = 0. \quad (2)$$

Note that in the Ampere's law used to derive this equation, there is no free current term. The factor of conductivity can be absorbed to the effective permittivity as below, although the following analysis will generally omit the damping factor γ_τ, assuming the form defined in (1). Therefore

$$\nabla \times \vec{H} = j\omega\tilde{\varepsilon}_e\vec{E} \approx j\omega \operatorname{Re}[\tilde{\varepsilon}_e]\vec{E}. \quad (3)$$

3.2. Bulk Plasma and Surface Plasmon. Considering a plane wave solution in the form of $A\exp(-j\vec{k}\cdot\vec{r})$ for (2), a dispersion relation can be obtained as

$$|k|^2 = k_0^2\varepsilon_e(\omega) = \left(\frac{\omega}{c_0}\right)^2 \varepsilon_e(\omega). \quad (4)$$

Equation (4) clearly shows that a metal with positive permittivity will support plane waves in the bulk medium. In other words, if the driving frequency goes above the bulk plasma

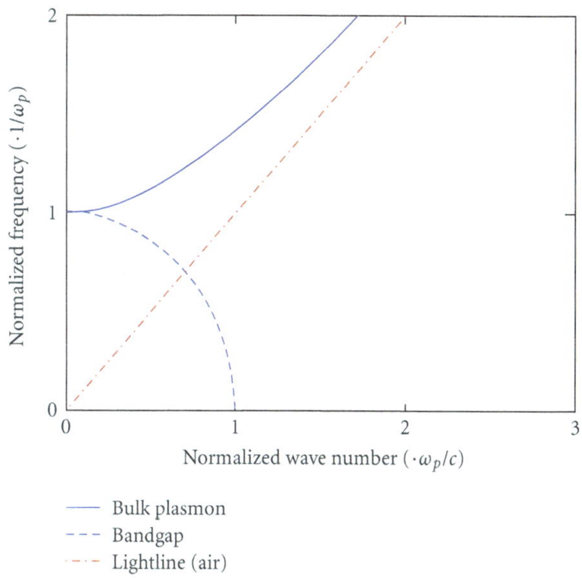

FIGURE 3: Dispersion relation of free electron gas (solid) compared to that of lightline of dielectric material (dash-dotted).

frequency, the plasma (or metal) becomes transparent to this external EM wave as

$$\beta = \sqrt{\varepsilon_e}k_0 = k_0\sqrt{1 - \frac{\omega_p^2}{\omega^2}}. \quad (5)$$

This dispersion relation of free-electron gas is shown in the solid blue curve in Figure 3.

On the other hand, a negative dielectric constant is obtained when the driving frequency of the incident optical beam falls below the plasma frequency, which leads to an evanescent wave with an imaginary wave vector, as given by

$$k_{\text{eva}} = \sqrt{\varepsilon_e}k_0 = \pm jk_0\sqrt{\frac{\omega_p^2}{\omega^2} - 1}. \quad (6)$$

This relation is shown in the dashed curve bearing a lower frequency cutoff related to plasma frequency ω_p in Figure 3. Physically, whatever direction the incident field comes from, the field will impinge into a small thickness known as skin depth, and then be fully shielded by the free electrons. The concept and the term of "skin depth" conveys an implication that only a small fraction of bulk metal could take part in the photon-material interaction, and very limited field-penetration or power-propagation could ever happen. External optical fields can only penetrate into metals for a couple of nanometers range among this spectrum, and are therefore trivial compared to the reflecting power radiated away from the metal surfaces.

The bulk plasma does not support EM waves with negative permittivity. However, it is also known that dispersion with negative permittivity has great potentials to be engineered or incorporated into complex composites for novel applications, as researchers have already done to artificially shaped structures such as photonic crystals [14] or metamaterials [15]. In this sense, the dispersion of plasmons described by the Drude model provides certainly an opportunity to manipulate electromagnetic waves and fields, as will be seen for the case of surface plasmons.

For surface plasmons, the dispersion curve will be more complex. Surface plasmon waves, which reveal the macroscopic motion of surface plasmon-polaritons (SPPs), are a coherent combination of electron gas oscillation (plasmons) and photons. For a simple case of a single metal-insulator interface in 2-dimensional (2D) view, a transverse-magnetic (TM) solution for (2) is available [1, 16]. A matched real propagation constant (β) is required for a maintained wave motion in both materials (ε_m for metal and ε_d for insulator). Therefore, the dispersion relation for SPPs at a single interface is

$$\beta = \frac{\omega}{c_0\sqrt{\varepsilon_m\varepsilon_d/\varepsilon_m + \varepsilon_d}}. \quad (7)$$

With propagation constant (β), it is feasible to define the skin depths δ as $\delta = 1/2\gamma$ (Figure 4).

With a new term of surface plasmon frequency ($\omega_{\text{sp}} = \omega_p/\sqrt{1+\varepsilon_d}$) being defined, (7) leads to a dispersion plot for SPPs. For the air-metal case, the surface plasmon frequency is $0.707\ \omega_p$.

In Figure 5, there is a bandgap between ω_{sp} and ω_p, when the curve only gives imaginary propagation constant (evanescent waves). When $\omega < \omega_{\text{sp}}$, SPP modes in TM polarization exist. A very large β value can be obtained at ω_{sp}, which denotes a point of resonance at that frequency point, namely the SP resonant oscillation. When $\omega > \omega_p$, $\varepsilon_m > 0$; so metal will exhibit dielectric behaviors with radiation unbounded to insulator since the mode lies above the lightline. The dispersion relation plotted above does not take damping factor into account, which will instead force a finite wave vector $\text{Re}[\beta]$ at ω_{sp} due to the damping of free-electron oscillations. In this case, the right-hand side of (7) becomes a complex value and the dispersion will also slightly differ from the case of ideal conductors [16].

The single-interface case introduced above suggests several important characters related to SPPs. Firstly, a sustained surface wave can be excited along metal surfaces accompanied by ohmic loss, although at the same frequency bulk metal remains "opaque" for EM waves. Secondly, a bounded solution exists even with a half-space metal-cladded insulator, which means the SPP mode is confined to a single interface of metal-insulator with small out-of-plane radiation. Thirdly, the propagation constant of SPP mode is larger than that of a lightline at the same frequency point when the driving field has less-than-ω_{sp} frequency, which indicates a smaller effective wavelength for SPP mode compared to optical mode at same frequency. These three features are the sources of most applications reviewed in this article, and are the fundamentals of many prior efforts.

3.3. Metal-Insulator-Metal (MIM) Waveguides. Although surface plasmons can be confined to the metal-insulator boundary, the mode spread of the SP waves depends on the configuration of the waveguides and also the frequency. The length of the evanescent tail in a waveguide medium can

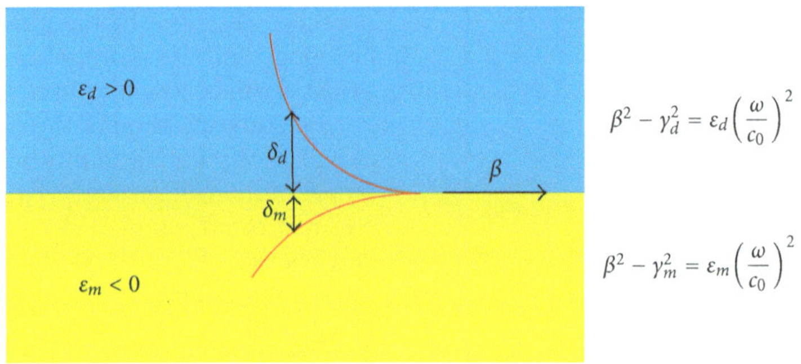

FIGURE 4: Definition of skin depths and their relation with propagation constant.

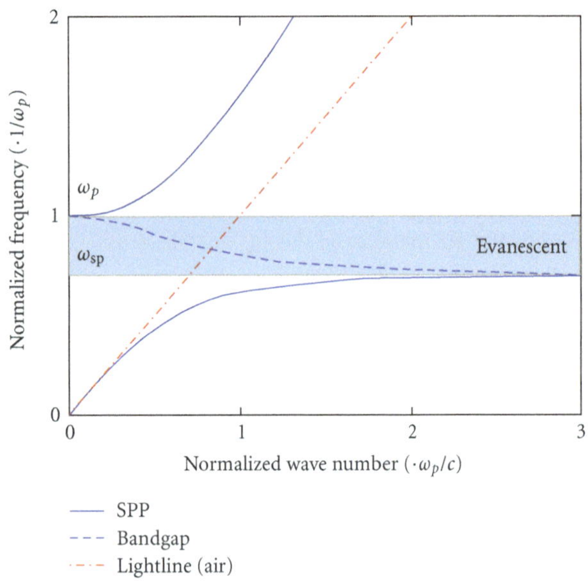

FIGURE 5: Dispersion relation of SPP (solid) in metal-insulator configuration. The curve for the frequency between 0.707 and 1 denotes imaginary propagation constant.

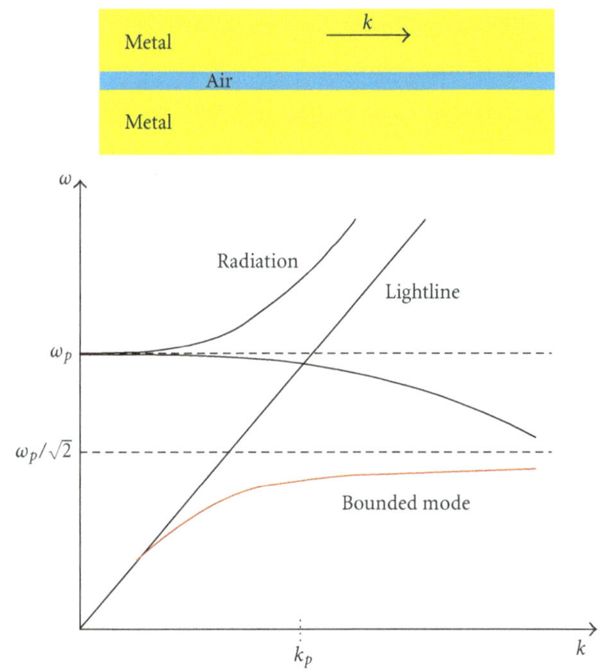

FIGURE 6: Dispersion plot for the MIM-layered structure adapted from [1].

be approximated by $\delta = 1/2\gamma$, when γ is defined from the conservation of momentum as follows:

$$\beta^2 - \gamma_d^2 = \varepsilon_d \left(\frac{\omega}{c_0}\right)^2 \quad (8)$$

$$\beta^2 - \gamma_m^2 = \varepsilon_m \left(\frac{\omega}{c_0}\right)^2. \quad (9)$$

Table 2 shows that for the metal-insulator (MI) waveguide, the evanescent tails can be very long compared to wavelength. For subwavelength and deep subwavelength mode confinement, it is necessary to explore different waveguide forms. The following section will start from an important plasmonic waveguide, namely metal-insulator-metal (MIM) waveguide.

The idea to incorporate surface plasmons to guided EM waves can be traced back to 1969, in Economou's paper [1] on theoretical study of EM field inside thin-films configuration. Note that it was about 30 years later, with the progress of advanced simulation tools and microscale fabrication technology, that people started to characterize plasmonic waveguides at optical frequencies.

A simple yet important form of plasmonic waveguide is based on a metal-insulator-metal (MIM) configuration (Figure 6). This waveguide, as its name indicates, consists of three layers with a dielectric core of either air or other insulator (usually with low optical loss and comparatively small refractive index), sandwiched by two metal claddings. In Economou's work [1], the dispersion relation for MIM configuration is analytically solved. In recent years, detailed analytical [17, 18] and experimental studies [19] of MIM are still performed and reported.

As verified by extensive numerical analysis, MIM waveguides possess notable tradeoff between acceptable propagation loss and subwavelength mode size. Veronis and Fan [20], along with Feigenbaum and Orenstein [21], have performed 3-dimensional (3D) simulation and mode analysis for MIM (sometimes called MDM for metal-dielectric-metal too)

TABLE 2: The lengths of evanescent tails at different wavelengths for metal-insulator boundary.

Wavelength (λ)	0.5 μm	1.55 μm	10.5 μm	300 μm
Frequency (c_0/λ)	600 THz	193 THz	28 THz	1 THz
ε_{Al}	$-36.5 + j9.4$	$-252 + j46$	$-8e3 + j5.2e3$	$-6e4 + j9e4$
ε_{Air}	1.0	1.0	1.0	1.0
Decay factor (γ/k_0)	0.1678	0.0631	0.0112	0.0041
Tail length (δ)	0.23 μm	1.95 μm	74.7 μm	5800 μm
Relative tail length (δ/λ)	0.47	1.26	7.12	19.49

TABLE 3: TM-mode size comparison for dielectric and plasmonic waveguides with λ = 1550 nm light. Data acquired from FDTD Solutions's integrated mode solver (http://www.lumerical.com/).

Waveguide type	Dielectric	MIM	IMI	MI
Materials	Silicon	Silver, air	Silver, air	Silver, air
Core thickness	300 nm	30 nm	30 nm	—
Total cladding thickness	—	300 nm	300 nm	—
n_{eff}	2.5060	1.5788	2.7104	1.0039
Loss (dB/cm)	—	2180.5	846.5	44.0
FWHM (H intensity)	208 nm	46 nm	148 nm	983 nm

structures. They successfully show that while the mode size of MIM can be pushed down to very close to the physical dimension of the guiding core in deep subwavelength regime, the acceptable propagation length keeps MIM waveguides competitive for on-chip signal transmission (Table 3). The 2D to 3D transition is verified by shrinking a 2D nanoslit [1] to a 3D nanospot [20, 22]. This feature brings in the possibility of using MIM structure as a device to guide propagating waves in ultrasmall dimension (Figure 7), and therefore an interesting platform for applications that require extreme field confinement or intensified light-matter interactions. The metal sandwich could possibly contain unwanted scattering loss from rough surfaces, which is another advantage of the MIM configuration.

From a transfer matrix perspective, it has been shown that if the fundamental TM mode is excited inside an MIM gap, the waveguide works well as a transmission line (TL) system, and shows reduced scattering loss from sharp bend [23], unlike traditional optical waveguides. The transmission line (TL) theory, adopted from the microwave realm, has then been applied to cascaded MIM systems [24] to study the reflection and transmission coefficient for complex or multilevel compositions of MIM waveguides. The formalism of TL theory based on eigenmode expansion, is found to be successful especially in calculating the coupling efficiency between SP waves propagated through several MIMs.

Along with the MIM waveguide, the insulator-metal-insulator (IMI) configuration has also found many applications. IMI is formed by burying a thin metal core in dielectric background, and if the symmetry condition is strictly met, a TM mode very similar to a dielectric mode can be supported. Devices incorporating IMI have been demonstrated for both passive [25] and active [26] cases, and because IMI's propagation loss is considerably smaller than MIM, it is frequently used for transmitting NIR optical power in longer distance above 10 μm's mark. The lack of mode confinement, however, hampers its usage in deep subwavelength scale, so does it with metal-insulator (MI) waveguides (Table 3).

Based on the MIM platform highlighted with extreme power confinement, several inventions or discoveries are noteworthy. Atwater and his coworkers have contributed significantly to this topic with their characterization of MIM plasmonic waveguides [27]. Chen [28] from Cornell University evaporated gold onto silicon ribs to form MIM waveguide, which also developed a novel way to effectively excite SP modes from photons. On the field of active plasmonics, nanocavities formed in MIM waveguide are proposed for fast modulation [29, 30], and have exhibited great potentials for development of future small-print modulators.

It is also interesting to look at the MIM waveguide from a different perspective. If an MIM waveguide is chopped for a very small propagation (less than one hundred nanometers, e.g.,), and at the same time a light beam illuminates right into the waveguide from free space, the situation will be very close to transmission through an optical head [31], which can be used to count the power getting through a subwavelength slit or a hole inscribed on a thin film. One important literature published on extra transmission related to SP waves is from Ebbesen et al. [9]. In this paper, the observance of at least 10-folds of transmission enhancement is reported (Figure 8), after facilitated features of surface corrugations are applied. Ironically, such an important literature later is cited by the authors themselves that the reported enhancement is greatly exaggerated because of a problematic measurement. The role of plasmonics in extra transmission is therefore under debate, as [32] and especially [33] have reviewed. One of the authors from the original nature paper, Tineke Thio, converted to be one of the most radical opponents to the role of surface plasmons in enhanced transmission. A diffraction theory [32], from her point of view, is powerful enough to explain most phenomena observed in extratransmission.

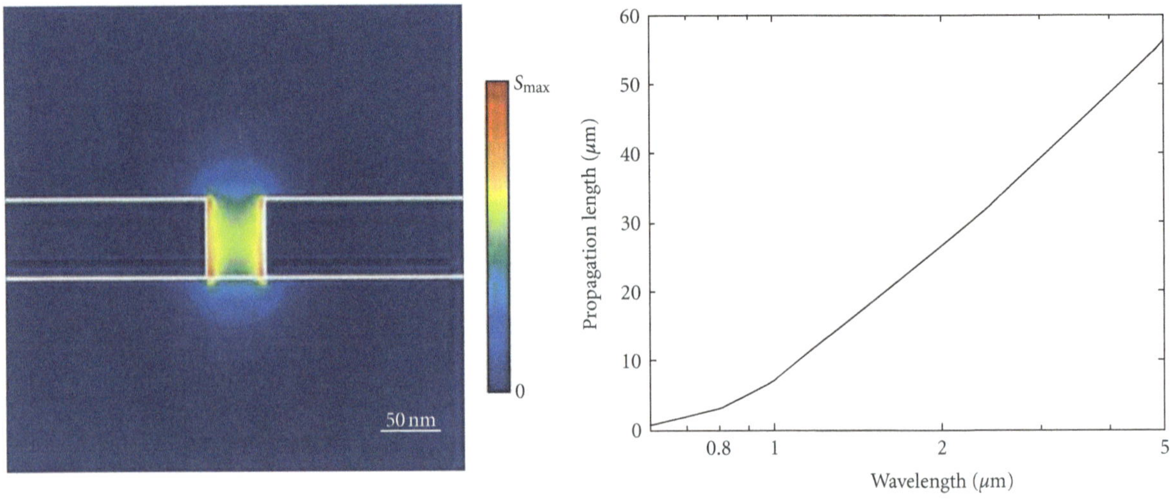

FIGURE 7: (a) A typical MIM-mode profile for 1550 nm light inside a 50 nm-by-50 nm air gap. The substrate and superstrate are silicon dioxide and air, respectively. (b) Propagation distance of the MIM mode for optical to NIR frequency [20].

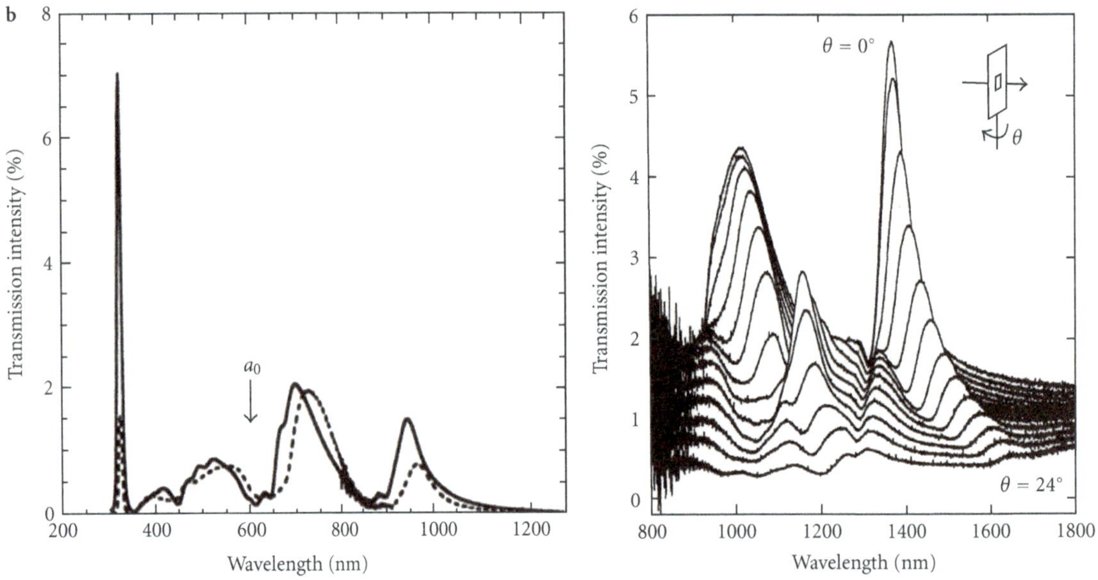

FIGURE 8: Transmission through subwavelength holes patterned in Ag as a function of wavelength (a) and illumination angle (b) [9].

Despite all the debates, it is then impossible to halt the eagerness of researchers on utilizing SP waves to accomplish higher transmission through subwavelength holes or slits. For example, Pacifici et al. [34] have used a near-field detector to characterize the field profile when corrugated surfaces are illuminated by plane waves. They managed to observe subwavelength focusing of light and have given their own physical model to explain it. A carefully designed mask [35] is found to be able to achieve a subwavelength transmission *beyond* near-field with radiationless interference. This discovery might be more interesting than the previous one for far-field applications such as imaging and nanolithography. More and more theoretical calculations on EM waves among periodic surface features or slit diffraction such as [36, 37] or [38] are still being elaborated and published, providing fresh approaches or understandings into this topic.

The extratransmission experiments introduced above are mostly based on incident plane waves from free space or uniform dielectric half space. In 2007, Veronis and Fan published simulation results [39] studying the behavior of NIR light when a dielectric waveguide is butt-connected to an MIM waveguide. They find that although the mode profiles between dielectric and MIM waveguides differ greatly, a transmission of ~70% can be achieved. There are quite a few previous unknowns mentioned by this theoretical analysis. Firstly, the mode overlap is not found as a key issue for coupling, as the "funneling" effect has been verified by numerical simulation performed recently [40]. Secondly, the direct coupling only works for TM polarization, which is different compared to the classical case of an aperture antenna loaded on an infinite ground plane [41]. Therefore, a different physics unlike the aperture antenna theory must exist. Thirdly, the way that optical modes couple to SP modes are different to TL formalism. A matched propagation constant here does not necessarily guarantee an optimal coupling efficiency, as the eigenmodes in plasmonic waveguides are different in nature regarding the orthonormality compared to conventional rectangular waveguides. As all being said, direct coupling remains an interesting topic to the plasmonics community and has always been appreciated as a practical scheme with plasmonic devices.

3.4. Subwavelength Waveguiding by Periodic Structures. Surface plasmons are grouped oscillations of free electrons near the surface. The coupling of plasmons and photons could happen on the boundary between a metal and an insulator. As the "donor" of plasmons, metals are usually designed to share a flat boundary with insulators to avoid unwanted out-of-plane scattering. Rough boundaries or surface corrugations, on the other hand, are occasionally introduced [16] for excitations or observations of SPPs. Theoretically, surface plasmons exist for middle infrared (MIR) or even far infrared (FIR) regime. We can recall the lengths of evanescent tails in the dielectric side are given by $1/2\gamma_d$, while the term γ_d is given by

$$\gamma_d = \sqrt{\beta^2 - \varepsilon_d \left(\frac{\omega}{c}\right)^2}. \tag{10}$$

For EM waves with frequencies much smaller than optical spectrum, ε_m is a negative term with large absolute value, thus the value of γ_d can be very small. For a metal-insulator case with $|\varepsilon_m| \gg |\varepsilon_d|$, and referring to the dispersion relation in (7), we can see β^2 will approach $\varepsilon_d(\omega/c)^2$, which is the right side of (8). Therefore, γ_d will likely be very small and leads to a very long evanescent tail. This type of surface wave looks close to the so-called Zenneck or Sommerfeld wave [42] and the field confinement will be very poor (Table 3).

Obviously, to achieve tightly bounded surface wave beyond optical spectrum such as terahertz (THz) or even gigahertz (GHz) regime, a flat surface is not enough. On the other hand, the lower frequency (or the longer wavelength) makes it possible to make comparatively small corrugation on metallic surfaces. If the considered corrugation is small enough compared to wavelength, it can be legitimately characterized by using the effective medium formalism. This approach is called "spoof plasmonics" by Pendry et al. [43], in their attempt to achieve arbitrary "spoof" plasma frequency via drilling holes on the perfect electric conductor (PEC).

When the corrugation is uniformly distributed in xy direction (Figure 9(a)), the bulk PEC can be regarded as a uniaxial crystal, a typical anisotropic metamaterial. The dispersion calculation yields an upper cutoff frequency and an extremely slow light towards this cut-off, same as the typical SPP behaves. More importantly, when an effective medium is used for analysis, the theory of conventional plasmonics can be applied to the structured surface. Therefore, the concept of surface plasmon frequency can also be extended to describe the property of this metamaterial. Instead of the canonical value, the new (or effective) surface plasmon frequency is determined by the dimension of surface corrugations. Therefore, the optical property of this metamaterial can be well controlled by tweaking the physical perturbations applied to the bulk material.

The spoof plasmonic approach has inspired many discoveries based on surface corrugations especially in THz region as metals are almost lossless and compact THz waveguides are in high demand. Maier et al. [44] designs a THz waveguide for subwavelength confinement in 2D working at 1 THz. Martin-Cano et al. have designed metal gratings as THz waveguide platform, which can integrate multiple passive devices at the same time [45]. In GHz region, Zhao et al. [46] used designer surface plasmon approach to demonstrate the same concept (Figure 10).

Another platform for spoof plasmonics is based on stratified medium, which finds vast applications in the optical regime. Multilayered composites with periodic layer-unit have been widely implored as negative index materials (NIMs) for superlensing. Treated as an effective medium, its anisotropy is well approximated by straightforward mixing rules [47, 48] if each repeated layer is thin enough compared to wavelength. Along with the superlensing effect [49], multilayered composites have also been designed to support surface-plasmon-like waves [50, 51] and used for nanolithography. Figure 11 shows the schematic electric field diagram of a multilayer structure, when a composition of surface waves is formed [48].

4. Plasmonic Materials and Fabrication Technique

The progress of experimental plasmonics is closely related to the advancement of material research and fabrication technology. Plasmonic phenomena have been observed and investigated for more than 40 years, originally on light scattering from metal particles in early years, gradually including light guiding by particle arrays and layered structures. Without the advanced fabrication methods to process plasmonic materials and integrate plasmonic structures into optical circuitry, it is impossible to expect the vast applications of plasmonic components in sensing, imaging, storage, and communications these days. In this section, the most important plasmonic materials and their processing methods will be reviewed.

In solid-state physics, surface plasmons are treated as grouped oscillation of electron gas. The electron gas model fits materials with a high volume of free electrons, which is inherently more accurate for simple metals (alkali metals). Metals found in semiconductor technology at room temperature have an electron concentration close to the order

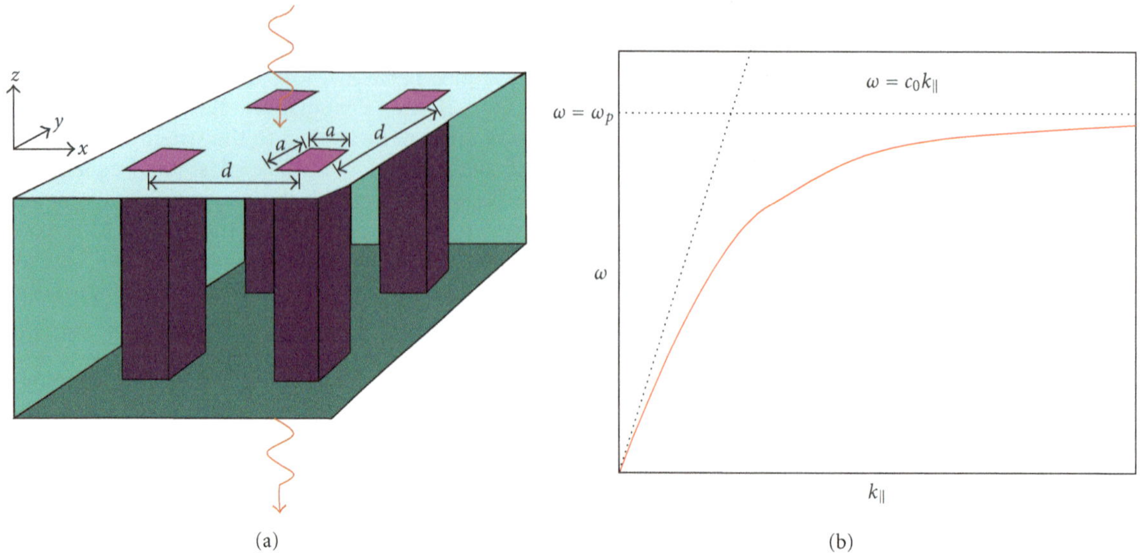

FIGURE 9: Spoof plasmonics in PEC shown by dimension (a) and dispersion (b) [43].

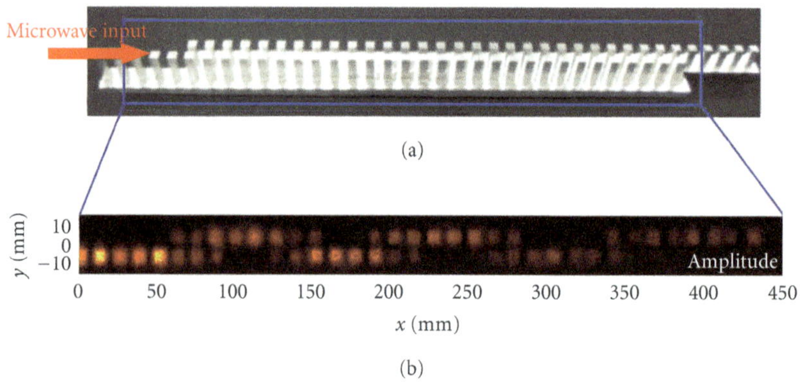

FIGURE 10: A spoof plasmonic waveguide guiding GHz waves [46], forming a directional coupler.

of 10^{22} per cm^3, while semiconductors or doped materials widely used in this industry have a carrier concentration of less than 10^{17} per cm^3. As a result, most plasmonic structures, especially those for optical spectrum, will involve metals at some level. The overwhelmingly used materials in plasmonics have remained to be those that are stable in room temperature, especially aluminum (Al) and noble metals such as gold (Au), silver (Ag), and copper (Cu) (Table 1). Alternative plasmonic materials known to contain abundant free carriers include metallic alloys, metallic compounds, and graphene.

4.1. Metals. Metals have been studied and understood through the free-electron model, in which the electrons inside metals are regarded as moving almost free from striking with each other. In the Drude formalism for 3D electron gas, the mean relaxation time is described by the damping frequency term γ_τ. For the noble metals frequently used in plasmonics, the total damping comes from not only the interactions between conduction electrons, but also from interband transitions [10, 16]. Wherever the damping of free-electron oscillation actually comes from, the propagation loss in classical electromagnetic problems for plasmonic metals can be evaluated from the imaginary parts of their dielectric constants and the electric field intensity, treated as dissipated or lost in the form of ohmic heat. Therefore, a straightforward figure of merit for choosing plasmonic material is its resistivity. By reading the optical resistivity data or the dielectric constants of various metals [52], it is possible to spot candidates of suitable materials for different frequency ranges. In NIR spectrum, silver has the smallest imaginary part of permittivity and the best conductivity, making it a highly favored material. The easy oxidation somehow limits its applications, and makes gold a popular candidate in this frequency range. For the applications of UV or DUV light such as the 193 nm light for DUV optical lithography, however, aluminum is proved to be the most attractive candidate [12] (Table 1).

Processing metals (especially noble metals) is in general complex, especially compared with semiconductors used in microfabrication. The deposition of plasmonic metals can be achieved in vacuum by sputtering or evaporation. To avoid percolation, and to obtain acceptable uniformity and quality,

Figure 11: Multilayered, periodic metal-insulator stacks (a) and the schematic E-field profile (b) inside [48].

the deposition rate is usually limited to less than 1 nm per second. The small deposition rate is normally not a big issue as in most cases the required thickness for plasmonics is small. Electroplating gives faster metal growth rate but the quality of the surface might not be considered fine enough for plasmon waves.

The pattern transfer from resist to metals is extremely challenging for noble metals, as there are not many effective methods to etch copper, silver, or gold. Plasmonic metals such as aluminum can be etched by plasma etch in Cl_2-based ambient. Other than this, there are generally two approaches to pattern metals. The first one is through image reversal and lift-off process, which requires resist overhang after development. There have been photoresists specifically used for lift-off process such as nLOR from MicroChem [53], while researchers also use bilayer or multilayer compositions to achieve the crucial overhang [54]. When using bilayer resists for liftoff, it is conventionally required to coat a bottom resist layer for at least twice as thick as the lift-off thickness. Therefore, it is not easy to have high aspect ratio, standalone metal features fabricated. A second approach to pattern metal is through ion milling. Focused ion beam can be used to directly mill patterns into gold or silver film, while it usually introduces redepositions from the beam source. Argon ion beam is another option that can work with noble metals, which requires hard masks such as carbon and might greatly increase the number of steps and also the complexity of processing.

4.2. Compounds. Metal silicides are originally used to decrease the contact resistivity in VLSI circuits and especially MOSFET technology. As they have an intermediate conductivity and carrier concentration between metal and semiconductor, silicide compounds have also been investigated for plasmonic applications especially in THz frequency [55]. In the form of compounds consisting of silicon and metals (Ti, Co, W, Ni, Pd, etc.), silicides have a smaller plasmonic frequency compared to noble metals used in the near-infrared regime and in general higher loss. However, as its fabrication and electric property have been extensively studied for large-scale integrated circuit technology, the deposition of very thin layers and accurate patterning is highly reliable.

Transparent conducting oxide (TCO) materials are optically transparent and electrically conductive. As one of the most widely used TCO, Indium tin oxide (ITO) has been widely studied and used in display industry and optoelectronics as an optical coating. The frequency range for ITO to exhibit negative permittivity is also in NIR, in which ITO is reported to have competitively low loss even compared to silver. The deposition of ITO films can be achieved through sputter deposition [56]. The major concern about using ITO is its price, and the instability of its optical property for different growth conditions. Similar materials include aluminum zinc oxide (AZO), gallium zinc oxide (GZO), and so forth [57].

4.3. Graphene and Others. A few rarely-used yet interesting materials exist as potential candidates for plasmonic applications. Among them, graphene and polymer-based plasmonics have attracted more and more attention, because of the distinguished electronic and optical properties. Graphene's plasmonic property has been predicted in [58–60], showing promising tunable optical conductivity. Standard approaches for graphene layer deposition are limited at this time, although depositions based on transfer-printing method [61], CVD [62], and electrostatic method [63] have been reported.

5. Conclusion

Surface plasmon has been studied extensively for optical sensing and imaging, even long before the term "plasmonics" became widely accepted. The transition from SPR in standalone devices to plasmonics, assisted by advanced simulation and fabrication tools, emphasizes the integration of plasmonic features into subsystems for all sorts of optical communications and information exchange. With more and more significant discoveries based on surface plasmon waveguiding in subwavelength or deep subwavelength regime, plasmonics is expected to further impact the research and development of integrated optical communications, data storage, optical sensing, and imaging technology.

Acknowledgment

This paper is based upon work supported in part by the U. S. Army under Award no. W911NF-10-1-0153 and the National Science Foundation under Award no. ECCS-1057381. Acknowledgment is made to the Donors of the American Chemical Society Petroleum Research Fund for partial support of this project.

References

[1] E. N. Economou, "Surface plasmons in thin films," *Physical Review*, vol. 182, no. 2, pp. 539–554, 1969.

[2] R. H. Ritchie, "Surface plasmons in solids," *Surface Science*, vol. 34, no. 1, pp. 1–19, 1973.

[3] M. I. Stockman, "Nanoplasmonics: past, present, and glimpse into future," *Optics Express*, vol. 19, no. 22, pp. 22029–22106, 2011.

[4] K. S. Kim, "On the evolution of PON-based FTTH solutions," *Information Sciences*, vol. 149, no. 1–3, pp. 21–30, 2003.

[5] M. Rothschild, "A roadmap for optical lithography," *Optics and Photonics News*, vol. 21, no. 6, pp. 26–31, 2010.

[6] J. Eid, A. Fehr, J. Gray et al., "Real-time DNA sequencing from single polymerase molecules," *Science*, vol. 323, no. 5910, pp. 133–138, 2009.

[7] W. A. Challener, C. Peng, A. V. Itagi et al., "Heat-assisted magnetic recording by a near-field transducer with efficient optical energy transfer," *Nature Photonics*, vol. 3, no. 4, pp. 220–224, 2009.

[8] M. L. Brongersma and P. G. Kik, *Surface Plasmon Nanophotonics*, Springer, Dordrecht, The Netherlands, 2007.

[9] T. W. Ebbesen, H. J. Lezec, H. F. Ghaemi, T. Thio, and P. A. Wolff, "Extraordinary optical transmission through subwavelenght hole arrays," *Nature*, vol. 391, no. 6668, pp. 667–669, 1998.

[10] C. Kittel, *Introduction to Solid State Physics*, John Wiley & Sons, New York, NY, USA, 1989.

[11] B. E. A. Salen and M. C. Teich, *Fundamentals of Photonics*, Wiley-Interscience, Hoboken, NJ, USA, 2nd edition, 2007.

[12] M. A. Ordal, R. J. Bell, J. Alexander, L. L. Long, and M. R. Querry, "Optical properties of fourteen metals in the infrared and far infrared: Al, Co, Cu, Au, Fe, Pb, Mo, Ni, Pd, Pt, Ag, Ti, V, and W," *Applied Optics*, vol. 24, no. 24, pp. 4493–4499, 1985.

[13] E. D. Palik, *Handbook of Optical Constants of Solids*, Academic Press, Orlando, Fla, USA, 1985.

[14] J. D. Joannopoulos, S. G. Johnson, and J. N. Winn, *Photonic Crystals: Molding the Flow of Light*, Princeton University Press, New Jersey, NJ, USA, 2008.

[15] N. Engheta and R. W. Ziolkowski, *Metamaterials: Physics and Engineering Explorations*, Wiley, 2006.

[16] S. A. Maier, *Plasmonics: Fundamentals and Applications*, Springer, New York, NY, USA, 2007.

[17] R. Gordon, "Light in a subwavelength slit in a metal: propagation and reflection," *Physical Review B*, vol. 73, no. 15, Article ID 153405, 2006.

[18] J. A. Dionne, L. A. Sweatlock, H. A. Atwater, and A. Polman, "Plasmon slot waveguides: towards chip-scale propagation with subwavelength-scale localization," *Physical Review B*, vol. 73, no. 3, Article ID 035407, 9 pages, 2006.

[19] S. I. Bozhevolnyi, V. S. Volkov, E. Devaux, J.-Y. Laluet, and T. W. Ebbesen, "Channel plasmon subwavelength waveguide components including interferometers and ring resonators," *Nature*, vol. 440, no. 7083, pp. 508–511, 2006.

[20] G. Veronis and S. Fan, "Guided subwavelength plasmonic mode supported by a slot in a thin metal film," *Optics Letters*, vol. 30, no. 24, pp. 3359–3361, 2005.

[21] E. Feigenbaum and M. Orenstein, "Modeling of complementary (Void) plasmon waveguiding," *Journal of Lightwave Technology*, vol. 25, no. 9, pp. 2547–2562, 2007.

[22] R. Yang, M. A. G. Abushagur, and Z. Lu, "Efficiently squeezing near infrared light into a 21nm-by-24nm nanospot," *Optics Express*, vol. 16, no. 24, pp. 20142–20148, 2008.

[23] G. Veronis and S. Fan, "Bends and splitters in metal-dielectric-metal subwavelength plasmonic waveguides," *Applied Physics Letters*, vol. 87, no. 13, Article ID 131102, 3 pages, 2005.

[24] Ş. E. Kocabaş, G. Veronis, D. A. B. Miller, and S. Fan, "Modal analysis and coupling in metal-insulator-metal waveguides," *Physical Review B*, vol. 79, no. 3, Article ID 035120, 2009.

[25] R. Charbonneau, C. Scales, I. Breukelaar et al., "Passive integrated optics elements based on long-range surface plasmon polaritons," *Journal of Lightwave Technology*, vol. 24, no. 1, pp. 477–494, 2006.

[26] T. Nikolajsen, K. Leosson, and S. I. Bozhevolnyi, "Surface plasmon polariton based modulators and switches operating at telecom wavelengths," *Applied Physics Letters*, vol. 85, no. 24, pp. 5833–5835, 2004.

[27] J. A. Dionne, H. J. Lezec, and H. A. Atwater, "Highly confined photon transport in subwavelength metallic slot waveguides," *Nano Letters*, vol. 6, no. 9, pp. 1928–1932, 2006.

[28] L. Chen, J. Shakya, and M. Lipson, "Subwavelength confinement in an integrated metal slot waveguide on silicon," *Optics Letters*, vol. 31, no. 14, pp. 2133–2135, 2006.

[29] J. A. Dionne, K. Diest, L. A. Sweatlock, and H. A. Atwater, "PlasMOStor: a metal-oxide-si field effect plasmonic modulator," *Nano Letters*, vol. 9, no. 2, pp. 897–902, 2009.

[30] W. Cai, J. S. White, and M. L. Brongersma, "Compact, high-speed and power-efficient electrooptic plasmonic modulators," *Nano Letters*, vol. 9, no. 12, pp. 4403–4411, 2009.

[31] M. J. Levene, J. Korlach, S. W. Turner, M. Foquet, H. G. Craighead, and W. W. Webb, "Zero-mode waveguides for single-molecule analysis at high concentrations," *Science*, vol. 299, no. 5607, pp. 682–686, 2003.

[32] H. J. Lezec and T. Thio, "Diffracted evanescent wave model for enhanced and suppressed optical transmission through subwavelength hole arrays," *Optics Express*, vol. 12, no. 16, pp. 3629–3651, 2004.

[33] T. Thio, "A bright future for subwavelength light sources," *American Scientist*, vol. 94, no. 1, p. 40, 2006.

[34] D. Pacifici, H. J. Lezec, L. A. Sweatlock, R. J. Walters, and H. A. Atwater, "Universal optical transmission features in periodic and quasiperiodic hole arrays," *Optics Express*, vol. 16, no. 12, pp. 9222–9238, 2008.

[35] R. Merlin, "Radiationless electromagnetic interference: evanescent-field lenses and perfect focusing," *Science*, vol. 317, no. 5840, pp. 927–929, 2007.

[36] F. J. García-Vidal, S. G. Rodrigo, and L. Martín-Moreno, "Foundations of the composite diffracted evanescent wave model," *Nature Physics*, vol. 2, no. 790, pp. 262–267, 2006.

[37] L. Chen, J. T. Robinson, and M. Lipson, "Role of radiation and surface plasmon polaritons in the optical interactions between a nano-slit and a nano-groove on a metal surface," *Optics Express*, vol. 14, no. 26, pp. 12629–12636, 2006.

[38] D. Pacifici, H. J. Lezec, H. A. Atwater, and J. Weiner, "Quantitative determination of optical transmission through subwavelength slit arrays in Ag films: role of surface wave interference

and local coupling between adjacent slits," *Physical Review B*, vol. 77, no. 11, Article ID 115411, 2008.

[39] G. Veronis and S. Fan, "Theoretical investigation of compact couplers between dielectric slab waveguides and two-dimensional metal-dielectric-metal plasmonic waveguides," *Optics Express*, vol. 15, no. 3, pp. 1211–1221, 2007.

[40] R. A. Wahsheh, Z. Lu, and M. A. G. Abushagur, "Nanoplasmonic couplers and splitters," *Optics Express*, vol. 17, no. 21, pp. 19033–19040, 2009.

[41] C. A. Balanis, *Antenna Theory: Analysis and Design*, Wiley, New Jersey, NJ, USA, 2005.

[42] A. Sommerfeld, "Propagation of waves in wireless telegraphy," *Annals of Physics*, vol. 81, pp. 1367–1153, 1926.

[43] J. B. Pendry, L. Martín-Moreno, and F. J. Garcia-Vidal, "Mimicking surface plasmons with structured surfaces," *Science*, vol. 305, no. 5685, pp. 847–848, 2004.

[44] S. A. Maier, S. R. Andrews, L. Martín-Moreno, and F. J. García-Vidal, "Terahertz surface plasmon-polariton propagation and focusing on periodically corrugated metal wires," *Physical Review Letters*, vol. 97, no. 17, Article ID 176805, 4 pages, 2006.

[45] D. Martin-Cano, M. L. Nesterov, A. I. Fernandez-Dominguez, F. J. Garcia-Vidal, L. Martin-Moreno, and E. Moreno, "Domino plasmons for subwavelength terahertz circuitry," *Optics Express*, vol. 18, no. 2, pp. 754–764, 2010.

[46] W. Zhao, O. M. Eldaiki, R. Yang, and Z. Lu, "Deep subwavelength waveguiding and focusing based on designer surface plasmons," *Optics Express*, vol. 18, no. 20, pp. 21498–21503, 2010.

[47] D. J. Bergman, "The dielectric constant of a composite material-a problem in classical physics," *Physics Reports*, vol. 43, no. 9, pp. 377–407, 1978.

[48] B. Wood, J. B. Pendry, and D. P. Tsai, "Directed subwavelength imaging using a layered metal-dielectric system," *Physical Review B*, vol. 74, no. 11, Article ID 115116, 2006.

[49] W. Cai, D. A. Genov, and V. M. Shalaev, "Superlens based on metal-dielectric composites," *Physical Review B*, vol. 72, no. 19, Article ID 193101, 4 pages, 2005.

[50] J. Elser, A. A. Govyadinov, I. Avrutsky, I. Salakhutdinov, and V. A. Podolskiy, "Plasmonic nanolayer composites: coupled plasmon polaritons, effective-medium response, and subdiffraction light manipulation," *Journal of Nanomaterials*, vol. 2007, Article ID 79469, 8 pages, 2007.

[51] R. Yang, X. Huang, and Z. Lu, "Arbitrary super surface modes bounded by multilayered metametal," *Micromachines*, vol. 3, no. 1, pp. 45–54, 2012.

[52] D. R. Lide, *CRC Handbook of Chemistry and Physics*, CRC Press, 2004.

[53] MicroChem Corp, March2012, http://www.microchem.com.

[54] Photolithography Resist Processes and Capabilities (CNF), March 2012, http://www.cnf.cornell.edu/cnf_process_photo_resists.html.

[55] R. A. Soref, S.-Y. Cho, W. Buchwald, and R. E. Peale, "Silicon plasmonic waveguides," in *Silicon Photonics for Telecommunications and Biomedical Applications*, B. Jalali and S. Fathpour, Eds., Taylor & Francis, London, UK, 2010.

[56] "PVD 75 Sputter Deposition from Kurt J. Lesker by CNF," March 2012, http://www.cnf.cornell.edu/cnf5_tool.taf?_function=detail&eq_id=156.

[57] P. R. West, S. Ishii, G. V. Naik, N. K. Emani, V. M. Shalaev, and A. Boltasseva, "Searching for better plasmonic materials," *Laser & Photonics Reviews*, vol. 4, no. 6, pp. 795–808, 2010.

[58] D. R. Andersen, "Graphene-based long-wave infrared TM surface plasmon modulator," *Journal of the Optical Society of America B*, vol. 27, no. 4, pp. 818–823, 2010.

[59] F. H. L. Koppens, D. E. Chang, and F. J. Garciia de Abajo, "Graphene plasmonics: a platform for strong Light-Matter interactions," *Nano Letters*, vol. 11, no. 8, pp. 3370–3377, 2011.

[60] J. T. Kim and S.-Y. Choi, "Graphene-based plasmonic waveguides for photonic integrated circuits," *Optics Express*, vol. 19, no. 24, pp. 24557–24562, 2011.

[61] X. Li, Y. Zhu, W. Cai et al., "Transfer of large-area graphene films for high-performance transparent conductive electrodes," *Nano Letters*, vol. 9, no. 12, pp. 4359–4363, 2009.

[62] A. Reina, X. Jia, J. Ho et al., "Large area, few-layer graphene films on arbitrary substrates by chemical vapor deposition," *Nano Letters*, vol. 9, no. 1, pp. 30–35, 2009.

[63] A. N. Sidorov, M. M. Yazdanpanah, R. Jalilian, P. J. Ouseph, R. W. Cohn, and G. U. Sumanasekera, "Electrostatic deposition of graphene," *Nanotechnology*, vol. 18, no. 13, Article ID 135301, 2007.

17

Development of Automatic 3D Blood Vessel Search and Automatic Blood Sampling System by Using Hybrid Stereo-Autofocus Method

Eiji Nakamachi, Yusuke Morita, and Yoshifumi Mizuno

Department of Biomedical Engineering, Doshisha University, 1-3 Miyakodani, Tatara, Kyotanabe, Kyoto 610-0394, Japan

Correspondence should be addressed to Eiji Nakamachi, enakamac@mail.doshisha.ac.jp

Academic Editor: Nanguang Chen

We developed an accurate three-dimensional blood vessel search (3D BVS) system and an automatic blood sampling system. They were implemented into a point-of-care system designed for medical care, installed in a portable self-monitoring blood glucose (SMBG) device. The system solves problems of human error caused by complicated manual operations of conventional SMBG devices. We evaluated its accuracy of blood-vessel position detection. The 3D BVS system uses near-infrared (NIR) light imaging and the stereo and autofocus hybrid method to determine blood vessel locations accurately in three dimensions. We evaluated the accuracy of our 3D BVS system using a phantom of human skin, blood vessels, and blood. Additionally, we established an automatic blood sampling system for SMBG and assessed its performance in relation to punctures, blood suction, transport, and discharge on an enzyme sensor. The 3D BVS and automatic blood sampling system are adequate for use in a portable SMBG device.

1. Introduction

Recently, the rapidly increasing number of diabetic patients has [1] become a major social problem throughout the world. Diabetic patients must independently measure blood glucose concentrations using a self-monitoring blood glucose (SMBG) device and self-inject insulin several times each day [2, 3]. Diabetes patients and candidates might constitute up to one-fifth of the world population [1]. Therefore, SMBG devices are expected to play an increasingly important role in daily medical care. The commercial SMBG devices of various kinds that are on the market today can be categorized as either noninvasive or minimally invasive. Noninvasive devices adopt infrared spectroscopy [4], FT-IR-ATR method [5], photoacoustic spectroscopy [6], and so on. However, these methods present severe problems of low accuracy and unreliability. In contrast, minimally invasive devices [7, 8] can measure blood glucose concentrations with high accuracy [9], but puncture with a needle or knife edge is painful. In addition, such devices frequently induce human error because of their complicated manual operations during puncture, treatment of small blood samples, self-measurement, and so on. Therefore, many improvements are necessary for SMBG devices. Several groups have developed minimally invasive SMBG devices equipped with a blood vessel search system to reduce human error. For example, Yamakoshi et al. developed a device to detect the positions of blood vessels in the arm using reflected near-infrared (NIR) light [10], but the blood vessel depth could not be determined accurately. Another example is a bio-medical-chip developed by Horiike et al., which can detect the position of a blood vessel in three dimensions using electric signals [11]. However, their chip showed low accuracy of blood vessel location detection, and the device entailed operations that were time-consuming and painful for patients. Although Kandani et al. developed a three-dimensional blood vessel search (3D BVS) system using the stereo method [12], the value of the standard deviation of measured positions was higher for deeper blood vessel locations. In this study, we set our goal as integration of a 3D BVS system and an automatic blood sampling system. To accomplish that goal, we combined the autofocus method

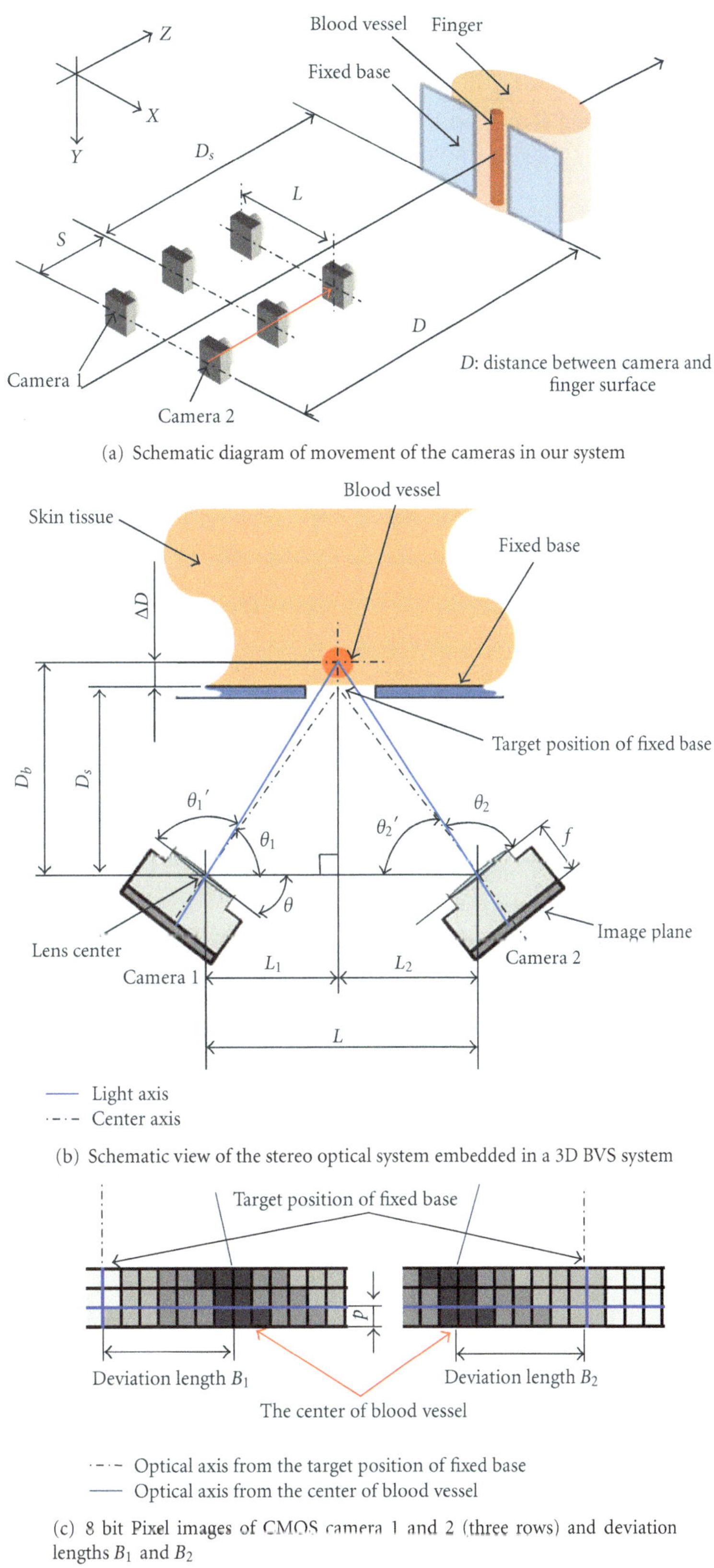

FIGURE 1: Schematic diagram of the autofocus and stereo hybrid methods embedded in a 3D BVS system.

with the original stereo method to establish a hybrid method. We set the allowed value of 100 μm for the targeted blood vessel, which had diameter of 500–1000 μm. An automatic blood sampling system for collection and transport of the blood for blood glucose measurements was connected with 3D BVS to produce the portable medical care system. For patients with diabetes, this system is expected to improve quality of life to a considerable degree.

As described in this paper, we first discuss the hybrid stereo-autofocus method for 3D BVS systems by introducing images processed with NIR light. Then we examine the accuracy of the 3D BVS system. Secondly, to evaluate our newly developed automatic blood sampling system for blood glucose measurement, we verify its capabilities of (1) accurate puncture, (2) stable blood transport, and (3) blood discharge on a glucose enzyme sensor. In light of those results, we assess the performance of this integrated portable medical care system.

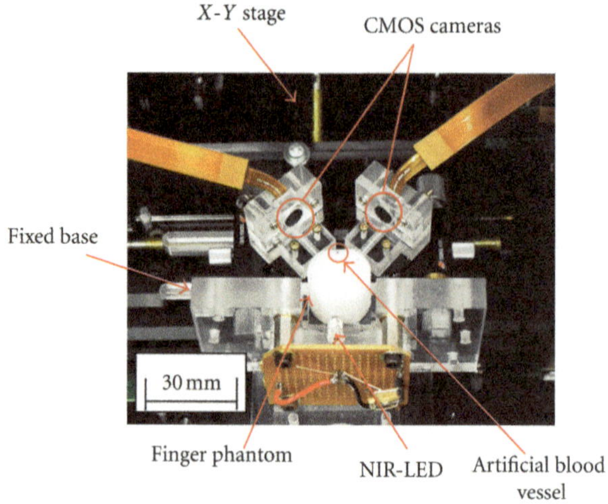

FIGURE 2: Photograph of the 3D BVS system.

2. Hybrid Stereo-Autofocus Method and Blood Vessel Visualization

2.1. Autofocus Method and Stereo Method. We used a hybrid stereo-autofocus method to detect the position of a blood vessel located between the first and second joints of the finger using transmitting NIR light. Our previous studies, which were done using a stereo method, did not determine the blood vessel depths accurately [12]. The measurement error was high for blood vessels located deep in the finger. Therefore, we adopted an autofocus method to obtain a very clear image of blood vessels using the template search method in the region of blood vessel [13]. This method improved the accuracy considerably. As presented in Figure 1(a), we took blood vessel images using different camera positions in relation to the finger based on the autofocus method. D means the length between the initial set position and the fixed base, and S the slide length as indicated in Figure 1(a). Fixed base means the upper surface of the glass plate, which is located at the position D_s and on which the finger was set. We evaluated the blood vessel image sharpness based on the template-matching algorithm and selected the clearest image at the camera position D_s as shown in Figure 1(a) [13]. In the auto focus procedure, we fix the incline angle θ as indicated in Figure 1(b).

Next, as shown in Figure 1(b), we measured blood vessel positions using the stereo method [12]. In Figure 1(b), L stands for the distance between the two CMOS cameras, f represents the focal length, D_b signifies the vertical distance between the center of the blood vessel and the line drawn between two center points of the lenses, and D_s denotes the vertical distance between the reference surface of the fixed base and the line drawn between respective center points of the lenses. In Figure 1(b), ΔD is the depth of the blood vessel center from the reference surface of fixed base. Figure 1(b) shows θ_1 and θ_2 as the angles of the light axes from the center of the blood vessel against the horizontal line between the respective centers of the two lenses. As shown in Figure 1(c), the center of the blood vessel can be detected from the luminance distribution of the blood vessel images obtained using the two CMOS cameras. Furthermore, θ_1 and θ_2 are calculable using trigonometric functions. The size of image plane pixel, 5.55 μm, was used to evaluate the blood vessel center position with the same length ratio as the pixel. Therefore, we obtain the deviation lengths B_1 and B_2 as indicated in Figure 1(c), which can be calculated as a number of deviated pixels times by pixel size 5.55 μm. Finally, D_b is expressed as

$$D_b = \frac{\tan\theta_1 \cdot \tan\theta_2}{\tan\theta_1 + \tan\theta_2} L, \tag{1}$$

here, $\theta_1 = \pi - \theta - \theta_1'$, $\theta_2 = \pi - \theta - \theta_2'$, $\theta_1' = \tan^{-1}(f/B_1)$, $\theta_2' = \tan^{-1}(f/B_2)$, $L = L_1 + L_2 = (D_b/\tan\theta_1) + (D_b/\tan\theta_{21})$.

Length D_s is already determined by using the autofocus method as shown in Figure 1(a). Therefore, the blood vessel depth (ΔD) can be calculated as

$$\Delta D = D_b - D_s. \tag{2}$$

2.2. 3D BVS System and Blood Vessel Visualization. We incorporated the hybrid stereo-autofocus components into a 3D BVS system, which included four NIR LED lights (peak wavelength of 870 nm, KED871M51A; Kyosemi Corp.) as the light sources, two CMOS cameras (PPV404C; Asahi Denshi Inc.), and two image-processing units (PP-DEB-007; Asahi Electronics Laboratory Co., Ltd.). The NIR LEDs were set behind the finger to irradiate the finger with NIR light. We used two CMOS cameras to take images using the NIR transmitted light. Details of the CMOS cameras were the following: 5.32 mm focal length, 8 × 8 × 5.55 mm size, and 5.55 × 5.55 μm pixel size. Images were of 8-bit digital grayscale values ranging from black (0) to white (255) for one pixel. The 640 × 480 pixel images were stored as JPEG image files. The images were processed by an image processing unit (AT401; Asahi Denshi Inc.) to calculate the blood vessel depth using the autofocus and stereo algorithms. Figure 2 presents a photograph of our 3D BVS system. The respective centers of CMOS cameras and LEDs are set on the same

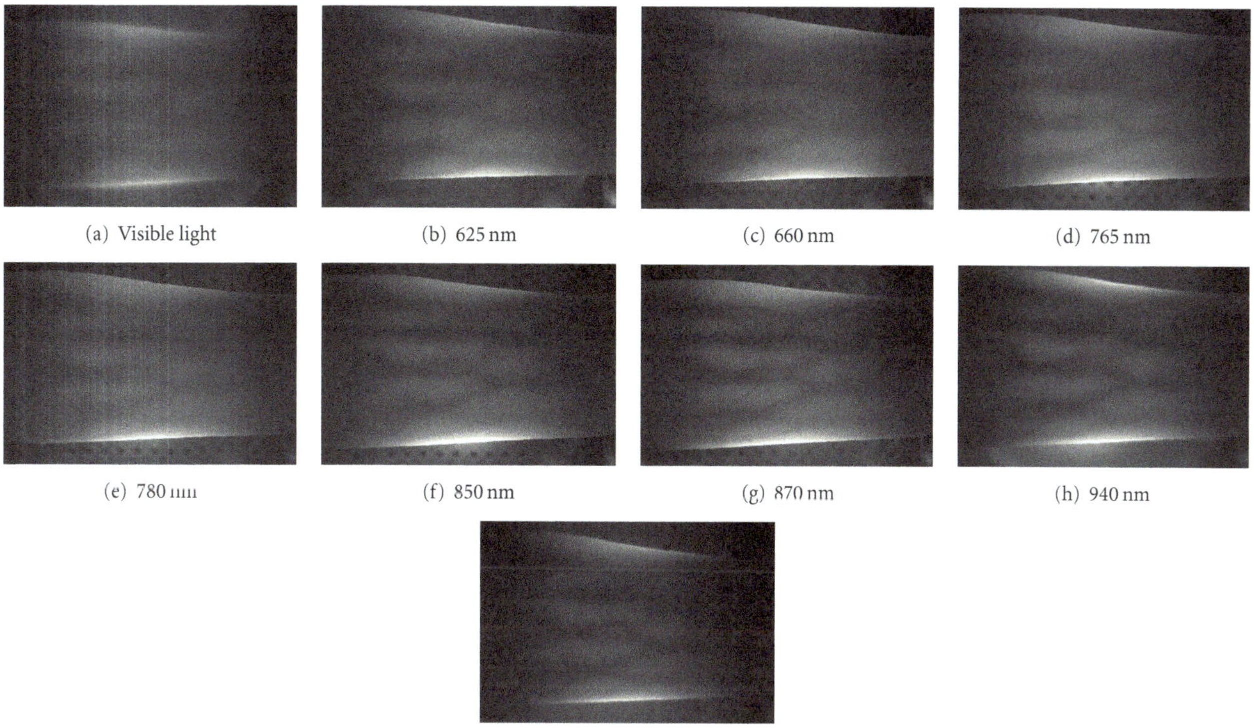

(a) Visible light (b) 625 nm (c) 660 nm (d) 765 nm
(e) 780 nm (f) 850 nm (g) 870 nm (h) 940 nm
(i) 950 nm

FIGURE 3: Blood vessel photos in cases of NIR lights, wavelengths 625–950 nm, and visible light.

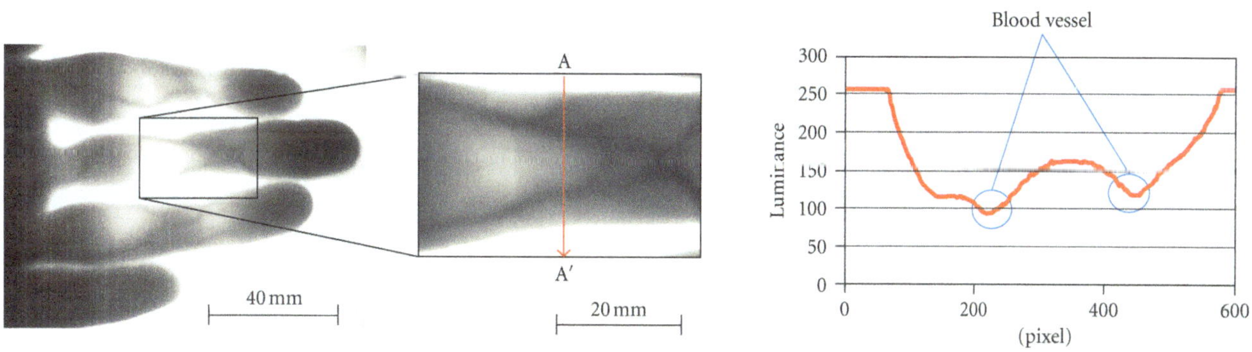

(a) Blood vessel image of the hand (b) Blood vessel image of the finger hand (c) Luminance distribution of image on line A-A′

FIGURE 4: Image of the blood vessel in the human finger. The NIR light wavelength is 870 nm.

horizontal plane. The length between the cameras (L value: Figure 1(b)) was set to 30 mm. The length between the object and the camera lens (D_s value: Figure 1) was set to 18.8 mm. The CMOS cameras were inclined at 33.5 deg using a rotating stage (KSP-256; Sigma Koki Co., Ltd.). The size of our 3D BVS device became 40 × 30 × 10 mm. Therefore, our system had size and features that allowed its installation into our new portable SMBG device. We used the transmitted NIR light to visualize blood vessels located in the human finger [14–16]. We selected small veins as our measurement targets and employed NIR light with wavelengths of 650–950 nm. The NIR light was transmitted through the finger and was recorded in images by the cameras in the positions presented in Figures 1(b) and 2. Figure 3 shows images obtained using

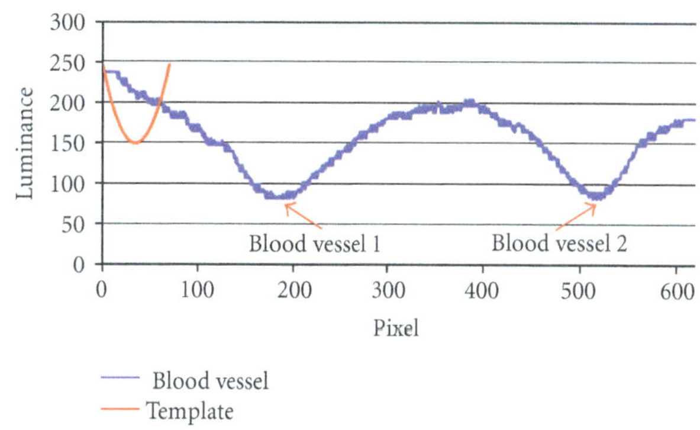

FIGURE 5: Luminance distribution of the image including the blood vessel and the template to search the blood vessels.

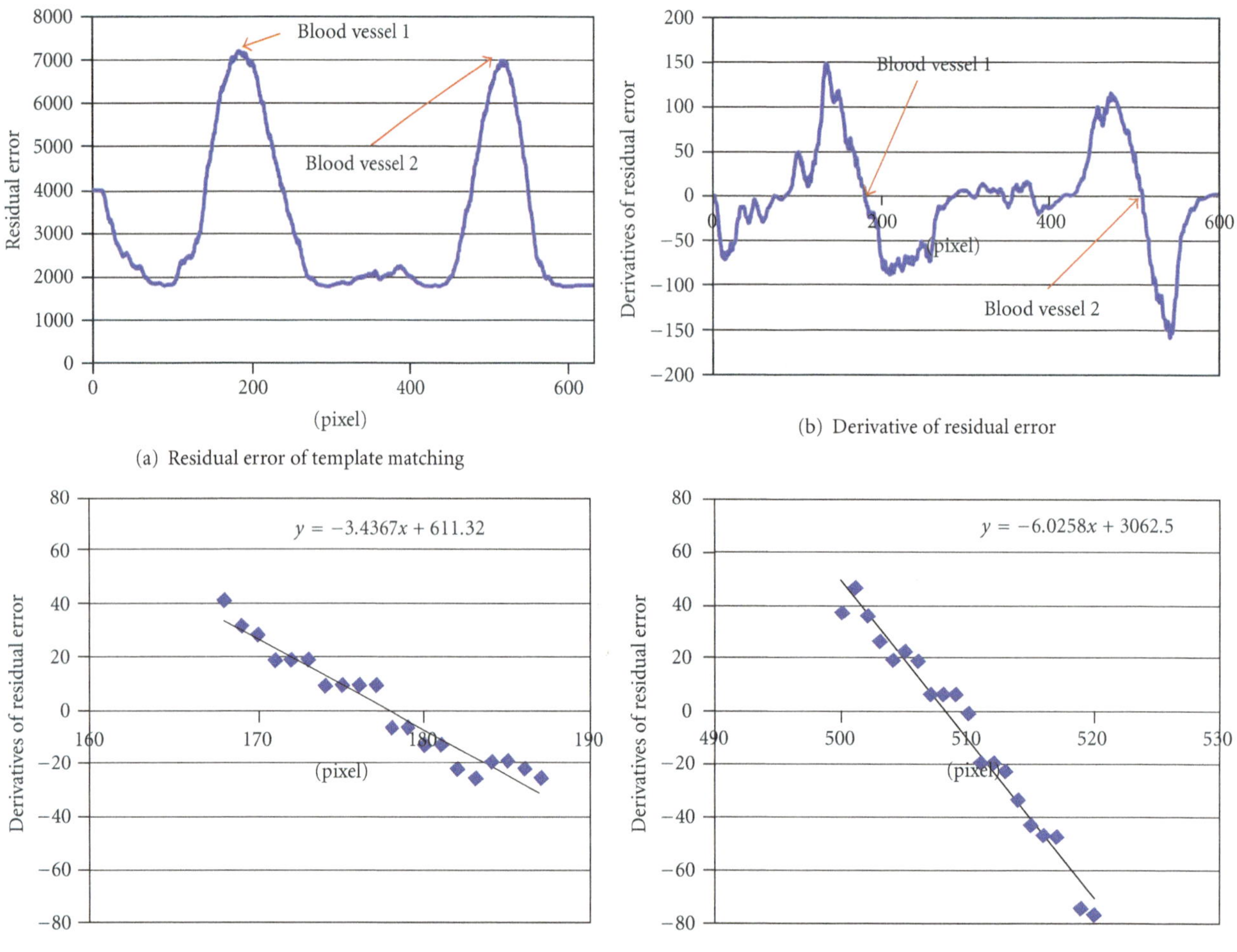

(a) Residual error of template matching

(b) Derivative of residual error

(c) Derivative of residual error around blood vessel 1

(d) Derivative of residual error around blood vessel 2

FIGURE 6: Residual errors calculated by template matching algorithm and its derivatives distributions to determine the sharpness.

the visible light and NIR light of the wavelengths shown in the figures. According to the steepest gradient search of the luminance distribution, especially in the close area to the blood vessel, finally we chose the wavelength of 870 nm [11, 13]. Figure 4 presents blood vessel images and the luminance distribution along line A–A' as indicated in Figure 4(b). We concluded that our imaging system can provide a clearer image of the blood vessel to determine its location in three dimensions.

Next we discuss the image processing associated with blood vessel detection. We adopted template matching to determine the image sharpness. The templates were produced on the assumption that the luminance distributions of blood vessels are convex-downward quadratic curves, as shown by the red curve in Figure 5, where the luminance distribution is portrayed as a blue line. Figures 6(a)–6(d) show results of template matching. Figure 6(a) presents the distribution of residual error obtained by template matching calculation. It shows two local maximum points, which correspond to the blood vessel position. Figure 6(b) portrays the distribution of the derivative of residual error. Figures 6(c) and 6(d) show the point at which the derivative value becomes zero, which is the center of the blood vessel. Moreover, we assumed the inclination of the diagram of derivative values to be the rate of luminance change. Therefore, we confirmed that the blood vessel can be detected clearly as the rate of luminance change. The sharpness can be evaluated using the angle of inclination of derivatives, as shown in Figures 6(c) and 6(d). Results show that we can select a clear image of a blood vessel taken by the camera located at distance D_s shown in Figure 1(a).

3. Evaluation of Blood Vessel Position Detection

We evaluated the accuracy of our 3D BVS system by carrying out experiments using finger phantoms containing artificial blood vessels. The photograph and the finger phantom size are depicted in Figure 7. Small veins of a human finger, with diameters of 0.5–1.0 mm, exist at depths of 0.5–2.0 mm from the surface. Therefore, we used the diameter of 0.75 mm for artificial blood vessels. We generated finger phantoms of seven types with artificial blood vessels embedded at different depths of 0.5–2.0 mm with 0.25 mm intervals. The finger phantoms were made of the polyacetal resin (TPS-POM; Toyo Plastic Seiko Co., Ltd.), which had optical

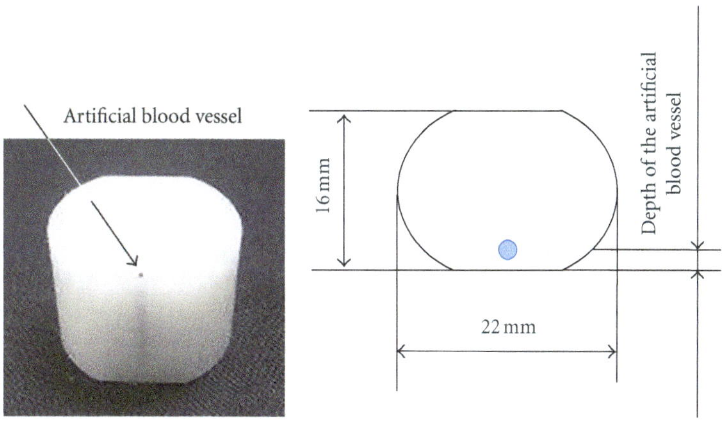

(a) Photograph of the finger (b) Size of finger phantom

Figure 7: Finger phantom image and its dimensions.

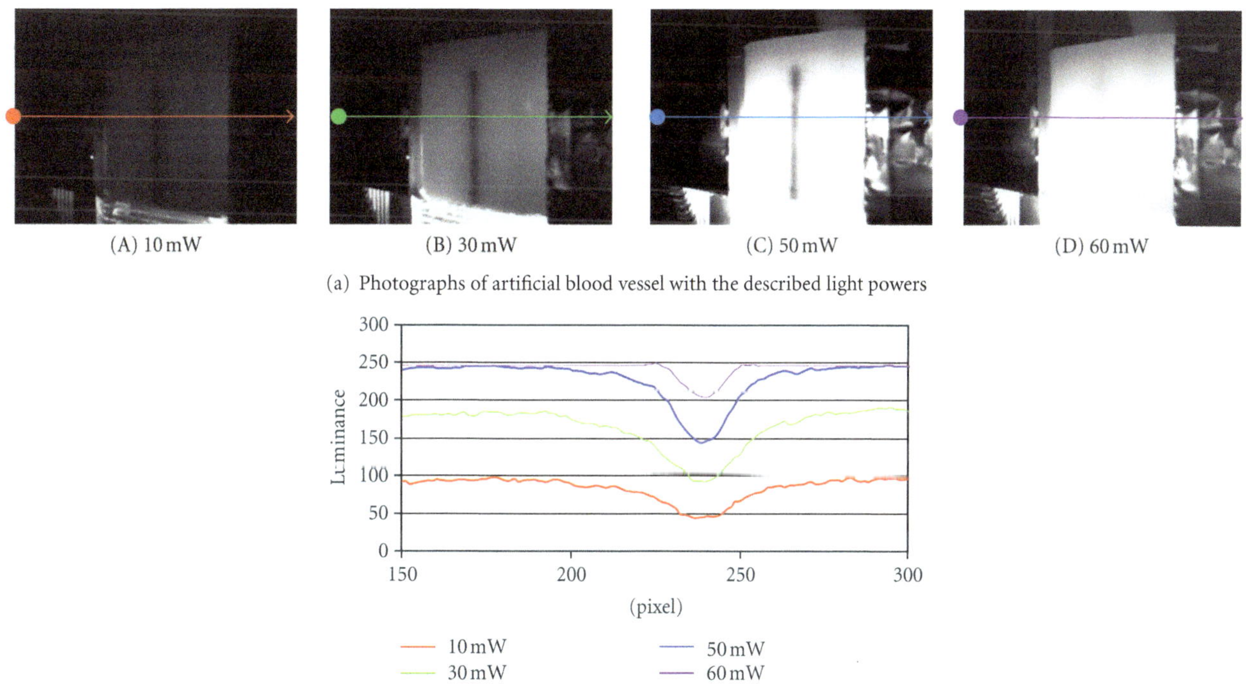

(a) Photographs of artificial blood vessel with the described light powers

(b) Luminance distributions of artificial blood vessel for the described input light powers

Figure 8: Influence of light power on the luminance distributions.

characteristics similar to those of human skin [17]. We included colored water in the artificial blood vessels.

We first increased the input power of NIR light to reduce the influence of scattering and thereby obtain a clear image. The scattering of NIR light reduces both the intensity of the transmitted light and the contrast between the artificial blood vessel and the tissue. We adopted four levels of input power of the NIR light: 10, 30, 50, and 60 mW. Figures 8(a) and 8(b) show photographs of blood vessels and luminance distributions of finger phantom images with the four input powers. Through comparison with the slope and concentration of the luminance distribution, we finally selected 50 mW to overcome the scattering and provide the clearest image of artificial blood vessels.

We measured the depths of the artificial blood vessels using our 3D BVS system, which is based on the hybrid stereo-autofocus method. To avoid the influence of light from the environment, we covered the whole system with a light-proof cloth. First, we analyzed the images and determined the vertical distance between the reference surface of the fixed base and the line drawn between respective center points of the lenses as D_s, presented in Figure 1(a), using autofocus methods. Secondly, we measured the position of the artificial blood vessel based on the stereo method.

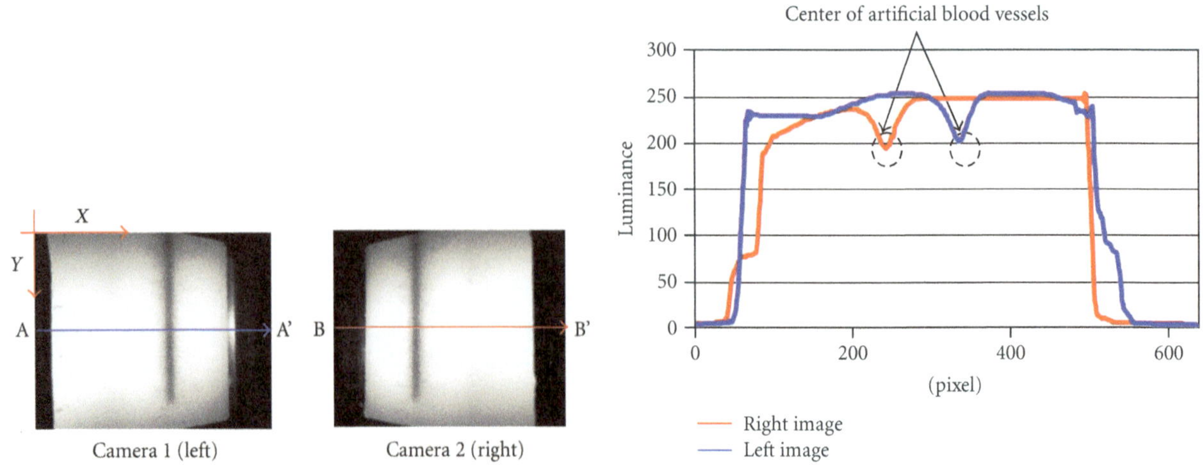

(a) Stereo images of the artificial blood vessel

(b) Luminance distributions on lines A-A' and B-B'

FIGURE 9: Stereo images and luminance distributions obtained using two cameras, wavelength of 870 nm.

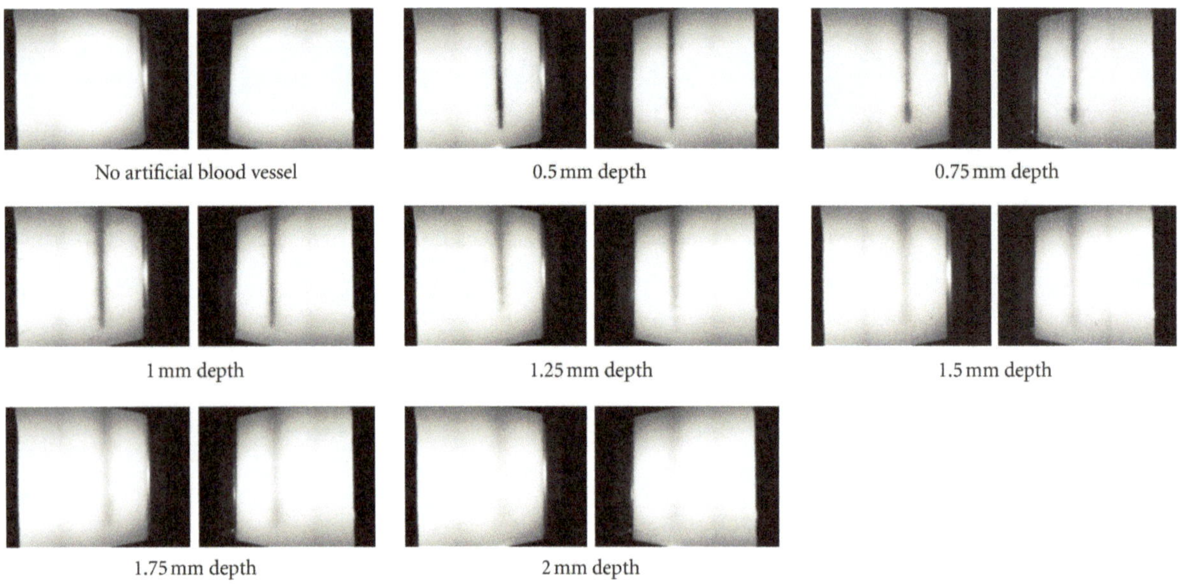

FIGURE 10: Stereo images of artificial blood vessels embedded in the finger phantom at the indicated depth.

Figure 9(a) shows stereo images of the artificial blood vessels. The rows of pixels on the center lines represented as A-A' and B-B', as portrayed in Figure 9(a), were used to determine the center and the blood vessel depth. Pixels located on the line (X, Y) from value (0, 240)–(640, 240) provide luminance according to the eight-byte digital number, where X-Y coordinates were defined as portrayed in Figure 9(a). Figure 9(b) shows that a rapid decrease in brightness is observed at the area of the blood vessel. Therefore, we are able to determine the artificial blood vessel position. The center of the artificial blood vessel position of camera 1 was set to the 327th pixel and that of camera 2 to the 316th pixel. According to Figures 1(b) and 1(c), the deviations between the center position of blood vessel and the fixed base plate in the image planes of CMOS camera 1 and 2 were 7 pixels and 4 pixels, respectively. Because the pixel size was 5.55 μm, the actual deviation lengths were, respectively, 64.8 μm in camera 1 and 93.6 μm in camera 2. Next, the angles of light axes, θ_1 and θ_2, as indicated in Figure 1(b), were obtained using these deviation length B_1 and B_2 and focal length f as shown in Figure 1(c). We then calculated the blood vessel depth ΔD using (1) and (2). Figure 10 shows stereo images of the artificial blood vessels embedded at the indicated depths, where the left image was taken by camera 1 and the right image by camera 2,

$$\theta = \sin^{-1}\left(\frac{\sin\theta}{n}\right),$$
$$\phi = \tan^{-1}\left(\frac{f}{s}\right), \qquad (3)$$
$$d = d'\frac{\tan\theta}{\tan\phi}.$$

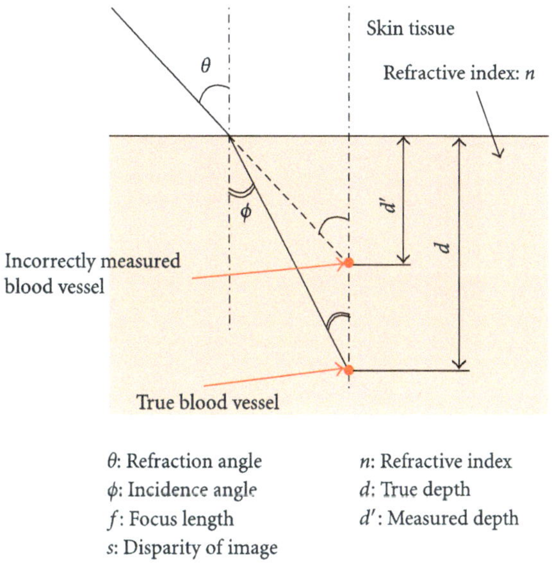

FIGURE 11: Refraction of light in the skin tissue and depth correction.

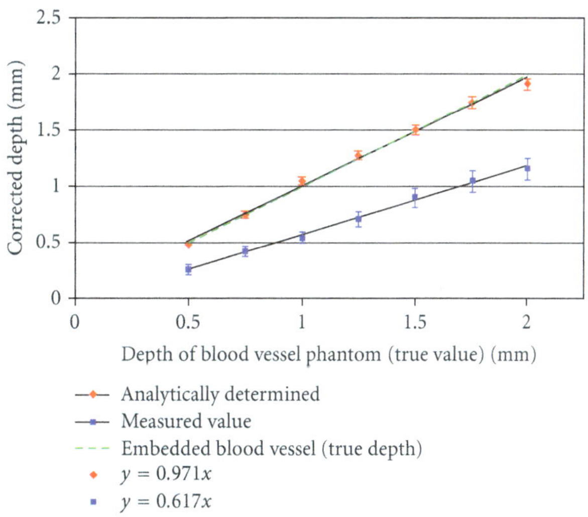

FIGURE 12: Relation between the estimated depths and true depth of artificial blood vessel.

Next, to assess the position detection accuracy of this 3D BVS detection device, we strove to eliminate the influence of the refraction of NIR light in the visualization procedure. Because the refractive index of polyacetal differs from that of air, NIR light is refracted at the interface and the difference occurs between the blood vessel position of the image and that of the actual value. Therefore, we corrected the result of the measurements based on Snell's law using the refractive index of the polyacetal as 1.48. Figure 11 shows the trajectory of NIR light from the finger phantom to the camera. Figure 11 shows that the measured depth of the blood vessel becomes smaller than the actual depth because of the refraction of NIR light. Formulas for the incidence angle, refraction angle, and the blood vessel depth are shown in (3). Using these values, we corrected the measured depth obtained using (1) and (2). Figure 12 shows the relation between the corrected depth and the actual one, which we named "true" depth. We calculated the standard deviations in seven cases of "true" depths. In each "true" depth, we measured more than ten times and calculated the standard deviation by using the differences between the "true" depth and measured depths. It shows that the deeper the depth, the larger the standard deviation. The maximum standard deviation was 110 μm, and the average value of standard deviations was 63 μm. The average of all deviations was 29.8 μm, which means that average value of differences between measured and true depths in seven cases. Good correlation ($R^2 = 0.946$) was found between the corrected depths and the actual depths. These errors were less than the allowance value of 150 μm for the targeted blood vessel with diameter greater than 500 μm.

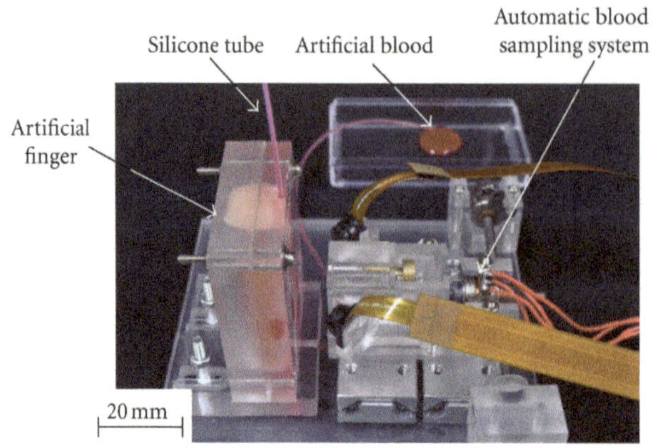

FIGURE 13: Automatic blood sampling system equipped with artificial finger and blood vessel.

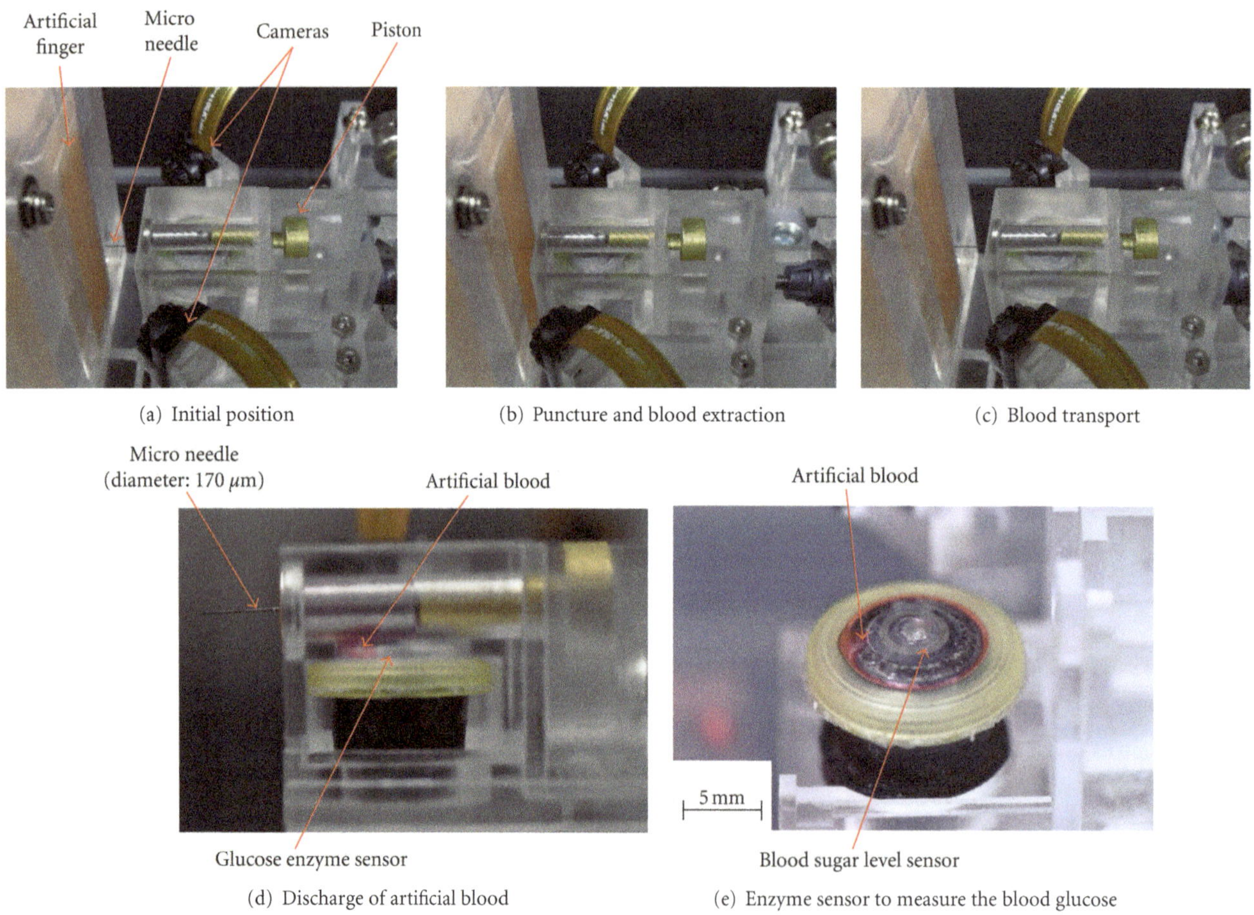

FIGURE 14: Photos of automatic blood sampling procedure and the enzyme sensor used to measure the blood glucose.

4. Automatic Blood Sampling System

We fabricated an automatic blood sampling system that had been designed to be built in a portable SMBG device for use in conjunction with the 3D BVS system. In addition to two small stepping motors, we used a glucose enzyme sensor made by Horiba, Ltd. for accurate and reliable blood glucose measurement purposes [18], which was expected to perform accurate and quick blood collection. Figure 13 shows a prototype of the newly developed automatic blood sampling system, with size of 47 × 30 × 20 mm. Regarding performance evaluation, we used an artificial finger made of silicone rubber (RTV2 VP7550; Asahi Chemical Ind. Co. Ltd.), which had similar elastic properties to those of human skin. The artificial blood vessel diameter was set as 1.0 mm; the blood vessel was embedded to 1.0 mm depth in the artificial finger. We used a 50% glycerin solution, which had similar viscosity to that of human blood, as artificial blood.

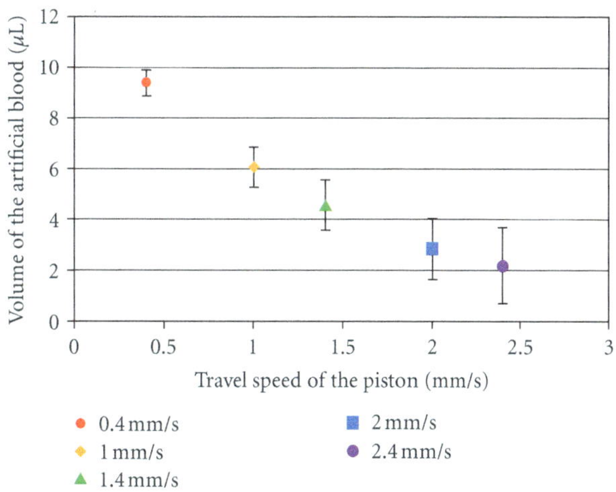

FIGURE 15: Relation between the volume of artificial blood on the sensor and the piston travel speed (±SD).

This artificial blood was kept in the artificial blood vessel. We examined the capabilities for puncture, blood extraction, transport, and discharge on the enzyme sensor. The glucose enzyme sensor required at least 5.0 μL of blood for blood glucose measurements. Figures 14(a)–14(e) show four stages of blood sampling procedures, such as puncture, extraction, transport, discharge, and measurement. We analyzed blood sampling and measurement capabilities to provide 5.0 μL of blood on the enzyme sensor. Figure 15 presents the relation between the artificial blood volume on the enzyme sensor and the piston speed for blood transport and discharge. As the piston speed decreases, the amount of artificial blood increases. We concluded that an adequate piston speed to provide over 5.0 μL of blood for measurement was 1.0 mm/s.

5. Conclusions

We developed a 3D BVS system and an automatic blood sampling system to fabricate a portable SMBG device. We adopted the hybrid stereo-autofocus method to detect the position of blood vessels in three dimensions. We also examined the performance of our automatic blood sampling system, which features puncture, blood extraction, transport, and discharge for an enzyme sensor to measure the blood glucose. Results of the evaluation of our 3D BVS and automatic blood sampling devices are summarized as follows. (1) Accuracy evaluation of 3D BVS system was confirmed. The maximum standard deviation was 111 μm and the average of the standard deviation was 63 μm when detecting 0.5 mm-diameter blood vessels embedded at depths of 0.5–2.5 mm with 0.25 mm intervals. Results show very good correlation with actual measurements: $R^2 = 0.946$. (2) Blood sampling performance evaluation was done with the piston speed for blood suction and discharge selected as 1.0 mm/s to provide 0.5 μL artificial blood on the glucose enzyme sensor. That amount was sufficient for measuring the blood glucose concentration. Our integrated system, combined with 3D BVS and automatic blood sampling systems, can be used effectively as a portable medical care system.

References

[1] Matsushita Co. Ltd., "Blood sugar level measurement system," *Matsushita Technical Journal*, vol. 51, no. 3, pp. 248–253, 2005 (Japanese).

[2] M. Miyazaki, "Reliability related to small size and simple blood sugar weighing device and the usages," *Journal of Clinical Laboratory Instruments and Reagents*, vol. 28, no. 4, pp. 291–304, 2005 (Japanese).

[3] N. Ozeki, Y. Kamata, Y. Okuzumi et al., "Examination related to measurement accuracy of small electrode type blood sugar measuring instrument Antisense β," *Journal of Medicine and Pharmaceutical Science*, vol. 53, no. 3, pp. 361–370, 2005 (Japanese).

[4] K. Jungheim and T. Koschinsky, "Risky delay of hypoglycemia detection by glucose monitoring at the arm," *Diabetes Care*, vol. 24, no. 7, pp. 1303–1306, 2001.

[5] K. Fujita, K. Tamura, K. Wataru, H. Ishizawa, and E. Toba, "Development of noninvasive blood glucose sensor using the infrared spectroscopy," *Transactions of the Institute of Electrical Engineers of Japan C*, vol. 129, no. 9, pp. 1759–1765, 2004 (Japanese).

[6] T. Karatsu, D. Iwata, T. Nakajima et al., "Determination of the glucose in serum by the FT-IR-ATR method," *Japan Analyst*, vol. 54, no. 2, pp. 149–154, 2005.

[7] E. Nakamachi, S. Jinninn, Y. Uetsuji, K. Tsuchiya, and H. Yamamoto, "Sputter generating and characterization of a titanium alloy microneedle for applying to Bio-MEM," *Transactions of the Japan Society of Mechanical Engineers A*, vol. 72, no. 716, pp. 471–477, 2006.

[8] T. Asakura and H. Seino, "Basic study of a micro-tapered needle (TN-3305) used for insulin pre-filled products," *Japanese Journal of Pharmaceutical Health Care and Sciences*, vol. 30, no. 6, pp. 368–376, 2004 (Japanese).

[9] H. A. MacKenzie, H. S. Ashton, S. Spiers et al., "Advances in photoacoustic noninvasive glucose testing," *Clinical Chemistry*, vol. 45, no. 9, pp. 1587–1595, 1999.

[10] K. Yamakoshi, S. Ishimaru, A. Nakabayashi et al., "Development of a compact device for self-monitoring of venous blood glucose using miniature needle and infra-red CCD monitor," *Transactions of the Japanese Society for Medical and Biological Engineering*, vol. 42, p. 198, 2003 (Japanese).

[11] Y. Horiike, Y. Ogawa, M. Niihashi, A. Oki, M. Takai, and M. Nagai, "Development of a health-care-chip," *Chemical Devices*, pp. 81–85, 2004.

[12] H. Kandani, Y. Uetsuji, K. Tsuchiya, and E. Nakamachi, "Development of blood vessel searching system for HMS," in *Proceedings of the SPIE Optics and Photonics*, vol. 7055, SanDiego, Calif, USA, 2008.

[13] E. Nakamachi, T. Kuroda, K. Tsuchiya, Y. Uetsuji, and T. Uenoya, "Development of a three-dimensional blood vessel searching system by using the near infra-red light," *Transactions of the Japan Society of Mechanical Engineers C*, vol. 73, no. 731, pp. 2088–2094, 2007.

[14] M. Tamura, M. Kaneko, K. Shimizu, K. Yamamoto, and T. Mikami, "Basic study for visualization of blood vessels using near-infrared light," Tech. Rep. MBE89, IEICE, 1989.

[15] M. Kono, H. Ueki, and S. Umemura, "A new method for the identification of individuals by using of vein pattern matching of a finger," in *Proceedings of the 5th Symposium on Pattern Measurement*, pp. 9–12, Yamaguchi, Japan, 2000.

[16] E. Nakamachi, S. Jinninn, Y. Uetsuji, K. Tsuchiya, and H. Yamamoto, "Sputter generating and characterization of a titanium alloy microneedle for applying to Bio-MEM," *Transactions of the Japan Society of Mechanical Engineers A*, vol. 72, no. 716, pp. 471–477, 2006.

[17] N. Furuya, "3D measuring system for assembly robot using stereo camera. Camera model and parameter calibration by geometrical method," *Journal of the Japan Society for Precision Engineering*, vol. 65, no. 8, pp. 1116–1120, 1999.

[18] E. Nakamachi, "Development of automatic operated blood sampling system for portable type self-monitoring blood glucose device," in *Proceedings of the 32nd Annual International Conference of the IEEE Engineering in Medicine and Biology Society (EMBC '10)*, pp. 335–338, September 2010.

18

Effect of Fluorescent Particle Size on the Modulation Efficiency of Ultrasound-Modulated Fluorescence

Yuan Liu,[1,2] Baohong Yuan,[1,2] and Joseph Vignola[3]

[1] *Department of Bioengineering, The University of Texas at Arlington, Arlington, TX 76019, USA*
[2] *Joint Biomedical Engineering Program, The University of Texas at Arlington, and The University of Texas Southwestern Medical Center at Dallas, TX 75390, USA*
[3] *Department of Mechanical Engineering, The Catholic University of America, Washington, DC 20064, USA*

Correspondence should be addressed to Baohong Yuan, baohong@uta.edu

Academic Editor: Nanguang Chen

To investigate whether the size of fluorescent particles affects the modulation efficiency of ultrasound-modulated fluorescence (UMF), we measured UMF and DC (direct current) signals of the fluorescence emission from four different-sized fluorescent particles: (1) three carboxylate-modified fluorescent microspheres (FM) with diameters of 20 nm, 200 nm, and 1.0 μm and (2) streptavidin-conjugated Alexa Fluor 647 with a diameter of approximately 5 nm. The UMF and DC signals were simultaneously measured using a broadband lock-in amplifier and a narrowband amplifier, respectively. The ratio of the UMF strength to the DC signal strength is defined as the modulation efficiency. This modulation efficiency was then used to evaluate the effects of fluorophore size and concentration. Results show that the modulation efficiency was improved by approximately a factor of two when the size of the fluorescent particles is increased from 5 nm to 1 μm. In addition, the linear relationship between the UMF strength and ultrasound pressure (observed in our previous study) was maintained regardless of the fluorescent particle sizes.

1. Introduction

Tissue fluorescence imaging has been well developed and widely used because of its high sensitivity and specificity [1, 2]. Fluorescence techniques can provide unique tissue physiological information when compared with other noninvasive imaging modalities (ultrasound, magnetic resonance imaging, computed tomography, etc.) and are sensitive to tissue microenvironments, such as tissue pH, temperature, and gas/ion concentrations. Also, they are relatively cost efficient, flexible in imaging probes selection (from organic dyes, to quantum dots, and to nanoparticles or microparticles), highly sensitive to imaging probes (fM-nM, 10^{-15}–10^{-9} mole/liter), and nonionizing radiative [1–3]. Commonly used high-resolution fluorescence microscopy faces a fundamental challenge due to tissue's strong optical scattering, which typically limits penetration depth to a few hundred micrometers [3]. Techniques used to image deep tissue at ranges of millimeters or centimeters, such as fluorescence diffuse optical tomography (FDOT) [4], take advantage of diffused photons that have been scattered many times before being detected. These diffused photons can penetrate biological tissue up to tens of millimeters at the red or near infrared (NIR) region [3] at the expense of spatial resolution (limited to ~1–5 mm) [4].

Ultrasound-modulated fluorescence (UMF) has been proposed to increase spatial resolution while maintaining imaging depth [5]. This is possible because ultrasonic scattering is two to three orders of magnitude less than optical scattering in biological tissues [3]. The concept of UMF is similar to ultrasound-modulated optical tomography (UOT) [6–8], although their mechanisms may be different [9, 10]. In UMF, a focused ultrasound beam is used to tag the diffused fluorescence photons in the focal zone. By detecting and analyzing the tagged (or modulated) fluorescence photons, one can quantify the fluorescence properties in the ultrasound focal zone with ultrasonic spatial resolution that is usually much higher than the resolution of FDOT [11–15]. Recently, several research groups have investigated UMF both theoretically [5, 10] and experimentally [11–15]. The modulation efficiency of UMF is limited and has to be significantly improved for applications in biomedical imaging [11–15].

Kobayashi et al. reported a modulation efficiency in UMF adequate for tomographic imaging in a scattering medium using relatively large fluorescent microspheres (~ microns in diameters) [11]. However, their results seem to disagree with our recently reported experimental data collected using relatively small organic fluorescent dyes (nanometers in diameters) [12, 13] and that of Hall et al. [14, 15] (see the detailed discussion about the disagreements in [12]). Therefore, the question of whether relatively large fluorescent particles (~microns) can reliably and significantly improve the modulation efficiency of UMF is of interest.

The study of the effect of particle size on the modulation efficiency can also lead to a better understanding of the underlying UMF mechanisms. The modulation mechanisms of UMF are quite different from those of UOT [10] due to the intrinsic incoherent properties of fluorescence emission in tissue. While UOT has been well studied using coherent light [9], the modulation mechanisms of UMF are still poorly understood due to inconsistent data in literature [11–15]. Previously, we have proposed several possible mechanisms for UMF [10]. The most plausible one is that the UMF signal is caused by periodic oscillations of the fluorophore concentration that are generated by ultrasonic pressure wave [10]. We have reported a supporting evidence of this mechanism based on a recent experimental data [12, 13]. Although this mechanism can only provide a relatively small modulation efficiency (typically 10^{-4}–10^{-6}) in a scattering medium, it exists in all types of fluorophores from small nanoscale fluorescent molecules to relatively large nano-/microscale fluorescent particles [10, 12, 13]. Intuitively, one might think that it is easier to modulate large fluorescent microparticles with an ultrasound wave because larger particles may be ultrasonically oscillated more efficiently; besides, each microparticle can encapsulate more fluorescent molecules. Potentially, this could provide a means to improve the modulation efficiency of UMF. If this hypothesis is proved true, the next question would be how much improvement can be achieved and is it sufficient for biomedical imaging applications? To address these basic questions, UMF modulation efficiencies for four communally available fluorescing particles were measured and discussed in this study.

2. Methods

Our previous studies have shown that the UMF signal is difficult to quantify in a scattering medium due to its poor signal-to-noise ratio [13]. As a result, water was used as the medium in this study. Three fluorescent microspheres with diameters of 20 nm, 200 nm, and 1.0 μm were tested in this study. To further reduce the particle size, Alex Fluor 647 conjugated with streptavidin (~1 nm in diameter for the dye molecule and ~5 nm in diameter for the streptavidin [16, 17]) was selected as a sample in comparison. The UMF signal was measured using a broadband lock-in amplifier, and simultaneously the fluorescent DC signal was measured using a narrowband amplifier. With both measurements calibrated, the ratio of these two quantities is recognized as the modulation efficiency. A consistent experimental setup was used for all the samples in the study so that results for different samples could be compared. The modulation efficiency is evaluated as a function of the fluorescent particle size.

The experimental setup is shown in Figure 1. The excitation light was produced by a fiber-coupled diode laser (57ICS064/SP/HS, Melles Griot) with a wavelength of 638 nm. The monochromatic light is guided via a polarization maintaining fiber (57FTP602/FC/1.0 Melles Griot, core diameter: 4 μm) to a collimating lens (F240FC-B, Thorlabs). The power density of the laser was controlled as 411 mW/cm^2. A silicon tube (FT, Instech Laboratories, inner diameter: 0.79 mm, outer diameter: 2.39 mm) was positioned in a water-filled container (40 (x) $\times$ 120 (y) $\times$ 120 (z) mm) and orientated parallel to the y-axis. An ultrasound transducer (UST, V314-SU-F-1.00-IN-PTF, Olympus NDT, center frequency: 1 MHz) was focused on the intersection of the laser beam and the FT (see Figure 1). The UST was driven by a continuous sinusoidal signal with a frequency of 1.0 MHz that was generated from a function generator (FG, Agilent 33120A, Agilent Tech, Calif, USA) and amplified by a power amplifier (PA, 7600 M, Krohn-Hite, Mass, USA). The peak pressure at the transducer focus was measured by a hydrophone (HNP-0200, Onda Corporation, Calif, USA) and preamplifier (AH2010, Onda Corporation, Calif, USA). Two lenses (L1 and L2, AC254-030-A1, Thorlabs) were used to collect the emitted fluorescence. Long-pass emission filters (EmF, BLP01-635R, Semrock, cutoff wavelength: 650 nm) were employed to reject the excitation light. The emission photons were focused on two adjacent channels of a 16-channel photomultiplier tube (PMT, R5900U-01-L16, Hamamatsu, Japan) at the central area of this PMT to monitor the UMF and the fluorescence DC signals. The photocurrent from one channel was converted to a voltage signal using a transimpedance amplifier (TIA, 313A-1-1 pF, Analog Modules, Longwood, Florida), and its output was directed to a broadband lock-in amplifier (LIA, SR844, Stanford Research Systems, Calif, USA). The synchronized TTL (transistor-transistor logic) signal of the function generator was used as the reference signal for the LIA. The output of the LIA represents the UMF strength. The photocurrent from the second channel was directly connected to an oscilloscope (OS, TDS 3032B, Tektronix, Oregon). The input impedance of the oscilloscope was selected to be 1 Mega Ohms so that an appropriate gain can be obtained with a limited bandwidth. This signal was used to record the DC level of the fluorescence. Four sets of measurements were conducted for each sample using a new solution and tube. In each set of measurements, the UST was scanned for 3 times along the x-axis using a linear translation stage (PT1, Thorlabs, NJ, USA), and the DC and UMF signals were recorded at each selected position of the UST. The average and the standard derivation were calculated and displayed based on all the measurements.

Before any measurements were conducted, the tube was ultrasonically located by connecting the UST to a pulse generator/receiver (5077PR, Olympus NDT, Waltham, Mass, USA). The peak-to-peak voltage of the ultrasonic echoes were recorded at selected positions when scanning the UST along the x-axis. The background voltage, which was mainly caused by the electronic interference from the UST driven

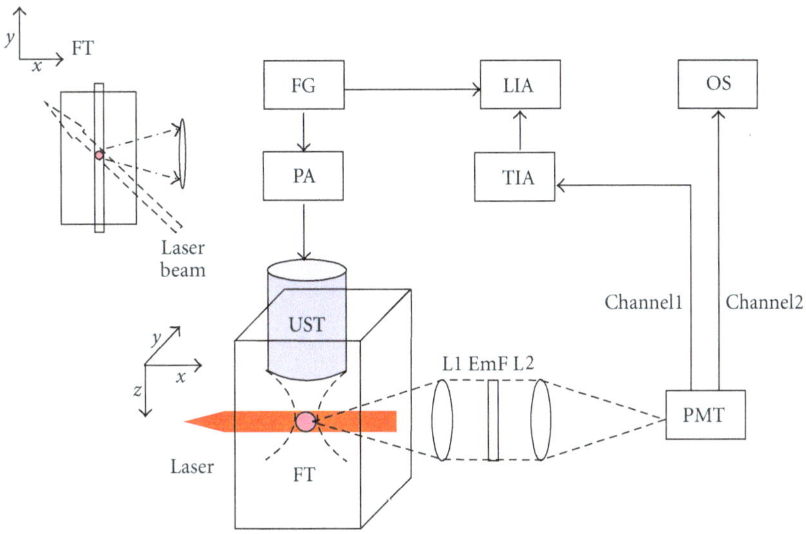

FIGURE 1: Schematic of the measurement system. FG: function generator; PA: power amplifier; UST: ultrasound transducer; FT: fluorescent tube; L1–L2: two lenses; EmF: emission filters; PMT: photomultiplier tube; TIA: transimpedance amplifier; LIA: lock-in amplifier; OS: oscilloscope. The inset represents the top view of the container and FT.

signal, was recorded when the excitation laser beam was blocked. This background voltage is independent of the UST position and can be subtracted from the measured UMF signals.

The UMF signal and the fluorescent DC signal were measured as a function of the fluorophore concentration. The concentrations of the Alexa Fluor 647 solutions (S21374, Invitrogen) were 0.025, 0.05, 0.1, and 0.2 mg/mL, and the concentrations for the fluorescent microsphere solutions (FM, F8782, F8806, F8816, FluoSpheres, Invitrogen) were 0.5, 1.0, 2.0, and 4.0 mg/mL. New clear tubes were used for each fluorescent sample to avoid the contamination from the previous sample. The effect of the tube autofluorescence is undetectable at the 638 nm excitation wavelength and can be ignored. This has been verified by detecting the UMF signal when filling the tube with water. Our results showed that the detected signal is almost a constant and independent of the location of the ultrasound transducer, which implies that no UMF signal can be detected when no fluorescent solution is present. Ideally, the linear fits of the experimental data should pass through the origin, implying that both the UMF and the fluorescence DC signals should be zero when the concentration is zero. In practice, this is not always true because of the existence of background interference and/or instrument bias. Therefore, the intercepts of the linearly fitted data were subtracted during the data processing. The ratio of the intercept-subtracted UMF signal to the intercept-subtracted DC signal is defined as the modulation efficiency. The absolute value of the modulation efficiency (the ratio) may not represent the true physical meaning of the modulation depth because the gains of the two channels may not be the same. However, because the measurement system and all experimental parameters were maintained the same for all the samples, the relative values of the defined modulation efficiency for all the samples can be compared.

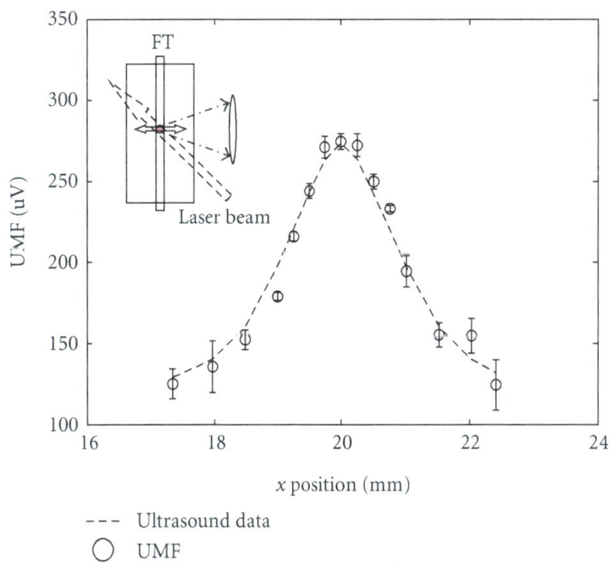

FIGURE 2: The circles with error bars represent the measured UMF signals from one of the fluorescent microsphere solutions (with the diameter of 200 nm and the concentration of 1.0 mg/mL) as a function of the x position of the UST. The dashed line shows the rescaled ultrasound signal. The inset represents the measurement configuration, and the doubled arrow shows the UST scan line. $P_{\mathrm{fspp}} = 180$ KPa.

3. Results and Discussion

Figure 2 displays the measured UMF signals (the circles with error bars) from the tube filled with 200 nm diameter FM solution (1.0 mg/mL) after scanning the UST along the x-axis. The inset shows the measurement configuration. The ultrasound data was linearly rescaled as to display both the ultrasound data and the UMF data in the same figure. The origin of the x-axis is defined as the left edge of the

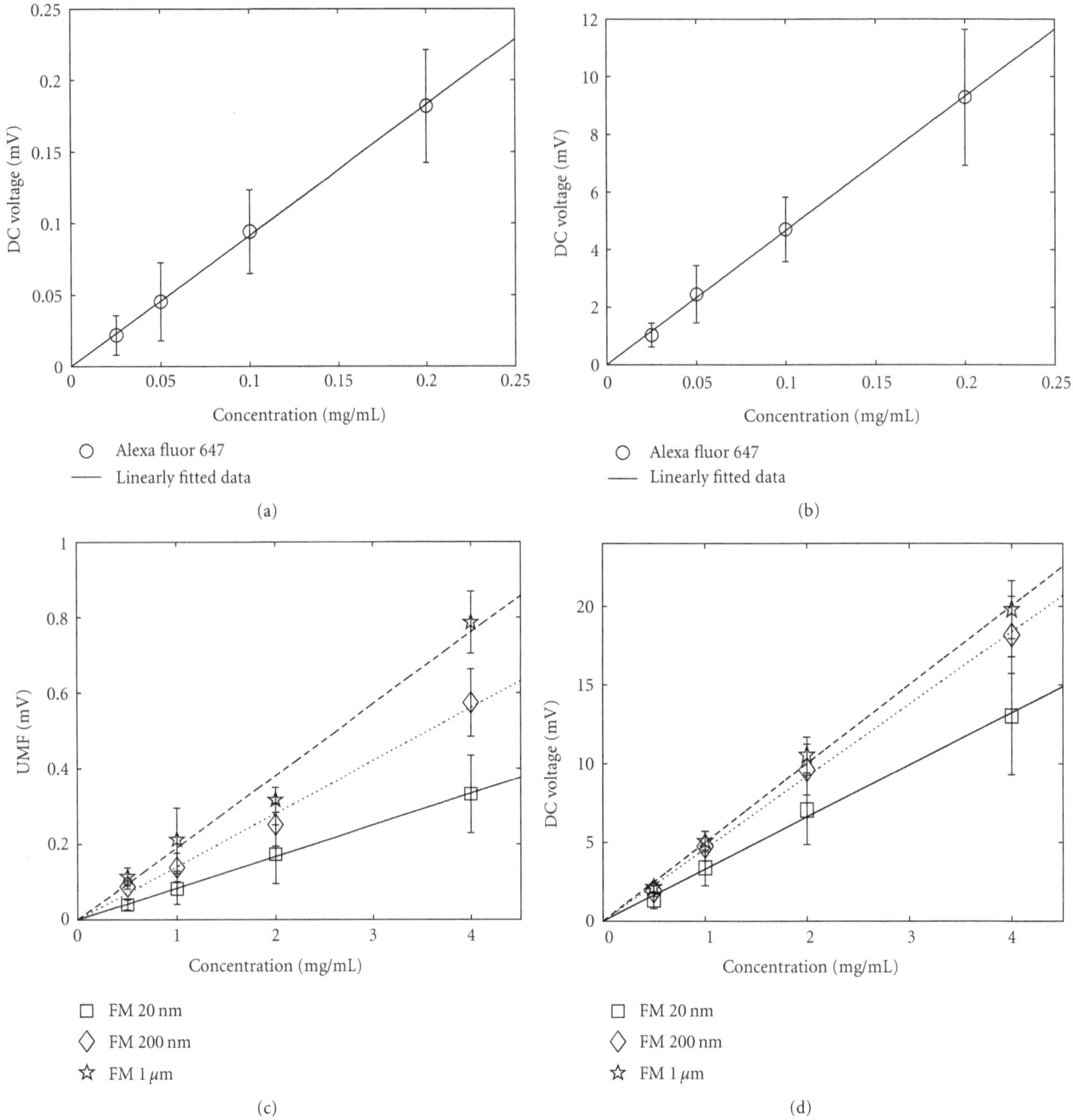

FIGURE 3: UMF and DC signals as a function of fluorophore concentration when $x = 20\,\text{mm}$, $y = 0\,\text{mm}$, and $P_{\text{fspp}} = 180\,\text{KPa}$. (a) and (b) show the UMF signal and DC signal of streptavidin-conjugated Alexa Fluor 647, respectively. The circles with error bars represent the experimentally measured data, and the solid line is the linearly fitted data. (c) and (d) show the UMF signal and DC signals from the fluorescent microsphere solutions, respectively. The squares, diamonds, and stars represent the experimentally measured data. The solid, dotted, and dashed lines are the linearly fitted data.

tank, and the tube is located at the center of the tank ($x = 20$ and $y = 0\,\text{mm}$). The peak pressure applied at the transducer focus (P_{fspp}) is 180 KPa. Figure 2 indicates a correlation between the UMF signal and the ultrasound. The modulation efficiency of different fluorophores is based on the peak values of the UMF signals.

Figure 3 shows the peak values of the measured UMF and DC signals from the four fluorescent particle solutions as a function of fluorophore concentration when $P_{\text{fspp}} = 180\,\text{KPa}$. Figures 3(a) and 3(b) show the UMF signal and the DC signal from the streptavidin-conjugated Alexa Fluor 647, respectively. The circles with error bars represent the experimentally measured data, and the solid lines are the linearly fitted data. Similarly, Figures 3(c) and 3(d) show the UMF signal and the DC signals from the three fluorescent microsphere solutions, respectively. The squares, diamonds, and stars represent the experimentally measured data. The solid, dotted, and dashed lines represent linear fits

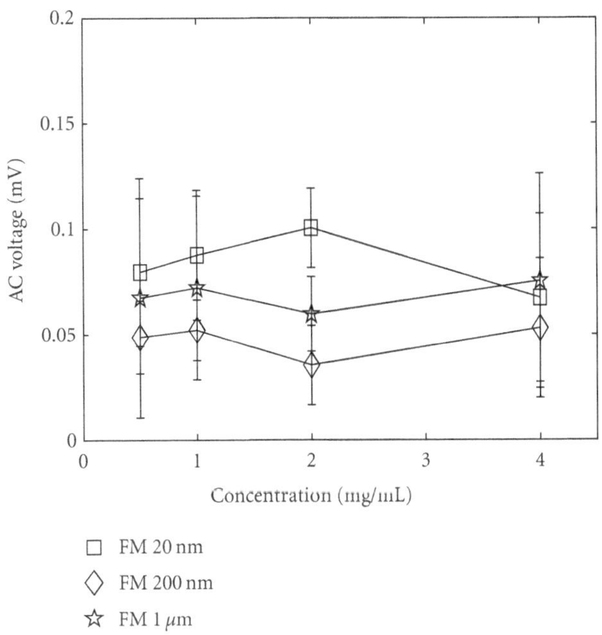

Figure 4: Ultrasound-modulated signal (AC voltage) of the scattered excitation light from the microsphere solutions as a function of concentration, when $x = 20\,\text{mm}$, $y = 0\,\text{mm}$, and $P_{\text{fspp}} = 180\,\text{KPa}$. The squares, diamonds, and stars represent the experimentally measured data.

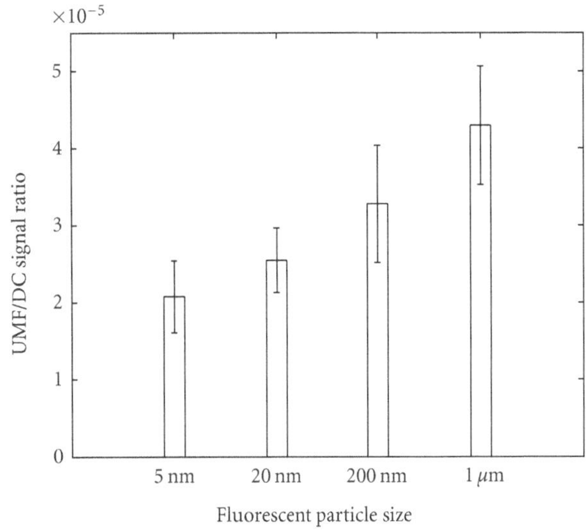

Figure 5: Modulation efficiencies of the four types of samples with diameters of 5 nm (as an approximation to streptavidin-conjugated Alexa Fluor 647), 20 nm, 200 nm, and 1.0 μm.

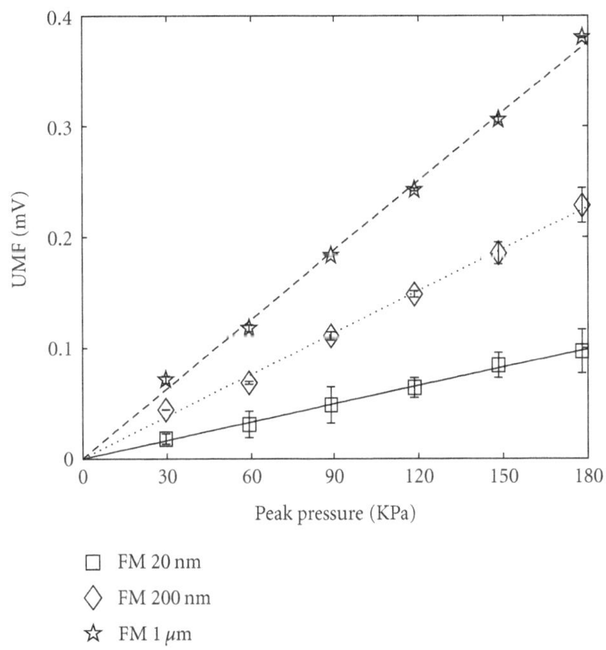

Figure 6: UMF signals from the fluorescent microsphere solutions with the concentration of 1.0 mg/mL as a function of P_{fspp} (after the subtraction of the electronic interference) when $x = 20\,\text{mm}$ $y = 0\,\text{mm}$. The squares, diamonds, and stars with error bars represent the experimentally measured data from the three different-sized microsphere solutions (20 nm, 200 nm, and 1.0 μm), respectively. The solid, dotted, and dashed lines are the linear fits.

of the data. Figures 3(b) and 3(d) show a linear relationship between the DC signal and fluorophore concentrations from all the four samples. These linear relationships imply that the concentrations chosen in this study are appropriate and the inner filter effect or self-quenching is negligible.

To investigate the possibility of ultrasound-induced modulation of excitation light, we also detected the AC signal of the excitation light by blocking the emission light. Figure 4 shows the ultrasound-modulated signal (AC voltage) of the excitation light as a function of microsphere concentration. The squares, diamonds, and stars represent the experimentally measured data. For all three microsphere solutions, the AC signals do not vary significantly with the concentrations in the adopted range. Also, the differences of signal strengths from different particle solutions are not significant. Interestingly, the signal from 1.0 micron particles is higher than that from 200 nm particles. However, both are lower than that of 20 nm particles solution. These results indicate that modulation of the excitation light is complicated and may originate from various mechanisms [9]. Although the dominant modulation mechanism is difficult to be judged from the data shown in Figure 4, a few conclusions can be drawn based on our current and previous studies [5, 10, 12, 13]. First, the modulated fluorescence signal can be clearly observed in a solution without scatterers, such as rhodamine B aqueous solution where the fluorophores are smaller than 1 nm in diameters and do not contribute to light scattering [13]. It implies that the scatters are not necessary to generate UMF signal. Secondly, the modulation of the excitation light is not well correlated with fluorophore concentration and particle size. In contrast, the observed UMF does. This result may indicate that the modulation of the fluorophore concentration is dominant in which the unmodulated excitation light excites the modulated fluorophores.

Figure 5 shows the calculated modulation efficiencies (the ratio of the UMF signal to the fluorescence DC signal) of three FM and the streptavidin-conjugated Alexa Fluor 647 fluorescent probes. The modulation efficiency increases with increasing particle size for fluorescent microsphere solutions. The modulation efficiency of the fluorescent microsphere sample with a diameter of 1.0 μm is ~1.3 times higher than that of the 200 nm diameter sample, ~1.7 times higher than that of 20 nm diameter sample, and ~2 times higher than that of the sample of streptavidin-conjugated Alexa Fluor 647 (~5 nm diameter). Unfortunately, the factor of 2 improvement in the modulation efficiency is limited compared to the factor of 200 increase in the size of fluorescent particles (from 5 nm to 1 micron). This result implies that the observed increases in modulation efficiencies from different research groups mentioned in the introduction section cannot be justified based solely on the size differences of the adopted fluorescent particles. The finding of the size-dependent modulation efficiency in this study implies that large fluorescent particles (microns) are easier to modulate compared with small fluorescent particles (nanometers). If one assumes that the major modulation mechanism is the modulation of fluorophore concentration as discussed in our previous studies [10, 13], the current results imply that for the same concentration of the fluorophore the larger microspheres can be modulated relatively more efficiently. Note that the upper limit of the size being effective in the modulation has not been specifically studied. However, the particle size is usually limited to be smaller than capillary diameter for biomedical imaging applications. Therefore, we limited our particle size to below one micron in this study.

The peak UMF signals from the three-sized fluorescent microsphere solutions (with the concentration of 1.0 mg/mL) were measured as a function of the peak pressure applied at the transducer focus (P_{fspp}). The background electronic interference was measured at each P_{fspp} value and subtracted from the corresponding UMF signal. Figure 6 shows a the linear relationships between the UMF and the P_{fspp} for all the three samples. This result is consistent with our previous studies [10, 13] and indicates that the fluorescent particle size does not affect the linear relationship between the UMF strength and the ultrasound pressure [13].

4. Summary

In summary, modulation efficiencies from four different-sized fluorescent particles solutions (from 5 nm to 1 um scale) were measured and compared. The modulation efficiency increases by approximately a factor of two when increasing the fluorescent particle size from 5 nanometers to 1 micron. This size-dependent modulation efficiency indicates that large fluorescent particles can be slightly more efficiently modulated. The mechanism may be that larger particles are ultrasonically oscillated more efficiently and larger particle can encapsulate more fluorescent molecules. However, the improvement is very limited and may not be considered as a strategy to significantly improve the modulation efficiency of UMF for biological imaging applications. This study also indicates that new modulation mechanisms should be explored to gain further improvements in modulation efficiency.

Acknowledgments

The authors acknowledge the funding support of National Institute of Health (7R15EB012312-02), CPRIT (RP120052) and seed grant support of Research Enhancement Program from University of Texas at Arlington

References

[1] W. Cai and X. Chen, "Multimodality molecular imaging of tumor angiogenesis," *Journal of Nuclear Medicine*, vol. 49, supplement 2, pp. 113S–128S, 2008.

[2] J. R. Lakowicz, *Principles of Fluorescence Spectroscopy*, Speinger, Berlin, Germany, 3rd edition, 2006.

[3] L. V. Wang, "Multiscale photoacoustic microscopy and computed tomography," *Nature Photonics*, vol. 3, no. 9, pp. 503–509, 2009.

[4] J. Culver, W. Akers, and S. Achilefu, "Multimodality molecular imaging with combined optical and SPECT/PET modalities," *Journal of Nuclear Medicine*, vol. 49, no. 2, pp. 169–172, 2008.

[5] B. Yuan, "Theoretical study of ultrasound-modulated fluorescence photon waves in turbid media," in *Diffuse Optical Tomography and Fluorescence Diffuse Optical Tomography, PhD Dissertation*, chapter 6, University of Connecticut, Connecticut, RI, USA, 2006.

[6] Y. Li, H. Zhang, C. Kim, K. H. Wagner, P. Hemmer, and L. V. Wang, "Pulsed ultrasound-modulated optical tomography using spectral-hole burning as a narrowband spectral filter," *Applied Physics Letters*, vol. 93, no. 1, Article ID 011111, 2008.

[7] T. W. Murray, L. Sui, G. Maguluri et al., "Detection of ultrasound-modulated photons in diffuse media using the photorefractive effect," *Optics Letters*, vol. 29, no. 21, pp. 2509–2511, 2004.

[8] G. Rousseau, A. Blouin, and J.-P. Monchalin, "Ultrasound-modulated optical imaging using a powerful long pulse laser," *Optics Express*, vol. 16, no. 17, pp. 12577–12590, 2008.

[9] L. V. Wang, "Mechanisms of ultrasonic modulation of multiply scattered coherent light: an analytic model," *Physical Review Letters*, vol. 87, no. 4, Article ID 043903, 2001.

[10] B. Yuan, J. Gamelin, and Q. Zhu, "Mechanisms of the ultrasonic modulation of fluorescence in turbid media," *Journal of Applied Physics*, vol. 104, no. 10, Article ID 103102, 2008.

[11] M. Kobayashi, T. Mizumoto, Y. Shibuya, M. Enomoto, and M. Takeda, "Fluorescence tomography in turbid media based on acousto-optic modulation imaging," *Applied Physics Letters*, vol. 89, no. 18, Article ID 181102, 2006.

[12] B. Yuan, Y. Liu, P. M. Mehl, and J. Vignola, "Microbubble-enhanced ultrasound-modulated fluorescence in a turbid medium," *Applied Physics Letters*, vol. 95, no. 18, Article ID 181113, 2009.

[13] B. Yuan and Y. Liu, "Ultrasound-modulated fluorescence from rhodamine B aqueous solution," *Journal of Biomedical Optics*, vol. 15, no. 2, Article ID 021321, 2010.

[14] D. J. Hall, U. Sunar, and S. Farshchi-Heydari, "Quadrature detection of ultrasound-modulated photons with a gain-modulated, image-intensified, CCD camera," *The Open Optics Journal*, vol. 2, pp. 75–78, 2008.

[15] D. J. Hall, M. J. Hsu, S. Esener, and R. F. Mattrey, "Detection of ultrasound-modulated photons and enhancement with ultrasound microbubbles," *Proceedings of SPIE*, vol. 7190, article 71900L, 2009.

[16] J. H. Shin, Y. Jiang, B. Grabowski, J. Hurwitz, and Z. Kelman, "Substrate requirements for duplex DNA translocation by the

eukaryal and archaeal minichromosome maintenance helicases," *The Journal of Biological Chemistry*, vol. 278, no. 49, pp. 49053–49062, 2003.

[17] H. Li, S. H. Park, J. H. Reif, T. H. LaBean, and H. Yan, "DNA-templated self-assembly of protein and nanoparticle linear arrays," *Journal of the American Chemical Society*, vol. 126, no. 2, pp. 418–419, 2004.

19

Bragg-Scattering Four-Wave Mixing in Nonlinear Fibers with Intracavity Frequency-Shifted Laser Pumps

Katarzyna Krupa,[1] Michela Bettenzana,[2] Alessandro Tonello,[1] Vincent Couderc,[1] Philippe Di Bin,[1] Stefan Wabnitz,[2] and Alain Barthélémy[1]

[1] *XLIM, UMR CNRS 6172, Université de Limoges, 123 av A. Thomas, 87060 Limoges, France*
[2] *Dipartimento di Ingegneria dell'Informazione, Università di Brescia, Via Branze 38, 25123 Brescia, Italy*

Correspondence should be addressed to Alessandro Tonello, alessandro.tonello@unilim.fr

Academic Editor: Christophe Finot

We experimentally study four-wave mixing in highly nonlinear fibers using two independent and partially coherent laser pumps and a third coherent signal. We focus our attention on the Bragg-scattering frequency conversion. The two pumps were obtained by amplifying two Intracavity frequency-shifted feedback lasers working in a continuous wave regime.

1. Introduction

Over the last decades, the nonlinear phenomena of parametric four-wave mixing (FWM) have been intensively studied in view of their great potential to provide a large variety of all-optical functionalities for ultrafast signal processing [1–4]. In particular, optical parametric amplification, frequency conversion, phase conjugation, and nonlinear switching have been widely explored. The main interest was developed for FWM with coherent pumps, since this technique may provide high conversion efficiencies with appropriate dispersion curves [5]. Nevertheless, in recent years, FWM in optical fibers with incoherent or partially coherent pumps has attracted great attention, in view of its polarization-independent gain and increased resilience to stimulated Brillouin scattering (SBS) [6, 7]. Let us remember that SBS limits the direct use of intense and narrowband lasers in nonlinear fiber. For the specific case of highly nonlinear fibers (hereafter HNLF) the Brillouin threshold may be in the order of few tens of milliwatt for a fiber length of 500 m. The common way to increase the Brillouin threshold requires the adoption of phase modulation of pumps. For the case of two-pump parametric devices, optimal architectures of phase modulations have been proposed to suppress SBS. These solutions are based on phase modulators, generally driven by a pseudorandom bitsequence generator [4, 5, 8, 9] at GHz repetition frequency. With pure phase modulation the pumps can be spectrally broadened without affecting the time domain intensity.

The adoption of incoherent pumps may represent a cost-effective solution to study FWM in different scenarios and may open a novel kind of applications, even if this may take place at the expense of low conversion efficiency and spectral broadening of the converted signal. The interplay between four-wave mixing processes with mixed coherent-incoherent pumps may also open entirely new features as reported very recently in [10]. Till now, most of investigations on nonlinear effects with incoherent pumping have been conducted for single pump FWM processes and so far, little attention has been paid to a specific type of FWM generally called Bragg-scattering FWM (BS-FWM) [6, 10]. The low-noise nature of the BS-FWM is a very attractive feature [8, 11], which makes this process an excellent candidate for light-by-light manipulation with possible extension to the case of very weak signals. These optical functionalities may be of interest to manipulate optical quantum states in view of frequency conversion in quantum networks.

In this paper, we study the case of BS-FWM with partially coherent pumps in HNLF. For this purpose we used two broadband intra-cavity frequency-shifted feedback lasers

(IFSFLs) that operated in the continuous wave (CW) regime. Although these pumps are independent and exhibit low coherence time, we experimentally observed a clear signature of frequency conversion. In addition, we discuss the competition between BS-FWM and other types of FWM simultaneously occurring in HNLFs. In particular, we experimentally demonstrate the competition with degenerate (i.e., single pump) FWM.

2. Experimental Setup and Pump Sources

In the case of a parametric frequency conversion, the coherence time of the pump t_C should be compared to the characteristic evolution time t_P to establish such a nonlinear interaction. In HNLF and for powers in the order of 50 mW one could estimate t_P from the effective nonlinear length and group speed [12]. Whereas t_C from our pumps is instead associated to the bandwidth of at least 50 GHz of our pumps, so we may expect that the nonlinear interplay becomes affected by the partial coherence of pumps.

To study the case of BS-FWM frequency conversion in such a different regime we propose the experimental setup presented in Figure 1. Instead of filtering amplified spontaneous emission (ASE) sources (often employed in previous studies of incoherent FWM [6]) in our experiment we used a pair of intra-cavity frequency-shifted feedback lasers (IFSFLs) that we independently amplified to obtain two intense pumps.

The coherence properties of IFSFLs have been widely investigated and exploited in metrology and in astronomical imaging [13–15]. This kind of laser has also been proposed to increase by several orders of magnitude the SBS threshold in [16] allowing supercontinuum generation from an amplified IFSFL with several Watts of input power. However BS-FWM requires two distinct pumps. Thus, we developed two identical tunable fiber IFSFLs that operated at different wavelengths. Each IFSFL has been separately amplified by an erbium-doped fiber amplifier (EDFA) and protected against unwanted back reflections by an isolator. To reduce the ASE noise we put one bandpass filter at the input and two bandpass filters at the output of each EDFA (DiCon thin-film filters with 5 nm bandwidth each). These sets of filters carve the spectrum of each pump to reduce the background spectral level without significantly affecting the 3 dB bandwidth of the sources. Next, the two pumps were combined by a 50/50 fiber coupler and multiplexed with a third signal from a commercially available tunable extended cavity laser (laser power of 4 mW) through a 95/5 "tap" coupler. Pumps and signal were injected into a 480 m long off-the-shelf segment of HNLF with nominal zero dispersion wavelength (ZDW) of 1545 nm.

The spectrum at the output end of the fiber was analyzed by an optical spectrum analyzer (OSA) and by a 1 GHz photodiode connected either to an oscilloscope (OSC) or to a radio frequency (RF) analyzer. In what follows we will refer to a set of measurements in which the average powers of pump 1, 2, and signal at the input end of HNLF were 43 mW, 62 mW, and 0.12 mW, respectively.

The two central elements of our experimental setup are the fiber IFSFLs. We present their synoptic scheme in Figure 2. Each laser cavity consists of a 20 m long polarization maintaining (PM) erbium-doped fiber pumped by a laser diode emitting at 1480 nm and an inline PM acousto-optic frequency shifter (AOFS) controlled by a fixed RF driver at 110 MHz. The cavity mirrors are provided by a fiber FC-PC connector (see Figure 2, on the right) to introduce a natural Fresnel-type back reflection; the laser is connected to an output fiber with a FC-APC connector though a mating sleeve containing an air gap. The second mirror is represented on the left-hand side of Figure 2 by a blazed grating in Littrow configuration working at the first diffraction order. The microlens has a focal length $f' = 8$ mm while the grating has density of 600 grooves per millimeter. The central wavelength of the laser can be freely tuned all over the bandwidth permitted by the erbium gain by a simple rotation of the grating. The two IFSFLs are identical copies, and they are both based on this scheme: they were tuned to the upper and lower spectrum edge allowed by such lasers (1540 nm and 1563 nm for pump 1 and 2, resp.) so as to obtain the largest possible difference between their carrier frequencies. The corresponding frequency difference in this case is of 2.87 THz.

The free-space optics section reported in Figure 2 provides the transfer function of intra-cavity frequency dependent loss. In the case of perfect collimation, it turns out that the filtered bandwidth is inversely proportional to the focal length f' of the microlens. The fiber end close to the lens was cleaved with an angle of 8° to reduce any undesired back reflection. Under these conditions the low numerical aperture of the fiber provides the filtering mechanism. In order to smooth the optical spectrum, we used the AOFS inside the fiber cavity. A dynamical equilibrium can be found between frequency shifting, tight frequency-dependent loss, and broadband gain. However, a more detailed analysis of the RF spectrum at the output of one of the IFSFL unveils its complex behavior, as reported on the left-hand side of Figure 3.

We observe the presence of a comb of frequencies regularly spaced by 3.6 MHz, which compares well with the reciprocal of the round trip time of the fiber cavity. We compare these results with the direct numerical calculation coming from a model described in [17] (see Figure 3(b)). To implement equation (3) of [17] we assumed a fiber length of 27 m (in fact both the doped fiber and the AOFS are spliced to PM standard monomode fiber patchcords on both ends), a group index of 1.45, a frequency shift of 2×110 MHz (the AOFS is seen two times per round trip), and a random phase for each frequency comb component. Note that in our case the AOFS is driven by a radio frequency which is much higher than the inverse of the round trip time of the cavity.

The dynamical evolution of the IFSFLs is also a key point for applications in nonlinear optics. Indeed we observed as many other authors that our IFSFLs may operate in a CW or in a pulsed regime (see, e.g., [18]), depending on the level of feedback in the cavity. However, it would be desirable that the IFSFLs work close to a CW mode to observe BS-FWM as one would expect that a nonsynchronized pulsed regime

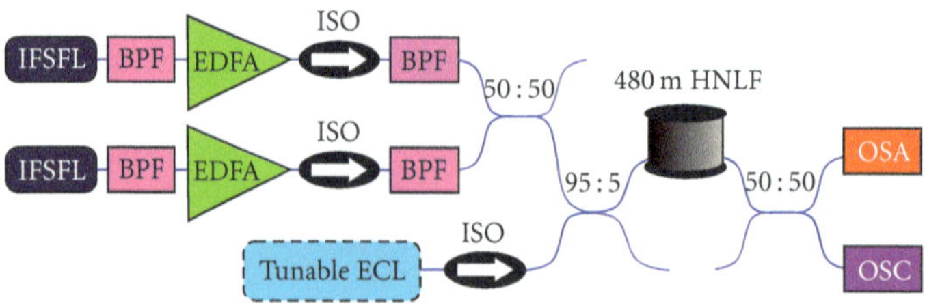

Figure 1: Experimental setup; IFSFL: intra-cavity frequency-shifted feedback laser, BPF: bandpass filter, EDFA: erbium-doped fiber amplifier, ISO: isolator, ECL: extended cavity tunable laser, HNLF: highly non-linear fiber, OSA: optical spectrum analyzer, and OSC: oscilloscope.

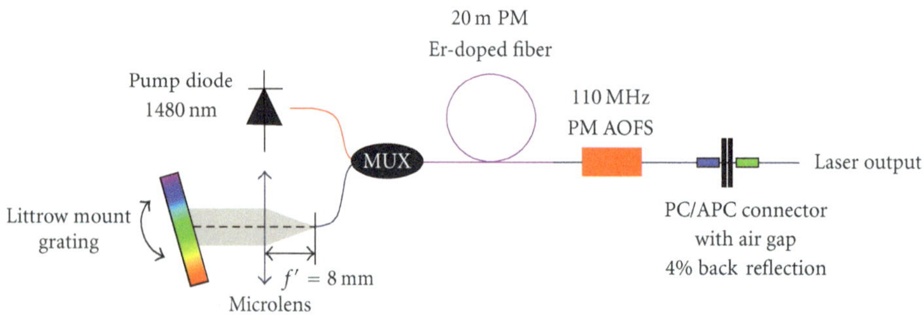

Figure 2: Setup of each IFSFL; MUX: multiplexer, PM: polarization maintaining, and AOFS: acousto-optic frequency shifter.

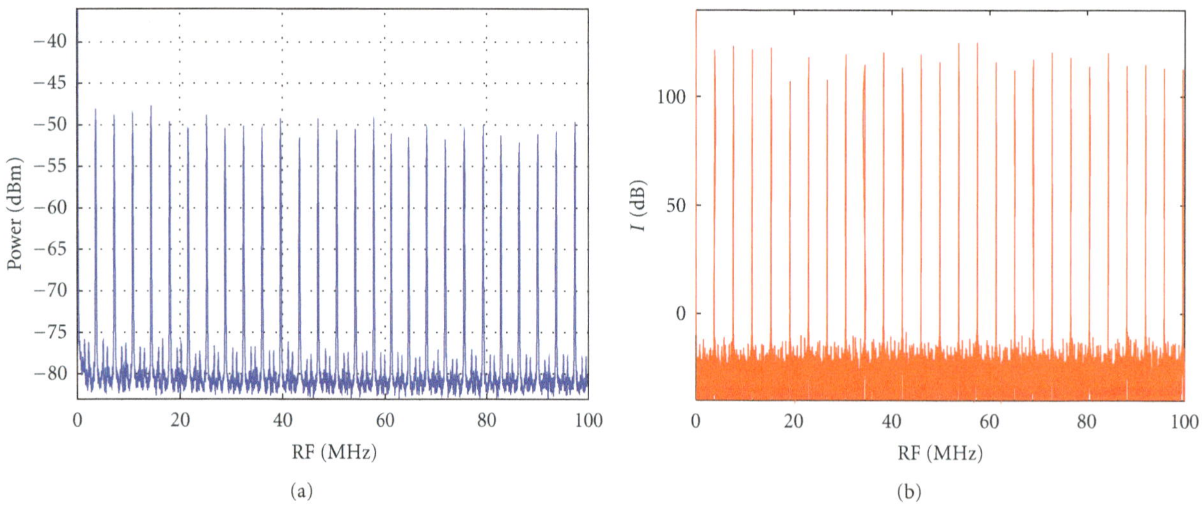

Figure 3: (a) Experimental RF spectrum for IFSFL 1 with a preamplified RF analyzer: the frequency comb free spectral range is of 3.6 MHz; (b) numerical simulation of RF photodiode current (spectrum of equation (3) of [17]).

between the pumps may drastically degrade the conversion efficiency in a dual pump process. According to [18], the bifurcation point between CW and pulsed solution can be influenced by the pump diode power. To keep the IFSFLs operating in the CW regime, we have then set their diode currents to a low value (65 mA for both lasers). For higher values of diode current one observes a rapid transition to time-periodic solutions [18] by direct observation of the laser output through a photodiode and an oscilloscope. The working points of our lasers were finely adjusted by verifying the absence of relaxation oscillations with an oscilloscope and a 1 GHz bandwidth InGaAs photodetector.

3. Background Theory

Four-wave mixing is usually characterized as 3 distinct processes depending on the relative positions of the pumps, signal, and idler frequencies (see Figure 4(a)) [4, 8]: modulation instability (MI), phase conjugation (PC), and Bragg-scattering (BS). All of them may be simultaneously

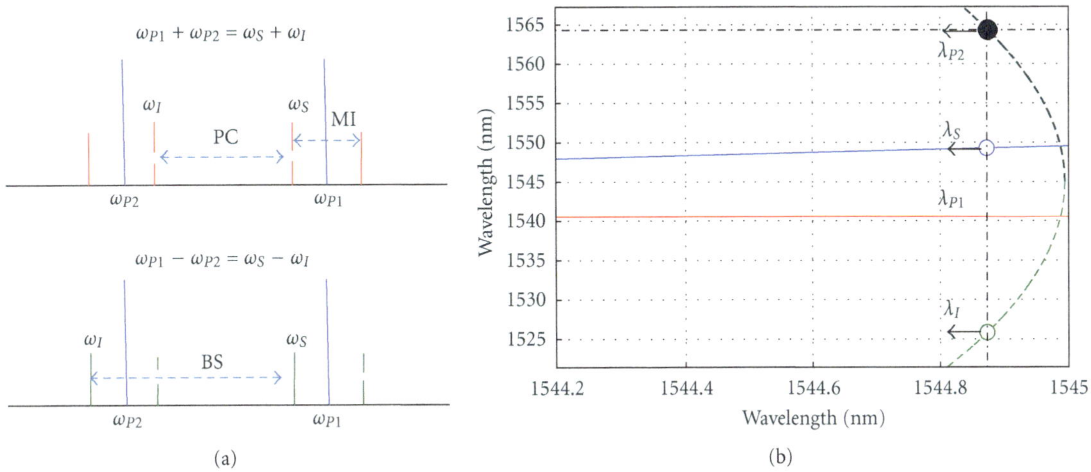

FIGURE 4: (a) Four-wave mixing processes, from [4]; (b) example of BS-FWM phase-matching diagram: thick solid (dashed) line (curve) for pump 1 (pump 2); thin solid (dashed) line (curve) for signal (idler).

observable [5] especially for low dispersion values, as it will be shown at the end of Section 4. MI requires one pump process. PC and BS require instead two distinct pumps. BS-FWM has a low noise nature as the frequency conversion may take place at phase matching in the absence of exponential gain.

For BS-FWM energy and momentum conservation give for angular frequencies ω_j and wave-vector k_j ($j = P1, P2, S,$ for the pump 1, pump 2, signal and idler resp.) [4, 5, 11]:

$$\omega_I = \omega_S + (\omega_{P1} - \omega_{P2}), \quad (1)$$

$$k_I = k_S + (k_{P1} - k_{P2}). \quad (2)$$

The dependence of wave vectors on angular frequency leads to the determination of the best conditions for frequency conversion. In close proximity to the ZDW, the dispersion coefficient of an HNLF is generally approximated by a linear function of wavelength. Under this approximation, if one fixes for instance the wavelength of the pump 1, one can build a chart of optimization for the remaining wavelengths combining (1) and (2) [4, 5]. An example for such a diagram is given in Figure 4(b) for a fiber with ZDW of 1545 nm and dispersion slope of 0.018 ps/(km nm^2). The horizontal axis represents a common reference wavelength: its corresponding angular frequency is the mean value between the frequency of the idler and that of pump 2. One can start to choose a value for the wavelength for pump 2 by covering the corresponding thick dashed curve: the wavelength for pump 2 will be the vertical coordinate of the selected point. Then from the chart one can automatically predict the best signal wavelength and hence the position of the corresponding idler. One should draw first a vertical line that matches the selected choice of pump 2. The wavelengths for signal and idler will be read at the intersection points with the other curves. More specifically the signal will be read at the intersection with the thin solid line while the idler will come from the intersection with the thin dashed curve. In practice, it is likely that the ZDW may fluctuate by about 1 nm along the fiber length, according to the analysis presented in [19] for similar types of fibers. This fact may degrade the overall conversion efficiency [5].

4. Experimental Results and Discussion

Let us first focus on the spectral region between 1520 nm and 1535 nm, where the BS-FWM appears whenever the signal is varied between 1547.3 nm and 1554.3 nm. The collection of these experimental results is shown in Figure 5(a). Note that the BS-FWM sideband varies from 1524.5 nm to 1531.5 nm and keeps a fixed distance of 2.87 THz from the signal. BS-FWM is then a follower of the signal frequency, being the frequency distance between signal and idler in BS-FWM decided by the pumps frequencies. We better show this effect in Figure 5(b) where the experimental results of signal and idler carrier wavelength compare well with the energy conservation law of (1). However, in the same spectral range we observe a competition with the degenerated FWM driven by pump 1. Hereafter we will indicate this effect as modulation instability (MI) independently from the presence or absence of parametric gain. Note that we work with very low dispersion values and one of these sidebands is seeded by the input signal itself. Differently from BS-FWM, the sidebands of MI-FWM follow a symmetry law around the pump. Such symmetry in spectrum is very well known and can be understood again from the energy conservation law that reads as $\omega_I = 2\omega_{P1} - \omega_S$. If we assume that the angular frequency of signal is detuned from that of a pump $P1$ as $\omega_S = \omega_{P1} - \Omega$, then it turns out that $\omega_I = \omega_{P1} + \Omega$.

Therefore if we keep fixed the spectral position of pumps and we gradually reduce the angular frequency ω_S (i.e., we increase the signal wavelength) we can observe two opposite behaviors: the angular frequency ω_I of degenerated FWM will increase whereas that of BS-FWM will decrease. In our experiment this fact leads even to a superposition of the two idlers at 1528.75 nm, when the signal is tuned precisely at $\lambda_S = 1551.42$ nm. This crossing effect upon signal wavelength

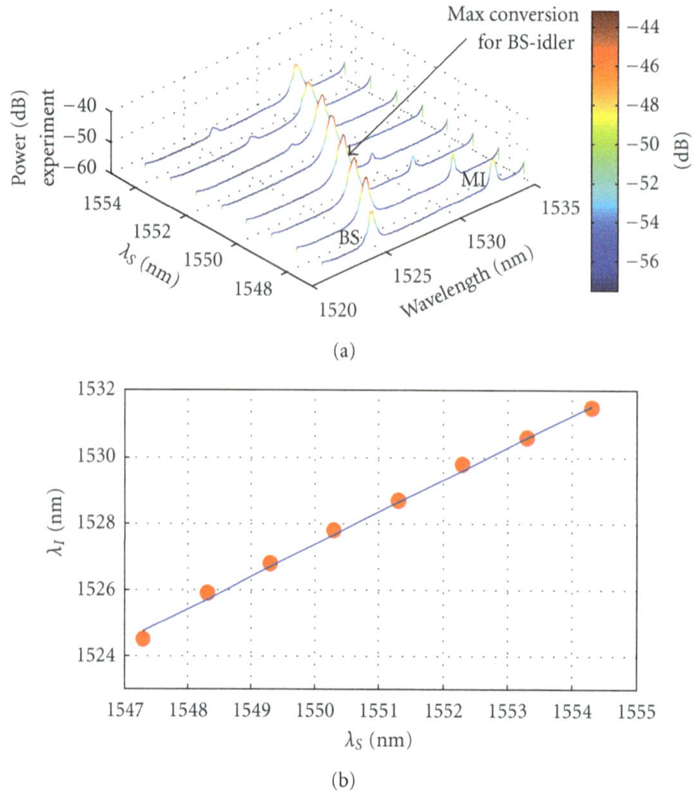

Figure 5: (a) Collection of experimental spectra upon different input signal wavelengths; (b) BS-FWM idler wavelength upon signal wavelength; dots: experimental results; solid line: equation (1).

is also shown in Figure 5(a). The coexistence of these two parametric processes and their tiny frequency separation should be then considered carefully in applications. Note that pump 1 is very close to the ZDW and a selection rule of parametric processes, based on phase mismatch may become less effective in practice.

As we show in Figure 6(a), we obtained in our experiment 16 dB from the background noise level at the idler central wavelength for the case of maximally efficient BS-FWM. Note that in our spectral measurements we set to 1 kHz the video bandwidth digital filter of the optical spectrum analyzer. The maximum conversion is found for a signal at 1550.3 nm. The discrepancy from phase matching (1549.8 nm) may be due to uncertainties in the dispersion parameters of the fiber and to the fluctuation in ZDW along the beam propagation in the fiber [5], as we have already mentioned. We obtained a depletion of the signal of about 1 dB and a ratio between the spectral peak of the idler and that of the signal of the signal of −20 dB: this result is much lower than that reported in [5] for the case of coherent pumps with direct phase modulation. This is mainly due to the difference in fiber parameters and choice of wavelengths. Additionally note that the experiments show that the idler is spectrally broadened as it comes from a combination of broadband pumps and a narrowband signal [6]. In the experiments we estimated the 3 dB bandwidths of pumps as 20 GHz and 44 GHz for IFSL 1 and 2, respectively, (our spectrum analyzer has a minimum resolution of 0.05 nm which correspond to 6.24 GHz at these wavelengths). We experimentally obtained a 57 GHz wide BS-idler at 1527.8 nm and for a signal wavelength λ_S = 1550.3 nm.

Note that in our experiment we can shift the signal within the relatively wide spectral range and still obtain efficient BS-idler generation (see Figure 5(b)). This effect may be potentially applicable for the simultaneous conversion of a large spectrum. In our case, the conversion efficiency BS-FWM varies by less than 3 dB when the signal is tuned by 9 nm.

Let us now report on the case of the two external sidebands (see a and b in Figure 6(b)). By using the word "spontaneous" in the label, we mean that these sidebands are directly generated by the two pumps and they do not require a third signal to appear. Their spectral position is entirely due to the spectral positions of pumps: with respect to pump 1 (pump 2), sideband a (b) is at the same frequency detuning of pump 2 (pump 1). Because of the low dispersion of the fiber and relatively high intensity of pumps, sidebands a and b have large amplitudes. Note that the amplitude of the left-side spontaneous FWM idler (see a in Figure 6(b)) is larger than that of the right-side one (see b in Figure 6(b)). There are also some additional frequency conversion effects. Since their appearance is strongly connected to the fine adjustment of the working points of IFSFLs, we report our best observation in Figure 7. More specifically we observed the signature of MI-FWM around pump 2 and PC-FWM.

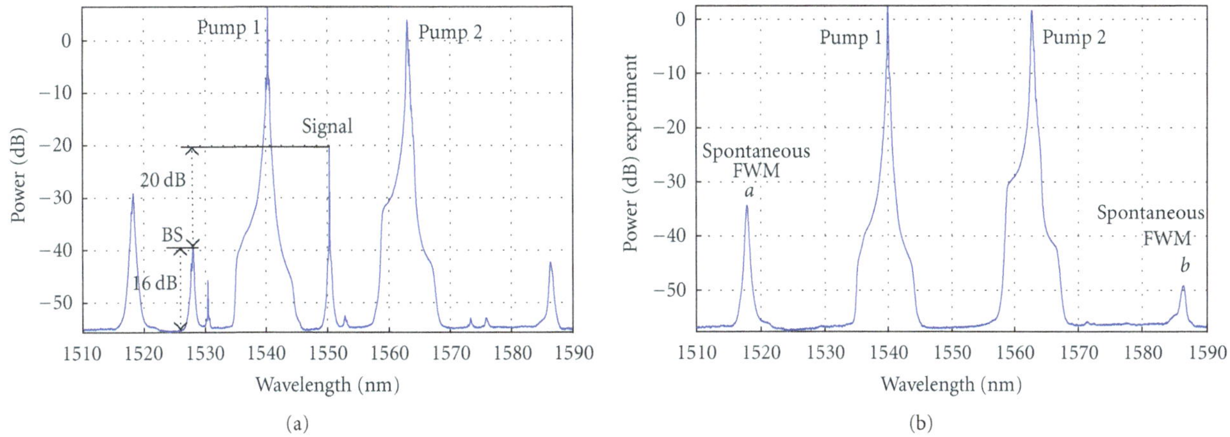

FIGURE 6: (a) experimental output spectrum obtained for the signal wavelength at 1550.3 nm: maximal conversion of BS-FWM; phase-matching conditions for BS-FWM. (b) experimental output with two pumps only.

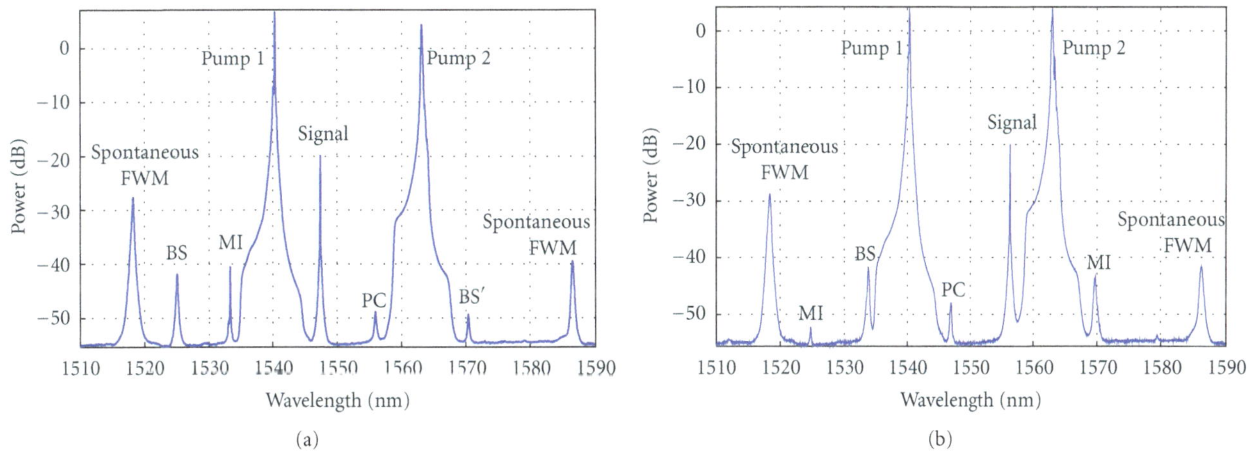

FIGURE 7: Experimental output spectrum obtained for signal wavelength at 1547.3 nm (a) and 1556.3 nm (b); existence of multiple simultaneous FWM processes.

We also observed another kind of BS-FWM that comes from the subtraction instead of the sum of the beating frequency of 2.87 THz (BS′). The conversion efficiencies of these last processes remain low, probably due to the large phase mismatch. We show in Figure 7 two output spectra taken with a signal at 1547.3 nm (a) and 1556.3 nm (b).

5. Conclusion

In this paper, we studied BS-FWM in HNLF with a novel cost-effective and partially coherent pumping scheme. We experimentally demonstrated efficient and tunable BS-FWM despite the large bandwidth of the pumps and their complete statistical independence. The spectral peak of the converted BS-FWM is 20 dB below that of the original signal and 16 dB above the background noise level. We noted coexistence of multiple independent FWM processes. In particular, we observed a relevant competition BS-FWM and MI-FWM. Our measurements agree well with a theoretical analysis based on the phase-matching diagram for the case of monochromatic waves. In conclusion, nonlinear fiber optics frequency converters based on broadband pumps appear to be promising for their implementation in all-optical signal processing devices.

Acknowledgments

The authors acknowledge the financial support from the Region Limousin and from the French National Research Agency under Grant ANR-08-JCJC-0122 PARADHOQS. They thank Alain Dexet for his assistance in manufacturing some of the mechanical parts used in the experiment.

References

[1] C. J. McKinstrie, S. Radic, and A. R. Chraplyvy, "Parametric amplifiers driven by two pump waves," *IEEE Journal on Selected Topics in Quantum Electronics*, vol. 8, no. 3, pp. 538–547, 2002.

[2] R. Slavík, F. Parmigiani, J. Kakande et al., "All-optical phase and amplitude regenerator for next-generation telecommunications systems," *Nature Photonics*, vol. 4, no. 10, pp. 690–695, 2010.

[3] J. Hansryd, P. A. Andrekson, M. Westlund, J. Li, and P. O. Hedekvist, "Fiber-based optical parametric amplifiers and their applications," *IEEE Journal on Selected Topics in Quantum Electronics*, vol. 8, no. 3, pp. 506–520, 2002.

[4] D. Méchin, R. Provo, J. D. Harvey, and C. J. McKinstrie, "180-nm wavelength conversion based on Bragg scattering in an optical fiber," *Optics Express*, vol. 14, no. 20, pp. 8995–8999, 2006.

[5] R. Provo, S. Murdoch, J. D. Harvey, and D. Méchin, "Bragg scattering in a positive β_4 fiber," *Optics Letters*, vol. 35, no. 22, pp. 3730–3732, 2010.

[6] Y. Yan and C. Yang, "Four-wave mixing between coherent signal and incoherent pump light in nonlinear fiber," *Journal of Lightwave Technology*, vol. 27, no. 22, pp. 4954–4959, 2009.

[7] D. Nodop, D. Schimpf, J. Limpert, and A. Tünnermann, "SBS suppression in high power fiber pulse amplifiers employing a superluminescence diode as seed source," in *Proceedings of the Conference on Lasers and Electro-Optics (CLEO EUROPE '11)*, Munich, Germany, May 2011, paper CJ7.6 THU.

[8] A. H. Gnauck, R. M. Jopson, C. J. McKinstrie, J. C. Centanni, and S. Radic, "Demonstration of low-noise frequency conversion by Bragg Scattering in a Fiber," *Optics Express*, vol. 14, no. 20, pp. 8989–8994, 2006.

[9] S. Radic, R. M. Jopson, A. Gnauck, C. J. McKinstrie, J. C. Centanni, and A. R. Chaplyvy, "Stimulated-brillouin-scattering supression using a single modulator in two-pump parametric architectures," in *Proceedings of the Optical Fiber Communication Conference (OFC '05)*, Anaheim, Calif, USA, March 2005, paper OWN5.

[10] J. Schröder, A. Boucon, S. Coen, and T. Sylvestre, "Interplay of four-wave mixing processes with a mixed coherent-incoherent pump," *Optics Express*, vol. 18, no. 25, pp. 25833–25838, 2010.

[11] H. J. McGuinness, M. G. Raymer, C. J. McKinstrie, and S. Radic, "Quantum frequency translation of single-photon states in a photonic crystal fiber," *Physical Review Letters*, vol. 105, no. 11, Article ID 093604, 2010.

[12] Y. Yan and Y. Changxi, "Coherent light wave generated from incoherent pump light in nonlinear kerr medium," *Journal of the Optical Society of America B*, vol. 26, no. 11, pp. 2059–2063, 2009.

[13] H. G. Chatellus and J.-P. Pique, "Coherence properties of modeless lasers," in *Proceedings of the Science, Quantum of Quasars Workshop*, Grenoble, France, December 2009.

[14] L. P. Yatsenko, B. W. Shore, and K. Bergmann, "Coherence in the output spectrum of frequency shifted feedback lasers," *Optics Communications*, vol. 282, no. 2, pp. 300–309, 2009.

[15] L. P. Yatsenko, B. W. Shore, and K. Bergmann, "An intuitive picture of the physics underlying optical ranging using frequency shifted feedback lasers seeded by a phase-modulated field," *Optics Communications*, vol. 282, no. 11, pp. 2212–2216, 2009.

[16] P. A. Champert, V. Couderc, and A. Barthélémy, "1.5-2.0-μm multiwatt continuum generation in dispersion-shifted fiber by use of high-power continuous-wave fiber source," *IEEE Photonics Technology Letters*, vol. 16, no. 11, pp. 2445–2447, 2004.

[17] K. Nakamura, T. Miyahara, and H. Ito, "Observation of a highly phase-correlated chirped frequency comb output from a frequency-shifted feedback laser," *Applied Physics Letters*, vol. 72, no. 21, pp. 2631–2633, 1998.

[18] M. Stellpflug, G. Bonnet, B. W. Shore, and K. Bergmann, "Dynamics of frequency shifted feedback lasers: Simulation studies," *Optics Express*, vol. 11, no. 17, pp. 2060–2080, 2003.

[19] B. Auguié, A. Mussot, A. Boucon, E. Lantz, and T. Sylvestre, "Ultralow chromatic dispersion measurement of optical fibers with a tunable fiber laser," *IEEE Photonics Technology Letters*, vol. 18, no. 17, pp. 1825–1827, 2006.

20

Experiments on Linear and Nonlinear Localization of Optical Vortices in Optically Induced Photonic Lattices

Daohong Song,[1] Cibo Lou,[1] Liqin Tang,[1] Zhuoyi Ye,[1] Jingjun Xu,[1] and Zhigang Chen[1,2]

[1] *The Key Laboratory of Weak-Light Nonlinear Photonics, Ministry of Education and TEDA Applied Physics School, Nankai University, Tianjin 300457, China*
[2] *Department of Physics & Astronomy, San Francisco State University, San Francisco, CA 94132, USA*

Correspondence should be addressed to Zhigang Chen, biglasers@gmail.com

Academic Editor: Jan Masajada

We provide a brief overview on our recent experimental work on linear and nonlinear localization of singly charged vortices (SCVs) and doubly charged vortices (DCVs) in two-dimensional optically induced photonic lattices. In the nonlinear case, vortex propagation at the lattice surface as well as inside the uniform square-shaped photonic lattices is considered. It is shown that, apart from the fundamental (semi-infinite gap) discrete vortex solitons demonstrated earlier, the SCVs can self-trap into stable gap vortex solitons under the normal four-site excitation with a self-defocusing nonlinearity, while the DCVs can be stable only under an eight-site excitation inside the photonic lattices. Moreover, the SCVs can also turn into stable surface vortex solitons under the four-site excitation at the surface of a semi-infinite photonics lattice with a self-focusing nonlinearity. In the linear case, bandgap guidance of both SCVs and DCVs in photonic lattices with a tunable negative defect is investigated. It is found that the SCVs can be guided at the negative defect as linear vortex defect modes, while the DCVs tend to turn into quadrupole-like defect modes provided that the defect strength is not too strong.

1. Introduction

Vortices and vortex solitons are ubiquitous in many branches of sciences such as hydrodynamics, superfluid, high-energy physics, laser and optical systems, and Bose-Einstein condensates [1, 2]. In optics, optical vortices are phase singularities embedded in electromagnetic waves, and the intensity in a singularity point vanishes due to the undetermined phase. The field phase changes by $2\pi m$ (m is an integer called topological charge) along any closed loop around the singularity point. The study of linear and nonlinear propagation dynamics of optical vortex beams in continuum media attracted great attentions for decades [1]. In general, a vortex dark ring beam will expand during linear propagation due to the diffraction. However, nonlinearity can compensate the diffraction, leading to the formation of nonlinear localized states called spatial vortex solitons. It is found that an optical vortex can self-trap into a vortex soliton in nonlinear media with a *self-defocusing* nonlinearity [3–7], even the vortex is nested in a partially incoherent optical beam [8]. However, with a *self-focusing* nonlinearity, optical vortices tend to breakup into filaments due to the azimuthally modulation instabilities [9].

Several different physical mechanisms have been proposed for suppression of the azimuthally instability [10–12]. Guiding the vortices in periodic structures such as photonic crystals or photonic lattices opened a new avenue to stabilize optical vortices in nonlinear media. Generally speaking, the vortices propagation in periodic structures is dominated by the lattice discreteness, nonlinearity, and the phase singularity. Therefore, the localization and stabilization of such vortices exhibit many new phenomena which do not exist in continuous medium. In fact, discrete SCV solitons have been both theoretically predicated and successfully observed experimentally in square photonic lattices with self-focusing nonlinearity under four-site excitation [13–16]. Singly charged high-band vortex solitons are also realized in photonic lattices [17]. On the other hand, DCV propagation in isotropic square photonic lattice leads to periodic charge flipping with a quadrupole-like mediate states [18].

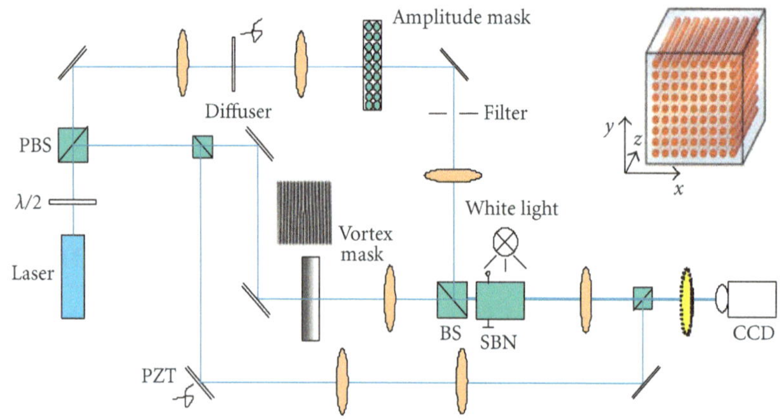

FIGURE 1: Experimental setup for optical induction of photonic lattices in a biased photorefractive crystal by amplitude modulation of a partially coherent beam. PBS: polarizing beam splitter; SBN: strontium barium niobate. Top path is the lattice-inducing beam, the middle path is the probing vortex beam, and the bottom path is the reference beam. The top right insert illustrates the scheme of induced waveguide arrays in the otherwise uniform SBN nonlinear crystal.

Stable SCV, DCV, and multivortex solitons in hexagonal lattices have also been experimentally demonstrated [19–21]. Besides square-like and hexagonal lattices, it has also been proposed that discrete vortex solitons with higher topological charges even exist in photonic quasicrystals [22].

In this paper, we present a brief overview of our recent work on both linear and nonlinear propagation of SCV and DCV beams in 2D photonic lattices in bulk photorefractive material by optical induction. In the nonlinear cases, vortex beam propagation both inside and at the surface of photonic lattices is considered. We show that under appropriate nonlinear conditions both SCV and DCV beams can stably self-trapped into a gap vortex soliton in photonic lattices with self-defocusing nonlinearity. While a SCV beam can turn into a stable gap vortex soliton under four-site excitation, a DCV beam tends to turn into a quadrupole gap soliton which bifurcates from the band edge. However, under geometrically extended eight-site excitation, the DCV can evolve into a stable gap vortex soliton. At the interface between photonics lattice and a homogenous media, under appropriate conditions, the SCV beam can turn into a stable surface discrete vortex soliton under four-site excitation with self-focusing nonlinearity, while it self-traps into a quasi-localized surface state under single site excitation. The stability of such nonlinear localized states is monitored by the numerical simulation to long propagation distance. In the linear case, the SCV beam can also evolve into a linear localized vortex defect mode in photonic lattice with a negative defect, while the DCV beam turns again into a quadrupole-like defect mode, provided that the defect strength is not too deep. Our results may prove to be relevant to the studies of similar phenomena in superfluids and Bose-Einstein condensates.

2. Experimental Setup

The experimental setup for our study is similar to those we used earlier for observation of discrete solitons [15, 23, 24], as shown in Figure 1, except that the input probe beam here can be either a SCV or DCV beam. Our experiments are typically performed in a biased SBN:60 ($5 \times 5 \times 10\,\text{mm}^3$) photorefractive crystal illuminated by a laser beam with wavelength 488 nm from the Argon ion lasers. The biased crystal provides a self-focusing or defocusing non-instantaneous nonlinearity by simply reversing the external biased field. To generate a 2D-waveguide lattice, we use an amplitude mask to spatially modulate the otherwise uniform incoherent beam after the diffuser. The rotating diffuser turns the laser beam into a partially spatially incoherent beam with a controllable degree of spatial coherence. The mask is then imaged onto the input face of the crystal; the periodic intensity pattern is a nearly nondiffracting pattern throughout the crystal after proper spatial filtering. The lattice beam is ordinarily polarized, so it will induce a nearly linear waveguide array that remains invariant during propagation. We can create not only uniform lattices but also lattices with negative defects and lattices with sharp surfaces by simply changing the specially designed amplitude mask. The extraordinarily polarized beam passes through a computer generated hologram to create a ring vortex beam, and then it is focused by a circular lens. The vortex beam is then used as our probe beam or *soliton-forming* beam, propagating collinearly with the *lattice-inducing* beam. In our experiment, the linear propagation of the vortex beam can be realized either by keeping the vortex beam at low intensity or by taking advantage of the noninstantaneous of the photorefractive nonlinearity. A weak reference beam is introduced to interfere with the vortex probe beam after it exits the crystal (bottom path of the Figure 1).

A piezotransduced (PZT) mirror is added to the optical path of the reference beam which can be used to actively vary the relative phase between the plane wave and the vortex beam. In addition, a white-light background beam illuminating from the top of the crystal is typically used for fine-tuning the photorefractive nonlinearity [8, 15].

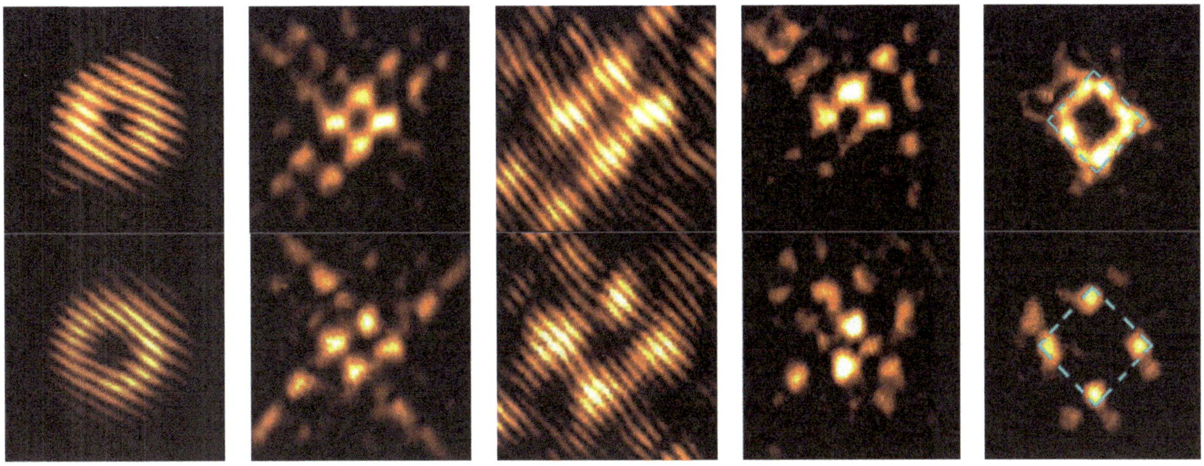

FIGURE 2: Experimental results of self-trapping of SCVs (top) and DCVs (bottom) in a defocusing photonic lattice. (a) Interferograms showing the phase of the input vortex beams, (b) intensity patterns of self-trapped vortex beams at lattice output, (c, d) interferograms between (b) and a tilted plane wave (c) and an on-axis Gaussian beam (d), respectively, and (e) the k-space spectra of (b) where the dash squares mark the first Brillouin zone of the square lattice. (a, c) are zoomed in with respect to (b, d) for better visualization.

3. Nonlinear Localization of Optical Vortices in Photonic Lattices

3.1. Discrete Gap Vortex Solitons. First, we study the SCV and DCV beams propagating inside optically induced photonic lattices under the self-defocusing nonlinearity. We find that a donut-shaped SCV beam can self trap into a stable gap vortex soliton, while the DCV beam evolves into a quadruple soliton under the four-site excitation. Spectrum measurement and numerical analysis suggest that the gap vortex soliton does not bifurcate from the edge of the Bloch band, quite different from previously observed spatial gap solitons.

In our experiment, the lattice beam induces a "backbone" lattice with a self-defocusing nonlinearity where an intensity minimum corresponds to an index maximum [25]. The off-site excitation scheme is used so that the vortex core is on an index minimum, while the donut-like vortex beam covers four adjacent index maxima. Under proper conditions, self-trapping of the vortex beam can be established. Typical experimental results are presented in Figure 2, for which the lattice spacing is about 27 μm, the intensity ratio of the vortex beam to the lattice beam is about 1 : 4, and the bias field is about −1.2 kV/cm. The interferograms of the input vortex beam with an inclined plane wave are shown in Figure 2, where the central fork in the interference fringes indicates the phase singularity ($m = 1$ for top panels and $m = 2$ for bottom panels) of the vortex beam. When the self-trapped state is established in the nonlinear regime, both $m = 1$ and $m = 2$ vortices break up into four filaments to cover the initially excited four lattice sites, just as in the self-focusing case [15, 16] except with longer tails along the two principal axes. Although the intensity patterns of self-trapped $m = 1$ and $m = 2$ vortices look somewhat similar, significant differences can be found in their phase structure and spatial spectrum. Two different interference techniques are used to identify the phase structure of self-trapped vortices. One is to send a titled broad beam (quasiplane wave) to interfere with the output vortex beam (Figure 2(c)). For the limited propagation distance of our crystal length (10 mm), it appears that the vortex singularity (manifested by the central forks in the interferograms) persists after the nonlinear propagation through the crystal, although it seems that charge-flipping (reversal of forks) is associated with the DCV at crystal output. However, from numerical simulations to longer propagation distances, the singularity can maintain only for the $m = 1$ but not for the $m = 2$ vortices [26]. In fact, our theoretical analysis shows that a "true" doubly charged gap vortex soliton does not exist under this excitation condition, and a quadrupole-like soliton structure is found instead for the $m = 2$ vortex [26]. The other technique is to send a coaxial broad Gaussian beam as an interference beam. We can clearly see that the phase structures for self-trapped $m = 1$ and $m = 2$ vortices are different (Figure 2(d)). The two diagonal spots are out-of-phase for the $m = 1$ vortex but in-phase for the $m = 2$ vortex.

We also measure the spatial spectrum of self-trapped vortices by using the Brillouin Zone (BZ) Spectroscopy technique [27]. Again, dramatic difference between $m = 1$ and $m = 2$ vortices can be seen in the k-space spectra, indicating quite different physical pictures for self-trapping. For the $m = 1$ vortex, most of the power is located alongside the first BZ, but it would not concentrate just to the four corner points (corresponding to four high-symmetry M points) which mark the edge of the first Bloch band and where the diffraction is anomalous. However, for the $m = 2$ vortex, the nonlinear spectrum reshaping makes the power spectrum settle into the M-points quickly, similar to those of the fundamental discrete gap solitons and gap soliton trains [28]. Numerical simulations show that such spectrum difference remains for long propagation distance. These results suggest

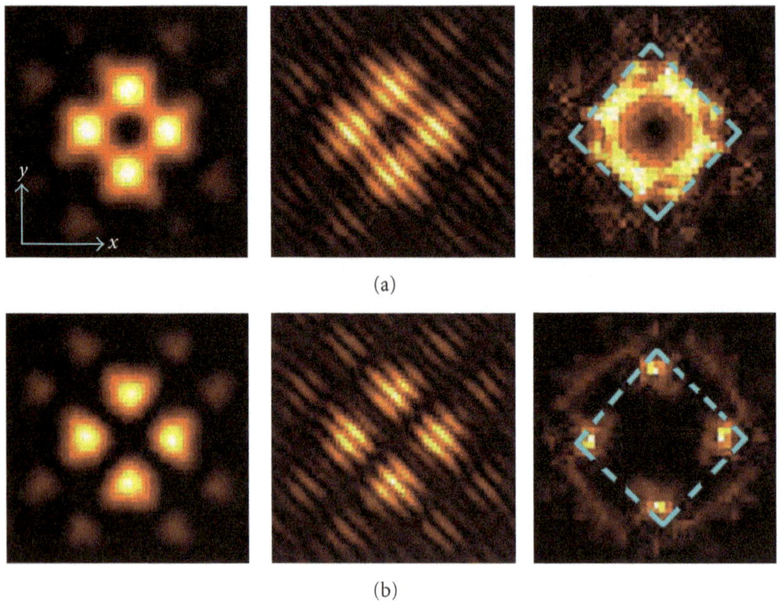

Figure 3: Simulation results of singly charged (a) and doubly charged (b) vortex beams propagating to a longer distance of 40 mm. Shown are the output transverse (x-y) intensity pattern (left), its interferogram with a tilted plane wave (middle), and its k-space spectrum (right) in both (a) and (b). Notice that the vortex singularity maintains in (a) but disappears in (b).

that although the $m = 1$ vortex can evolve into a gap vortex soliton, it does not bifurcate from the edge of the first Bloch band, quite different from all previously observed fundamental, dipole, or quadrupole-like gap soliton in self-defocusing lattices [28, 29]. On the other hand, the $m = 2$ vortex can evolve into a quadrupole-like localized state, which does seem to bifurcate from the edge of the first Bloch bands as confirmed by numerical analysis [26, 30].

The experimental observations are also compared with the numerical results obtained using beam propagation method with initial conditions similar to that for the experiment. The numerical model is a nonlinear wave equation with a 2D square lattice potential under saturate self-defocusing nonlinearity. Excellent agreement can be found for the propagation distance of 10 mm (i.e., our crystal length) for both $m = 1$ and $m = 2$ vortices [26]. To test whether the relevant gap soliton structure can persist for longer propagation distance, simulations are performed with propagation distance up to 40 mm while all other parameters are unchanged. It is found that the phase singularity can maintain for the singly charged vortex, while for the doubly charged vortex at long propagation distance, it turns into a quadrupole-like structures. The simulation results are shown in Figure 3. Notice that the interference patterns of the nonlinear output with a tilted plane wave show that the fork is still in the center for the $m = 1$ vortex, but this is not the case for the $m = 2$ vortex which turns into an out-of-phase quadrupole soliton. The power spectrum is also similar to that of 1 cm propagation in experiment: the spectrum for the $m = 1$ gap vortex soliton does not evolve into the vicinity of the first-band M points, whereas that for the $m = 2$ vortex does. Intuitively, the nonedge bifurcation of the singly charged gap vortex may be attributed to the nontrivial helical phase structure of the vortex, which cannot be expressed as a simple superposition of linear Bloch modes near the band edge. Such subtle difference between the SCV and DCV gap solitons is also confirmed by theoretical analysis [26].

It is interesting to explore whether stable doubly charged gap vortex soliton exists in *self-defocusing* square photonic lattices without "decaying" into a quadrupole soliton, although such stable high-order discrete vortex solitons have been observed in hexagonal photonic lattices under a *self-focusing* nonlinearity [20]. Indeed, we find that when the input vortex ring is expanded to cover eight lattice sites as shown in Figure 4(a), under appropriate nonlinear conditions, the phase singularity can maintain and it can evolve into a "true" stable gap vortex soliton. Typical experimental results are shown in Figure 4, where the results were obtained when the lattice spacing is 25 μm, the intensity ratio of the vortex beam to the lattice beam is 1 : 4, and the bias field is -1.4 kV/cm. When the nonlinear self-trapped state is realized, most of its energy is localized in the initially excited eight lattices (Figure 4(c)). More importantly, the interference pattern between an inclined plane wave and the nonlinear output clearly shows that the two forks along with the bifurcation direction still preserve compared with those of the input vortex shown in Figure 4(b). Again, in such a case, the spectrum is not located in the four M points, different from the nonlinear spectrum under four-site excitation shown in Figure 2(e). Such high-order vortex gap solitons do not bifurcate from the band edge either! We want to mention that the DCV beams can also stably self-trap into solitons in the semi-infinite gap under self-focusing nonlinearity with 8-site excitation. Such geometrical extended stabilization mechanism has also been confirmed by theoretical analysis [31].

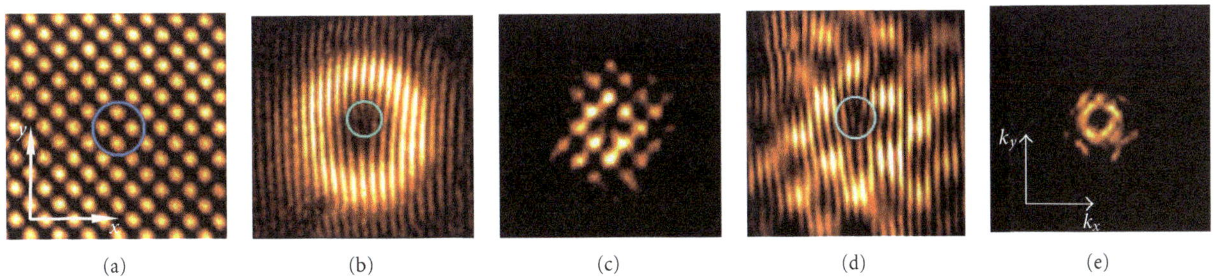

FIGURE 4: Experimental observation of an extended $m = 2$ gap vortex soliton in photonic lattices with self-defocusing nonlinearity. (a) lattice beam, where the circle shows the location of the input vortex, (b) the interferogram of the input vortex with a tilted plane wave to show the phase singularity, (c) self-trapped vortex pattern, (d) the zoom-in interferogram of (c) with a tilted plane wave, and (e) the Fourier-space spectrum of the $m = 2$ vortex soliton. The circles in (b) and (d) show the position of the phase singularity.

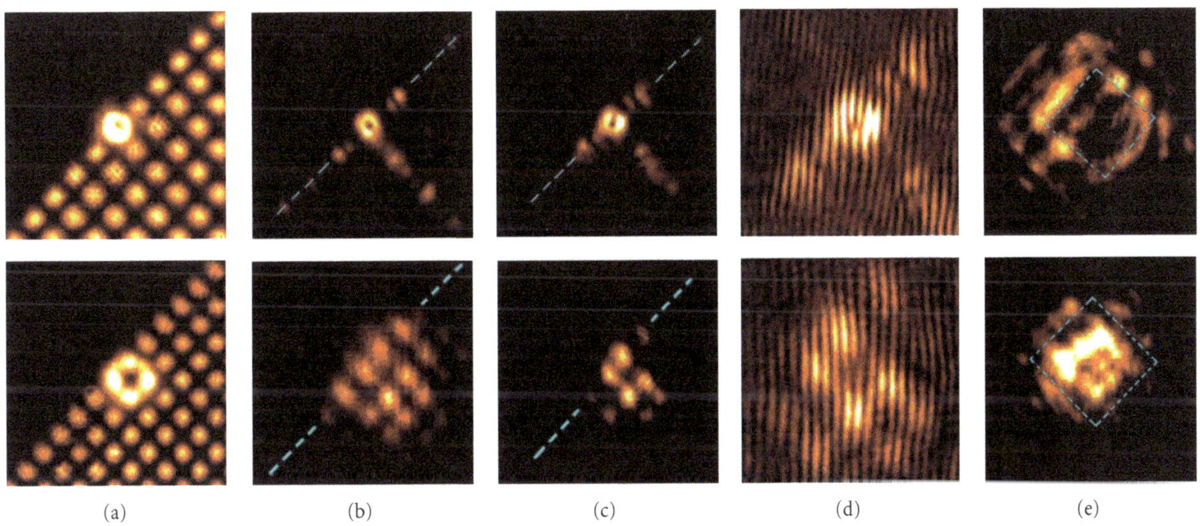

FIGURE 5: Experimental results of vortex self-trapping under single site surface excitation (top) and four-site surface excitation (bottom). (a) Lattice beam superimposed with the vortex beam, (b) linear and (c) nonlinear output of the vortex beam, where blue dashed line indicates the surface location, (d) zoom-in interferogram of (c) with an inclined plane wave, and (e) k-space spectrum of (c). The dashed square in (e) marks the boundary of the first Brillouin zone.

3.2. Discrete Surface Vortex Solitons. Discrete surface solitons form an important family of discrete solitons that exist at interface between the semi-infinite photonic lattices and a homogenous media [32]. Now, let us use the SCV beam to excite at the surface of the photonic lattice rather than inside the lattice under the self-focusing nonlinearity. Surface vortex solitons have also been theoretically considered [33, 34]. Here, we present our experimental results of discrete surface vortex solitons with different excitation conditions. Both single-site and four-site excitation schemes are considered. We find that a SCV beam under the four-site excitation can evolve into a stable discrete surface vortex soliton in the semi-infinite gap, while under the single-site excitation, it evolves into a unstable quasilocalized surface state in the first photonic bandgap. The stability of such localized surface states depending on their initial excitation has been studied numerically for longer propagation distance [35].

First, we use the single-site excitation so the vortex ring covers only one lattice site at the surface (Figure 5(a) top). The vortex beam exhibits asymmetric discrete diffraction when its intensity is low but self-traps at the surface site when its intensity is high while keeping the bias field (2.2 kV/cm) unchanged. Typical experimental results are presented in Figure 5. The lattice spacing is about 30 μm. When the nonlinearity is absent (the intensity ration of vortex-to-lattice is less than 1 : 10), the vortex beam diffracts as shown in Figure 5(b). The diffraction tails have double peak intensity at each lattice site, indicating the excitation of second-band Bloch modes. The tail going into the lattice is much longer than that at the lattice surface, owing to surface enhanced reflection [36, 37]. As we increase the intensity ratio to 1 : 3, self-trapping of the vortex at the lattice surface occurs when the vortex confines more at the initial excitation site, maintaining a donut-shaped pattern (Figure 5(c) top). To verify its phase singularity, a tilted plane wave is introduced to interfere with the surface vortex beam. The single fork remains in the interferogram, indicating that the phase singularity is preserved during the 1 cm experimental propagation length. Since the input vortex beam is tightly focused, its initial k-space spectrum extends beyond the first BZ (Figure 5(e) top).

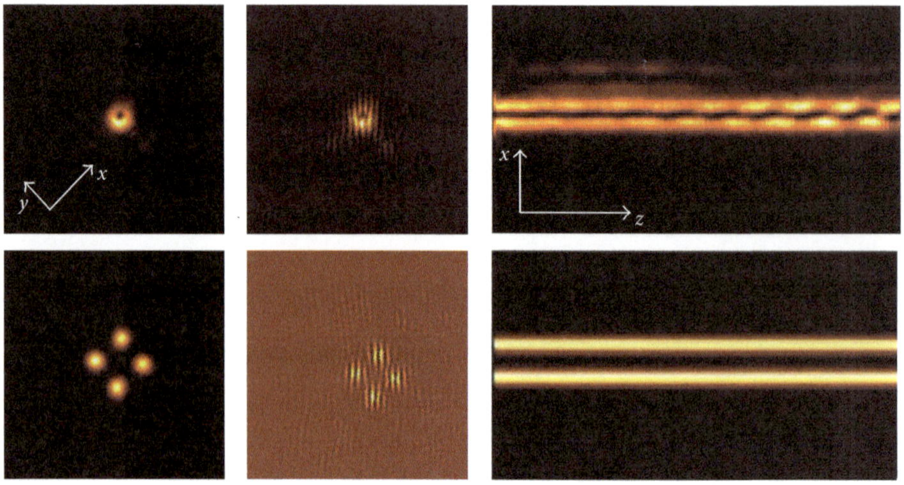

FIGURE 6: Numerical simulation of single-site (top) and four-site (bottom) surface excitation of a SCV after 4-cm of nonlinear propagation. Shown are (a) vortex output intensity pattern and (b) its corresponding interferogram, and (c) side-view of the vortex propagation to $z = 4$ cm.

Such input condition excites high-band Bloch modes. In the case of the self-focusing nonlinearity, this leads to that the power spectrum is concentrated at the normal diffraction regions outside the first BZ but close to the high-symmetry X-points, as shown in Figure 5(e) top. Apparently, such self-trapped surface vortex states arise from excitation of Bloch modes near the X-points of the second Bloch band. However, numerical results to longer propagation distance (4 cm) shows that these quasilocalized surface states residing in the first photonic bandgap are unstable and thus would break up after long-distance propagation [35].

Next, we use the four-site excitation so the vortex ring is expanded to cover four nearest lattice sites at the surface (the vortex core locates in the index minimum as off-site excitation, as shown in Figure 5(a) bottom). Under appropriate nonlinear conditions, the SCV self-traps into a discrete surface vortex soliton. Typical experimental results are presented in the bottom panels of Figure 5. At a low nonlinearity, the vortex beam exhibits discrete diffraction (Figure 5(b)). Notice that in this case, the diffracted intensity pattern does not have fine features of double-peak in each lattice site, different from that of single-site excitation. This suggests that, under this condition, the expanded vortex excites the first-band Bloch modes. At a high nonlinearity, self-trapping of the vortex beam into a four-spot pattern is achieved (Figure 5(c)).

Due to that the induced surface is slightly deformed, the four-spot does not match a perfect square pattern. To monitor the phase structure of the self-trapped vortex, an inclined plane wave is again introduced for interference. The vortex singularity persists clearly in the nonlinear output (Figure 5(d) bottom). The nonlinear spectrum is also different from that of single-site excitation shown in Figure 5(e), since now most of the energy is concentrated in the central region of the normal diffraction within the first BZ. This indicates that the surface vortex soliton is formed in the semi-infinite gap with modes primarily from near the Γ-point of the first Bloch band. As shown below, this four-site surface vortex soliton can be stable and thus remains invariant for long distance propagation.

Experimentally, it is difficult to test the stability of the self-trapped surface vortex structures due to the limited crystal length (i.e., 1 cm). Thus, we study the stability of the self-trapped vortex states by numerical simulation with parameters similar to the experimental conditions, but to much longer propagation distance (4 cm). For the four-site excitation vortex in the semi-infinite gap is stable, but for the single-site vortex in the Bragg reflection gap is unstable as shown in Figure 6. The nonlinear output of the SCV under single-site excitation broke up after 4 cm propagation, while the four-site surface soliton is nearly invariant. This stability analysis is also confirmed theoretically [35].

4. Linear Localization of Optical Vortices in Photonic Lattices with Negative Defects

In the previous experiments, the localization of the vortex beam inside and at the surface of the photonic lattice is related to the nonlinearity-induced defect by the vortex beam itself. These self-trapped nonlinear localized topological states can be considered as self-induced *nonlinear* defect modes (DMs). A nature question arises whether vortices can be localized in a fabricated defect as *linear* defect modes within photonic lattices? In fact, band gap guidance of a Gaussian beam in photonic lattice with a negative defect has been realized in optically induced and fs-laser written waveguide arrays [38, 39].

Different from light guided in higher refractive-index region due to the total internal reflection, light guidance in negative defects results from the Bragg reflection in periodic structures. Here, we show that both a SCV and DCV beams

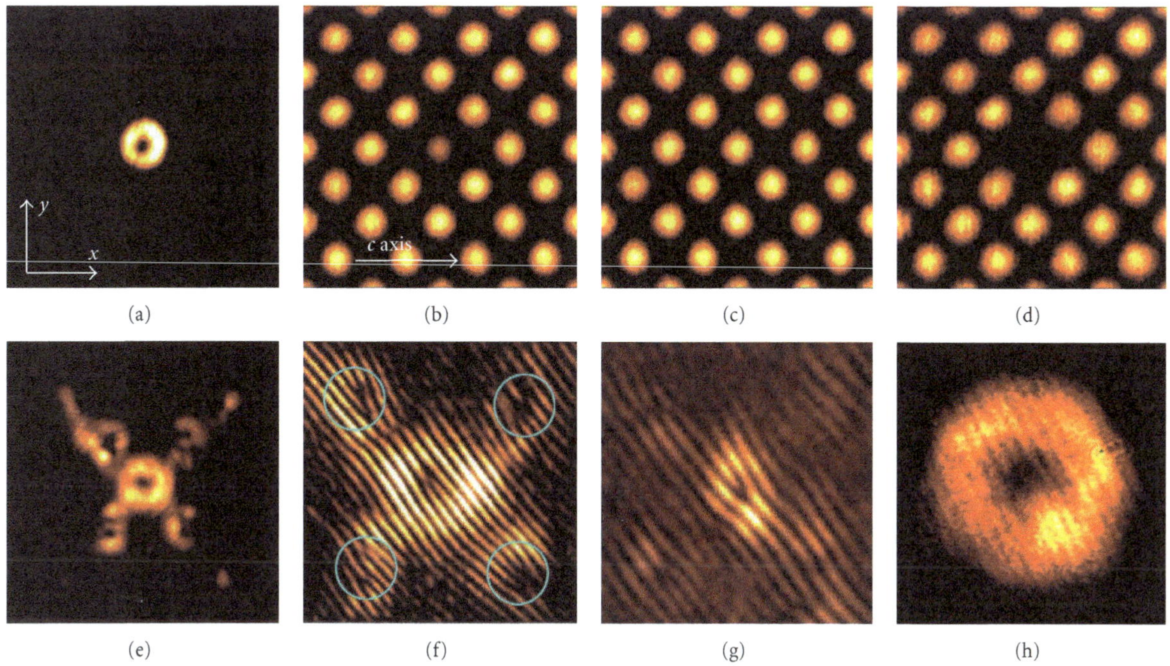

FIGURE 7: Experimental results of bandgap guidance of a vortex beam in a tunable negative defect. (a) Vortex at input. (b)–(d) Induced lattices with nonzero-intensity defect, no defect, and zero-intensity defect, respectively. (e), (f) Vortex output from the defect in (b) and its zoom-in interferogram. The circles in (f) mark the location of the vortex pairs. (g) Interferogram when the vortex is excited at nondefect site. (h) Vortex diffraction output when lattice is absent.

can be guided in the negative defects in 2D photonic lattices provided the defect strength is not too deep. The experimental setup is similar to Figure 1, except that the ordinarily-polarized beam is split again into two lattice-inducing beams: one has an uniform periodic intensity pattern as shown in Figure 7(c) (peak intensity I_c), and the other has a zero-intensity defect site on otherwise uniform periodic pattern as shown in Figure 7(d) [peak intensity I_d]. Superimposing the two lattice-inducing beams of equal lattice spacing results in a nonzero but tunable defect [an example obtained at $I_d/I_c = 1/2$ is shown in Figure 7(b)]. The tunability of the defect strength is realized by varying the intensity ratio I_d/I_c (e.g., at $I_d/I_c \gg 1$, the defect is deep or close to zero intensity, but at $I_d/I_c \ll 1$, the defect is shallow or close to washout).

Typical experimental results are presented in the bottom panels of Figure 7. Under proper experimental conditions, the input vortex beam (20 μm in diameter) is guided into the defect channel, maintaining a donut-shaped pattern in the defect site with "tails" extending along two principal axes of the square lattice. This result was obtained at 42 μm lattice spacing, 2.3 kV/cm, and $I_d/I_c = 1/2$. For comparison, the same vortex beam diffracts dramatically when the defect is removed, showing no self-action of the vortex beam under the same bias condition (Figure 7(h)). Fine features can be seen in the interferogram obtained by interference between the output vortex beam and an inclined broad beam (quasiplane wave). While the phase singularity is maintained in the vortex center, additional vortex pairs are evident along the four "tails" away from the defect site, manifested by two close but separated fork fringes in the interferogram (Figure 7(f)). These fine structures in the "tails" provide us a way to identify the properties of the vortex DMs. By comparing with the theoretical results [40], it is apparent that the input SCV beam has evolved into a high-gap DM, whose propagation constant resides between the second and third Bloch bands. We emphasize that the localization of the vortex beam arises from the linear bandgap guidance, quite different from stationary propagation of second-band vortex solitons where the vortex beam itself creates a positive defect with self-focusing nonlinearity [17]. For comparison, the interferogram from the vortex beam exiting a normal lattice site is shown in Figure 7(g), where a single vortex (instead of vortex pair) is observed along each "tail."

The formation of the vortex DMs depends on the defect strength. A series of experiments is performed to illustrate the influence of the defect strength. To do so, we keep the bias field and I_c fixed but tune the defect strength by gradually varying I_d. We found that the vortex can be guided only when the defect is not too deep (or the intensity in the defect site is not close to zero). Examples of varying the intensity ratio I_d/I_c or the bias field (both control the induced index change at the defect site) are shown in Figure 8. When the biased field is set at 2.6 kV/cm, guidance of the vortex beam is clearly observed at $I_d/I_c = 1$ (Figure 8(a)) (which corresponds to a half-intensity defect). However, once the uniform lattice beam is removed ($I_c = 0$), the vortex beam cannot be confined in the defect site under the same bias condition (Figure 8(b)). At this zero-intensity defect, the vortex DM cannot form, even at a decreased bias

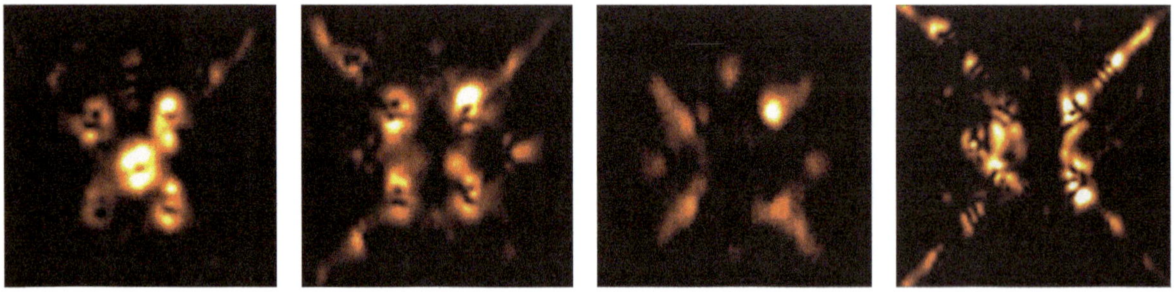

(a) (b) (c) (d)

Figure 8: Experimental results of SCV beam output after propagating through (a) a half-intensity ($I_d/I_c = 1/1$) defect and (b) a zero-intensity defect at a fixed bias field of 2.6 kV/cm. (c), (d) Output of the vortex beam through the zero-intensity defect when the bias field is decreased to 1.4 kV/cm and increased to 3.2 kV/cm, respectively.

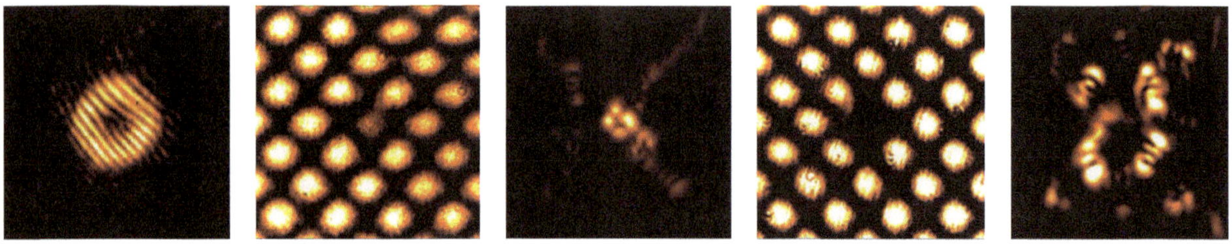

(a) (b) (c) (d) (e)

Figure 9: Experimental results of bandgap guidance f a DCV beam in a tunable negative defect. (a) interference pattern of input vortex with an inclined plane wave (zoomed in), (b) lattice beam with a nonzero-intensity defect, (c) linear output of the doubly charged vortex beam exiting the defect channel in the lattice (b), (d) lattice beam with a zero-intensity defect, and (e) linear output of the DCV through the defect in lattice (d).

field of 1.4 kV/cm (Figure 8(c)) or an increased bias field of 3.2 kV/cm (Figure 8(d)). Thus, bandgap guidance of the vortex clearly occurs in the half-intensity defect but deteriorates in the zero-intensity defect. There might be a threshold for the defect strength that would support the vortex DM, but experimentally it is difficult to determine such a threshold. Theoretically, it is found that, when the defect gets deeper the propagation constant of the DM moves deeper, toward a higher band-gap. Because of the limited number of fully opened higher gaps, the propagation constant of the DM may move into a higher Bloch band and eventually localized DM disappears [41].

Finally, a DCV beam is used as a probe beam sent into the tunable negative defect to check whether a high-order optical vortex can be guided in it. We find the DCV cannot be guided and it tends to break up again into a quadruple-like structure when the defect strength is not too deep. Typical experimental results are shown in Figure 9. These results were obtained when the lattice spacing is 42 μm, the biased field is about 3.4 kV/cm, and the intensity ratio is $I_d/I_c = 2$. As before with nonlinear propagation of DCVs, the linear output is no longer in a donut shape, but rather turns into a quadrupole pattern with the symmetry similar to the lattice as shown in Figure 9(c). However, when the defect strength is very strong (i.e., intensity at the defect site is close to zero as in Figure 9(d), even the quadrupole-like pattern cannot be maintained (Figure 9(e)). Under this condition, there is no light guided at the defect site at all when the bias field or the lattice intensity is increased.

5. Summary and Conclusions

We have studied experimentally the dynamical propagation of optical vortices in optically induced square photonic lattices under different settings. Inside the uniform photonic lattices, we find that the SCV beams self-trap into stable gap vortex solitons under four-site excitation with a defocusing nonlinearity. However, the DCV beams tend to break up and turn into quadrupole gap solitons. With geometrically extension eight-site excitation, the DCV beams can evolve into stable gap vortex solitons. Both the SCV and DCV gap solitons do not bifurcate from modes at the band edges, which is different from all the previously observed gap solitons. At the lattice surface, the SCV beams turn into stable surface vortex solitons in the semi-infinite gap under the four-site excitation with a self-focusing nonlinearity. However, they evolve into unstable quasivortex solitons with propagation constants in the first Bragg gap under the single-site excitation. In photonic lattices with negative defects, the SCV beams can be guided as linear vortex defect modes under appropriate defect conditions due to the bandgap guidance, while the DCV beams are unstable and evolve into quadrupole-like defect modes. Our results about such linear and nonlinear localized vortex states may prove to be relevant to similar vortex phenomena in other discrete systems beyond optics.

Acknowledgments

This work was supported in part by the 973 Program (2007CB613203) and NSFC (10904078) in China and by US National Science Foundation (NSF) and Air Force Office of Scientific Research (AFOSR). The authors thank J. Yang, P. G. Kevrekidis, K. Law, and X. Wang for helpful discussions and assistance.

References

[1] A. S. Desyatnikov, Y. S. Kivhsar, and L. Torner, "Optical vortices and vortex solitons," in *Progress in Optics*, E. Wolf, Ed., vol. 47, pp. 219–319, North-Holland, Amsterdam ,The Netherlands, 2005.

[2] M. R. Dennis, Y. S. Kivshar, M. S. Soskin, and G. A. Swartzlander Jr., "Singular optics: more ado about nothing," *Journal of Optics A*, vol. 11, no. 9, Article ID 090201, 2009.

[3] G. A. Swartzlander and C. T. Law, "Optical vortex solitons observed in Kerr nonlinear media," *Physical Review Letters*, vol. 69, no. 17, pp. 2503–2506, 1992.

[4] J. Christou, V. Tikhonenko, Y. S. Kivshar, and B. Luther-Davies, "Vortex soliton motion and steering," *Optics Letters*, vol. 21, no. 20, pp. 1649–1651, 1996.

[5] G. Duree, M. Morin, G. Salamo et al., "Dark photorefractive spatial solitons and photorefractive vortex solitons," *Physical Review Letters*, vol. 74, no. 11, pp. 1978–1981, 1995.

[6] Z. Chen, M. Segev, D. W. Wilson, R. E. Muller, and P. D. Maker, "Self-trapping of an optical vortex by use of the bulk photovoltaic effect," *Physical Review Letters*, vol. 78, no. 15, pp. 2948–2951, 1997.

[7] Z. Chen, M. F. Shih, M. Segev, D. W. Wilson, R. E. Muller, and P. D. Maker, "Steady-state vortex-screening solitons formed in biased photorefractive media," *Optics Letters*, vol. 22, no. 23, pp. 1751–1753, 1997.

[8] Z. Chen, M. Mitchell, M. Segev, T. H. Coskun, and D. N. Christodoulides, "Self-trapping of dark incoherent light beams," *Science*, vol. 280, no. 5365, pp. 889–891, 1998.

[9] W. J. Firth and D. V. Skryabin, "Optical solitons carrying orbital angular momentum," *Physical Review Letters*, vol. 79, no. 13, pp. 2450–2453, 1997.

[10] C. C. Jeng, M. F. Shih, K. Motzek, and Y. Kivshar, "Partially incoherent optical vortices in self-focusing nonlinear media," *Physical Review Letters*, vol. 92, no. 4, pp. 439041–439044, 2004.

[11] I. Towers and B. A. Malomed, "Stable (2+1)-dimensional solitons in a layered medium with sign-alternating Kerr nonlinearity," *Journal of the Optical Society of America B*, vol. 19, no. 3, pp. 537–543, 2002.

[12] D. Briedis, D. E. Petersen, D. Edmundson, W. Krolikowski, and O. Bang, "Ring vortex solitons in nonlocal nonlinear media," *Optics Express*, vol. 13, no. 2, pp. 435–443, 2005.

[13] B. A. Malomed and P. G. Kevrekidis, "Discrete vortex solitons," *Physical Review E*, vol. 64, no. 2, pp. 266011–266016, 2001.

[14] J. Yang and Z. H. Musslimani, "Fundamental and vortex solitons in a two-dimensional optical lattice," *Optics Letters*, vol. 28, no. 21, pp. 2094–2096, 2003.

[15] D. N. Neshev, T. J. Alexander, E. A. Ostrovskaya et al., "Observation of discrete vortex solitons in optically induced photonic lattices," *Physical Review Letters*, vol. 92, no. 12, p. 123903, 2004.

[16] J. W. Fleischer, G. Bartal, O. Cohen et al., "Observation of vortex-ring "discrete" solitons in 2D photonic lattices," *Physical Review Letters*, vol. 92, no. 12, p. 123904, 2004.

[17] G. Bartal, O. Manela, O. Cohen, J. W. Fleischer, and M. Segev, "Observation of second-band vortex solitons in 2D photonic lattices," *Physical Review Letters*, vol. 95, no. 5, Article ID 053904, pp. 1–4, 2005.

[18] A. Bezryadina, E. Eugenieva, and Z. Chen, "Self-trapping and flipping of double-charged vortices in optically induced photonic lattices," *Optics Letters*, vol. 31, no. 16, pp. 2456–2458, 2006.

[19] B. Terhalle, D. Göries, T. Richter et al., "Anisotropy-controlled topological stability of discrete vortex solitons in optically induced photonic lattices," *Optics Letters*, vol. 35, no. 4, pp. 604–606, 2010.

[20] B. Terhalle, T. Richter, K. J. H. Law et al., "Observation of double-charge discrete vortex solitons in hexagonal photonic lattices," *Physical Review A*, vol. 79, no. 4, pp. 043821–043828, 2009.

[21] B. Terhalle, T. Richter, A. S. Desyatnikov et al., "Observation of multivortex solitons in photonic lattices," *Physical Review Letters*, vol. 101, no. 1, pp. 013903–013904, 2008.

[22] K. J. H. Law, A. Saxena, p. G. Kevrekidis, and A. R. Bishop, "Stable structures with high topological charge in nonlinear photonic quasicrystals," *Physical Review A*, vol. 82, no. 3, Article ID 035802, 2010.

[23] H. Martin, E. D. Eugenieva, Z. Chen, and D. N. Christodoulides, "Discrete solitons and soliton-induced dislocations in partially coherent photonic lattices," *Physical Review Letters*, vol. 92, no. 12, p. 123902, 2004.

[24] Z. Chen, H. Martin, E. D. Eugenieva, J. Xu, and A. Bezryadina, "Anisotropic enhancement of discrete diffraction and formation of two-dimensional discrete-soliton trains," *Physical Review Letters*, vol. 92, no. 14, p. 143902, 2004.

[25] N. K. Efremidis, J. Hudock, D. N. Christodoulides, J. W. Fleischer, O. Cohen, and M. Segev, "Two-dimensional optical lattice solitons," *Physical Review Letters*, vol. 91, no. 21, pp. 213906/1–213906/4, 2003.

[26] D. Song, C. Lou, L. Tang et al., "Self-trapping of optical vortices in waveguide lattices with a self-defocusing nonlinearity," *Optics Express*, vol. 16, no. 14, pp. 10110–10116, 2008.

[27] G. Bartal, O. Cohen, H. Buljan, J. W. Fleischer, O. Manela, and M. Segev, "Brillouin zone spectroscopy of nonlinear photonic lattices," *Physical Review Letters*, vol. 94, no. 16, Article ID 163902, 2005.

[28] C. Lou, X. Wang, J. Xu, Z. Chen, and J. Yang, "Nonlinear spectrum reshaping and gap-soliton-train trapping in optically induced photonic structures," *Physical Review Letters*, vol. 98, no. 21, Article ID 213903, 2007.

[29] L. Tang, C. Lou, X. Wang et al., "Observation of dipole-like gap solitons in self-defocusing waveguide lattices," *Optics Letters*, vol. 32, no. 20, pp. 3011–3013, 2007.

[30] J. Wang and J. Yang, "Families of vortex solitons in periodic media," *Physical Review A*, vol. 77, no. 3, Article ID 033834, 2008.

[31] K. J. H. Law, D. Song, p. G. Kevrekidis, J. Xu, and Z. Chen, "Geometric stabilization of extended S=2 vortices in two-dimensional photonic lattices: theoretical analysis, numerical computation, and experimental results," *Physical Review A*, vol. 80, no. 6, Article ID 063817, 2009.

[32] S. Suntsov, K. G. Makris, G. A. Siviloglou et al., "Observation of one- and two-dimensional discrete surface spatial solitons,"

Journal of Nonlinear Optical Physics and Materials, vol. 16, no. 4, pp. 401–426, 2007.

[33] H. Susanto, p. G. Kevrekidis, B. A. Malomed, R. Carretero-González, and D. J. Frantzeskakis, "Discrete surface solitons in two dimensions," *Physical Review E*, vol. 66, Article ID 046602, 2007.

[34] Y. V. Kartashov, A. A. Egorov, V. A. Vysloukh, and L. Torner, "Surface vortex solitons," *Optics Express*, vol. 14, no. 9, pp. 4049–4057, 2006.

[35] D. Song, C. Lou, K. J. H. Law et al., "Self-trapping of optical vortices at the surface of an induced semi-infinite photonic lattice," *Optics Express*, vol. 18, no. 6, pp. 5873–5878, 2010.

[36] S. Suntsov, K. G. Makris, D. N. Christodoulides et al., "Observation of discrete surface solitons," *Physical Review Letters*, vol. 96, no. 6, Article ID 063901, 2005.

[37] X. Wang, A. Bezryadina, Z. Chen, K. G. Makris, D. N. Christodoulides, and G. I. Stegeman, "Observation of two-dimensional surface solitons," *Physical Review Letters*, vol. 98, no. 12, Article ID 123903, 2007.

[38] I. Makasyuk, Z. Chen, and J. Yang, "Band-gap guidance in optically induced photonic lattices with a negative defect," *Physical Review Letters*, vol. 96, no. 22, Article ID 223903, 2006.

[39] A. Szameit, Y. V. Kartashov, M. Heinrich et al., "Observation of two-dimensional defect surface solitons," *Optics Letters*, vol. 34, no. 6, pp. 797–799, 2009.

[40] J. Wang, J. Yang, and Z. Chen, "Two-dimensional defect modes in optically induced photonic lattices," *Physical Review A*, vol. 76, no. 1, Article ID 013828, 2007.

[41] D. Song, X. Wang, D. Shuldman et al., "Observation of band-gap guidance of optical vortices in a tunable negative defect," *Optics Letters*, vol. 35, no. 12, pp. 2106–2108, 2010.

21

Volumetric Diffuse Optical Tomography for Small Animals Using a CCD-Camera-Based Imaging System

Zi-Jing Lin,[1] Haijing Niu,[2] Lin Li,[1] and Hanli Liu[1]

[1] Department of Bioengineering, Joint Graduate Program between University of Texas at Arlington and University of Texas Southwestern Medical Center, University of Texas at Arlington, Arlington, TX 76019, USA
[2] State key Laboratory of Cognitive Neuroscience and Learning, Beijing Normal University, Beijing 10875, China

Correspondence should be addressed to Hanli Liu, hanli@uta.edu

Academic Editor: Javier A. Jo

We report the feasibility of three-dimensional (3D) volumetric diffuse optical tomography for small animal imaging by using a CCD-camera-based imaging system with a newly developed depth compensation algorithm (DCA). Our computer simulations and laboratory phantom studies have demonstrated that the combination of a CCD camera and DCA can significantly improve the accuracy in depth localization and lead to reconstruction of 3D volumetric images. This approach may present great interests for noninvasive 3D localization of an anomaly hidden in tissue, such as a tumor or a stroke lesion, for preclinical small animal models.

1. Introduction

Over the last decade, diffuse optical tomography (DOT) has become an emerging medical imaging modality [1, 2], which permits near infrared (NIR) light (650–900 nm) to penetrate deep tissues in several centimeters noninvasively. This technique has been investigated for several clinical applications, particularly for monitoring of functional brain activities [3–5] and detection of breast cancer [6, 7]. The fundamental principle of DOT is that the detected NIR signals reflect optical properties of the underlying biological tissues [8, 9] and that the spatial distribution of such measured optical properties can be used to reconstruct optical tomographic images.

In the meantime, small animal models have been often utilized in preclinical research in order to investigate a variety of human diseases. Commonly used modalities to image small animals include magnetic resonance imaging (MRI) [10], computed tomography (CT) [11], positron emission tomography (PET) [12], and single photon emission computed tomography (SPECT) [13]. As compared to MRI and CT, DOT is able to provide functional sensitivity; as compared to CT, PET, and SPECT, DOT is radiation-free. More recently, DOT has become an imaging methodology in preclinical studies for certain diseases, such as stroke [14] and cancer [15].

To date, fiber-based and charge-coupled device- (CCD-) camera-based DOT techniques are well developed and widely performed in both human and animal studies. A fiber-based DOT system has several advantages, namely, being compatible with different geometry and shape of a measured organ and having low noise because of direct contact of fibers on the tissue surface. However, a limited surface space on small animals is often a constraint to place many fiber optodes and thus restricts the spatial resolution of reconstructed images. A CCD-camera-based DOT system eliminates such a problem since the CCD camera can serve as a detector array with possible thousands of virtual detectors and cover a wide field of view (FOV). Moreover, such a DOT system is simpler and more portable with lower cost, as compared to a fiber-based DOT system.

While DOT is a promising tool to image diseases noninvasively, one major limitation exists due to scattering nature of photons traveling in tissue. Namely, strong light scattering of biological tissue causes the detection sensitivity of DOT to attenuate exponentially with increased depth, resulting in poor depth localization in DOT. A variety of efforts have been made by different research groups to improve the accuracy in depth localization. A layer-based sigmoid adjustment (LSA) method was introduced to balance the sensitivity contrast in depth by directly adjusting the forward sensitivity matrix [16]. Spatially variant regularization (SVR) [17] was also introduced to compensate the decrease in measurement sensitivity with increase of depth by modifying the penalty term of regularization along depth. SVR has been utilized in the frequency-domain and CW-based DOT techniques for imaging human breast cancer and brain functions. Recently, we have developed a depth compensation algorithm (DCA) to significantly improve the accuracy of DOT in depth localization, as demonstrated by both laboratory phantom and human brain measurements [18, 19], using fiber-based DOT systems.

In this study, we wish to demonstrate the feasibility and improvement of three-dimensional (3D) volumetric DOT for small animal imaging by using CCD-camera-based DOT in combination with DCA. Our results are based on computer simulations and laboratory phantom experiments, confirming that the combination of a CCD camera and DCA is an effective approach for 3D volumetric images to be utilized in preclinical research.

2. Materials and Methodology

2.1. Implementation of a CCD-Camera-Based DOT System. We implemented a CCD-camera-based, NIR spectroscopic imaging system, which consisted of a broadband light source, a multiplexer, a 6-filter wheel, and a high-speed CCD camera (Figure 1(a)) with a pixel matrix of 496 × 656. Such a system allowed us to acquire the transmitted/reflected NIR signals from living tissues at specific wavelengths so as to obtain hemodynamic signals within a time scale of a hundred milliseconds. We performed CCD-camera calibration by utilizing laboratory tissue phantoms (Figure 1(b)) in order to (a) calibrate the CCD camera for its nonuniformity and to (b) examine the stability and linearity of the calibrated camera.

2.2. Correction of the CCD-Camera-Based DOT System. Numbers of factors affect the results of a CCD image. These factors include (1) small electric signals seen by the camera without any optical input, or dark current, (2) pixel-to-pixel sensitivity variations, and (3) nonuniform illumination of the CCD chip due to vignetting and dust. Flat field correction is a common technique [20] that can be used to obtain improved quality of CCD images. The flat field technique removes the effects due to dark current, pixel-to-pixel sensitivity variations, uneven illumination, and nonuniformity caused by vignetting and dust.

During the process of correction, we collected two types of images besides taking raw object images: (1) dark images without any optical input and (2) reference images with planer illumination. The dark images were needed to correct the electric current that flows through each pixel of the CCD array without any optical input. The planer-illuminated reference images were obtained by uniformly illuminating a highly reflecting reference plate, which has a reflectivity of 98% and a dimension of 200 × 200 mm^2. To approximately achieve a planer illumination, we used a white light lamp and placed it 1.5 meters away from the measurement site, with a good amount of light intensity, without saturating the camera.

The procedures used to perform flat field correction in this study are listed below: first, take a dark image by blocking the camera's entrance completely, followed by taking object images with a point-source illumination from one side of the object (see Figure 1(b)). The integration times were kept the same for both situations. Second, take a reference image with a planer illumination followed by the corresponding dark image, also using the same integration time for both. Then, corrected object images can be obtained with the following equation:

$$I = K\frac{O - D_O}{R - D_R}, \tag{1}$$

where I represents the corrected image, R is the reference image from the reference plate, O is the raw object image, D_O and D_R represent dark images taken with the corresponding integration times for the object and reference plate, respectively, and K is a constant and usually is equal to the intensity averaged over the area of reference image. To demonstrate the necessity of correcting a raw image, Figure 2 shows the cross-section profiles along X axis and Y axis of a CCD image before and after flat field correction. It is noticed that the center of FOV detects much higher light intensity than the edges before correction. After the correction, it is clearly shown that the uniformity on the edges is much improved while the correction is not perfect.

For quantitative comparison, we calculated maximal relative deviations between the central intensity and the intensity at the edge of the images for both the raw and corrected images, using the following expression:

$$\text{relative deviation} = \frac{\text{central intensity}}{\text{intensity at the edge}}, \tag{2}$$

The estimated maximal values of relative deviations were 40% and 10% for the raw image and corrected image, respectively. In most of the image regions, such deviations in the corrected images were less than 10%.

2.3. Stability and Linearity of the CCD Camera System. To check the stability of the CCD camera system, repeated measurements were taken at different times. We utilized a 1% Intralipid solution (Baxter Healthcare Corporation, Deerfield, IL) as a testing medium and took optical readings as the baseline images (I_B) at time "0", followed by other images (I_M) taken 1 minute, 5 minutes, 10 minutes, 30 minutes, 1 hour, and 1 day after the baseline measurement using the same experimental conditions (i.e., the same integration time). Each of the images was calibrated or corrected with the flat field correction and averaged over all the pixels per image to obtain an averaged light intensity for each respective image. Then, a ratio between the averaged intensity from each set of I_M and the averaged intensity from I_B (the baseline image) was obtained for all cases,

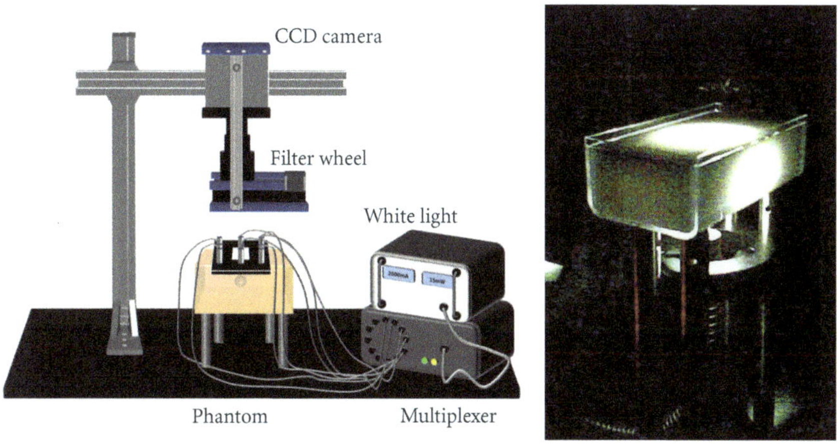

FIGURE 1: (a) Schematic diagram of a CCD-camera-based DOT system for laboratory measurement; (b) a tissue phantom which was sitting on the phantom supporter with the illumination light coming from the bottom.

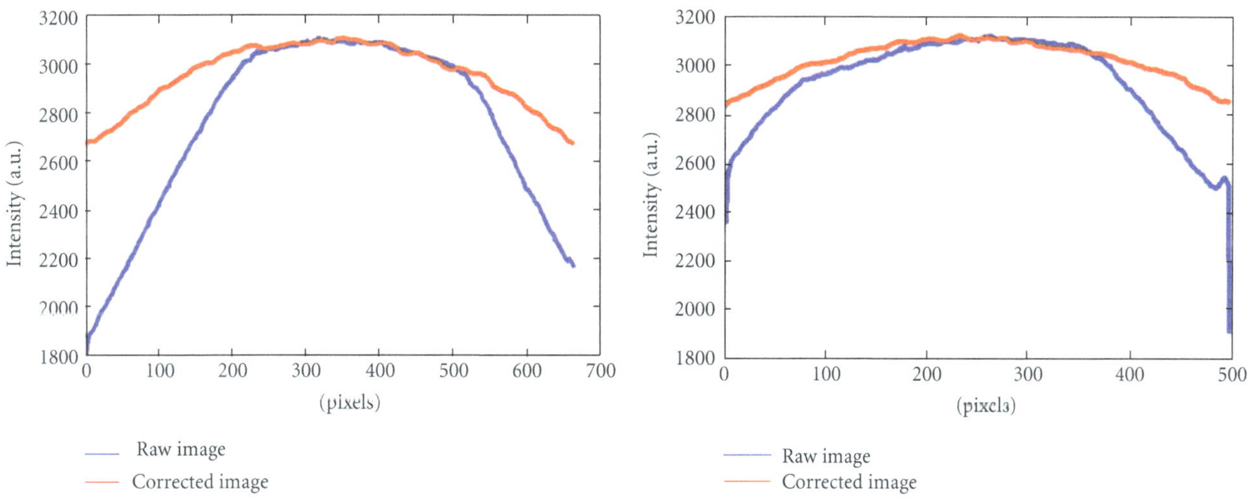

FIGURE 2: The cross-section profile of intensity from the raw and corrected image along (a) X axis and (b) Y axis.

as shown in Figure 3(a). It is very clear that the image intensities taken at different time intervals are fluctuated less than 1% of that of the baseline image, confirming good stability of the CCD camera system. Similar experiments were repeated using a 1.5% Intralipid solution to increase the light scattering property of the testing medium. The results are very consistent with the results from 1% Intralipid solution, demonstrating that a CCD camera has high stability with a fluctuation of intensity less than 1%.

In addition, we checked whether the intensity of light source could affect the results. We took numerous images using the same Intralipid concentration, but with different intensities of the light source. The averaged intensity of an image taken at light source level 5 was selected as the baseline, I_B; the ratios between the averaged intensities from other images, I_M, with different light source levels to the chosen baseline, I_M/I_B, are shown in Figure 3(b). This set of tests confirms that the level of the light source does not significantly affect the relative changes of our CCD camera output. Namely, the camera provides stable output images, independent of variable light levels.

Furthermore, to test the linearity of the CCD camera system, we performed two experiments: (1) one was to determine the image intensity changes versus absorption changes within the testing medium, and (2) the other one was to obtain the relationship between the image intensity measured versus changes in light scattering property within the testing medium. The tissue-mimic Intralipid solution was held in an acrylic box, which had a dimension of 10 cm × 5 cm × 3 cm (length × width × height). The acrylic box was wrapped by black tape except the top side of the box to prevent light from passing through the box wall.

For absorption changes, the experiments were performed using 1% Intralipid solutions with addition of 10 μL, 20 μL,

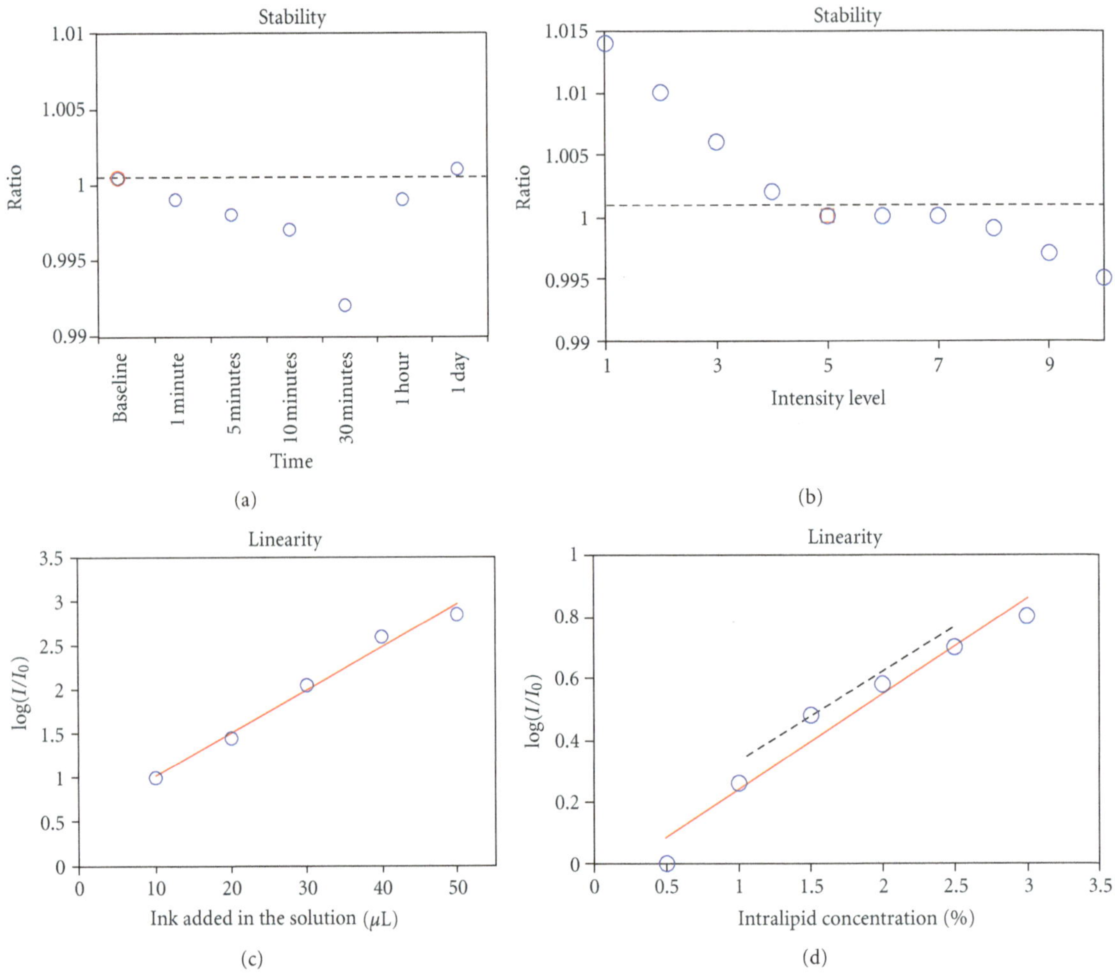

FIGURE 3: Ratio of the intensities (I_M/I_B) for a 1% Intralipid solution (a) at different times of measurement and (b) with different light intensities. The relationships between $\log(I_0/I)$ and (c) the volume of ink added into the testing solution for a 1% Intralipid solution, and (d) the concentration of Intralipid. Dashed line in (d) shows the linear relationship within Intralipid concentrations between 1% and 2.5%.

30 μL, 40 μL, and 50 μL of diluted ink. The CCD camera readings were taken throughout these Intralipid solutions, using the same integration time. We further plotted values of $\log(I_0/I)$ versus the volume of the ink added in the solution, where I_0 and I represented the averaged image intensities taken from the Intralipid solutions without ink and with different concentrations of ink. As seen in Figure 3(c), the data can be fitted well by a linear line. A similar behavior was seen for the 1.5% Intralipid case, confirming that a CCD camera has high-quality performance in its linearity with respect to light absorption.

For light scattering changes, the experiments were taken from multiple Intralipid solutions with concentrations of 0.5%, 1%, 1.5%, 2%, 2.5%, and 3% (using the same integration time). A plot of $\log(I_0/I)$ versus concentration of Intralipid is given in Figure 3(d), where I_0 and I represent the averaged image intensities taken from the Intralipid solutions at 0.5% and other concentrations. A good linear relationship is seen when Intralipid concentrations are between 1% and 2.5% (shown by the dashed line in Figure 3(d)). While changes of image intensity due to light scattering may non-linearly depend on the Intralipid concentration [21], such non-linearity is relatively small and can be nonsignificant within the scattering property range for biological tissues [22, 23]. Thus, the optical density measured from the CCD camera is linearly proportional to the light scattering changes, at the first-order approximation.

2.4. Optode Geometry for CCD-Camera-Based DOT. Instrument noise may appear during the measurements with a CCD camera that serves as multiple detectors; dark noise and nonuniformity of CCD chip sensitivity are two noise sources. Another problem often encountered is associated with the limited dynamic range of a CCD camera: when an individual pixel of the CCD array reaches its maximal intensity, the individual pixel well is occupied by electrons which can also overflow into the neighboring pixel wells. Such an overspill of electrons thereby distorts the actual signals not only at the saturated pixel but also at several adjacent pixels. In order to minimize the problem of signal saturations, the source fibers in this study were arranged around the FOV of the CCD camera, as shown in Figure 4. Such geometry can also be utilized while keeping the source fibers within the FOV.

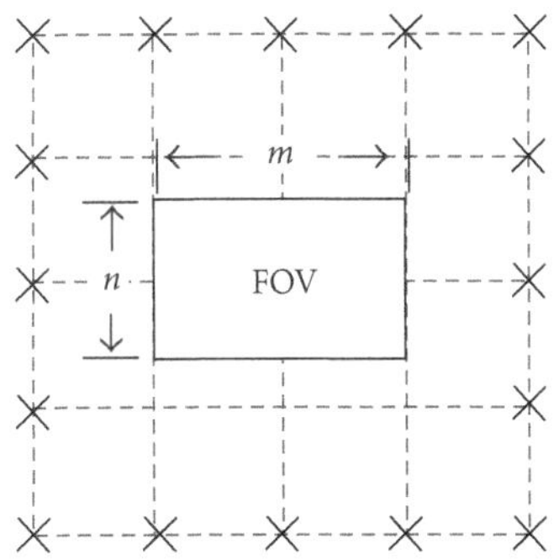

FIGURE 4: Optode geometry for CCD-camera-based DOT system. The dimension of FOV was $m \times n$ in mm^2. X represents possible locations of source fibers.

As shown in Figure 4, the dimension of FOV of the CCD camera was $m \times n\,\text{mm}^2$. Instead of using physical detectors, virtual detectors were used in this noncontact CCD camera imaging system. Virtual detectors refer to groups of CCD pixels combined within FOV of the CCD camera. The signal-to-noise ratio (SNR) at selected locations can be enhanced by averaging the signals of all pixels selected. For example, if we selected 10×10 pixels to be binned as a virtual detector, the entire FOV would have a total number of $\sim 50 \times 66$ virtual detectors for the DOT measurements. This implies that a CCD-camera-based DOT system may serve as a high-density imaging device with an easy-to-change detector array by grouping the pixels in different ways as detectors.

2.5. Theory of Image Reconstruction

2.5.1. Conventional Image Reconstruction in DOT. The photon propagation in a highly scattering medium follows the diffusion equation, and the analytic solution of the diffusion equation can be utilized to study optical signal changes due to absorption perturbation, as given by

$$\begin{aligned}\Delta\text{OD} &= -\log\left(\frac{\Phi_{\text{pert}}}{\Phi_0}\right) = -\frac{1}{2.3}\ln\left(\frac{\Phi_{\text{pert}}}{\Phi_0}\right) \\ &= \frac{1}{2.3}\int \Delta\mu_a(r)L(r)dr,\end{aligned} \tag{3}$$

where ΔOD is the change in optical density. Φ_0 is the photon fluence in a semi-infinite, homogeneous medium, and Φ_{pert} is the photon fluence with absorbers embedded. L is the effective pathlength of light through the tissue experiencing the absorption change. While (3) is known as the modified Beer-Lambert law, it can be generalized in a matrix form and written as $\mathbf{y} = \mathbf{Ax}$, where matrix $\mathbf{y}$ represents measured changes in optical density from all the measurements, matrix $\mathbf{x}$ corresponds to a volumetric vector of the unknown perturbation in absorption coefficient in the volume under study, and matrix $\mathbf{A}$ describes the sensitivity of measurements to the change in absorption within each voxel in the medium.

However, solving $\mathbf{y} = \mathbf{Ax}$ is nontrivial, involving an underdetermined, ill-posed inverse problem because the number of measurements is much fewer than the number of unknowns (i.e., voxels) to be reconstructed. In the current research field, to reconstruct a DOT image, the Tikhonov regularization [24] method has been employed; an image of $\mathbf{x}$ can be given by

$$\hat{x} = A^T\left(AA^T + \alpha S_{\max} I\right)^{-1} y, \tag{4}$$

where $\hat{x}$ is the reconstructed image of absorption perturbation; I is the identity matrix; $S_{\max}$ is the maximal singular value of matrix AA^T, and α is the regularization parameter.

2.5.2. Depth Compensation Algorithm. Because of the severe attenuation in light intensity as increase of propagation depth, the measurement sensitivity of DOT in deep tissue is significantly lower than that in superficial tissue. In order to overcome this problem, a new approach of DCA has been recently developed by introducing a weighted matrix $\mathbf{M}$, which has a pseudo-exponential magnitude with depth to counterbalance the loss of sensitivity of $\mathbf{A}$ in deep layers. The weighted matrix M is formed as

$$M = \text{diag}\left(M(A_L), M(A_{L-1}), \ldots, M(A_2), M(A_1)\right)^{\gamma}, \tag{5}$$

where $M(A_i)$ represents the maximum singular values for measurement sensitivities within the particular layer i which is decomposed from forward matrix A; γ is an adjustable power and can be varied between 0 and 3. The detailed derivation and validation of DCA can be found in references [18, 19]. The key component in DCA is the formation of a transformation matrix $\mathbf{A}^{\#}$, as defined by $\mathbf{A}^{\#} = \mathbf{AM}$. Then $\mathbf{A}^{\#}$ is used to replace the original sensitivity matrix, $\mathbf{A}$, in (4). The reconstructed image to be obtained using DCA can be written as follows:

$$\hat{x}_{\text{DCA}} = (AM)^T\left(AM(AM)^T + \alpha S'_{\max} I\right)^{-1} y, \tag{6}$$

where $S'_{\max}$ is the maximum eigenvalue of $AM(AM)^T$. Notice that $\hat{x}_{\text{DCA}}$ in (6) is not equal to $\hat{x}$ in (4) since matrix M is empirically introduced. It is known now that (6) yields minimum error in depth localization for reconstructed NIRS images, while the trade-off of DCA is an inaccurate quantification for absorption perturbation.

As demonstrated in our recent publications [18, 19], the power of depth compensation is controlled by the exponent parameter of γ in M. The larger the γ value is, the larger the weight is to compensate the measurement sensitivity toward a deeper layer, and vice versa. An inappropriate selection of γ could lead to overcompensation and give rise to inaccurate results. Our previous studies have suggested an optimal range of γ to be 1.0–1.6 for accurate depth localization using a fiber-based DOT system. In this study, we need to select and confirm a suitable range of γ for a CCD-camera-based DOT system.

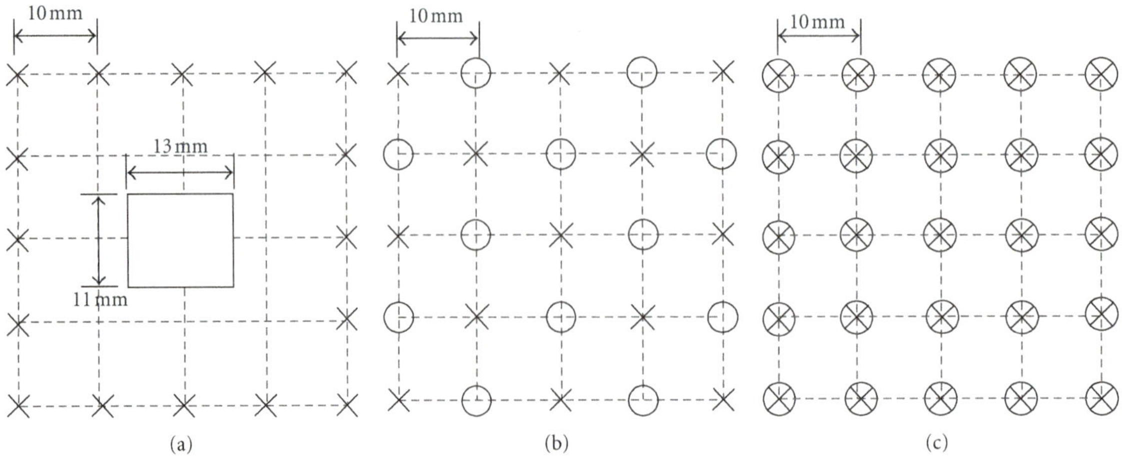

FIGURE 5: Scheme of optode geometries used in the simulation study. (a) to (c) are corresponded to G-I to G-III. Notice that the dimension of all dashed unit squares is $10 \times 10\,mm^2$. (x: source fibers; o: detector fibers; ⊗: bifurcated fibers).

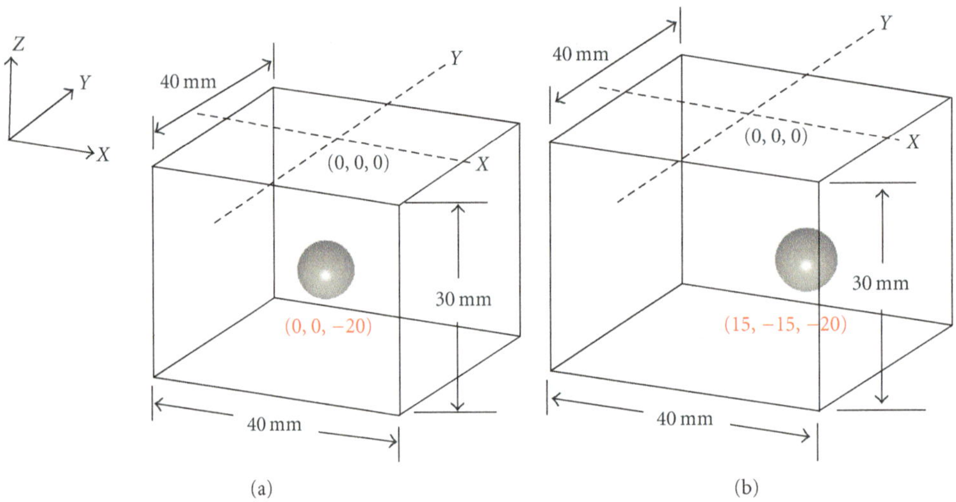

FIGURE 6: Schematic diagram of setups used in simulation study. (a) Setup I: the inclusion was located right below the center of measurement plane. (b) Setup II: the inclusion was located close to the right side corner of the measurement volume.

3. Results

3.1. Computer Simulations. To evaluate reconstruction performances using the source-camera geometry given in Section 2.4 (also replotted in Figure 5(a) and denoted as G-I), we performed computer simulations and compared the results with those that were derived using the source-detector geometries commonly seen in the fiber-based DOT system (shown in Figures 5(b) and 5(c)). Several sets of simulations were carried out under the conditions of (1) using different numbers of sources and virtual detectors, (2) without and (3) with utilization of DCA.

As shown in Figure 5(a), sixteen sources were used in G-I. In terms of the number of virtual detectors, we defined the dimension of FOV to be $13 \times 11\,mm^2$ and the dimension of each virtual detector to be $1\,mm^2$. Therefore, there were a total of $13 \times 11 = 143$ virtual detectors used in the simulations. The geometry shown in Figure 5(b) (G-II) is commonly used optode geometry with sources and detectors being alternately placed. The geometry given in Figure 5(c) (G-III) exhibits the optode setup with bifurcated fibers, that is, each fiber can serve as a source and detector.

The reconstruction performances were evaluated by two parameters: (1) the position error (PE) and (2) contrast-to-noise ratio (CNR) [25]. PE is the distance from the center of actual object to the center of the reconstructed object. The reconstructed object was defined by the pixels with values above half of the maximum value. CNR indicates whether the reconstructed object can be clearly distinguished from the background. A reconstructed image was divided into two regions: volume of interest (VOI) and volume of background (VOB). VOI was defined by the location and

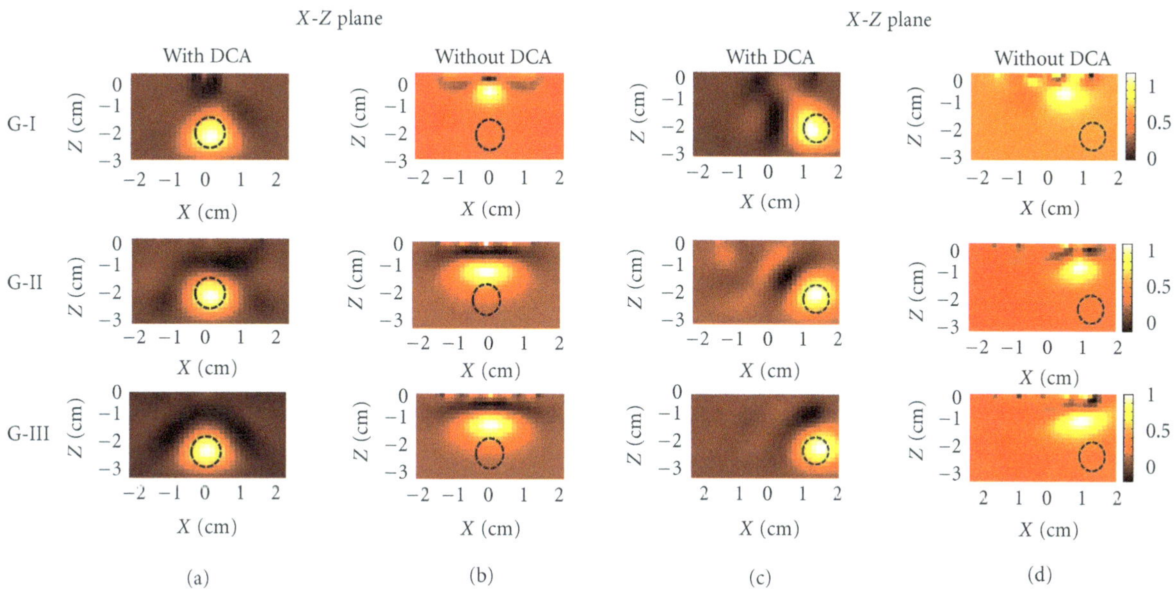

FIGURE 7: Reconstructed DOT images of a single object in X-Z plane: (a) and (c) with DCA ($\gamma = 1.3$), and (b) and (d) without DCA. Top-to-bottom rows were obtained from G-I to G-III (Figure 5).

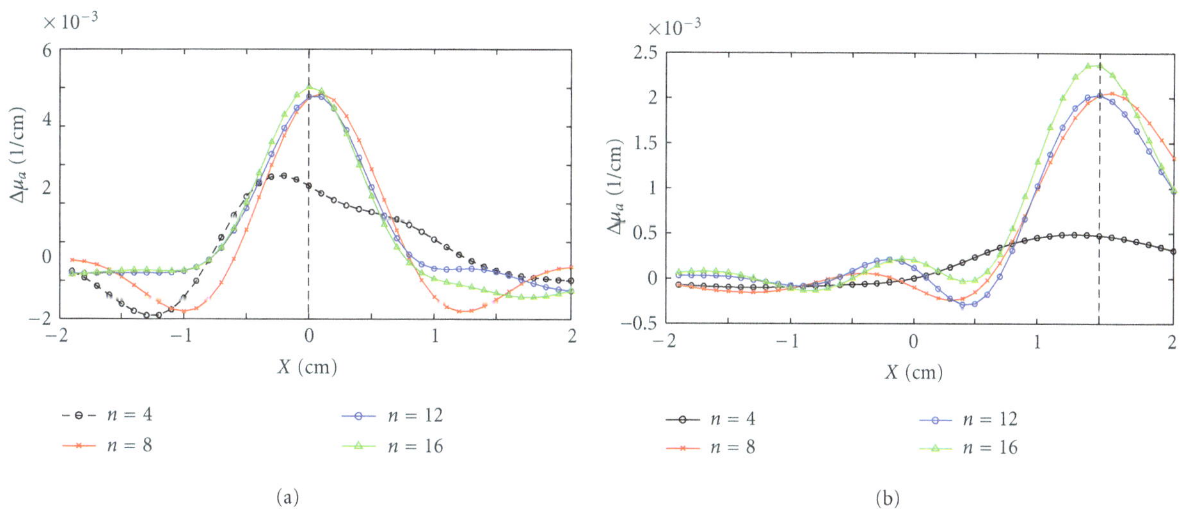

FIGURE 8: Profiles of reconstructed $\Delta\mu_a$ values along x-axis for (a) experimental Setup I and (b) experimental Setup II. Dashed vertical lines represent the actual location of the inclusion in x-axis.

size of the reconstructed object. VOB was defined by the remaining volume of the image. CNR can be calculated by

$$\mathrm{CNR} = \frac{\mu_{\mathrm{VOI}} - \mu_{\mathrm{VOB}}}{\left[w_{\mathrm{VOI}}\sigma_{\mathrm{VOI}}^2 + w_{\mathrm{VOB}}\sigma_{\mathrm{VOB}}^2\right]^{1/2}}, \tag{7}$$

where w_{VOI} and w_{VOB} are the weight factors of the VOI and VOB relative to the entire volume (i.e., VOI or VOB divided by the entire volume), μ_{VOI} and μ_{VOB} are the mean values of object and background volumes in a 3D reconstructed image, and σ_{VOI} and σ_{VOB} are the respective standard deviations.

3.1.1. Image Reconstruction of Embedded Objects. In the simulation, we considered a homogeneous semi-infinite medium with a volume of $40 \times 40 \times 30\,\mathrm{mm}^3$ (Figure 6). The origin of the coordinates was set at the center of top surface, as marked in the figure; this surface was also chosen to be the measurement surface in the simulation. As commonly used, a set of absorption and reduced scattering coefficients of $\mu_a = 0.1\,\mathrm{cm}^{-1}$ and $\mu_s' = 10\,\mathrm{cm}^{-1}$ were selected to simulate the background tissue.

Two different setups were utilized to evaluate the performances of optode geometries: Setup 1 is shown in Figure 6(a), where a spherical absorber with an 8 mm diameter was placed at $(X, Y, Z) = (0, 0, -20)$ (in mm), with a 3 : 1 contrast ratio in μ_a relative to the background μ_a. By looking at the optode geometry of the CCD-camera-based DOT imaging system, it is noticed that there are fewer overlapping measurements near four corners of FOV

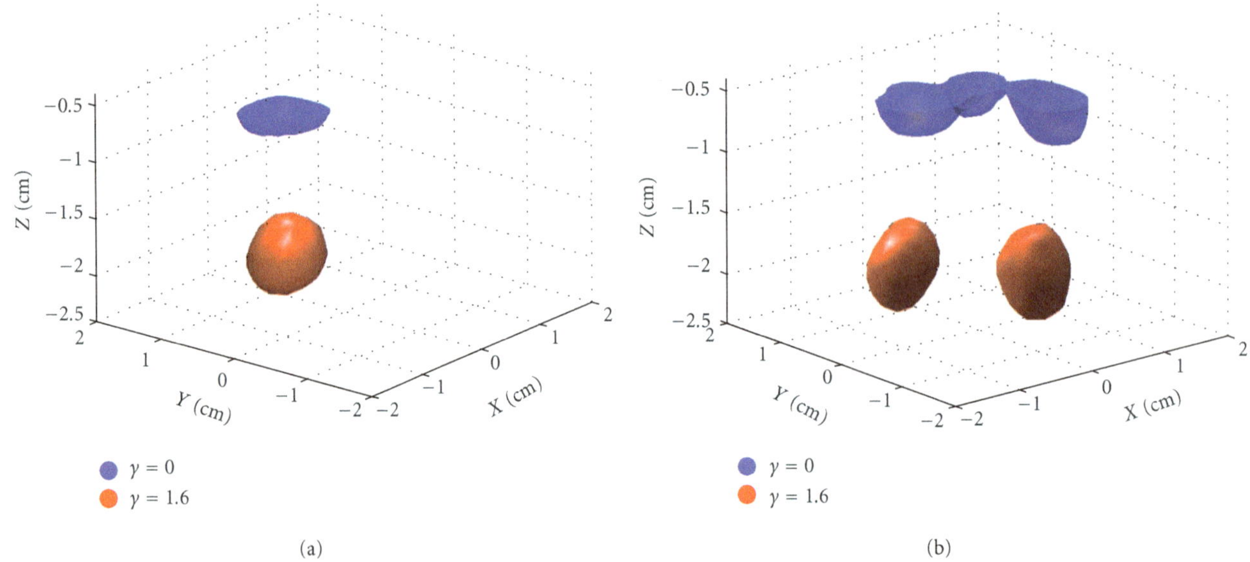

FIGURE 9: Three dimensional plots of reconstructed images for (a) one inclusion with DCA ($\gamma = 1.6$) and without DCA (i.e., $\gamma = 0$), and (b) two inclusions with DCA ($\gamma = 1.6$) and without DCA ($\gamma = 0$). Voxels with the reconstructed absorption values above 80% of the maximum value are selected as reconstructed objects.

of the camera, as compared to the center of FOV. This is because more overlapping measurements occur near the center of FOV when we pair most of the virtual detectors with given light sources in the current source-camera setup. It has also shown that a higher measurement density does not necessarily improve the spatial resolution significantly [26]. Thus, to evaluate and compare the effect of the number of overlapping measurements at the four corners, in Setup 2, one spherical absorber was placed near the right corner of the medium with respect to Setup 1, as shown in Figure 6(b). A random Gaussian noise of 1% in amplitude was added to the simulated measurement data. DCA was employed with a compensation power of $\gamma = 1.3$.

The reconstructed images in the X-Z plane from both setups are shown in Figure 7: Figures 7(a) and 7(b) are obtained from Setup 1 with the absorber placed at the center of measurement plane, while Figures 7(c) and 7(d) result from Setup 2. Also, the three rows from top to bottom in the figure correspond to the fiber-camera or fiber-fiber geometry of G-I to G-III, respectively. Figures 7(b) and 7(d) clearly show that the reconstructed images for all geometries are pulled towards the superficial surface along Z direction when DCA is not used or $\gamma = 0$. After DCA was employed with $\gamma = 1.3$, the reconstructed object was reconstructed at the expected depth (Figures 7(a) and 7(c)). By comparing the reconstructed images using the three geometries of G-I to G-III, we can see that the overall qualities of DOT images are all similar. However, in Figure 7(c), it is noticed that the DOT images with G-II and G-III exhibit a relatively better shape for the reconstructed objects, while G-I leads to a little distorted profile (tilted to the center) when the absorber was close to the edge of the imaging medium.

To further quantify the comparison among the three geometries, we have calculated respective values of CNR and PE from each geometry (Figure 5) and for two simulation setups (Figure 6), and the corresponding values are listed in Table 1. It is clearly seen that there are significant improvements of PE values with DCA, as we expected, for both setups, while the overall PE values in Setup 1 are better than those in Setup 2. Next, we focus on only the data with DCA: by close inspection on G-I to G-III for Setup 1, we observe that CNR values from all three optode geometries are not greatly different although G-III has a little higher CNR. A similar trend can be seen for Setup 2. Regarding PE values obtained only with DCA, no noticeable difference appears for all three geometries (less than 1 mm) in either of the setups.

Overall, the computer simulation results confirm that the geometry of G-I, which can be utilized as a CCD-camera-based DOT system, is able to deliver excellent DOT images at the same quality level as the other two fiber-based DOT systems do. Given the simplicity, low cost, and portability of such a system, it may be preferred in preclinical, small animal studies.

3.1.2. Optimization of the Number of Sources. It is common that the space available in preclinical studies around animals is restricted or limited to allow for many optodes. Thus, we next further investigate a possible optimal number of optodes that can still permit a reasonably good DOT image. In geometry G-I, only source fibers need to be arranged around FOV of the CCD camera. Therefore, several computer simulations were performed based on both setups (Figure 6) used in Section 3.1.1 in order to optimize the numbers of sources and virtual detectors. Specifically, the light sources were placed symmetrically around the origin to prevent the distortion of reconstructed images.

Figure 8 shows several cross-section profiles of the reconstructed $\Delta\mu_a$ values along x-axis with different numbers of sources while the absorber was placed at center (Figure 8(a))

TABLE 1: Summary of the CNR and PE values for three geometries.

	Setup 1 (absorber at center)						Setup 2 (absorber on side)					
	G-I		G-II		G-III		G-I		G-II		G-III	
	W	*W/O*	*W*	*W/O*	*W*	*W/O*	*W*	*W/O*	*W*	*W/O*	*W*	*W/O*
CNR	5.66	4.39	5.65	4.12	5.73	4.58	5.45	5.46	5.61	5.62	6.47	4.72
PE(cm)	0.20	0.82	0.20	0.75	0.20	0.72	0.36	0.81	0.35	0.79	0.35	0.78

* *W*: with DCA; *W/O*: without DCA.

TABLE 2: Summary of the CNR, PE, and CPU time for different numbers of sources.

	Setup I (absorber at center)				Setup II (absorber on side)			
Numbers of source	**4**	**8**	**12**	**16**	**4**	**8**	**12**	**16**
CNR	4.27	5.66	5.67	5.83	1.24	4.42	4.51	4.58
PE (cm)	0.41	0.20	0.20	0.20	0.88	0.35	0.32	0.32
CPU time (sec)	27.1	102.1	209.4	289.2	27.1	102.1	209.4	289.2

and close to the edge of medium (Figure 8(b)). It is clear from this figure that an increase in number of the sources can lead to an enhanced reconstructed contrast, and thus resulting in a clearer DOT image. However, no substantial difference is observed in absorption contrast while the number of sources is increased from 8 to 12 and further to 16 in both cases.

More quantitatively, we have calculated CNR and PE values for all four cases with the number of sources of $n = 4$, 8, 12, and 16. The respective values are given in Table 2. It shows unambiguously that no difference in PE and CNR appears when n has changed from 8 to 16. However, the computational time for $n = 16$ is ~2.5 times longer than that for $n = 8$. Hence, for practical use, we would suggest to utilize eight sources instead of twelve or more, which offer the same quality of DOT images but have less complexity of a DOT system.

3.2. Phantom Experiments and Results. Besides the computational investigation given above, we have also performed laboratory phantom studies to confirm what we found. A 1% Intralipid solution was used to generate the liquid tissue phantom (Figure 1(b)) with background optical properties of $\mu_a = 0.1\,\text{cm}^{-1}$ and $\mu_s' = 10\,\text{cm}^{-1}$. A spherical gelatin inclusion of 8 mm in diameter was embedded 20 mm below the liquid surface, with a 2.5 : 1 contrast ratio in absorption between the inclusion and background. A broad-band white light source (Illumination Technologies, Inc. Model 3900, quartz-tungsten halogen lamp) was used, and a 12-bit CCD camera (SamBa Q34 with Navitar Zoom 7000 lens) was employed as the detector. The FOV of the camera was $13 \times 11\,\text{mm}^2$, the same dimension as used in the simulations. The total FOV was divided into 143 (13×11) virtual detectors, meaning that each virtual detector had 38×38 pixels. In this study, 8 light sources were connected to and controlled by a multiplexer (Avantes Inc. Multiplex Channels 1×16), as shown in Figure 1(a). A total of 8×143 source-detector (virtual) pairs were pooled to perform DOT image reconstructions with utilization of DCA. The total volumetric imaging space was $40 \times 40 \times 30\,\text{mm}^3$. Moreover, another similar experiment was repeated with two identical inclusions embedded at 20 mm depth.

3D reconstructed images are shown in Figures 9(a) and 9(b) for the phantom experiments described above. To obtain 3D plots, we have selected the reconstructed absorption values at 80% of its maximum as a volume threshold in order to form the imaged object. From both Figures 9(a) and 9(b), it is clear that the embedded objects can be successfully reconstructed at a relatively accurate depth when DCA is applied. However, because of the diffuse nature of light in tissues, the volume ratio between the reconstructed and real object is around 2 : 1, which also depends on the threshold value selected and the depth of the embedded inclusion.

4. Conclusion

In this study, we present the feasibility of using a CCD-camera-based system combined with our newly developed DCA to form volumetric diffuse optical tomography. We performed both computer simulations and laboratory phantom experiments to confirm our findings. Although only a 12-bit CCD camera was utilized in our experiment setup, the optical signals from 20 mm depth could still be detected by grouping several pixels from the CCD-camera readings as virtual detectors. Overall, we are confident to state that because of its simplicity, low cost, and portability of such a system, a CCD-camera-based DOT may become a highly appropriated imaging tool to be applied in preclinical, small animal studies. For example, this approach may present a great promise for 3D localization of an embedded anomaly tissue, such as a tumor embedded in a small animal.

For future work, *in vivo* animal measurements are planned for both brain imaging and tumor detection to further validate volumetric imaging by combining a CCD-camera-based DOT system with DCA. In addition, optimization of the source-detector geometry will be further investigated since different combinations of source-detector arrangement may result in different qualities of reconstructed images.

Authors' Contribution

The first two authors contributed equally to the paper.

Acknowledgments

This work was supported in part by NINDS from the NIH (5R33NS052850-04). The authors also acknowledge the PMI DOT imaging toolbox.

References

[1] N. Iftimia and H. Jiang, "Quantitative optical image reconstruction of turbid media by use of direct-current measurements," *Applied Optics*, vol. 39, no. 28, pp. 5256–5261, 2000.

[2] A. M. Siegel, J. J. A. Marota, and D. A. Boas, "Design and evaluation of a continuous-wave diffuse optical tomography system," *Optics Express*, vol. 4, no. 8, pp. 287–298, 1999.

[3] D. A. Boas, A. M. Dale, and M. A. Franceschini, "Diffuse optical imaging of brain activation: approaches to optimizing image sensitivity, resolution, and accuracy," *NeuroImage*, vol. 23, no. 1, pp. S275–S288, 2004.

[4] B. R. White, A. Z. Snyder, A. L. Cohen et al., "Resting-state functional connectivity in the human brain revealed with diffuse optical tomography," *NeuroImage*, vol. 47, no. 1, pp. 148–156, 2009.

[5] A. Custo, D. A. Boas, D. Tsuzuki et al., "Anatomical atlas-guided diffuse optical tomography of brain activation," *NeuroImage*, vol. 49, no. 1, pp. 561–567, 2010.

[6] D. R. Leff, O. J. Warren, L. C. Enfield et al., "Diffuse optical imaging of the healthy and diseased breast: a systematic review," *Breast Cancer Research and Treatment*, vol. 108, no. 1, pp. 9–22, 2008.

[7] B. J. Tromberg, B. W. Pogue, K. D. Paulsen, A. G. Yodh, D. A. Boas, and A. E. Cerussi, "Assessing the future of diffuse optical imaging technologies for breast cancer management," *Medical Physics*, vol. 35, no. 6, pp. 2443–2451, 2008.

[8] G. Xu, D. Piao, C. H. Musgrove, C. F. Bunting, and H. Dehghani, "Trans-rectal ultrasound-coupled near-infrared optical tomography of the prostate–part I: simulation," *Optics Express*, vol. 16, no. 22, pp. 17484–17504, 2008.

[9] Z. Jiang, D. Piao, G. Xu et al., "Trans-rectal ultrasound-coupled near-infrared optical tomography of the prostate—part II: experimental demonstration," *Optics Express*, vol. 16, no. 22, pp. 17505–17520, 2008.

[10] J. P. Sacha, M. D. Cockman, T. E. Dufresne, and D. P. Trokhan, "Quantification of regional fat volume in rat MRI," in *Medical Imaging 2003: Physiology and Function: Methods, Systems, and Applications*, vol. 5031 of *Proceedings of SPIE*, pp. 289–297, February 2003.

[11] A. Laib, O. Barou, L. Vico, M. H. Lafage-Proust, C. Alexandre, and P. Rügsegger, "3D micro-computed tomography of trabecular and cortical bone architecture with application to a rat model of immobilisation osteoporosis," *Medical and Biological Engineering and Computing*, vol. 38, no. 3, pp. 326–332, 2000.

[12] D. Sorger, M. Patt, P. Kumar et al., "[18F]Fluoroazomycinarabinofuranoside (18FAZA) and [18F]Fluoromisonidazole (18FMISO): a comparative study of their selective uptake in hypoxic cells and PET imaging in experimental rat tumors," *Nuclear Medicine and Biology*, vol. 30, no. 3, pp. 317–326, 2003.

[13] C. Vanhove, T. Lahoutte, M. Defrise, A. Bossuyt, and P. R. Franken, "Reproducibility of left ventricular volume and ejection fraction measurements in rat using pinhole gated SPECT," *European Journal of Nuclear Medicine and Molecular Imaging*, vol. 32, no. 2, pp. 211–220, 2005.

[14] J. P. Culver, T. Durduran, D. Furuya, C. Cheung, J. H. Greenberg, and A. G. Yodh, "Diffuse optical tomography of cerebral blood flow, oxygenation, and metabolism in rat during focal ischemia," *Journal of Cerebral Blood Flow and Metabolism*, vol. 23, no. 8, pp. 911–924, 2003.

[15] G. Gulsen, O. Birgul, M. B. Unlu, R. Shafiiha, and O. Nalcioglu, "Combined diffuse optical tomography (DOT) and MRI system for cancer imaging in small animals," *Technology in Cancer Research and Treatment*, vol. 5, no. 4, pp. 351–363, 2006.

[16] Q. Zhao, L. Ji, and T. Jiang, "Improving depth resolution of diffuse optical tomography with a layer-based sigmoid adjustment method," *Optics Express*, vol. 15, no. 7, pp. 4018–4029, 2007.

[17] B. W. Pogue, T. O. McBride, J. Prewitt, U. L. Österberg, and K. D. Paulsen, "Spatially variant regularization improves diffuse optical tomography," *Applied Optics*, vol. 38, no. 13, pp. 2950–2961, 1999.

[18] H. Niu, Z. Lin, F. Tian, S. Dhamne, and H. Liu, "Comprehensive investigation of three-dimensional diffuse optical tomography with depth compensation algorithm," *Journal of Biomedical Optics*, vol. 15, no. 4, article 046005, 2010.

[19] H. Niu, F. Tian, Z. Lin, and H. Liu, "Development of a compensation algorithm for accurate depth localization in diffuse optical tomography," *Optics Letters*, vol. 35, no. 3, pp. 429–431, 2010.

[20] S. B. Howell, *Handbook of CCD Astronomy*, 2000.

[21] S. C. Kanick, U. A. Gamm, M. Schouten, H. J. Sterenborg, D. J. Robinson, and A. Amelink, "Measurement of the reduced scattering coefficient of turbid media using single fiber reflectance spectroscopy: fiber diameter and phase function dependence," *Biomedical Optics Express*, vol. 2, pp. 1687–1702, 2011.

[22] M. Johns, C. A. Giller, D. C. German, and H. Liu, "Determination of reduced scattering coefficient of biological tissue from a needle-like probe," *Optics Express*, vol. 13, no. 13, pp. 4828–4842, 2005.

[23] G. Zonios and A. Dimou, "Modeling diffuse reflectance from semi-infinite turbid media: application to the study of skin optical properties," *Optics Express*, vol. 14, no. 19, pp. 8661–8674, 2006.

[24] S. R. Arridge, "Optical tomography in medical imaging," *Inverse Problems*, vol. 15, no. 2, pp. R41–R49, 1999.

[25] Q. Zhao, L. Ji, and T. Jiang, "Improving performance of reflectance diffuse optical imaging using a multicentered mode," *Journal of Biomedical Optics*, vol. 11, no. 6, Article ID 064019, 2006.

[26] F. Tian, G. Alexandrakis, and H. Liu, "Optimization of probe geometry for diffuse optical brain imaging based on measurement density and distribution," *Applied Optics*, vol. 48, no. 13, pp. 2496–2504, 2009.

Permissions

The contributors of this book come from diverse backgrounds, making this book a truly international effort. This book will bring forth new frontiers with its revolutionizing research information and detailed analysis of the nascent developments around the world.

We would like to thank all the contributing authors for lending their expertise to make the book truly unique. They have played a crucial role in the development of this book. Without their invaluable contributions this book wouldn't have been possible. They have made vital efforts to compile up to date information on the varied aspects of this subject to make this book a valuable addition to the collection of many professionals and students.

This book was conceptualized with the vision of imparting up-to-date information and advanced data in this field. To ensure the same, a matchless editorial board was set up. Every individual on the board went through rigorous rounds of assessment to prove their worth. After which they invested a large part of their time researching and compiling the most relevant data for our readers. Conferences and sessions were held from time to time between the editorial board and the contributing authors to present the data in the most comprehensible form. The editorial team has worked tirelessly to provide valuable and valid information to help people across the globe.

Every chapter published in this book has been scrutinized by our experts. Their significance has been extensively debated. The topics covered herein carry significant findings which will fuel the growth of the discipline. They may even be implemented as practical applications or may be referred to as a beginning point for another development. Chapters in this book were first published by Hindawi Publishing Corporation; hereby published with permission under the Creative Commons Attribution License or equivalent.

The editorial board has been involved in producing this book since its inception. They have spent rigorous hours researching and exploring the diverse topics which have resulted in the successful publishing of this book. They have passed on their knowledge of decades through this book. To expedite this challenging task, the publisher supported the team at every step. A small team of assistant editors was also appointed to further simplify the editing procedure and attain best results for the readers.

Our editorial team has been hand-picked from every corner of the world. Their multi-ethnicity adds dynamic inputs to the discussions which result in innovative outcomes. These outcomes are then further discussed with the researchers and contributors who give their valuable feedback and opinion regarding the same. The feedback is then collaborated with the researches and they are edited in a comprehensive manner to aid the understanding of the subject.

Apart from the editorial board, the designing team has also invested a significant amount of their time in understanding the subject and creating the most relevant covers. They scrutinized every image to scout for the most suitable representation of the subject and create an appropriate cover for the book.

The publishing team has been involved in this book since its early stages. They were actively engaged in every process, be it collecting the data, connecting with the contributors or procuring relevant information. The team has been an ardent support to the editorial, designing and production team. Their endless efforts to recruit the best for this project, has resulted in the accomplishment of this book. They are a veteran in the field of academics and their pool of knowledge is as vast as their experience in printing. Their expertise and guidance has proved useful at every step. Their uncompromising quality standards have made this book an exceptional effort. Their encouragement from time to time has been an inspiration for everyone.

The publisher and the editorial board hope that this book will prove to be a valuable piece of knowledge for researchers, students, practitioners and scholars across the globe.

List of Contributors

Wei Jiang
Department of Electrical and Computer Engineering, Rutgers University, Piscataway, NJ 08854, USA
Institute for Advanced Materials, Devices and Nanotechnology, Rutgers University, Piscataway, NJ 08854, USA

Jun Tan and Weiwei Song
Department of Electrical and Computer Engineering, Rutgers University, Piscataway, NJ 08854, USA

Marco Masi
Institut des Nanotechnologies de Lyon, Universite de Lyon, INL-UMR5270, CNRS, INSA de Lyon, 69621 Villeurbanne, France

Mattia Mancinelli, Paolo Bettotti and Lorenzo Pavesi
Nano science Laboratory, Department of Physics, University of Trento, Via Sommarive 14, 38121 Trento, Italy

Martina Abb and Otto L .Muskens
SEPnet and the Department of Physics and Astronomy, University of Southampton, High field, Southampton SO17 1BJ, UK

Roberto Marani, Marco Grande, Vincenzo Petruzzelli and Antonella D'Orazio
Dipartimento di Elettrotecnica ed Elettronica, Politecnico di Bari, Via E. Orabona 4, 70125 Bari, Italy

Dibakar Roy Chowdhury, Ranjan Singh, Antoinette J. Taylor, Hou-Tong Chen and Abul K. Azad
Center for Integrated Nanotechnologies, Materials Physics and Applications Division, Los Alamos National Laboratory, Los Alamos, NM 87545, USA

Weili Zhang
School of Electrical and Computer Engineering, Oklahoma State University, Stillwater, OK 74078, USA

Gabrie lMartınez-Niconoff, G. Diaz-Gonzalez and J. Silva-Barranco
Departamento de ´Optica, Instituto Nacional de Astrofısica Optica y Electronica (INAOE), Apdo. Postal 51 y 216, Tonantzintla, PUE, Mexico

P.Martinez-Vara
Facultad de Ingenierıa, Benemerita Universidad Autonoma de Puebla (BUAP), Publa, PUE, Mexico

A. Carbajal-Domınguez
Universidad Juarez Autonoma de Tabasco, (UJAT), Cunduacan, Tab, Mexico

O. Vanvincq, A. Kudlinski, A. Betourne, A.Mussot, Y. Quiquempois and G. Bouwmans
Laboratoire PhLAM, UMR CNRS 8523, IRCICA, Universite Lille 1, 59655 Villeneuve dAscq Cedex, France

Sonia Boscolo
Photonics Research Group, School of Engineering and Applied Science, Aston University, Birmingham B4 7ET, UK

Christophe Finot
Laboratoire Interdisciplinaire Carnot de Bourgogne, UMR 6303 CNRS, Universite de Bourgogne, 21078 Dijon, France

Dharmadas Kumbhakar
Department of Electronics and Communication Engineering, Asansol Engineering College, Asansol, West Bengal 713305, India

G. Colas des Francs, S. Derom, R. Vincent, A. Bouhelier and A. Dereux
Laboratoire Interdisciplinaire Carnot de Bourgogne, UMR 6303, CNRS-Universite de Bourgogne, 9 Avenue A. Savary, BP 47 870, 21078 Dijon, France

Binlin Wu, M. Alrubaiee, W. Cai and S. K. Gayen
Department of Physics, The City College and The Graduate Center of The City University of New York, 160 Convent Avenue, New York, NY 10031, USA

M. Xu
Department of Physics, Fairfield University, 1073 North Benson Road, Fairfield, CT 06824, USA

Wei-Guo Jin and Tatsuya Minowa
Department of Physics, Faculty of Science, Toho University, Miyama 2-2-1, Funabashi, Chiba 274-8510, Japan

X.Wei, X. Z. Xie, W. Hu and J. F. Huang
School of Electromechanical Engineering, Guangdong University of Technology, Guangzhou 510090, China

Sunil Vyas, Rakesh Kumar Singh and P. Senthilkumaran
Department of Physics, Indian Institute of Technology Delhi, New Delhi 110016, India

Devinder Pal Ghai
Laser Science and Technology Centre, Metcalfe House, Delhi 110054, India

S. N. Kurilkina, V. N. Belyi and N. S. Kazak
B.I. Stepanov Institute of Physics of National Academy of Sciences of Belarus, 68 Nezavisimosti Avenue, 220072 Minsk, Belarus

Ruoxi Yang and Zhaolin Lu
Microsystems Engineering, Kate Gleason College of Engineering, Rochester Institute of Technology, Rochester, NY 14623, USA

Eiji Nakamachi, Yusuke Morita and Yoshifumi Mizuno
Department of Biomedical Engineering, Doshisha University, 1-3 Miyakodani, Tatara, Kyotanabe, Kyoto 610-0394, Japan

Yuan Liu and Baohong Yuan
Department of Bioengineering, The University of Texas at Arlington, Arlington, TX 76019, USA
Joint Biomedical Engineering Program, The University of Texas at Arlington, and The University of Texas Southwestern Medical Center at Dallas, TX 75390, USA

Joseph Vignola
Department of Mechanical Engineering, The Catholic University of America, Washington, DC 20064, USA

Katarzyna Krupa, Alessandro Tonello, Vincent Couderc, Philippe Di Bin and Alain Barthelemy
XLIM, UMR CNRS 6172, Universite de Limoges, 123 av A. Thomas, 87060 Limoges, France

Stefan Wabnitz and Michela Bettenzana
Dipartimento di Ingegneria dell'Informazione, Universita di Brescia, Via Branze 38, 25123 Brescia, Italy

Zhigang Chen
The Key Laboratory of Weak-Light Nonlinear Photonics, Ministry of Education and TEDA Applied Physics School, Nankai University, Tianjin 300457, China
Department of Physics & Astronomy, San Francisco State University, San Francisco, CA 94132, USA

Daohong Song, Cibo Lou, Liqin Tang, Zhuoyi Ye and Jingjun Xu
The Key Laboratory of Weak-Light Nonlinear Photonics, Ministry of Education and TEDA Applied Physics School, Nankai University, Tianjin 300457, China

Zi-Jing Lin, Lin Li and Hanli Liu
Department of Bioengineering, Joint Graduate Program between University of Texas at Arlington and University of Texas Southwestern Medical Center, University of Texas at Arlington, Arlington, TX 76019, USA

Haijing Niu
State key Laboratory of Cognitive Neuroscience and Learning, Beijing Normal University, Beijing 10875, China